# CODED-MODULATION TECHNIQUES FOR FADING CHANNELS

# THE KLUWER INTERNATIONAL SERIES IN ENGINEERING AND COMPUTER SCIENCE

## COMMUNICATIONS AND INFORMATION THEORY

*Consulting Editor*

**Robert Gallager**

***Other books in the series:***

**ELLIPTIC CURVE PUBLIC KEY CYRPTOSYSTEMS**, Alfred Menezes
ISBN: 0-7923-9368-6
**SATELLITE COMMUNICATIONS: Mobile and Fixed Services**, Michael Miller, Branka Vucetic and Les Berry
ISBN: 0-7923-9333-3
**WIRELESS COMMUNICATIONS: Future Directions**, Jack M. Holtzman and David J. Goodman
ISBN: 0-7923-9316-3
**DISCRETE-TIME MODELS FOR COMMUNICATION SYSTEMS INCLUDING ATM**, Herwig Bruneel and Byung G. Kim
ISBN: 0-7923-9292-2
**APPLICATIONS OF FINITE FIELDS**, Alfred J. Menezes, Ian F. Blake, XuHong Gao, Ronald C. Mullin, Scott A. Vanstone, Tomik Yaghoobian
ISBN: 0-7923-9282-5
**WIRELESS PERSONAL COMMUNICATIONS**, Martin J. Feuerstein, Theodore S. Rappaport
ISBN: 0-7923-9280-9
**SEQUENCE DETECTION FOR HIGH-DENSITY STORAGE CHANNEL**, Jaekyun Moon, L. Richard Carley
ISBN: 0-7923-9264-7
**DIGITAL SATELLITE COMMUNICATIONS SYSTEMS AND TECHNOLOGIES: Military and Civil Applications**, A. Nejat Ince
ISBN: 0-7923-9254-X
**IMAGE AND TEXT COMPRESSION**, James A. Storer
ISBN: 0-7923-9243-4
**VECTOR QUANTIZATION AND SIGNAL COMPRESSION**, Allen Gersho, Robert M. Gray
ISBN: 0-7923-9181-0
**THIRD GENERATION WIRELESS INFORMATION NETWORKS**, Sanjiv Nanda, David J. Goodman
ISBN: 0-7923-9128-3
**SOURCE AND CHANNEL CODING: An Algorithmic Approach**, John B. Anderson, Seshadri Mohan
ISBN: 0-7923-9210-8
**ADVANCES IN SPEECH CODING**, Bishnu Atal, Vladimir Cuperman, Allen Gersho
ISBN: 0-7923-9091-1
**SWITCHING AND TRAFFIC THEORY FOR INTEGRATED BROADBAND NETWORKS**, Joseph Y. Hui
ISBN: 0-7923-9061-X
**ADAPTIVE DATA COMPRESSION**, Ross N. Williams
ISBN: 0-7923-9085
**SOURCE CODING THEORY**, Robert M. Gray
ISBN: 0-7923-9048-2
**AN INTRODUCTION TO ERROR CORRECTING CODES WITH APPLICATIONS**, Scott A. Vanstone, Paul C. van Oorschot
ISBN: 0-7923-9017-2
**FINITE FIELDS FOR COMPUTER SCIENTISTS AND ENGINEERS**, Robert J. McEliece
ISBN: 0-89838-191-6
**AN INTRODUCTION TO CRYPTOLOGY**, Henk C. A. van Tilborg
ISBN: 0-89838-271-8

# CODED-MODULATION TECHNIQUES FOR FADING CHANNELS

*by*

**S. Hamidreza Jamali**
*Tehran University*

**Tho Le-Ngoc**
*Concordia University*

**KLUWER ACADEMIC PUBLISHERS**
**Boston / Dordrecht / London**

**Distributors for North America:**
Kluwer Academic Publishers
101 Philip Drive
Assinippi Park
Norwell, Massachusetts 02061 USA

**Distributors for all other countries:**
Kluwer Academic Publishers Group
Distribution Centre
Post Office Box 322
3300 AH Dordrecht, THE NETHERLANDS

**Library of Congress Cataloging-in-Publication Data**

Jamali, S. Hamidreza, 1955-
Coded-modulation techniques for fading channels / by S. Hamidreza Jamali, tho Le-Ngoc.
p. cm. -- (The Kluwer international series in engineering and computer science ; 268. Communications and information theory)
Includes bibliographical references and index.
ISBN 0-7923-9421-6 (alk. paper)
1. Digital modulation. 2. Error-correcting codes (Information theory) 3. Radio-Transmitters and transmission--Fading. I. Le, Ngoc Tho, 1954- . II. Title. III. Series: Kluwer international series in engineering and computer science ; SECS 268. IV. Series: Kluwer international series in engineering and computer science. Communications and information theory.
TK5103.7.J36 1994
621.382'23--dc20 94-2903
CIP



*Printed on acid-free paper.*

Printed in the United States of America

*To my mother, E. Mousavi,*
*and my father, S. B. Jamali*
**S. HAMIDREZA JAMALI**

*In loving memory of my father,*
*and to my mother,*
*my wife, Le-Chi,*
*and my daughters, Anh, Minh and Linh-Lan*
**THO LE-NGOC**

# Contents

## Chapter 3



## Chapter 4

## Chapter 5



## Chapter 6

## Chapter 7



## Chapter 8



## Chapter 9

# Appendix B



# Appendix C



# Appendix D

# Preface

During the past decade there has been a rapidly growing research in the area of bandwidth-efficient coded-modulation techniques. The primary advantage of these schemes over modulation schemes employing traditional error correcting codes is their ability to improve the performance of the communication system without bandwidth expansion. Such a property makes them a suitable choice for channels that are limited in both power and bandwidth. A typical example of such channels is mobile satellite channel, where it is desired to accommodate a large number of users in a given bandwidth with a power which is constrained by the physical size of the satellite and by the vehicle's antenna.

This book is intended to provide graduate students and practicing engineers with a sound background for the application of bandwidth-efficient coded-modulation techniques in fading channels. Our motivation is twofold. First, we wish to systematically present the materials of a field that has been grown rapidly in the last decade and to provide a solid frame or reference for further research in this area. Second, we wish to make the basic concepts and results accessible to practicing engineers. For an active research area like coded-modulation techniques, where new ideas are continually generated, this book is far from a complete treatise on the subject. Rather, it is intended as an educational primer. We have attempted to present the subject in a self-contained manner. However, it is assumed that the reader has some prior knowledge of probability and stochastic processes and some background of digital communications and error correcting codes, especially convolutional codes.

Chapter 1 of this book introduces the multipath fading channel and reviews its physical behaviour. Statistical models for some fading channels are considered and some related parameters are discussed.

In Chapter 2, the elements of a coded communication system in a fading channel are introduced. The random coding bound on the performance of digital communication systems over fading channels is then derived. The cut-off rate for different signal constellations and decoding strategies is discussed. This general discussion reveals the potential gains of coded-modulation schemes in bandwidth-limited fading channels.

Convolutional codes are discussed in Chapter 3. This chapter covers the construction, representation, distance properties and generating function of these codes. It also discusses maximum likelihood decoding and performance evaluation techniques. The aim of this chapter is to provide background materials for the following chapters on trellis-coded modulation.

The fourth chapter introduces the development of Trellis-Coded Modulation (TCM) as well as Block-Coded Modulation (BCM) schemes. Even though the emphasis of this book is on fading channels this chapter is devoted to the Additive White Gausian Noise (AWGN) channel in order to visualize the concept of these techniques and their historical development.

Chapter 5 treats the application of TCM schemes on fading channels. In this context the performance of these schemes over fading channels is evaluated. The parameters dominating the performance of these coded schemes over fading channels are extracted from this evaluation. The development of TCM schemes based on the fading channel design criteria is also addressed. It includes the performance of some TCM schemes when coherent demodulation is employed. The results are obtained from a combination of analysis and simulation.

Error correcting Reed-Solomon (RS) codes are introduced in Chapter 6. The goal of this chapter is to provide background material for the following chapters. Both soft-decision and hard-decision decoding techniques for these codes are treated in this chapter.

The results of Chapter 2 show that the gain in signal-to-noise (SNR) due to soft-decision decoding in fading channels is much higher than the gain in an AWGN channel. Applying this decoding technique for block-coded schemes, hence, is of great importance to fading channels. In Chapter 7, the issue of soft-decision decoding of RS-coded schemes is considered. This decoding technique is implemented using the Viterbi Algorithm (VA) for short RS codes. For the longer low rate RS coded schemes the Successive Erasure Minimum Distance Decoding (SEMDD) is discussed. The performance of these decoding strategies are evaluated and compared with the other techniques.

In Chapter 8, the performance evaluation of the coded-modulation schemes over fading channels is extended to multilevel differential encoding. Although differential detection degrades the performance of the system, it has an important advantage over coherent reception in that a means for extracting a carrier demodulation reference is not necessary at the receiver.

Chapter 9 represents the application of the trellis and RS-coded schemes in a shadowed Rician fading channel. The performance of these coded schemes is evaluated for ideal interleaving/de-interleaving. The effect of imperfect interleaving on the performance of the coded scheme is also investigated.

Finally, the last chapter is devoted to Quadrature Pulse-Overlapping Modulation (QPOM) techniques, which are attractive from the point of view of being a bandwidth and power efficient signaling method and being tolerant to nonlinear distortions. Although these schemes do not employ coding, the memory introduced by the overlapping pulse shapes makes them a non-redundant coded modulation scheme. In fact, the QPOM structure can be decomposed into a convolutional encoder followed by a memoryless modulator, which makes it equivalent to a coded-modulation structure. When soft-decision decoding is employed, QPOM schemes provide an increased time diversity, which improves their performance in fading channels. Chapter 10 covers the concept and spectral properties of these schemes and their performance in both AWGN and fading channels employing Maximum Likelihood Sequence Estimation (MLSE) techniques. It also discusses the performance improvement using a combination of nested coding and QPOM schemes.

In addition to numerous examples, a set of problems is introduced at the end of each chapter. Most of these problems are relatively straightforward applications of the material covered in that chapter. Some of the problems are designed to review background material, or to supplement and extend the results derived in the text. The 250 problems are a useful aid to learn the topics covered in this book.

This book can be used as a text for an introductory course on coded modulation techniques and their applications at a senior or graduate level and it is applicable to different course structures. For example, a one-semester course on the introduction to trellis-coded modulation can cover Chapters 1 through 5 while Chapters 9 and 10 can be used for further reading. For a one-semester course on both trellis and block coded modulation techniques, taught to the audience with a strong background in convolutional and block codes, all chapters can be covered with Chapters 3 and 6 considered as review materials. Alternatively, this book can be used for a two-semester sequence, with Chapters 1 through 5 on trellis-coded modulation taught in the first semester, and Chapters 6 through 10 on block coded modulation and applications taught in the second semester.

This book can also be used for short-courses or as a self-study guide for practicing engineers who want to learn the fundamentals of coded modulation and how these schemes can be applied to the design of coded systems in both AWGN and fading channels.

There are numerous acknowledgments which we would like to make. A book of this nature relies heavily on the pioneering work of the many researchers in this field. We have tried to adequately acknowledge this work in the references cited throughout the text.

This book, as well as our research contributions described therein, is made possible by the gracious and enriching environments of Concordia University and Tehran University. We would like to thank both universities for their support of our efforts in writing this book and for providing us with the proper facilities. We are grateful to the National Sciences and Engineering Research Council of Canada and the Fonds pour la Formation de Chercheurs et l'Aide à la Recherche of Quebec for their continuing support of our research.

We would like to express our deep gratitude to Slimane Ben Slimane for his valuable suggestions and contributions to the materials in Chapters 5, 9 and

10. In particular, Chapter 10 is based on his Ph.D dissertation [151].

Most of the manuscript and figures were prepared on LATEX by Hamid Reza Mehrvar and Slimane Ben Slimane. Their excellent work, dedication and patience with the innumerable changes are sincerely appreciated.

We specially thank Bonnie G. Weppler, Anader Benyamin-Seeyar, Yu-Dong Yao, Qiang Wang and two anonymous reviewers of the book proposal for their helpful comments and suggestions, many of which have been incorporated. We also wish to thank Mohammad R. Soleymani, Yousef R. Shayan and Vijay K. Bhargava for various discussions on the subject.

Finally, S. Hamidreza Jamali would like to take this opportunity to thank his parents for their moral support. Tho Le-Ngoc thanks his wife, Le-Chi for encouragement, continuing love and affection throughout this project and his daughters Anh, Minh, and Linh-Lan for playing cheerfully around him during the long writing process. Without them, his life would seem strangely disconnected.

**S. HAMIDREZA JAMALI**
**THO LE-NGOC**

# Chapter 1

# FADING CHANNEL MODELING

The purpose of this chapter is to review the physical behaviour and characteristics of fading channels. After giving a description of multipath fading channels, we develop a statistical model for these channels. This model and its related parameters will be used throughout the book. Finally we introduce computer models for various kinds of fading channels which are useful in computer simulation of communication systems over fading channels.

## 1.1 INTRODUCTION

The simplest model that is frequently used for a transmission channel is the *additive white Gaussian noise* (AWGN) channel model. In this model the received signal is the sum of the transmitted signal (attenuated in a known way) and Gaussian noise. This simple channel model has great theoretical and practical importance and is an accurate model for many communication channels, such as satellite and deep-space communication channels. In many communication systems, however, the channels are subjected to various impairments in addition to the additive noise. For these channels the simple model of AWGN is no longer valid and one must resort to more precise and complex channel models.

One of the channel types which frequently occurs in radio communication is the *fading channel*. There are numerous causes of fading. A few of these causes will be addressed in this introductory section. One common cause of fading is the multipath nature of propagation media. In this case, the transmitted signal arrives at the receiver from different paths with time-varying nature. These individual paths cannot be distinguished at the receiver's end and are all added together. Thus, the signal becomes a replica of the transmitted one with random amplitude and phase. This kind of fading occurs in ground-to-ground radio communication, and tropospheric and ionospheric scatter communication [105], [60].

The second kind of fading, which is most common in ionospheric high frequency (HF) radio propagation, is caused by the electron density variation versus altitude in the ionospheric layers. In HF radio communication the ionospheric layers are used as reflectors to communicate between two points which are not in line-of-sight or which are too far apart for ground wave propagation. The heights and electron densities of the ionospheric layers are closely related to solar radiation and these parameters vary due to the day-night changes or sun spot activity, among other causes. The dynamic changes of ionospheric layers varies the length of the transmission path, and consequently, the received signal experiences random variations in both amplitude and phase.

Another significant cause of fading is the relative motion of the transmitter and receiver in a static multipath environment. In mobile radio systems, the propagation between the transmitting antenna and the mobile unit antenna is over several paths, namely, the line-of-sight path and the paths due to the scattering caused by reflections from and diffractions around obstacles. These interfering signals produce a complex standing-wave pattern of varying field strength, with minima and maxima being of the order of a quarter wavelength apart. As a result of the vehicle movement through standing-wave pattern, the amplitude and phase variations are included in the received signal.

Two kinds of fading, namely, *short-term* and *long-term* fading are often considered in fading channels. In short-term fading the changes in channel characteristics occur within a time scale that ranges from fractions of a second to several seconds. In long-term fading the variations of the channel characteristics are in the range of minutes, tens of minutes, hours or even more. These variations are often related to solar or meteorological influences [106]. A channel may undergo both short-term and long-term fadings continually. However, from the signaling design point of view we are most concerned with short term fading [106] and we shall primarily consider this kind of fading henceforth. In the following sections we provide a statistical model for fading channels and derive some useful parameters which are essential in characterizing the channel behavior. Our treatment of the problem is brief and most of the material covered here may be found in more detail in [59, Chapter 9], [60, Chapter 7], and [106].

## 1.2 STATISTICAL CHARACTERIZATION OF MULTIPATH FADING CHANNELS

In this section we will consider the characterization of multipath fading channels. The multipath propagation for a scattering medium is illustrated in Figure 1.1 as an example. In this case a large number of scatterers are located at

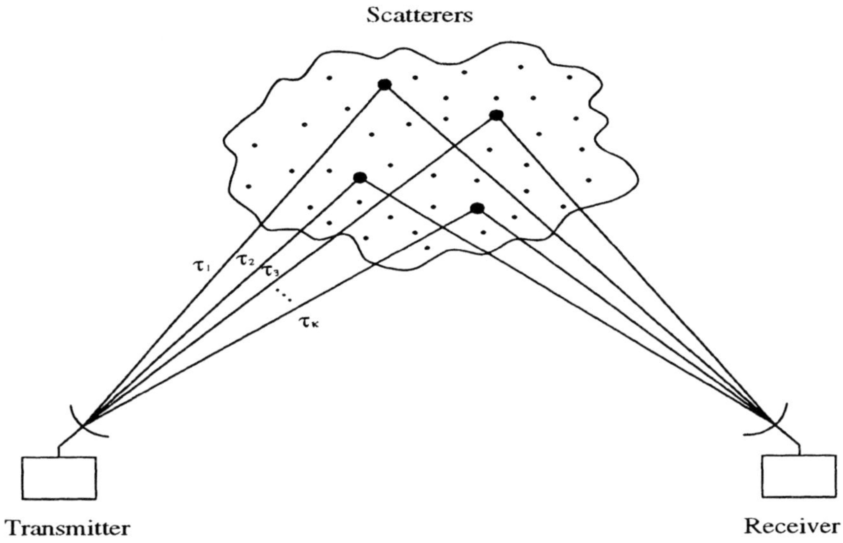


Figure 1.1: Multipath fading channel.

random points within the propagation path. The cloud of scattering particles causes the transmitted signal to be scattered and to arrive via several paths, with different time delays, at the receiver. If the time difference between time delays of these paths is smaller than the reciprocal of the transmitted signal bandwidth, the receiver sees only the superposition of the arriving signals without distinguishing several paths.

To visualize the behavior of fading channels and to develop a statistical model for these channels we first consider the propagation of an unmodulated carrier in a multipath fading environment. Consider a sinusoid $s(t) = \cos 2\pi f_c t$. Assume that $s(t)$ is transmitted and scattered by randomly moving particles. The scattered signals arrive at the receiver via different paths. Each path introduces a time delay and an attenuation factor into the transmitted signal. Because of random motion of the scattering particles the time delay and attenuation factor are time-variant. For the $i$th path the associated time delay and attenuation factor are denoted by $\tau_i(t)$ and $\alpha_i(t)$, respectively.

The received signal (apart from any additive noise) can be obtained by summing different signal components which are arriving via different paths at the receiver as shown below

$$\begin{aligned} r(t) &= \sum_i \alpha_i(t) \cos(2\pi f_c t - \theta_i(t)) \\ &= a_I(t) \cos(2\pi f_c t) + a_Q(t) \sin(2\pi f_c t), \end{aligned} \tag{1.1}$$

where

$$a_I(t) = \sum_i \alpha_i(t)\ \cos\theta_i(t)$$
$$a_Q(t) = \sum_i \alpha_i(t)\ \sin\theta_i(t) \qquad (1.2)$$

and $\theta_i(t) = 2\pi f_c \tau_i(t)$. To investigate the statistical properties of $a_I(t)$ and $a_Q(t)$ we assume that the samples of the random process $\alpha_i(t)$ are independent and identically distributed random variables. Also, we assume that the samples of the random process $\theta_i(t)$ are uniformly distributed random variables over $[0, 2\pi]$, each of which is statistically independent of others and of the $\alpha_i$'s. Under these assumptions the central limit theorem arguments lead to the conclusion that the in-phase and quadrature components, $a_I(t)$ and $a_Q(t)$, are two statistically independent Gaussian random processes [17, Page 531]. In fading channels with only a diffused multipath signal, these Gaussian processes have zero mean and the fading envelope, defined as

$$a = \sqrt{a_I^2 + a_Q^2},$$

has a *Rayleigh* distribution with a probability density function (pdf)

$$p_A(a) = \frac{a}{\sigma_a^2}\ e^{-\frac{a^2}{2\sigma_a^2}} \qquad (1.3)$$

where $\sigma_a^2$ represents the variance of the Gaussian random variables $a_I$ and $a_Q$. The average energy of the fading envelope is related to $\sigma_a^2$ as

$$E[a^2] = 2\sigma_a^2.$$

The fading phase defined as

$$\phi = \tan^{-1}\frac{a_Q}{a_I},$$

is a uniformly distributed random process with a pdf of

$$p_\Phi(\phi) = \frac{1}{2\pi}, \qquad 0 \le \phi < 2\pi.$$

In some fading channels, in addition to the diffused multipath fading process, there exists a dominant line-of-sight (direct) signal component. This occurs when there are fixed scatterers or signal reflectors in the medium in addition to randomly moving scatterers. Denoting the direct component by $A\cos 2\pi f_c t$[1] the received signal can be written as

$$r(t) = (A + a_I(t))\cos(2\pi f_c t) + a_Q(t)\sin(2\pi f_c t),$$

[1] We neglect the time delay of the line-of-sight (LOS) signal component. In this case the time delays of the diffused multipath signal components are relative values with respect to the LOS component.

where $a_I(t)$ and $a_Q(t)$, previously defined in (1.2), are zero mean Gaussian random processes, each with variance $\sigma_a^2$. Note that the sum $(A + a_I(t))$ also represents a Gaussian process with mean A and variance $\sigma_a^2$. Defining the fading envelope as

$$a = \sqrt{(A + a_I)^2 + a_Q^2},$$

it can be shown that the fading envelope has a *Rician* distribution with a pdf[2] (see Problem 1.1)

$$p_A(a) = 2a\ (1 + K)\ e^{-(K+a^2(1+K))}\ I_0(2a\sqrt{K(K+1)}) \tag{1.4}$$

where

$$I_0(x) = \sum_{k=0}^{+\infty} \left(\frac{x^k}{2^k k!}\right)^2$$

is the zero-order modified Bessel function of the first kind, and $K$ is the *Rician parameter* defined as the ratio of the energy of direct component to the energy of the diffused multipath component, that is

$$K = \frac{A^2}{2\sigma_a^2}.$$

In a Rician fading channel the fading phase defined as

$$\phi = \tan^{-1} \frac{a_Q}{A + a_I}$$

is no longer uniform and can be described by a pdf [3] (see Problem 1.1).

$$p_\Phi(\phi) = \frac{e^{-K}}{2\pi} + \frac{\sqrt{K}\ \cos\phi\ e^{-K\sin^2\phi}}{2\sqrt{\pi}}\ (2 - \mathrm{erfc}(\sqrt{\mathrm{K}}\cos\phi)), \quad |\phi| \le \pi \tag{1.5}$$

where $\mathrm{erfc(x)} = \frac{2}{\sqrt{\pi}} \int_{\mathrm{x}}^{\infty} \mathrm{e}^{-\mathrm{t}^2} \cdot \mathrm{dt}$ is the complementary error function. Note that the envelope and phase statistics of a Rician fading for $K = 0$ reduce to those of a Rayleigh fading (with normalized energy, $2\sigma_a^2 = 1$).

The multipath fading phenomenon may be most visualized by considering the received signal as the sum of a number of time-variant phasors. Using the complex representation, the received signal (for unmodulated transmission), given in (1.1), can be written as

$$r(t) = \mathrm{Re}\left[\sum_i \alpha_i(t)\ e^{j2\pi f_c(t-\tau_i)}\right],$$

[2] We assume that the energy of the fading process is normalized to unity, that is $E[a^2] = 1$.

[3] The effect of Doppler shift on the direct path is ignored. In fact the Doppler shift (which will be considered later) changes the carrier frequency and causes the phase of the direct path continually to vary with respect to the transmitted carrier.

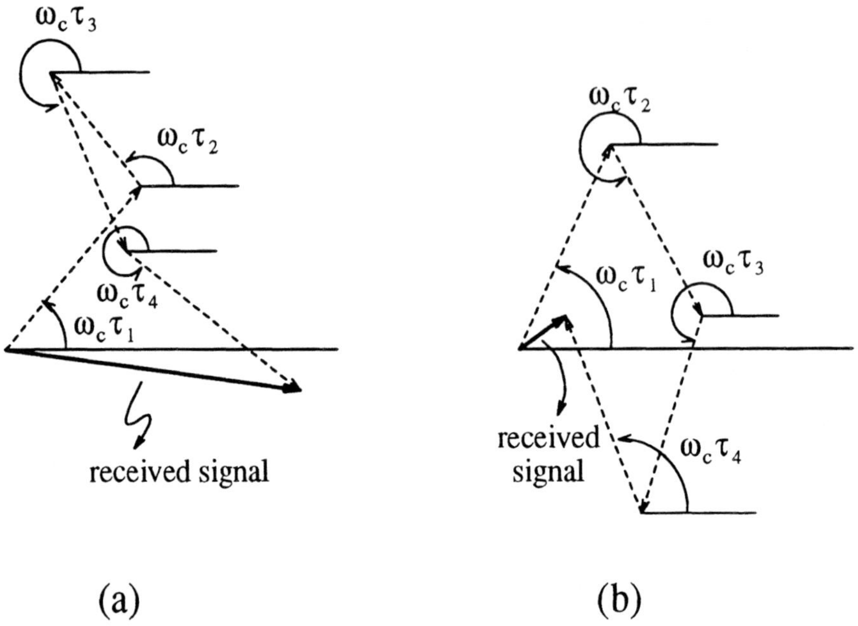


Figure 1.2: Phasor representation of a received signal in a multipath fading channel: (a) constructive addition (b) destructive addition.

where Re[·] shows the real part of a complex value. In phasor terms, the received signal is the sum of several phasors, $\{\alpha_i(t)\ e^{-j2\pi f_c\tau_i(t)}\}$, each of which contains randomly varying amplitudes and phases. For large values of the radio frequency $f_c$, relatively small motions of the scatterers may change the phase of each phasor over a full $(0,\ 2\pi)$ range even if the attenuation factor does not change significantly. As a consequence of phase changes, the phasors may add constructively (Figure 1.2(a)), or destructively (Figure 1.2(b)). The received signal $r(t)$ becomes very small when the phasors add destructively. The variation of the received signal envelope due to the constructive and destructive interference caused by the random phase changes is called *multipath fading*.

### 1.2.1 Linear Time-Variant Filter Model

It is possible to model a short-term fading channel by a linear filter whose parameters are time-variant [106]. Consider the effect of fading channels on a transmitted signal with complex representation,

$$s(t) = \text{Re}\ [\ u(t)\ e^{j2\pi f_c t}],$$

where $u(t)$ is the equivalent low-pass complex signal and $f_c$ is the carrier frequency. The equivalent low-pass signal $u(t)$ has an equivalent low-pass spectrum as

$$U(f) = \int_{-\infty}^{+\infty} u(t)\ e^{-j2\pi ft} dt. \tag{1.6}$$

Transmitting $s(t)$ over a multipath fading channel gives rise to a received signal,

$$\begin{aligned} r_c(t) &= \sum_i \alpha_i(t)\ s(t-\tau_i(t)) \\ &= \text{Re}\left[\left\{\sum_i \alpha_i(t)\ u(t-\tau_i(t))\ e^{-j2\pi f_c\tau_i(t)}\right\}\ e^{j2\pi f_c t}\right], \end{aligned}$$

where $\alpha_i(t)$ and $\tau_i(t)$ are the time-variant attenuation factor and time delay associated with the $i$th propagation path. The received signal can also be represented by its equivalent low-pass signal,

$$r(t) = \sum_i \alpha_i(t)\ u(t-\tau_i(t))\ e^{-j2\pi f_c\tau_i(t)}. \tag{1.7}$$

It is observed from (1.7) that the received signal is mathematically described as the convolution of the signal $u(t)$ and

$$h(\tau;\ t) = \sum_i \alpha_i(t)\ e^{-j2\pi f_c\tau_i(t)}\ \delta(\tau-\tau_i(t)).$$

Consequently, the multipath fading channel is modeled by a time-variant linear filter whose equivalent low-pass impulse response is given by $h(\tau;\ t)$.

In fading channels where the received signal consists of a continuum of multipath components the sum in (1.7) is replaced by an integral. In this case the received signal $r(t)$ becomes

$$r(t) = \int_{-\infty}^{+\infty} \alpha(\tau,\ t)\ u(t-\tau)\ e^{-j2\pi f_c\tau} d\tau, \tag{1.8}$$

where $\alpha(\tau,\ t)$ shows the attenuation factor associated with a $\tau$-delay path at time $t$. Observing (1.8) the equivalent low-pass impulse response of the channel is obtained as

$$h(\tau;\ t) = \alpha(\tau,\ t)\ e^{-j2\pi f_c\tau}.$$

Taking the Fourier transform of $h(\tau;\ t)$ relative to its $\tau$-dependence, the time-varying equivalent low-pass transfer function of the channel is obtained as

$$H(f;t) = \int_{-\infty}^{+\infty} h(\tau;t)\ e^{-j2\pi f\tau} d\tau. \tag{1.9}$$

To characterize $H(f;t)$ statistically, we interpret this time-varying transfer function as the response of the multipath fading channel to a transmitted sine wave of relative frequency $f$. The received signal $r(t)$ can then be rewritten in terms of $H(f;t)$ as

$$r(t) = \int_{-\infty}^{+\infty} U(f) \cdot H(f;\ t)\ e^{j2\pi ft} df, \tag{1.10}$$

where $U(f)$ is the equivalent low-pass spectrum of the transmitted signal $u(t)$, given in (1.6). Assume that the transmitted signal $u(t)$ is a sine wave at relative frequency $f'$. In this case $U(f) = \delta(f - f')$ implying that $r(t) = H(f';\ t)$. Thus, $H(f;\ t)$ describes exactly the response of the multipath fading channel to a pure sine wave at relative frequency $f'$. Considering the fact that the response of multipath fading channels to an unmodulated carrier is a complex Gaussian process, $H(f;\ t)$ and consequently $h(\tau;\ t)$ can also be characterized as complex Gaussian processes.

### 1.2.2 Autocorrelation Functions

The foregoing discussions lead to the conclusion that a multipath fading channel may be viewed as a time-variant linear filter whose impulse response and equivalent low-pass transfer function are complex Gaussian random processes. For short-term fading it is reasonable to assume that these processes are Wide-Sense Stationary (WSS), and hence, their statistics are not affected by a shift in the time origin.

To provide some useful terms concerning the characteristics of a fading channel we use the autocorrelation functions of its impulse response $h(\tau;t)$ and its equivalent low-pass transfer function $H(f;t)$. The autocorrelation function of $h(\tau;t)$ is defined by

$$R_h(\tau_1,\ \tau_2;\ \Delta t) = \frac{1}{2}\ E[h(\tau_1;\ t)\ h^*(\tau_2;\ t + \Delta t)], \tag{1.11}$$

where $h(\tau_1;\ t)$ and $h(\tau_2;\ t + \Delta t)$ are the impulse responses corresponding to two path delays $\tau_1$ and $\tau_2$. Assume that the scattering at two paths, with corresponding delays $\tau_1$ and $\tau_2$, is uncorrelated. This is called *uncorrelated scattering* and it is a reasonable assumption for most fading channels.

For uncorrelated scattering the autocorrelation function, defined in (1.11), can be expressed as

$$R_h(\tau_1,\ \tau_2;\ \Delta t) = R_h(\tau_1;\ \Delta t)\ \delta(\tau_1 - \tau_2), \tag{1.12}$$

where $R_h(\tau;\ \Delta t)$ represents the autocorrelation with time lag $\Delta t$ for a path delay $\tau$. The value of $R_h(\tau;\ \Delta t)$ at $\Delta t = 0$, that is $R_h(\tau;\ 0)$, describes the average output power of the channel at relative path delay $\tau$. The function $R_h(\tau;\ 0)$ is called the *multipath profile* [106] or the *delay power-spectrum* [60, Chapter 7] of the fading channel.

Returning to $H(f;\ t)$, we now define the autocorrelation function of the equivalent low-pass transfer function of the fading channel as

$$R_H(f_1, f_2; \Delta t) = \frac{1}{2}\ E[H(f_1;\ t)\ H^\star(f_2; t + \Delta t)], \tag{1.13}$$

where we have implicitly assumed that $H(f;\ t)$ is a WSS process. The Fourier transform relationship between $h(\tau;\ t)$ and $H(f;\ t)$ implies that the two autocorrelation functions $R_h(\tau_1,\ \tau_2;\ \Delta t)$ and $R_H(f_1,\ f_2;\ \Delta t)$ are also related by a Fourier transform.

**Example 1.1** To show that $R_h(\tau_1, \tau_2; \Delta t)$ and $R_H(f_1, f_2; \Delta t)$ are related by a Fourier transform we substitute (1.9) into (1.13)

$$\begin{aligned}
R_H(f_1, f_2; \Delta t) &= \frac{1}{2}E\left[\int_{-\infty}^{+\infty}\int_{-\infty}^{+\infty} h(\tau_1; t)\ h^\star(\tau_2; t)\ e^{-j2\pi f_1 \tau_1} \cdot e^{j2\pi f_2 \tau_2} d\tau_1\ d\tau_2\right] \\
&= \int_{-\infty}^{+\infty}\int_{-\infty}^{+\infty} \frac{1}{2}E[\ h(\tau_1; t)\ h^\star(\tau_2; t)]\ e^{-j2\pi f_1 \tau_1}\ e^{j2\pi f_2 \tau_2} d\tau_1\ d\tau_2 \\
&= \int_{-\infty}^{+\infty}\int_{-\infty}^{+\infty} R_h(\tau_1, \tau_2; t)\ e^{-j2\pi f_1 \tau_1}\ e^{j2\pi f_2 \tau_2}\ d\tau_1\ d\tau_2.
\end{aligned}$$

Using (1.12), for the uncorrelated scattering, the above relationship can be written as

$$\begin{aligned}
R_H(f_1, f_2; \Delta t) &= \int_{-\infty}^{+\infty}\int_{-\infty}^{+\infty} R_h(\tau_1; \Delta t)\ \delta(\tau_1 - \tau_2)\ e^{-j2\pi f_1 \tau_1}\ e^{j2\pi f_2 \tau_2} d\tau_1\ d\tau_2 \\
&= \int_{-\infty}^{+\infty} R_h(\tau_1; \Delta t)\ e^{-j2\pi (f_1 - f_2)\tau_1}\ d\tau_1 \\
&= \int_{-\infty}^{+\infty} R_h(\tau_1; \Delta t)\ e^{-j2\pi \Delta f \tau_1}\ d\tau_1 \\
&= R_H(\Delta f; \Delta t).
\end{aligned}$$

The above example shows that for uncorrelated scattering the autocorrelation function $R_H(f_1, f_2; \Delta t)$ depends only on the frequency difference $\Delta f = f_1 - f_2$. The function $R_H(\Delta f; \Delta t)$ is called the *spaced-tone autocorrelation* function [106].

### Coherence Bandwidth

The value of the spaced-tone autocorrelation function at $\Delta t = 0$, that is $R_H(\Delta f; 0)$, characterizes the *selectivity* of the fading process. If $R_H(\Delta f; 0)$ is nearly constant over a bandwidth $B_c$ then the two signals with frequency $f_1$ and $f_2$, for which $f_1 - f_2 < B_c$, are affected nearly the same by the fading channel. The bandwidth $B_c$ is called the *coherence bandwidth* of the fading channel.

For transmitted signals with a bandwidth less than $B_c$, all signal components are equally faded. In this case the fading is said to be *nonselective* or *flat fading*. On the other hand, if the bandwidth of the transmitted signal is more than $B_c$, then the signal components, with frequency difference more than $B_c$, are affected differently by the channel and the fading is called *selective fading*.

#### Multipath Spread

A closely related parameter to the coherence bandwidth of the fading channel is the *multipath spread*. The multipath spread, $T_M$, is the range of values of the path delay $\tau$ for which the multipath profile, $R_h(\tau;\ 0)$, is nearly constant. The Fourier transform relationship between $R_H(\Delta f;\ 0)$ and $R_h(\tau;\ 0)$ implies that the multipath spread is inversely reciprocal to the coherence bandwidth, $B_c$, that is $T_M \approx 1/B_c$. The selectivity of the fading channel, hence, can also be characterized by the value of the multipath spread $T_M$. Considering that, for simple symbol waveforms with duration $T$, the nominal bandwidth is approximately $1/T$ and the fading is nonselective if $T \gg T_M$.

#### Flat Fading

In a nonselective or flat fading environment the channel acts as an all-pass filter within the coherence bandwidth. In this case the time-varying channel gain is independent of the frequency and can be written as

$$H(f;t) = g(t),$$

where $g(t)$ is a complex Gaussian process. Using (1.10) the output of such a channel can be expressed in terms of its input and the complex channel gain as

$$r(t) = g(t)u(t),$$

apart from any additive Gaussian noise process. As it is seen, the flat fading channel imposes a multiplicative distortion $g(t)$ on the transmitted signal. Because of this multiplicative distortion flat fading has also been called *multiplicative fading*. This kind of fading channel is considered throughout the book and whenever we refer to "fading channel" flat fading is assumed.

### 1.2.3 Scattering Function

Besides the frequency (or path delay) dependence of the autocorrelation functions $R_H(\Delta f;\Delta t)$ (or $R_h(\tau;\Delta t)$), these functions are $\Delta t$-dependent. In fact, the $\Delta t$-dependence of any of these functions results from dynamic changes in the channel characteristics that can be related directly to the moving scatterers or reflectors in the medium. The time variations in the channel cause a relative frequency change which is called the *Doppler shift*. In order to relate the Doppler shift to the time variations of the channel we introduce the following transforms.

Considering the $\Delta t$-dependence of $R_h(\tau;\Delta t)$ a further Fourier transform of this function with respect to the variable $\Delta t$ is defined as

$$S_h(\tau;\lambda) = \int_{-\infty}^{+\infty} R_h(\tau;\Delta t)\; e^{-j2\pi\lambda\Delta t}\; d(\Delta t), \tag{1.14}$$

where $\lambda$ denotes the frequency variable (Doppler frequency) corresponding to the $\Delta t$-dependence of the autocorrelation function. The function $S_h(\tau;\lambda)$ is called the *scattering function*.

Similarly we can define the Fourier transform of $R_H(\Delta f;\Delta t)$ with respect to the variable $\Delta t$ as

$$S_H(\Delta f;\lambda) = \int_{-\infty}^{+\infty} R_H(\Delta f;\Delta t)\; e^{-j2\pi\lambda\Delta t}\cdot d(\Delta t). \tag{1.15}$$

Clearly, because of the Fourier transform relationship between $R_h(\tau;\Delta t)$ and $R_H(\Delta f;\Delta t)$, the scattering functions defined in (1.14) and (1.15) are Fourier transform pairs.

**Example 1.2** To show that $S_H(\Delta f;\lambda)$ and $S_h(\tau;\ \Delta t)$ are related by a Fourier transform we use the result of Example 1.1

$$\begin{aligned}
S_H(\Delta f;\lambda) &= \int_{-\infty}^{+\infty} R_H(\Delta f;\Delta t)\; e^{-j2\pi\lambda\Delta t} d(\Delta t) \\
&= \int_{-\infty}^{+\infty}\int_{-\infty}^{+\infty} R_h(\tau;\Delta t)\; e^{-j2\pi\Delta f\tau} e^{-j2\pi\lambda\Delta t}\; d(\Delta t)\; d\tau \\
&= \int_{-\infty}^{+\infty}\left[\int_{-\infty}^{+\infty} R_h(\tau;\ \Delta t) e^{-j2\pi\lambda\Delta t}\; d(\Delta t)\right] e^{-j2\pi\Delta f\tau} d\tau \\
&= \int_{-\infty}^{+\infty} S_h(\tau;\ \lambda) e^{-j2\pi\Delta f\tau} d\tau.
\end{aligned}$$

## Doppler Spread

Suppose a single tone (unmodulated sine wave) at relative frequency $f$ transmitted over a fading channel. In this case the envelope of the received signal is $H(f;t)$ and has the autocorrelation function $R_H(0;\Delta t)$ at relative frequency $f$. Setting $\Delta f$ to zero in (1.15) the Fourier transform of $R_H(0;\Delta t)$ becomes

$$\begin{aligned} S_H(0;\lambda) &= \int_{-\infty}^{+\infty} R_H(0;\Delta t)e^{-j2\pi\lambda\Delta t}d(\Delta t) \\ &\equiv S_H(\lambda), \end{aligned} \tag{1.16}$$

which indicates the power spectrum of the received signal. It is observed that, for a sine wave transmitted over a time-variant fading channel, the channel spreads the spectrum of the received signal over a range of frequency around $f$. Note that for a time-invariant channel $R_H(0;\Delta t)$ is a constant, and hence, $S_H(\lambda)$ becomes a delta function indicating that there is no spectral spreading observed in the transmission of a single tone.

The function $S_H(\lambda)$, defined in (1.16), is called the *Doppler power spectrum* of the channel and its nominal width is termed the *Doppler spread* $B_D$. A closely related parameter to the Doppler spread is the *coherence time*. The coherence time is the time span $\Delta t$ for which $R_H(0;\Delta t)$ remains nearly constant. Once again, because of the Fourier transform relationship between $S(\lambda)$ and $R_H(0;\Delta t)$, it is apparent that the coherence time is reciprocally related to the Doppler spread $B_D$.

**Spread factor**

The multipath spread (or equivalently coherence bandwidth) and Doppler spread (or equivalently coherence time) are main parameters which characterize the appropriate channel model for the specified signal. To be more specific, consider a symbol waveform of duration $T$. The required bandwidth for such waveform is roughly equal to $\frac{1}{T}$. The coherent processing of this waveform requires that the complex gain is essentially fixed over at least one signaling interval. This condition holds if

$$T \ll \frac{1}{B_D}. \tag{1.17}$$

In such a case the channel is said to be a *slow-fading channel.*

As noted earlier, in order for the fading to be nonselective (flat), the symbol duration $T$ must be much more than the multipath spread $T_M$, that is

$$T \gg T_M. \tag{1.18}$$

The conditions of slow and nonselective fading, given in (1.17) and (1.18), lead to the following double inequality

$$T_M \ll T \ll \frac{1}{B_D},$$

implying that for a slow flat-fading channel the product $B_D T_M$ must satisfy $B_D T_M \ll 1$. The product $B_D T_M$ is called the *spread factor* of the channel. The channels satisfying $B_D T_M \ll 1$ are said to be *underspread.*

**Example 1.3** Suppose that for a multipath fading channel the scattering function $S_h(\tau;\ \lambda)$ is uniform for the range of values $0 \leq \tau \leq 1$ msec and $-0.5 \leq \lambda \leq 0.5$ Hz. The symbol duration of the transmitted waveforms is selected to be $T = 10$ msec.

**i)** Determine the Doppler power spectrum $S_H(\lambda)$ and the autocorrelation function $R_H(0; \Delta t)$.

**ii)** Determine the spread factor of the channel.

**iii)** Determine whether the channel is selective or nonselective (flat) fading.

**iv)** Determine whether the channel is slow or fast fading.

**i)** Suppose that $S_h(\tau; \lambda) = A$ for the range of values $0 \leq \tau \leq 1$ msec and $-0.5 \leq \lambda \leq 0.5$ Hz. Using the result of Example 1.2 we have

$$\begin{aligned} S_H(\lambda) &\equiv S_H(0;\ \lambda) = \int_{-\infty}^{+\infty} S_h(\tau;\ \lambda)\, d\tau \\ &= \int_0^{10^{-3}} A d\tau = 10^{-3} A. \end{aligned}$$

Thus

$$S_H(\lambda) = \begin{cases} 10^{-3} A\ ; & -0.5 \text{ Hz} \leq \lambda \leq 0.5 \text{ Hz}, \\ 0\ ; & \text{Otherwise}. \end{cases}$$

Since the autocorrelation function $R_H(0; \Delta t)$ is the inverse Fourier transform of $S_H(\lambda)$, we have

$$\begin{aligned} R_H(0; \Delta t) &= \int_{-\infty}^{+\infty} S_H(0; \lambda) e^{j2\pi\lambda\Delta t} d\lambda \\ &= \int_{-0.5}^{0.5} 10^{-3} A e^{j2\pi\lambda\Delta t} d\lambda \\ &= 10^{-3} A \frac{\sin(\pi \Delta t)}{\pi \Delta t}. \end{aligned}$$

**ii)** The multipath spread $T_M$ is

$$T_M = 1\text{msec}.$$

The Doppler spread $B_D$ can be represented by the bandwidth of the rectangular spectrum $S_H(\lambda)$, i.e.

$$B_D = 1 \text{ Hz}.$$

In this case the spread factor of the channel is

$$B_D T_M = 10^{-3}.$$

Since $B_D T_M \ll 1$ the channel is underspread.

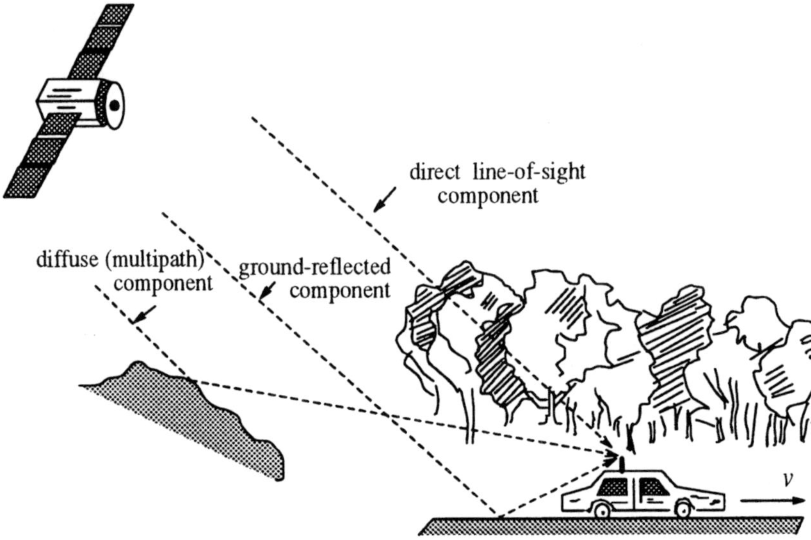


Figure 1.3: Mobile Radio Channel model.

**iii)** Since $T_M = 1$ msec $\ll T = 10$ msec the channel is nonselective (flat).

**iv)** Since $T = 10$ msec $\ll \frac{1}{B_D} = 1$ sec then the fading is slow-fading. Note that $R_H(0; \Delta t) \approx 10^{-3} A$ (constant) for $\Delta t \leq T$.

# 1.3 MOBILE RADIO CHANNEL CHARACTERISTICS

In mobile radio channels the received signal power varies due to signal *shadowing* and multipath fading. The shadowing of the received signal, which is caused by obstacles such as buildings, hills, trees, etc., changes the local mean of the received signal as the vehicle moves. In the following we characterize the statistics of the mobile radio channels and introduce some known statistical models for such channels.

A typical propagation environment for a mobile-satellite channel is shown in Figure 1.3. In general, the received signal consists of the sum of three components: the direct *line-of-sight* (LOS) component, the *ground-reflected* component, and the *diffuse* (multipath) component. Because of the received antenna pattern and the negative elevation angle of the ground-reflected component, the received antenna attenuates this reflection substantially [108]. Hence, the ground-reflected component can be either ignored or lumped into the direct LOS component. In this case the received signal can be viewed as the sum of only the direct line-of-sight and diffuse (multipath) components.

The direct LOS component is subjected to the shadowing effects caused by hills, trees, and various man-made structures. The amount of shadowing depends on various parameters such as the elevation angle and the area in which the vehicle moves. For example, in a rural (suburban) region with sparse tree cover, the line-of-sight component experiences a *light* shadowing, while in a region with a dense tree cover the shadowing is *heavy*.

As noted earlier (see Section 1.2), the multipath component of the received signal is characterized by a Rayleigh distributed envelope and a uniformly distributed phase.

Besides the random variation in the received signal envelope and its phase, a Doppler shift is associated with each signal component depending on the carrier frequency, the mobile speed, and the arrival angle. This Doppler shift is expressed as (see Problem 1.2)

$$f_d = \frac{v}{c} f_c \cos \alpha$$

where $v$ is the mobile speed, $f_c$ is the carrier frequency, $c$ is the speed of the light and $\alpha$ shows the angle that the propagation path makes with the mobile velocity vector. For $\alpha = 0$ the maximum value of the Doppler shift is obtained as

$$B \equiv (f_d)_{max} = \frac{v}{c} f_c.$$

For an omnidirectional mobile antenna and received plan waves uniformly distributed in arrival angle, the Doppler power spectrum $S_H(\lambda)$ and autocorrelation function $R_H(0; \Delta t)$ of a land-mobile radio channel can be represented as (see Problem 1.3)

$$S_H(\lambda) = \frac{A}{\pi\sqrt{B^2 - \lambda^2}}, \qquad |\lambda| < B, \tag{1.19}$$

and

$$R_H(0; \Delta t) = A J_0(2\pi B \Delta t) \tag{1.20}$$

where $A$ is a constant and $J_0(x) = \sum_{n=0}^{+\infty}(-1)^n \left[\frac{x^n}{2^n n!}\right]^2$ is the zero-order Bessel function of the first kind.

## 1.3.1 Shadowed Rician Channel

The probability functions used to describe the land-mobile satellite channels are the Rayleigh and Rician functions for multipath effects and the lognormal functions for shadowing. When the received signal consists of a multipath component, together with a direct (unshadowed) line-of-sight component its envelope is Rician distributed (see (1.4)). In this case the distribution of the received signal phase is given by (1.5). Under the condition of no line-of-sight component or complete blockage, only the multipath component is present and the channel is characterized by Rayleigh statistics.

As stated earlier, in a mobile radio channel the direct line-of-sight component undergoes shadowing effect. The shadowing of the LOS component is caused by the absorption and scattering of the incident direct wave by roadside trees or other obstacles as the wave is transmitted between the satellite and the vehicle. There are various types of statistical models available to describe the channel in which the LOS component is subjected to shadowing [138], [110], [112], [113], [114], and [115]. We only describe Loo's model [138], [110] but refer interested readers to the above-mentioned references for further reading.

A suitable model for the land mobile-satellite channel assumes that [138] the received signal is the sum of a line-of-sight component, $\gamma$, which is lognormally distributed, and a Rayleigh distributed multipath component. When the line-of-sight component is temporarily kept constant, then a conditional Rician distribution of the signal envelope holds, that is

$$p_{A|\Gamma}(a|\gamma) = \frac{a}{\sigma_a^2} \cdot e^{-(a^2+\gamma^2)/2\sigma_a^2} I_0(\frac{a\gamma}{\sigma_a^2}), \tag{1.21}$$

where $\sigma_a^2$ shows the average of the scattered power due to the multipath. Also, $I_0(\cdot)$ represents the zero-order modified Bessel function of the first kind. The probability density function (pdf) of the fading envelope, $a$, is obtained by averaging the conditional pdf of (1.21) with respect to the lognormally distributed process $\gamma$. The pdf of $\gamma$ is

$$P_\Gamma(\gamma) = \frac{1}{\sqrt{2\pi}\sigma_\gamma \cdot \gamma} e^{-(\ln\gamma - m_\gamma)^2/2\sigma_\gamma^2}, \tag{1.22}$$

where $\sigma_\gamma$ and $m_\gamma$ are the standard deviation and mean, respectively. Thus,

$$\begin{aligned} P_A(a) &= \int_0^\infty p_{A|\Gamma}(a \mid \gamma)\, p_\Gamma(\gamma)\, d\gamma \\ &= \frac{a}{\sqrt{2\pi}\,\sigma_a^2\sigma_\gamma} \int_0^\infty \frac{1}{\gamma} e^{-\left[\frac{(\ln\gamma - m_\gamma)^2}{2\sigma_\gamma^2} + \frac{(a^2+\gamma^2)}{2\sigma_a^2}\right]} \cdot I_0(\frac{a\gamma}{\sigma_a^2})\, d\gamma. \end{aligned} \tag{1.23}$$

The parameters $\sigma_a$, $\sigma_\gamma$, and $m_\gamma$ determine the degree of shadowing. As an example, Table 1.1 shows the values of these parameters used for the Canadian Mobile Satellite (MSAT) channel [116]. In this table the terms light, average, and heavy are used to represent an increasing effect of the lognormal process or the degree of shadowing.

The probability density function of the fading amplitude, given in (1.23), can be approximated for large and small values of $a$. It is shown that [138]

$$p_A(a) \approx \begin{cases} \frac{a}{\sigma_a^2} e^{-(\frac{a^2}{2\sigma_a^2})} & \text{for} \quad a \ll \sigma_a, \\ \frac{1}{\sqrt{2\pi}\,\sigma_\gamma a} e^{-(\ln a - m_\gamma)^2/2\sigma_\gamma^2} & \text{for} \quad a \gg \sigma_a. \end{cases}$$

Table 1.1: Channel Model Parameters for a Mobile Shadowed Rician Fading Channel.

| Degree of Shadowing | $\sigma_a^2$ | $m_\gamma$ | $\sigma_\gamma$ |
|---|---|---|---|
| Light | 0.1580 | 0.115 | 0.115 |
| Average | 0.1260 | -0.115 | 0.161 |
| Heavy | 0.0631 | -3.910 | 0.806 |

Hence, for small values of $a$ the fading amplitude is nearly Rayleigh distributed and for large values of $a$ the fading amplitude is nearly lognormal. For intermediate values of $a$ the fading amplitude pdf, $p_A(a)$, can be found by numerical integration.

## 1.4 FADING CHANNEL SIMULATOR

The performance evaluation of various modulation and coding strategies over a fading channel can be performed using a channel simulator. Various techniques on hardware and software simulators for different types of fading have been reported in the literature [117], [118], [119], and [111]. In the following, we describe a software channel simulator that allows different types of fading statistics to be modeled.

A general block diagram of a flat fading channel simulator is shown in Figure 1.4. The complex channel gain is generated by three independent zero mean Gaussian random variables (GRV) as shown in Figure 1.4. These random variables are filtered by a low-pass filter to shape the fading spectrum produced by the simulator. Depending on the Doppler power spectrum of the desired fading process an appropriate transfer function is chosen for low-pass filters.

Three types of fading channels, i.e., Rayleigh, Rician, and shadowed Rician fading channels, can be simulated using the general block diagram of Figure 1.4. This is shown schematically by the key S in this figure. If the key S is placed in position 2, then the complex channel gain consists of only in-phase and quadrature components. These two components are zero mean Gaussian random processes with variance $\sigma_a^2$ and shaped power spectrum corresponding to the Doppler power spectrum of the desired channel. For a normalized Rayleigh fading channel $\sigma_a^2$ is set to 1/2.

The Rician fading simulator is obtained by setting the key S to position 3. In this case a constant value of $\sqrt{2K\sigma_a^2}$ is added to the complex Gaussian gain,

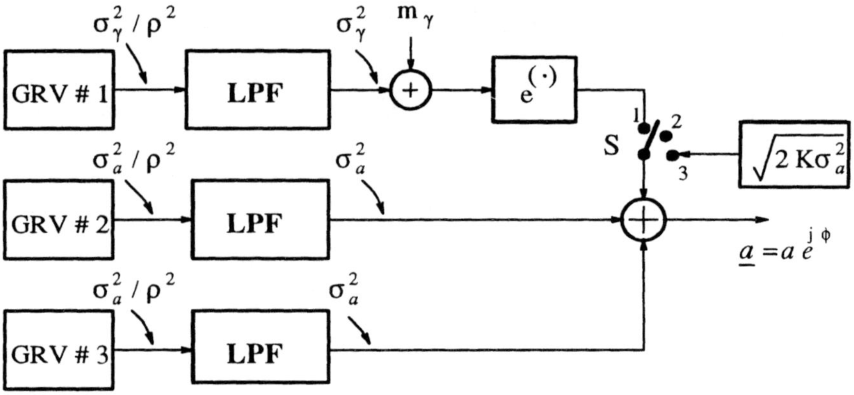


1 : Rician Shadowed Fading Channel
2 : Rayleigh Fading Channel
3 : Rician Fading Channel

Figure 1.4: The block diagram of the fading channel simulator.

where $K$ is the Rician parameter (see Section 1.2).

Setting the key S to position 1 simulates the shadowed Rician fading channel, introduced in Section 1.3. In this case, the parameters $\sigma_a^2$, $m_\gamma$, and $\sigma_\gamma^2$ are chosen according to Table 1.1.

## Problems

**1.1** The Gaussian variables $X$ and $Y$ are independent with zero-mean and equal variance $\sigma^2$. Show that

**a)** the random variable $Z = \sqrt{X^2 + Y^2}$ has a Rayleigh distribution with a probability density function

$$p_Z(z) = \frac{z}{\sigma^2} e^{-z^2/2\sigma^2}$$

**b)** the random variable $W = \sqrt{(X+a)^2 + Y^2}$ has a Rician distribution with a probability density function

$$p_W(w) = 2w(1+K)e^{-(K+(1+K)w^2)} I_0\left(2w\sqrt{K(K+1)}\right)$$

where $a$ is a constant, $K = a^2/2\sigma^2$, and $I_0(x) = \frac{1}{2\pi}\int_0^{2\pi} e^{x\cos t} dt$.

**c)** the random variable $\Phi = \tan^{-1}(Y/X)$ is uniformly distributed with a pdf of

$$p_\Phi(\phi) = \frac{1}{2\pi}, \qquad 0 \leq \phi < 2\pi.$$

**d)** the random variable $\Omega = \tan^{-1}[Y/(X+a)]$ has a pdf of

$$p_\Omega(w) = \frac{1}{2\pi} e^{-K} + \frac{\sqrt{K}\cos w e^{-K\sin^2 w}}{2\sqrt{\pi}} \left[2 - \mathrm{erfc}(\sqrt{\mathrm{K}}\cos\omega)\right], \qquad |w| \leq \pi.$$

where

$$\mathrm{erfc(x)} = \frac{2}{\sqrt{\pi}} \int_{\mathrm{x}}^{+\infty} \mathrm{e}^{-t^2} \mathrm{dt}.$$

and $K$ is defined as in **b)** above.

**1.2** The incident wave in Figure P1.2 is an unmodulated sinusoidal signal represented as

$$s(t) = \mathrm{Re}\left\{a e^{j2\pi f_c t}\right\}$$

where $a$ is a constant.

**a)** Show that the signal received by the vehicle can be represented as

$$r(t) = \mathrm{Re}\left\{b e^{j[2\pi(f_c - f_d)t + \phi]}\right\}$$

where $f_d = \frac{v}{c} f_c \cos\alpha$,
$c$ is the speed of the light,
$v$ is the velocity of the vehicle.

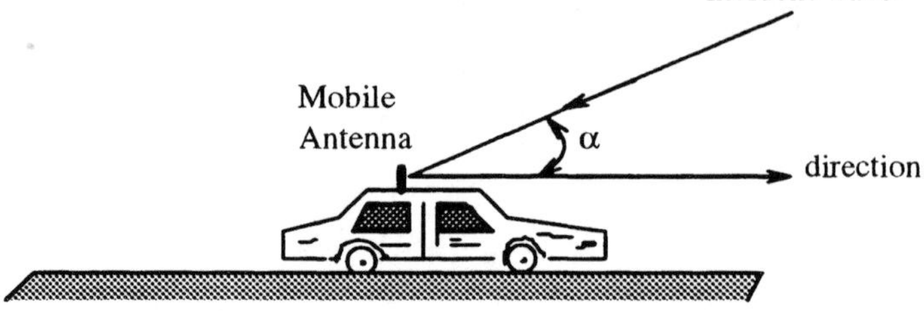


**Figure P1.2**

**b)** Show that the equivalent low-pass transfer function of the channel can be represented as

$$H(f_d;t) = de^{-j(2\pi f_d t + \phi)}$$

where $d$ is a constant.

**c)** Assume that the velocity of the vehicle is a random variable $V$ with a pdf of $p_V(v)$. Show that the Doppler power spectrum of the channel is

$$S_H(\lambda) = \frac{d^2}{2} p_V\left(\frac{c\lambda}{f_c}\cos\alpha\right)$$

**d)** Prove that the motion of the vehicle causes the spectral spreading.

**1.3** Consider a land-mobile radio channel with all reflected diffuse waves traveling in a plane. The incident wave at an arrival angle $\alpha$ (Figure P1.2) has a power angular density of $P(\alpha)$. The gain of the mobile antenna is $G(\alpha)$.

**a)** Prove that for a given Doppler frequency shift $\lambda = \frac{v}{c} f_c \cos\alpha$ the Doppler power spectrum $S_H(\lambda)$ satisfies the following relation

$$|S_H(\lambda)d\lambda| = [G(\alpha)P(\alpha) + G(-\alpha)P(-\alpha)]d\alpha$$

where $d\alpha$ and $d\lambda$ are the differential angle and differential frequency respectively.

**b)** Show that

$$S_H(\lambda) = \frac{G(\alpha)P(\alpha) + G(-\alpha)P(-\alpha)}{\sqrt{B^2 - \lambda^2}}$$

where $B = \frac{v}{c} f_c$.

c) Prove that for an omnidirectional mobile antenna and received plan waves uniformly distributed in arrival angle the Doppler power spectrum can be represented as

$$S_H(\lambda) = \frac{A}{\pi\sqrt{B^2 - \lambda^2}}, \qquad |\lambda| < B,$$

where $A$ is a constant.
Show that the autocorrelation function $R_H(0; \Delta t)$ has the following form

$$R_H(0; \Delta t) = A J_0(2\pi B \Delta t)$$

where $J_0(.)$ is the zero-order Bessel function of the first kind.

**1.4** a) Show that the zero-order Bessel function of the first kind $J_0(x) = \sum_{n=0}^{+\infty}(-1)^n \left[\frac{x^n}{2^n n!}\right]^2$ is a solution of the following differential equation

$$x\frac{d^2y}{dx^2} + \frac{dy}{dx} + xy = 0$$

b) Show that

$$\begin{aligned} J_0(x) &= \frac{1}{2\pi}\int_0^{2\pi} e^{jx\sin\theta}d\theta = \frac{1}{\pi}\int_{-\pi/2}^{\pi/2} e^{jx\sin\theta}d\theta \\ &= \frac{1}{\pi}\int_0^{\pi} e^{jx\cos\theta}d\theta \\ &= \frac{2}{\pi}\int_0^{\pi/2} \cos(x\sin\theta)d\theta \\ &= \frac{2}{\pi}\int_0^{\pi/2} \cos(x\cos\theta)d\theta \\ &= \frac{2}{\pi}\int_0^{1} \frac{\cos xt}{\sqrt{1-t^2}}dt. \end{aligned}$$

c) Show that the zero-order modified Bessel function of the first kind $I_0(x) = \sum_{n=0}^{+\infty}\left[\frac{x^n}{2^n n!}\right]^2$ is a solution of the following differential equation

$$x\frac{d^2y}{dx^2} + \frac{dy}{dx} - xy = 0$$

d) Show that

$$\begin{aligned} I_0(x) &= J_0(jx) \\ &= \frac{1}{2\pi}\int_0^{2\pi} e^{x\cos\theta}d\theta \end{aligned}$$

and

$$I_0(x) \approx \frac{e^x}{\sqrt{2\pi x}}, \qquad x >> 1.$$

**1.5** Other shapes of Doppler spectra of practical interest have been summarized by Mason [100]. They are

**Gaussian** (for aeronautical channels [129])

$$S_H(\lambda) = \frac{A}{\sqrt{2\pi}B} exp\left(-\frac{\lambda^2}{2B^2}\right)$$

**First-Order Butterworth**

$$S_H(\lambda) = \frac{A}{\pi B\left(1 + \frac{\lambda^2}{B^2}\right)}$$

**Second-Order Butterworth**

$$S_H(\lambda) = \frac{A}{1 + 16\frac{\lambda^4}{B^4}}$$

where $A$ and $B$ are constant. $B$ represents the Doppler spread of the channel.

**a)** Derive the autocorrelation function $R_H(0; \Delta t)$ for these channels.

**b)** Consider the coherence time as the time span $\Delta t$ for which $R_H(0; \Delta t)$ remains nearly constant with a variation of no more than 0.1%. Derive the coherence time of the above channels in terms of B.

**1.6** Consider the received signal $r(t)$ in a scattering medium discussed in Section 1.2. It can be represented as

$$r(t) = x(t)\cos w_c t + y(t)\sin w_c t$$

Assume that $x(t)$ and $y(t)$ are two correlated Gaussian processes with zero mean, equal variance $\sigma^2$ and a correlation coefficient $\rho$. Show that the pdf of the envelope of $r(t)$ is

$$p_Z(z) = \frac{z}{(1-\rho^2)\sigma^2} exp\left(\frac{-z^2}{2(1-\rho^2)\sigma^2}\right) I_0\left(\frac{\rho z^2}{2(1-\rho^2)\sigma^2}\right)$$

Verify that for $\rho = 0$, i.e., $x(t)$ and $y(t)$ are independent, the envelope of $r(t)$ has a Rayleigh probability distribution.

**1.7 Nakagami Distribution (m-Distribution)**

In general, for a channel with only diffused multipath components the fading amplitude is Nakagami distributed with a pdf [143]

$$p_A(a) = \frac{2}{\Gamma(m)} \left(\frac{m}{2\sigma^2}\right)^m a^{2m-1} e^{-ma^2/2\sigma^2}$$

where $\Gamma(x)$ is the Gamma function defined as

$$\Gamma(x) = \int_0^{+\infty} y^{x-1} e^{-y} dy, \qquad x > 0,$$

and $m$ is the Nakagami parameter ($m \geq 0.5$).

**a)** Show that the Gamma function has the following properties:

i) $\Gamma(1/2) = \sqrt{\pi}$
ii) $\Gamma(3/2) = \sqrt{\pi}/2$
iii) $\Gamma(x+1) = x\Gamma(x)$
iv) $\Gamma(n+1) = n!$ if n is an integer $\geq 0$
v) $(2N+1)!! \equiv (2N+1)(2N-1)\cdots 5.3.1$
$= \frac{\Gamma(2N+2)}{2^N \Gamma(N+1)}$, for integer $N \geq 0$

**b)** Show that $m$ is equal to the ratio

$$\frac{E^2\{A^2\}}{E\{(A^2 - E\{A^2\})^2\}},$$

i.e., $m$ represents the ratio of the square of the mean of the received energy to the variance of the received energy.

*Hint*: see Problem 1.9.

**c)** Verify that the Rayleigh pdf is a special case of the Nakagami pdf with $m = 1$. Show that the $k^{th}$ moment of a Rayleigh-distributed fading amplitude $A$ is

$$E\{A^k\} = (2\sigma^2)^{k/2}\Gamma\left(1 + \frac{k}{2}\right)$$

and the variance of $A$ is

$$\sigma_A^2 = E\{(A - E\{A\})^2\} = \left(2 - \frac{\pi}{2}\right)\sigma^2$$

Note that the Rayleigh-distributed fading amplitude $A$ can be considered as

$$A = \sqrt{X^2 + Y^2}$$

where $X$, $Y$ are statistically independent, identitically distributed, Gaussian variables with zero mean and variance $\sigma^2$.

## 1.8 Central Chi-Square Distribution

Consider the random variable $Z$ defined as

$$Z = \sum_{n=1}^{N} X_n^2$$

where $X_n$'s are statistically independent and identically distributed Gaussian random variables with zero mean and variance $\sigma^2$. Show that

**a)** the characteristic function of $X_n$ is

$$\Omega_{X_n}(jw) \equiv E\{e^{jwX_n}\} = e^{-w^2\sigma^2/2}$$

(Note: See Problem 5.23 for more discussions on the characteristic function).

**b)** the probability density function of $Z_n = X_n^2$ is

$$p_{Z_n}(z) = \frac{1}{\sqrt{2\pi\sigma^2 z}} e^{-z/\sigma^2}$$

**c)** the characteristic function of $Z_n$ is

$$\Omega_{Z_n}(jw) = \frac{1}{\sqrt{1 - j2\sigma^2 w}}$$

**d)** the characteristic function of $Z$ is

$$\Omega_Z(jw) = (1 - j2\sigma^2 w)^{-N/2}$$

**e)** the probability density function of $Z$ is

$$p_Z(z) = \frac{1}{\Gamma(N/2)(2\sigma^2)^{N/2}} z^{\frac{N}{2}-1} e^{-z/2\sigma^2}, \quad z \geq 0,$$

where $\Gamma(\cdot)$ is the Gamma function defined in Problem 1.7.
(Note: This pdf is called a (central) chi-square (or Gamma) pdf with $N$ degrees of freedom).

**f)** the first two moments of $Z$ are

$$E\{Z\} = N\sigma^2$$
$$E\{Z^2\} = (2 + N))N\sigma^4$$

and the variance of $Z$ is

$$\sigma_Z^2 = 2N\sigma^4$$

**1.9 Chi-distribution**

Consider the random variable

$$A = \sqrt{\sum_{n=1}^{N} X_n^2}$$

where $X_n$'s are statistically independent and identically distributed Gaussian random variables with zero mean and variance $\sigma^2$

**a)** Using the results of Problem 1.8, show that the pdf of $A$ is

$$p_A(a) = \frac{a^{N-1}}{2^{(N-2)/2}\Gamma(\frac{N}{2})\sigma^N} e^{-a^2/2\sigma^2}, \quad a \geq 0$$

where $\Gamma(\cdot)$ is the Gamma function defined in Problem 1.7 and the $k$th moment of $A$ is

$$E\{A^k\} = (2\sigma^2)^{k/2} \frac{\Gamma([N+k]/2)}{\Gamma(N/2)}, \quad k \geq 0.$$

Find the variance of $A$.

**b)** Show that, for $N = 2$, $A$ is a Rayleigh-distributed random variable.

**c)** Show that, for $N = 2m$, the pdf of $A$ is a special case of the Nakagami pdf discussed in Problem 1.7 with $m$ being an integer $\geq 1$, i.e., the pdf of $A$ can be rewritten as

$$p_A(a) = \frac{2}{\Gamma(m)} \left(\frac{m}{2\sigma_m^2}\right)^m a^{2m-1} e^{-ma^2/2\sigma_m^2}$$

where $\sigma_m^2 = \sigma^2/m$. Show that

$$m = \frac{E^2\{A^2\}}{E\{(A^2 - E^2\{A^2\})^2\}},$$

i.e., $m$ represents the inverse of the normalized variance of $A^2$.

# Chapter 2

# DIGITAL TRANSMISSION OVER FADING CHANNELS

Our objective in this chapter is to show the potential gain of bandwidth-efficient coded modulation schemes on bandwidth-limited fading channels. The treatment of this problem is general and does not include any specific coded scheme. With this objective in mind a general coded transmission system over fading channels will be described. To study the performance of the coded signaling schemes over fading channels the random coding bounding technique will be employed. In this approach the error bound is evaluated for an ensemble of coded signaling schemes rather than for a specific one. This yields an appropriate parameter, namely *cut-off rate*, which can be used for comparing different signaling schemes over a *Discrete Memoryless Channel* (DMC). We will evaluate this parameter for some two-dimensional signaling schemes, for different decoding strategies, on a Rayleigh fading channel and discuss the results. We will conclude this discussion by showing that the potential coding gain for bandwidth-efficient coded-modulation schemes over a Rayleigh fading channel is much more than that of the Additive White Gaussian Noise (AWGN) channel.

## 2.1 ELEMENTS OF DIGITAL COMMUNICATION SYSTEMS

The basic elements of a coded digital communication system are shown by the general block diagram in Figure 2.1. A sequence of digital information is produced by the data source at a rate of $\mathcal{R}$ information bits/sec. The channel

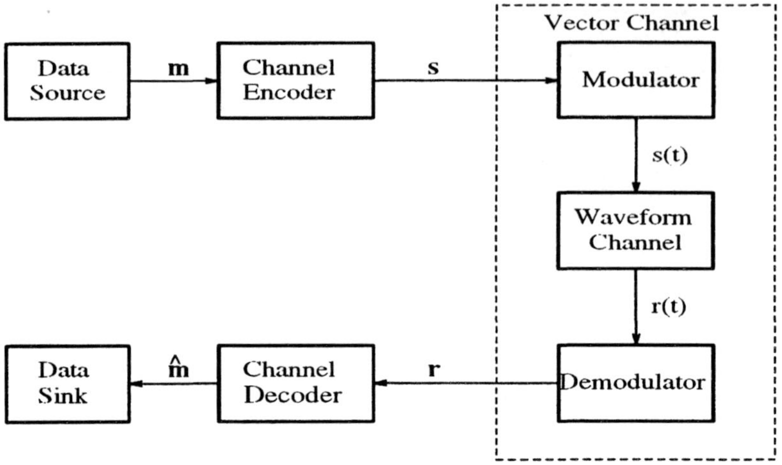


Figure 2.1: Block diagram a coded digital communication system.

encoder adds some redundancy to the incoming bits and delivers the encoded message to the modulator in the form of a signal vector s of $N$ components. Each component $s_n, n = 0, 1, \cdots, N-1$, ranges over the transmitter alphabet of $q$ elements $\{a_l\}$. The modulator transforms the signal vector to a waveform using a one-to-one mapping. In each time interval of $T$ sec the modulator transmits one of $M = 2^{\mathcal{R}T}$ possible waveforms $s_i(t)$, $i = 0, 1, \cdots, (M-1)$, corresponding to the input signal vector. This is called $M$-ary modulation. The transmitted waveforms are subjected to channel noise and distortion. In an additive Gaussian noise channel, only Gaussian-distributed white noise is added, while in a multiplicative fading channel, the amplitude and phase of the transmitted waveforms are distorted as well.

At the receiver the demodulator produces the corrupted vector **r** of $N$ components. The demodulator can quantize each received component into $q$ levels. In such a case, we say that the demodulator makes a *hard decision*. As an alternative, the demodulator may make $Q$-ary decision, where $Q > q$. In this case it is said to make *soft decision*. In its extreme case, $Q \to \infty$, no quantization is performed. Based on the quantized or unquantized information furnished by the demodulator, the decoder attempts to reconstruct the original transmitted message from the knowledge of the code used by the encoder and the redundancy contained in the received data.

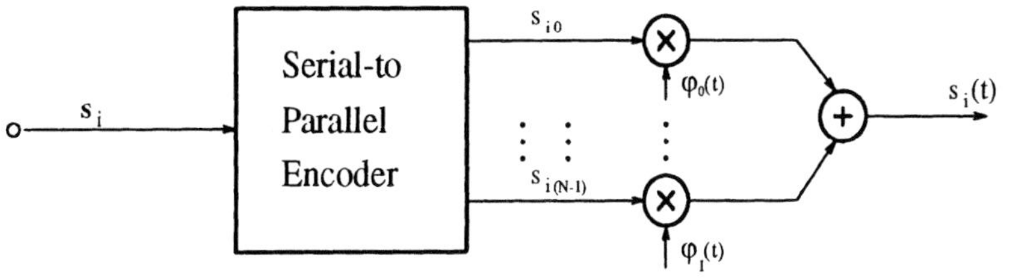


Figure 2.2: Block diagram of a modulator.

## 2.2 SIGNAL REPRESENTATION AND SIGNAL SPACE

From the analysis point of view, the function of the encoder-modulator can be divided into separate discrete-time and continuous-time operations using the *Gram-Schmidt orthogonalization* procedure [3, Appendix 2A], [17, Appendix 4A] (see Problem 2.1). This procedure permits us to expand any one of the M finite-energy time functions $s_i(t)$, $i = 0, 1, \cdots, M-1$, as linear combinations of $N \leq M$ *orthonormal basis functions* over the finite time interval $0 \leq t \leq T$, i.e.,

$$s_i(t) = \sum_{n=0}^{N-1} s_{in}\phi_n(t), \qquad i = 0, 1, \cdots, M-1, \tag{2.1}$$

where $\phi_n(t)$, $n = 0, 1, \cdots, N-1$, are N orthonormal functions defined as

$$\int_0^T \phi_l(t)\ \phi_j^*(t)dt = \delta_{lj} = \begin{cases} 1 & l = j, \\ 0 & l \neq j, \end{cases}$$

and the coefficient $s_{in}$ shows the projection of $s_i(t)$ onto the function $\phi_n(t)$ over the time interval $0 \leq t \leq T$, i.e.,

$$s_{in} = \int_0^T s_i(t) \cdot \phi_n^*(t)\ dt. \tag{2.2}$$

Using (2.1) the function of the modulator is shown schematically in Figure 2.2.

The orthogonalization procedure provides us with a tool to represent the waveforms $\{s_i(t)\}$ as the points of an *N-dimensional signal space* formed by the orthonormal basis, $\{\phi_i(t)\}$. The idea of signal space is of fundamental importance, for it visualizes the design of communication systems from an abstract point of view using the language of geometry. Each of the waveforms, $s_i(t)$, $i = 0,\ 1,\ \cdots,\ M-1$, are completely determined by the vector of their coefficients

$$\mathbf{s}_i = (s_{i0}, s_{i1}, \cdots, s_{i(N-1)}), \qquad i = 0, 1, \cdots, M-1,$$

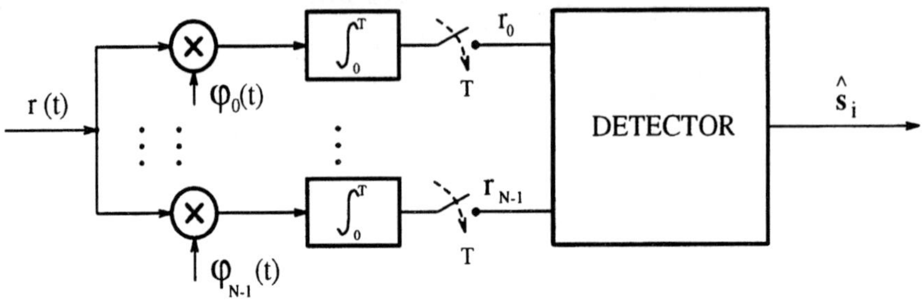


Figure 2.3: Block digram of a demodulator.

and their energies are interpreted as the norms of the corresponding vectors

$$E_s \equiv \int_0^T |s_i(t)|^2 \, dt = \sum_{n=0}^{N-1} |s_{in}|^2 \equiv |s_i|^2.$$

The distance between two signal waveforms $s_i(t)$ and $s_j(t)$ can be viewed as the *Euclidean distance* between the corresponding signal points in $N$-dimensional signal space

$$d_E^2 = \int_0^T |s_i(t) - s_j(t)|^2 \, dt = |s_i - s_j|^2.$$

To recover the transmitted vectors from the received waveforms the virtue of the orthonormality of the $\{\phi_i(t)\}$ is used at the receiver. This is done by multiplying the received signal by each of the orthonormal basis function and integrating the product over the interval $0 \leq t \leq T$. The implementation of this procedure is shown schematically in Figure 2.3.

For a multiplicative fading channel, the input to the receiver is a random process expressed as

$$r(t) = a(t)s_i(t) + n(t), \qquad 0 \leq t \leq T, \tag{2.3}$$

where $n(t)$ is the additive white Gaussian noise with a two-sided spectral density of $N_0/2$ and $a(t)$ is the fading process. Assume that $a(t)$ is slow enough to have a constant envelope $a$ and a constant phase[1] $\phi$ over the interval $0 \leq t \leq T$. In this case the integrator outputs, i.e.,

$$r_j \equiv \int_0^T r(t) \, \phi_j(t) dt, \qquad j = 0, 1, \cdots, N-1,$$

[1] It is assumed that the phase $\phi$ is known at the receiver, and that the receiver compensates the phase variations of the received signal by multiplying it by $\exp(-\phi)$. This kind of receiver is called *coherent receiver* and is considered in this chapter. Also we assume that the fading envelopes are independent random variables. This is achieved by using theoretically infinite size interleaver and de-interleaver in the coded communication system.

are random variables that constitute the components of a random vector

$$\mathbf{r} \equiv (r_{i0}, r_{i1}, \cdots, r_{i(N-1)}).$$

Considering (2.3) we have

$$\mathbf{r} = a\mathbf{s}_i + \mathbf{n}, \tag{2.4}$$

where

$$\mathbf{n} \equiv (n_{i0}, n_{i1}, \cdots, n_{i(N-1)})$$

is a random vector whose components, i.e.,

$$n_j \equiv \int_0^T n(t)\phi_j(t)dt, \qquad j = 0, 1, \cdots, N-1, \tag{2.5}$$

are independent zero-mean Gaussian random variables with variances $N_0/2$.

**Example 2.1** In order to verify that $\{n_j\}$, given in (2.5), are independent zero mean Gaussian random variables with variances $N_0/2$ we can write

$$\begin{aligned} E[n_j] &= E[\int_0^T n(t)\ \phi_j(t)\ dt] \\ &= \int_0^T E[n(t)]\ \phi_j(t)\ dt \\ &= 0. \end{aligned}$$

To find the variance of $n_j$ we have

$$\begin{aligned} \sigma_{n_j}^2 &= \frac{1}{2}\ E[|n_j|^2] \\ &= \frac{1}{2}E[\int_0^T \int_0^T n(t)\ n^*(u)\ \phi_j(t)\phi_j^*(u)dtdu] \\ &= \int_0^T \int_0^T R_n(t-u)\phi_j(t)\phi_j^*(u)dtdu \end{aligned}$$

where $R_n(t-u) = \frac{1}{2}\ E[n(t)\ n^*(u)]$ is the autocorrelation of the noise process $n(t)$. For a white Gaussian process with a double-sided spectral density of $\frac{N_0}{2}$ we have

$$R_n(t-u) = \frac{N_0}{2}\ \delta(t-u). \tag{2.6}$$

Using (2.6) the variance of $n_j$ is obtained as

$$\begin{aligned} \sigma_{n_j}^2 &= \int_0^T \int_0^T \frac{N_0}{2}\ \delta(t-u)\ \phi_j(t)\ \phi_j^*(u)\ dt\ du \\ &= \frac{N_0}{2} \int_0^T |\phi_j(t)|^2\ dt \\ &= \frac{N_0}{2}. \end{aligned}$$

To show that the components $n_i$ and $n_j$ $(i \neq j)$ are independent we have

$$\begin{aligned}
\frac{1}{2}E[n_i\, n_j^\star] &= \frac{1}{2}E[\int_0^T \int_0^T n(t)\, n^\star(u)\, \phi(t)\, \phi_j^\star(u) dt du] \\
&= \int_0^T \int_0^T R_n(t-u)\, \phi_i(t)\, \phi_j^\star(u)\, dt\, du \\
&= \int_0^T \int_0^T \frac{N_0}{2}\, \delta(t-u)\, \phi_i(t)\, \phi_j^\star(u)\, dt\, du \\
&= \frac{N_0}{2} \int_0^T \phi_i(t)\, \phi_j^\star(t)\, dt \\
&= 0,
\end{aligned}$$

which, since the variables $n_i$ and $n_j$ are Gaussian, implies that they are also independent.

Assuming that $n(t)$, hence $\mathbf{n}$, is statistically independent of $\mathbf{s}_i$, it is apparent that the vector $\mathbf{r}$ is a random vector whose components, $\{r_j\}$, are statistically independent Gaussian random variables with mean

$$E[r_j] = a \cdot s_{ij},$$

and variances $N_0/2$. It follows that the conditional density of $\mathbf{r}$, given the fading envelope $a$ and the transmitted vector $\mathbf{s}_i$, is

$$\begin{aligned}
p_N(\mathbf{r}|a, \mathbf{s}_i) &= \prod_{j=0}^{N-1} p(r_j|a, s_{ij}) \\
&= \prod_{j=0}^{N-1} \frac{1}{\sqrt{\pi N_0}} e^{-|r_j - a s_{ij}|^2/N_0}. \qquad (2.7)
\end{aligned}$$

## 2.3 MAXIMUM LIKELIHOOD RECEIVER

Based on the observation of the noisy received signal, the receiver decides which one of the signal points in the signal space is transmitted. A popular criterion for this decision is minimizing the probability of error.

Suppose the receiver observes $\mathbf{r}$ as the received signal and makes the decision $\hat{\mathbf{s}} = \mathbf{s}_i$ as the transmitted signal. The conditional error probability of such a decision, given the received signal $\mathbf{r}$, is

$$P(\mathcal{E}|\mathbf{r}) = Pr\{s_i \text{ not transmitted}|\mathbf{r}\}.$$

In this case the conditional probability of correct decision, given the received signal $\mathbf{r}$, is

$$P(\mathcal{C}|\mathbf{r}) = Pr\{s_i \text{ transmitted}|\mathbf{r}\}.$$

It is clear that, to minimize the error probability, the optimum receiver should choose $\hat{\mathbf{s}} = \mathbf{s}_i$ for which the conditional probability $P(\mathcal{C}|\mathbf{r})$ is maximum. Thus the optimum decision rule, based on minimizing the error probability, can be formulated as

$$\hat{\mathbf{s}} = \mathbf{s}_i \quad \text{if} \quad Pr\{\mathbf{s}_i \text{ transmitted } |\mathbf{r}\} \geq Pr\{\mathbf{s}_j \text{ transmitted } | \mathbf{r}\}, \quad \forall_{i \neq j}. \tag{2.8}$$

Such a receiver is called *Maximum A Posteriori Probability* (MAP) receiver.

Representing the a posteriori probability Pr{ $\mathbf{s}_i$ transmitted$|\mathbf{r}$} simply as $P(\mathbf{s}_i|\mathbf{r})$ and using the Bayes's rule we can write

$$P(\mathbf{s}_i|\mathbf{r}) = \frac{P(\mathbf{s}_i) p_{R|\mathbf{S}}(\mathbf{r}|\mathbf{s}_i)}{p_R(\mathbf{r})}, \tag{2.9}$$

where $P(\mathbf{s}_i)$ and $p_{R|\mathbf{S}}(\mathbf{r} \mid \mathbf{s}_i)$ show the a priori probability of the transmitted signal $\mathbf{s}_i$ and the *channel transition probability density function*, respectively. Substituting (2.9) into (2.8) the MAP decision rule can also be expressed as

$$\hat{\mathbf{s}} = \mathbf{s}_i \qquad P(\mathbf{s}_i) p_{R|\mathbf{S}}(\mathbf{r}|\mathbf{s}_i) \geq P(\mathbf{s}_j)\, p_{R|\mathbf{S}}(\mathbf{r} \mid \mathbf{s}_j), \quad \forall_{i \neq j}, \tag{2.10}$$

where $p_R(\mathbf{r})$ is ignored since it is independent of the index $i$.

Finally, for equally likely transmitted signals $P(\mathbf{s}_i) = P(\mathbf{s}_j)$, (2.10) is simplified to

$$\hat{\mathbf{s}} = \mathbf{s}_i \qquad p_{R|\mathbf{S}}(\mathbf{r} \mid \mathbf{s}_i) \geq p_{R|\mathbf{S}}(\mathbf{r} \mid \mathbf{s}_j), \quad \forall_{i \neq j}. \tag{2.11}$$

The decision rule, given in (2.11), and corresponding receiver are then referred to as *Maximum Likelihood* (ML). The channel transition probability $p_{R|\mathbf{S}}(\mathbf{r}|\mathbf{s}_i)$ is also called the *likelihood function*.

### Decision Regions

Using the MAP or ML rules the signal space can be partitioned into disjoint regions, each containing a signal point. These regions are called the decision regions. The decision region $\Lambda_i$ for a signal point $s_i$ is the set of all points such that whenever the received signal falls into $\Lambda_i$ the optimum receiver sets $\hat{s} = s_i$.

The concept of decision regions can be best visualized by considering an AWGN channel. Assuming that $\mathbf{r} = (r_0, r_1, \cdots, r_{N-1})$ is the received signal the likelihood function $p_{R|\mathbf{S}}(\mathbf{r}|\mathbf{s}_i)$ can be written as (see(2.7))

$$p_{R|\mathbf{S}}(\mathbf{r} \mid \mathbf{s}_i) = (\frac{1}{\pi N_0})^{N/2} \cdot e^{-\frac{1}{N_0}\sum_{k=0}^{N-1} |r_k - a s_{ik}|^2}, \tag{2.12}$$

where $a$ represents the channel gain. For the AWGN channel $a$ is constant and can be set to one for convenience. Substituting (2.12) into (2.11) the ML rule for the AWGN channel can be expressed as

$$\hat{\mathbf{s}} = \mathbf{s}_i \qquad d_E^2(\mathbf{r}, \mathbf{s}_i) \leq d_E^2(\mathbf{r}, \mathbf{s}_j), \qquad \forall_{i \neq j}, \tag{2.13}$$

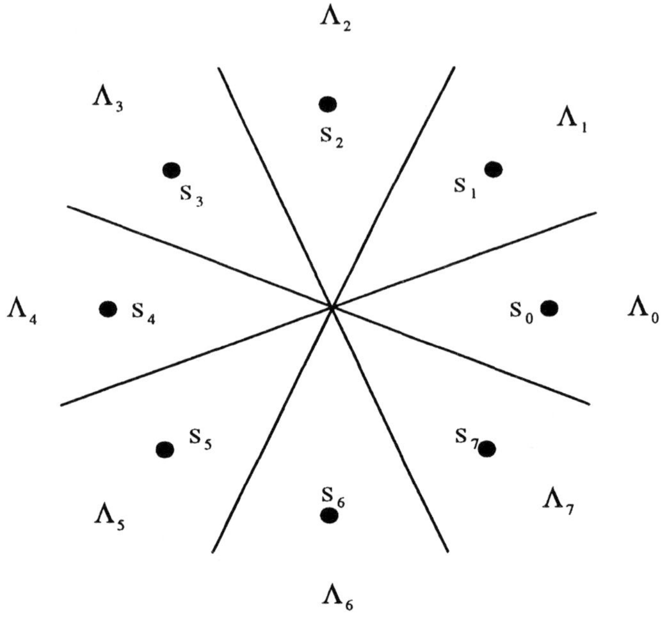


Figure 2.4: Decision regions for a signal set with 8 signals.

where $d_E^2(\mathbf{r},\ \mathbf{s}_i) = \sum_{k=0}^{N-1} |r_k - as_{ik}|^2$ is the Euclidean distance between the received signal $\mathbf{r}$ and the transmitted signal $\mathbf{s}_i$. Thus, in the AWGN channel the decision region $\Lambda_i$ for the signal point $\mathbf{s}_i$ is the set of all points that are closer to $\mathbf{s}_i$ than to any other signal point of the signal set. The decision regions for a two-dimensional signal set with eight signal points are shown in Figure 2.4 as an example.

## 2.4 PERFORMANCE OF SOME MODULATION TECHNIQUES IN AN AWGN CHANNEL

In this section we evaluate the performance of the coherent detection of some known signaling techniques in an AWGN channel.

### 2.4.1 Binary Modulation Techniques

In a binary modulation technique the modulator generates one of the two waveforms $s_0(t)$ or $s_1(t)$ corresponding to "0" or "1", respectively. In the following we evaluate the performance of two binary modulation techniques, namely, BPSK and BFSK, in an AWGN channel.

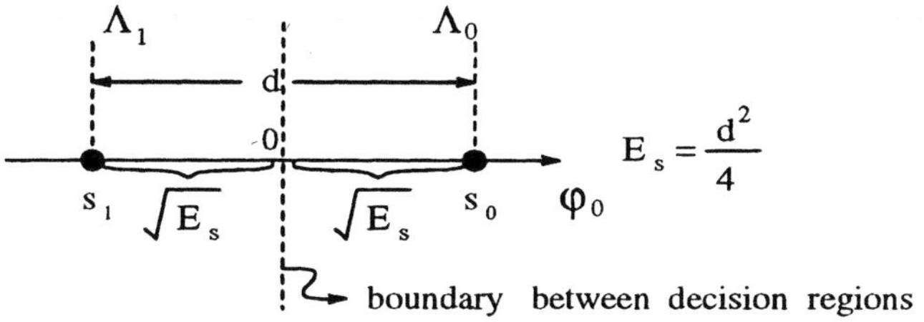


Figure 2.5: Signal space representation of BPSK signals.

### Binary Phase-Shift Keying (BPSK)

For the binary phase-shift keying the signals $s_0(t)$ and $s_1(t)$ are chosen to be *antipodal*, that is

$$s_1(t) = -s_0(t).$$

These two signals can be represented in one-dimensional space. Figure 2.5 shows the signal space and the signal points corresponding to BPSK. Representing the Euclidean distance of signals $s_0$ and $s_1$ by $d$, the average energy of the signal set is

$$E_s = \frac{1}{2}\frac{d^2}{4} + \frac{1}{2}\frac{d^2}{4} = \frac{d^2}{4}.$$

Assuming an ML receiver the boundary of the decision regions is a vertical line passing the origin. To compute the probability of error $P_e$ we have

$$\begin{aligned} P_e &= \frac{1}{2}P_{e|s_0} + \frac{1}{2}P_{e|s_1} \\ &= P_{e|s_0}, \end{aligned}$$

where $P_{e|s_i}, i = 0, 1$, is the probability of error when $s_i, i = 0, 1$, is transmitted.

Next we derive an expression for $P_{e|s_0}$. Considering that $\mathbf{r}$ is a Gaussian random variable with mean $d/2$ and variance $N_0/2$ the probability of error when $s_0$ is transmitted is obtained as

$$\begin{aligned} P_{e|s_0} &= \int_{-\infty}^{0} p(r \mid s_0)\, dr \\ &= \int_{-\infty}^{0} \frac{1}{\sqrt{\pi N_0}} e^{-(r-d/2)^2/N_0}\, dr \\ &= Q\left(\frac{d}{\sqrt{2N_0}}\right), \end{aligned}$$

where

$$Q(x) \equiv \frac{1}{\sqrt{2\pi}} \int_x^{\infty} e^{-t^2/2} dt.$$

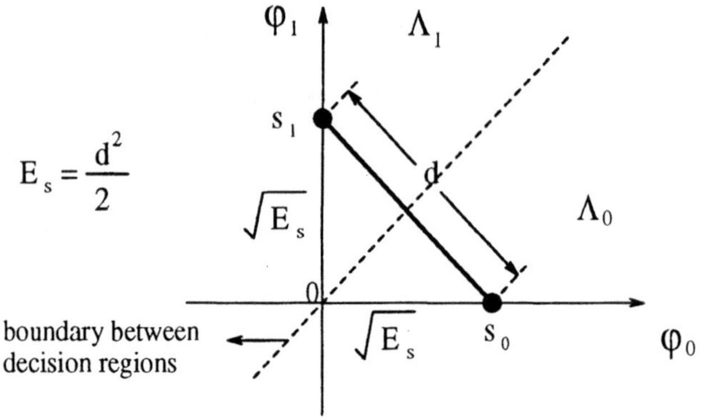


Figure 2.6: Signal space representation of BFSK signals.

Note that

$$\mathrm{erfc}(x) = \frac{2}{\sqrt{\pi}} \int_x^{+\infty} e^{-t^2} dt = 2Q(x\sqrt{2}) \quad \text{(see Problem 2.9)}.$$

Substituting $d = 2\sqrt{E_s}$ in the above equation yields the probability of error for BPSK

$$P_e = Q(\sqrt{\frac{2E_s}{N_0}}). \tag{2.14}$$

**Binary Frequency-Shift Keying (BFSK)**

Binary frequency shift keying (BFSK) is a special case of orthogonal signaling. In BFSK signaling the two waveforms $s_0(t)$ and $s_1(t)$ are two sine functions with frequencies $f_0$ and $f_1$, respectively. Using the orthogonality condition, that is,

$$\int_0^T s_0(t)\ s_1(t)\ dt = 0,$$

it can be shown that the orthogonality for $s_0(t)$ and $s_1(t)$ is obtained when $f_1 - f_0$ is equal to a multiple of $1/2T$ (see Problem 2.3).

The BFSK signals are represented in a two-dimensional signal space (see Figure 2.6). The average energy of this signal set can be obtained as

$$E_s = \frac{d^2}{2},$$

where $d$ is the Euclidean distance between signal points $s_0$ and $s_1$.

To compute the error probability of BFSK signaling we use the fact that the probability of error in the AWGN channel is unaffected by the signal set rotation or translation [17, Chapter 4]. This follows from the fact that the probability density function of the additive noise is spherically symmetric and independent of the transmitted signal. Thus, using the results of the probability of error of BPSK signal, the probability of error for BFSK signals is obtained as (see Problem 2.3)

$$\begin{aligned} P_e &= Q(\frac{d}{\sqrt{2N_0}}) \\ &= Q(\sqrt{\frac{E_s}{N_0}}). \end{aligned}$$

### 2.4.2 Non-Binary Modulation Techniques

The non-binary modulation techniques consist of waveforms $s_0(t), s_1(t), \cdots, s_{M-1}(t)$, that modulate $m = \log_2 M$ bits at a time into the signaling waveform. Among various types of these signaling schemes we consider those which have the throughput (bandwidth efficiency) more than 1 bit/s/Hz and are used in bandwidth limited channels.

#### M-ary Amplitude Modulation (M-AM)

In M-ary amplitude modulation the waveforms $s_i(t), i = 0, 1, \cdots, M-1$, are represented as

$$s_i(t) = A_i\ \phi_0(t), \qquad i = 0, 1, \cdots, M-1,$$

where $\{A_i\}$ are amplitudes having M discrete values

$$A_i = (M - 1 - 2i)\ \frac{d}{2}, \qquad i = 0, 1, \cdots, M-1.$$

Figure 2.7 shows the 4-AM and 8-AM signals as an illustrative example.

The average energy of M-AM signal set can be obtained as

$$\begin{aligned} E_s &= \frac{1}{M}\sum_{i=0}^{M-1} A_i^2 \\ &= \frac{d^2}{4M}\sum_{i=0}^{M-1}(M-1-2i)^2 \\ &= \frac{M^2-1}{12}\ d^2. \end{aligned}$$

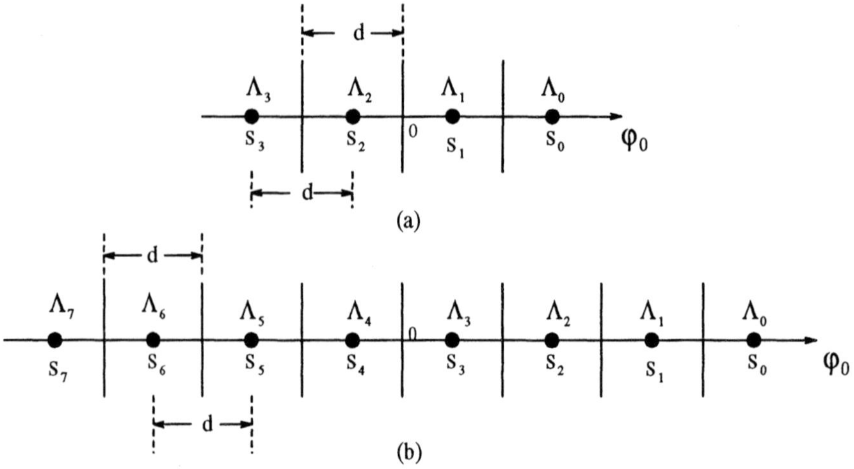

Figure 2.7: Signal space representation of M-AM signals: a) 4-AM, b) 8-AM.

Considering the fact that in uncoded M-ary signaling each signal carries $m = \log_2 M$ information bits, the average energy per information bit, $E_b$, can be related to $E_s$ as

$$E_b = E_s/m = E_s/\log_2 M. \tag{2.15}$$

Next we derive an expression for probability of error. In this case it is easier to compute the probability of correct detection first. Referring to Figure 2.7 we observe that the probability of correct decision when each of the inside signals, that is, $s_i$, $i = 1, \cdots, M-2$, is transmitted is the same. However, the probability of correct decision for either of the two outside signals, that is, $s_0$ and $s_{M-1}$, is different from the inside ones. On the basis that all transmitted signal are equiprobable, we can write

$$\begin{aligned} P_c &= \frac{M-2}{M} P_{c|s_1} + \frac{2}{M} P_{c|s_0} \\ &= \frac{M-2}{M} \int_{-d/2}^{d/2} \frac{1}{\sqrt{\pi N_0}} e^{-\frac{x^2}{N_0}} \cdot dx + \frac{2}{M} \int_{-d/2}^{\infty} \frac{1}{\sqrt{\pi N_0}} e^{-\frac{x^2}{N_0}} \cdot dx \\ &= 1 - \frac{2(M-1)}{M} Q(\frac{d}{\sqrt{2N_0}}). \end{aligned}$$

The probability of error, $P_e$, is obtained as

$$\begin{aligned} P_e &= 1 - P_c \\ &= \frac{2(M-1)}{M} Q(\frac{d}{\sqrt{2N_0}}). \end{aligned}$$

Representing $d$ in term of $E_s$ in the above equation yields

$$\begin{aligned} P_e &= \frac{2(M-1)}{M} Q\left(\sqrt{\frac{6}{(M^2-1)} \cdot \frac{E_s}{N_0}}\right) \\ &= \frac{2(M-1)}{M} Q\left(\sqrt{\frac{6\log_2 M}{(M^2-1)} \cdot \frac{E_b}{N_0}}\right) \end{aligned} \tag{2.16}$$

### M-ary Quadrature Amplitude Modulation (M-QAM)

The general form of M-ary quadrature amplitude modulation waveforms is

$$s_i(t) = A_i\phi_0(t) + B_i\phi_1(t), \qquad i = 0, 1, \cdots, M-1,$$

where $\{A_i\}$ and $\{B_i\}$ have discrete values and $\phi_0(t)$ and $\phi_1(t)$ are two orthonormal functions. In Figure 2.8 the signal set for some M-QAM signals is shown as an illustrative example. When $m = \log_2 M$ is an even number the M-QAM signal set can be constructed by choosing two coordinates $A_i$ and $B_i$ independently from a $2^{m/2}$-point M-AM signal set. For example, the 16-QAM signals, shown in Figure 2.8 can be considered as two 4-AM signals transmitted independently in two dimensions $\phi_0$ (in-phase) and $\phi_1$ (quadrature). This kind of MQAM signals is called *strict-sense QAM* (SSQAM).

The probability of error for an M-ary strict-sense QAM can be easily obtained from that of a $\sqrt{M}$-ary AM. In this case the probability of correct detection of M-ary SSQAM is expressed as

$$P_{c-MQAM} = P^2_{c-\sqrt{M}AM},$$

which yields

$$\begin{aligned} P_e &= 1 - P_{c-MQAM} \\ &= P_{e-\sqrt{M}AM}(2 - P_{e-\sqrt{M}AM}). \end{aligned} \tag{2.17}$$

Using (2.16), $P_{e-\sqrt{M}AM}$ can be obtained as

$$P_{e-\sqrt{M}AM} = \frac{2(\sqrt{M}-1)}{\sqrt{M}} Q\left(\sqrt{\frac{3}{(M-1)} \cdot \frac{E_s}{N_0}}\right), \tag{2.18}$$

where we have used the fact that the average energy in each $\sqrt{M}$-AM signal set is one-half of that in the QAM signal set. Substituting (2.18) into (2.17) yields

$$P_e = \frac{4(\sqrt{M}-1)}{\sqrt{M}} \cdot Q\left(\sqrt{\frac{3}{(M-1)} \cdot \frac{E_s}{N_0}}\right).$$

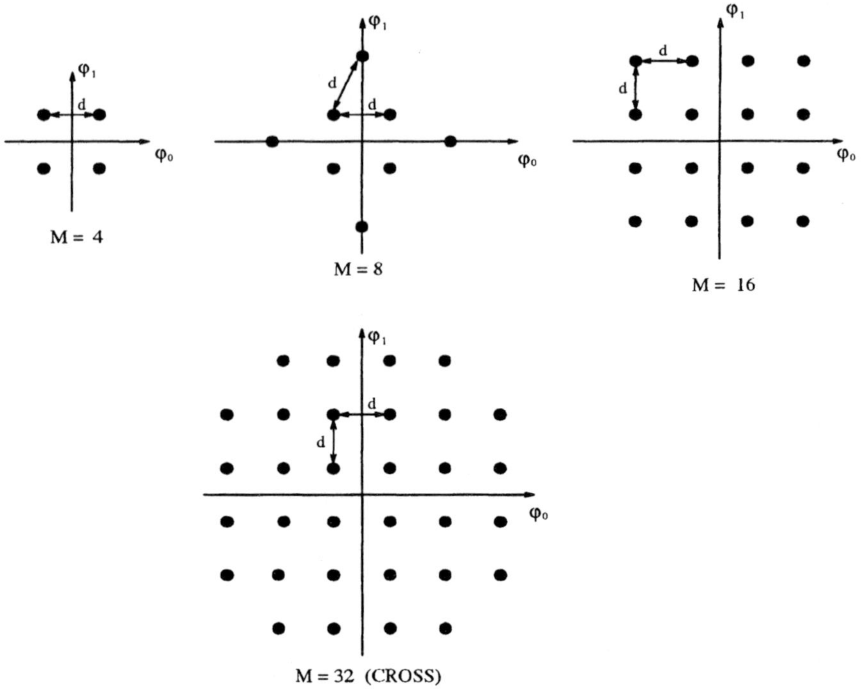


Figure 2.8: Some M-QAM signal sets.

$$\left[1 - \frac{(\sqrt{M}-1)}{\sqrt{M}}\, Q\left(\sqrt{\frac{3}{(M-1)} \cdot \frac{E_s}{N_0}}\right)\right], \qquad (2.19)$$

which can be expressed in terms of $\frac{E_b}{N_0}$ by using (2.15). For M-QAM signals in the general case an upper bound on $P_e$ can be derived (see Problem 2.7)

$$P_e \leq 4Q\left(\sqrt{\frac{3}{(M-1)} \cdot \frac{E_s}{N_o}}\right).$$

## M-ary Phase Shift-Keying Modulation (MPSK)

The general representation of MPSK waveforms is

$$s_i(t) = A\ \cos(\frac{2\pi}{M}i+\theta)\ \phi_0(t)\ +\ A\ \sin(\frac{2\pi}{M}i+\theta)\ \phi_1(t), \qquad i = 0,\ 1,\ \cdots,\ M-1,$$

where $\theta$ is a constant phase and $\phi_0(t)$ and $\phi_1(t)$ are two orthonormal functions. The MPSK waveforms have equal energy, and hence, in the signal space representation all the signal points are located on a circle with radius $A = \sqrt{E_s}$.

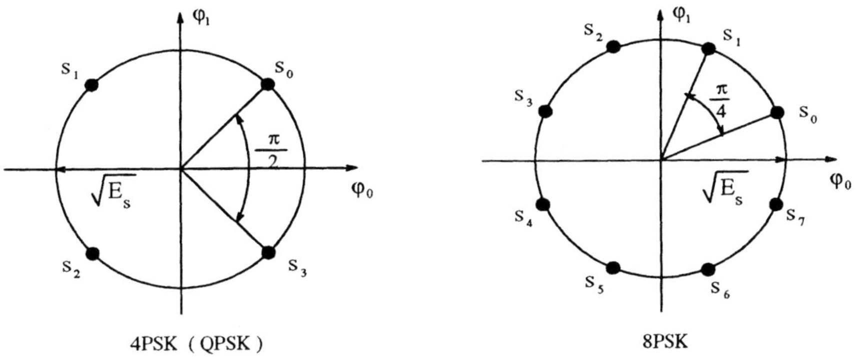


Figure 2.9: Some MPSK signal sets.

The signal sets of two examples of MPSK signals, i.e., QPSK and 8PSK, are shown in Figure 2.9. Note that QPSK scheme is the same as 4-QAM, and hence, its performance in an AWGN channel can be obtained from (2.19) by setting $M = 4$ (see Problem 2.8).

To evaluate the performance of MPSK signaling in the AWGN channel we derive upper and lower bounds on $P_e$ [120]. Considering equiprobable MPSK signal the probability of error $P_e$ is expressed as

$$P_e = P_{e|s_0},$$

where $P_{e|s_0}$ is the probability that $s_0$ is transmitted and the received signal falls outside of the decision region $\Lambda_0$, shown in Figure 2.10. A lower bound for $P_{e|s_0}$ can be obtained by calculating the probability that the received signal falls into the region $I_1$ (see Figure 2.10(a))

$$\begin{aligned} P_{e|s_0} &= Pr\{r \in \overline{\Lambda_0} \mid s_0\} \\ &> Pr\{r \in I_1 \mid s_0\} \\ &= Q(\frac{d}{\sqrt{2N_0}}) = Q\left(\sqrt{\frac{2E_s}{N_0}} \sin(\frac{\pi}{M})\right). \end{aligned}$$

Referring to Figure 2.10(b), the upper bound on $P_e$ is calculated as

$$\begin{aligned} P_{e|s_0} &= Pr\{r \in \overline{\Lambda_0} \mid s_0\} \\ &< Pr\{r \in I_1 \mid s_0\} + Pr\{r \in I_2 \mid s_0\} \\ &= 2Q(\frac{d}{\sqrt{2N_0}}) = 2Q\left(\sqrt{\frac{2E_s}{N_0}} \sin(\frac{\pi}{M})\right). \end{aligned}$$

The results of the foregoing discussions can be summarized as

$$Q\left(\sqrt{\frac{2E_s}{N_0}} \sin(\frac{\pi}{M})\right) < P_e < 2Q\left(\sqrt{\frac{2E_s}{N_0}} \sin(\frac{\pi}{M})\right).$$

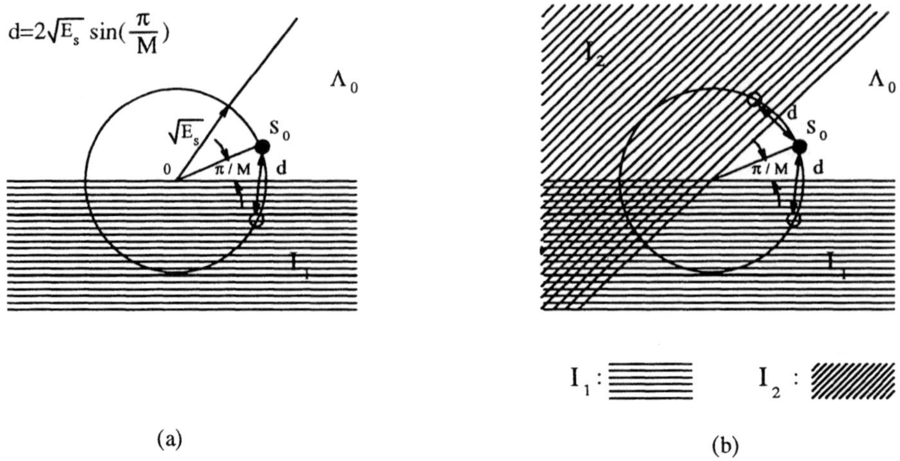


Figure 2.10: Illustration of decision regions used for computing error bounds of MPSK signals: a) lower bound, b)upper bound.

As the number of MPSK signals increases the error probability tends to the upper bound, i.e., $2Q\left(\sqrt{\frac{2E_s}{N_0}}\ \sin(\frac{\pi}{M})\right)$.

## 2.5 PERFORMANCE OF SOME MODULATION TECHNIQUES ON FADING CHANNELS

Using the results of the preceding section, the performance of the modulation schemes in a flat fading channel can be obtained. In this evaluation we assume that the fading is slow enough that the multiplicative channel gain, $a$, can be regarded as a constant during one signaling interval. In this case, assuming $s_i$ is transmitted, the received signal is expressed as (2.4).

To compute the error probability we first obtain the error probability conditioned on the fading amplitude $a$. This can be done easily by substituting $a^2E_s$ instead of $E_s$ in the expressions that we previously derived for the error probability in the AWGN channel. Averaging the conditional error probability over the probability density function of the fading amplitude yields the error probability in the flat fading channel.

**Example 2.2** Consider the BPSK signaling over a Rayleigh flat fading channel. The probability of error for this signaling scheme in the AWGN channel is given in (2.14). Using this equation the conditional error probability is expressed as

$$P_{e|a} = Q(a\sqrt{\frac{2E_s}{N_0}}).$$

To obtain the error probability we have

$$\begin{aligned} P_e &= \int_0^\infty p_A(a)\ Q(a\sqrt{\frac{2E_s}{N_0}})\ da \\ &= \int_0^\infty 2a\ e^{-a^2}\ Q(a\sqrt{\frac{2E_s}{N_0}})\ da \\ &= \frac{1}{2}\left(1-\sqrt{\frac{E_s/N_0}{1+E_s/N_0}}\right). \end{aligned}$$

For large signal-to-noise ratios, that is $\frac{E_s}{N_0} \gg 1$, the probability of error for BPSK signaling over a Rayleigh flat fading channel can be approximated as

$$P_e \approx \frac{1}{4E_s/N_0}.$$

**Example 2.3** The probability of error for BFSK signaling over a Rayleigh flat fading channel can be obtained by averaging the conditional error probability

$$P_{e|s_0} = Q\left(a\sqrt{\frac{E_s}{N_0}}\right),$$

over the pdf of the fading amplitude. The result is

$$P_e = \frac{1}{2}\left(1-\sqrt{\frac{E_s/N_0}{2+E_s/N_0}}\right),$$

which at high SNR's can be expressed as

$$P_e \approx \frac{1}{2E_s/N_0}.$$

The results of the foregoing examples show that the error probability in a fading channel decreases inversely with $E_s/N_0$. This is in contrast to the AWGN channel in which the error probability decreases exponentially with $E_s/N_0$ (for example, see (2.14)).

In order to reduce the error probability on a fading channel, the high error probability of a deep fade on a single transmission must be circumvented. This is accomplished by means of *diversity transmission* [17], [60], and [106]. One form of diversity transmission, called *time diversity*, involves sending a symbol $L_t$ times, in the hope that not all of the transmitted symbols will be subjected to deep fades. In other words, by transmitting the same information signal over independently fading channels, the probability that all the signal components will fade simultaneously is reduced considerably. In this technique the received signal performs some averaging to achieve an error performance that decreases exponentially with $E_s/N_0$ [17, Page 550]

$$P_e < e^{-0.149\ L_t(E_s/N_0)}.$$

From a coding point of view the time diversity technique can be regarded as a repetition (block) code of rate $1/L_t$. Since a repetition code is a simple form of coding, one might expect that selection of more efficient types of codes maintains the benefit of the time diversity technique. The order of time diversity provided by a code can be related to its minimum Hamming distance if maximum likelihood soft-decision decoding is considered (see Chapters 5 and 7).

## 2.6 ERROR BOUND AND CUT-OFF RATE OF CODED SCHEMES

To evaluate the error performance of a specific coded scheme, knowledge of the specific code and signal set is required. Exact expressions for the probability of error, in general, involve multidimensional integrals, which are complex to calculate. Upper bounds are hence developed [17], [3], which are applicable to any signal set. However, evaluation of these bounds for a specific signal set, other than a few simple cases, is essentially cumbersome. This becomes particularly prohibitive as the size of the signal set M and the dimensionality N becomes large.

These difficulties can be circumvented by bounding the probability of error for an *ensemble* of coded systems, rather than for a specific one. Strangely enough, it is much easier to find the error bounds not for just one communication system, but rather for a whole collection of communication systems, each consisting of an encoder, vector channel, and decoder (Figure 2.1). Since such an upper bound is the average of the probability of error over the entire ensemble of coded systems, it is obvious that at least one coded system must have a probability of error which is no greater than the ensemble average. Hence, the ensemble average is an upper bound of the probability of error for the optimum coded system with signal set of $M$ signal vectors of dimensionality N. This technique was first introduced by Shannon [121] and now is referred to as *random coding* bounding in information theory.

Using this technique Wozencraft and Jacobs show that the average error probability for an ensemble of the block coded system over a discrete memoryless channel is upper bounded as follows [17, page 392] (see Problem 2.13)

$$\overline{P_e} < 2^{-N(R_0 - R)}, \tag{2.20}$$

where $R$ is the information rate in bits per symbol.

Note that $M = 2^{NR} = 2^{\mathcal{R}T}$. The parameter $R_0$ is called the *cut-off rate* and is derived as [17, page396] (see Problem 2.13)

$$R_0 = -\log_2 \sum_{n=0}^{Q-1} \left[ \sum_{l=0}^{q-1} p_l \sqrt{p_{lh}} \right]^2, \tag{2.21}$$

where $q$ and $Q$ denote the size of transmitter alphabet $\{a_l\}$ and the receiver quantizer alphabet $\{b_h\}$, respectively. Referring to Figure 2.1 and discussions in Section 2.1, $p_l$ denotes the probability of transmitting symbol $a_l$, i.e.,

$$p_l \equiv P_r\{s_j = a_l\}.$$

$p_{lh}$ represents the channel transition probability, i.e.,

$$p_{lh} = P_r\{r_j = b_h | s_j = a_l\}.$$

We will use (2.21) to evaluate the cut-off rates of some known two-dimensional signaling schemes which employ hard-decision or erasure decoding techniques. In the following we extend the above results to fading channels with soft-decision decoding. The cut-off rate for such a case is derived in a manner similar to [17] for an ensemble of block coded systems.

Consider an ensemble of block codes with block length (dimensionality) of N and symbols chosen from a signal alphabet $\{a_l\}$ with cardinality of $q$. We assume that the codewords (signal vectors) of this ensemble are pairwise statistically independent, i.e.,

$$P(\mathbf{s}_i, \mathbf{s}_k) = P(\mathbf{s}_i) \cdot P(\mathbf{s}_k), \forall i \neq k, \tag{2.22}$$

and furthermore, the constituent components of each codeword have the same property[2]

$$P(\mathbf{s}_i) = \prod_{j=0}^{N-1} P(s_{ij}), \quad i = 0,\ 1,\ \cdots,\ M-1, \tag{2.23}$$

where M is the total number of codewords in the code.

To derive the upper bound of (2.20) for such an ensemble of codes we start with union bound. Assuming that the $i$th codeword $\mathbf{s}_i$ is transmitted, the probability of erroneous detection of this codeword is upper bounded by invoking the union bound,

$$P_{e|\mathbf{s}_i} \leq \sum_{\substack{k=0 \\ k \neq i}}^{M-1} P_2(\mathbf{s}_i, \mathbf{s}_k),$$

where $P_2(\mathbf{s}_i, \mathbf{s}_k)$ is called *pairwise error probability* and represents the probability of error when $\mathbf{s}_i$ is transmitted and $\mathbf{s}_k \neq \mathbf{s}_i$ is detected instead. By virtue of (2.22) and (2.23) the statistical average of $P_2(\mathbf{s}_i, \mathbf{s}_k)$ over the ensemble of coded system is independent of subscripts $i$ and $k$. Thus,

$$\overline{P_2(\mathbf{s}_i, \mathbf{s}_k)} = \overline{P_{2e}} \quad \forall i, k \neq i,$$

[2] In [17, Chap. 6] it is proved that these assumptions are true for an ensemble of parity check codes.

and

$$\overline{P_e} = \overline{P_{e|\mathbf{s}_i}} < M\overline{P_{2e}}, \tag{2.24}$$

where the bar sign denotes the averaging operation over all codes in the ensemble. As previously discussed, the number of codewords in the ensemble, $M$, is related to $R$ as

$$M = 2^{NR}. \tag{2.25}$$

Substituting (2.25) into (2.24) results in

$$\overline{P_e} < 2^{NR}\,\overline{P_{2e}}.$$

Hence, to obtain the desired bound of (2.20) we need only show that

$$\overline{P_{2e}} \leq 2^{-NR_0}. \tag{2.26}$$

### 2.6.1 Calculation of the Pairwise Error Probability

In fading channels, the decoding process uses a metric of the form $m(\mathbf{r}, \mathbf{s}, \mathbf{a})$ if an estimate of fading amplitude, $\mathbf{a}$, is available at the receiver, and $m(\mathbf{r}, \mathbf{s})$ if it is not. The estimate of the fading amplitude is called *Channel State Information* (CSI) [51]. Whichever metric is used, it is desirable, from the view point of simplifying the decoding process, that it has an additive property, namely, that the total metric for a codeword is the sum of metrics for the constituent components, i.e.,

$$m(\mathbf{r},\ \mathbf{s}) = \sum_{i=0}^{N-1} m(r_i,\ s_i).$$

If we consider the maximum likelihood criterion for the optimum decoder, then the ML metric is defined as

$$m(\mathbf{r}, \mathbf{s}) = \ln p_N(\mathbf{r}|\mathbf{s}, \mathbf{a}),$$

when CSI is available, and

$$m(\mathbf{r},\ \mathbf{s}) = \ln\ p_N(\mathbf{r}|\mathbf{s}),$$

if it is not. Note that for memoryless channels the above defined metrics have the additive property.

Assume that $\mathbf{s}$ is transmitted and $\mathbf{r}$ is received. Between the two codewords $\mathbf{s}$ and $\hat{\mathbf{s}}$ the optimum decoder incorrectly decides $\hat{\mathbf{s}}$ if and only if[3]

$$m(\mathbf{r},\ \hat{\mathbf{s}}) \geq m(\mathbf{r},\ \mathbf{s}),$$

[3] We drop subscripts $i$ and $k$ and use $\mathbf{s}$ and $\hat{\mathbf{s}}$ instead of $\mathbf{s}_i$ and $\mathbf{s}_k$ to show the fact that the pairwise error probability is independent of subscripts $i$ and $k$.

or equivalently

$$\sum_{i=0}^{N-1} m(r_i,\ \hat{s}_i) \geq \sum_{i=0}^{N-1} m(r_i,\ s_i).$$

Then the pairwise error probability is given by

$$P_2(\mathbf{s},\ \hat{\mathbf{s}}) = E_a\ [\ P_2(\mathbf{s},\ \hat{\mathbf{s}} \mid \mathbf{a})\ ], \tag{2.27}$$

where

$$P_2(\mathbf{s}, \hat{\mathbf{s}}|\mathbf{a}) = Pr\{m(\mathbf{r}, \hat{\mathbf{s}}) \geq m(\mathbf{r}, \mathbf{s})|\mathbf{s}, \mathbf{a}\} \tag{2.28}$$

is the pairwise error probability conditioned on the fading amplitude vector **a**, and E shows the statistical expectation operation. The conditional probability of (2.28) is upper-bounded using the Chernoff bound techniques [17], [4] (see Problem 2.11)

$$P_2(\mathbf{s},\ \hat{\mathbf{s}} \mid \mathbf{a}) \ \leq\ E_{\mathbf{r}|\mathbf{s}}\ [\exp(\lambda[m(\mathbf{r},\ \hat{\mathbf{s}})\ -\ m(\mathbf{r},\ \mathbf{s})])\ ], \tag{2.29}$$

where $\lambda \geq 0$ is the Chernoff parameter to be optimized for the tightest bound. Substituting (2.29) into (2.28) yields

$$P_2(\mathbf{s},\ \hat{\mathbf{s}}) \ \leq\ E_{\mathbf{a}}\left[E_{\mathbf{r}|\mathbf{s}}\ [\exp(\lambda[m(\mathbf{r},\ \hat{\mathbf{s}})\ -\ m(\mathbf{r},\ \mathbf{s})])\ ]\right]. \tag{2.30}$$

Defining the Chernoff bound between the codewords s and ŝ as

$$C(\mathbf{s},\ \hat{\mathbf{s}}, \lambda) \ \equiv\ E_{\mathbf{a}}\left[E_{\mathbf{r}|\mathbf{s}}\ [\exp(\lambda[m(\mathbf{r},\ \hat{\mathbf{s}})\ -\ m(\mathbf{r},\ \mathbf{s})])\ ]\right],$$

the pairwise error probability, given in (2.27), can be written as

$$P_2(\mathbf{s},\ \hat{\mathbf{s}}) \ \leq\ C(\mathbf{s},\ \hat{\mathbf{s}}, \lambda). \tag{2.31}$$

Averaging (2.31) over all codes in the ensemble and recalling that the constituent components of each codeword are statistically independent, we can write

$$\begin{aligned}\overline{P_{2e}} \ &\leq\ \overline{C(\mathbf{s}, \hat{\mathbf{s}}, \lambda)} \\ &=\ \prod_{i=0}^{N-1} \overline{C(s_i, \hat{s}_i, \lambda)} \\ &=\ [\ \overline{C(s, \hat{s}, \lambda)}]^N,\end{aligned}$$

where $C(s, \hat{s}, \lambda)$ is called the Chernoff factor of signal *components* $s$ and $\hat{s}$. Defining

$$R_0(\lambda) \equiv -\log_2 \overline{C(s, \hat{s}, \lambda)},$$

$\overline{P_{2e}}$ can be expressed as

$$\overline{P_{2e}} \ \leq\ 2^{-NR_0(\lambda)}, \qquad \lambda \geq 0. \tag{2.32}$$

The bound given in (2.32) is valid for any $\lambda \geq 0$. The parameter $\lambda$ should be chosen in such a way that the bound is as tight as possible. For this optimum $\lambda$, $R_0(\lambda)$ has the largest possible value. Defining

$$R_0 \equiv \max_{\lambda} R_0(\lambda) = R_0(\lambda_{opt}), \tag{2.33}$$

the desired result of (2.26) is obtained.

### 2.6.2 Calculation of $\overline{C(\mathrm{s}, \hat{\mathrm{s}}, \lambda)}$

Averaging $C(s, \hat{s}, \lambda)$ over all possible symbols $s$ and $\hat{s}$, we obtain

$$\overline{C(s, \hat{s}, \lambda)} = \sum_{i=0}^{q-1} \sum_{j=0}^{q-1} p(s_i)\, p(s_j)\, C(s_i, \hat{s}_j, \lambda).$$

Note that $p(s_i)$ and $p(s_j)$ are equivalent to $p_l$'s in (2.21). Then

$$R_0 = -\log_2 \min_{\lambda} \left\{ \sum_{i=0}^{q-1} \sum_{j=0}^{q-1} p(s_i) p(s_j) C(s_i, \hat{s}_j, \lambda) \right\}. \tag{2.34}$$

Assuming that the symbols are equally probable, i.e., $p(s_i) = p(s_j) = 1/q$, (2.34) turns out to be

$$R_0 = 2 \log_2 q - \log_2 \min_{\lambda} \left\{ \sum_{i=0}^{q-1} \sum_{j=0}^{q-1} C(s_i, \hat{s}_j, \lambda) \right\}. \tag{2.35}$$

Similar expression for cut-off rate is obtained in [55], [52] using the ensemble of convolutional coded schemes.

### 2.6.3 Cut-off Rate of Some Two-Dimensional Signal Sets

The cut-off rate, $R_0$, is the appropriate criterion for comparing different modulation schemes on a discrete memoryless channel [36]. The bigger that $R_0$ is for a given signal energy-to-noise ratio, the better the modulation system is. In the following, we calculate the cut-off rate for some known two-dimensional signal sets, and compare them based on the $R_0$ criterion over a Rayleigh fading channel.

To evaluate the cut-off rate of (2.35) the expression for the Chernoff factor is required. The Chernoff factors for fading channels with and without channel state information have been evaluated in [51] with an emphasis on MPSK signal sets. A procedure for evaluating these factors for general two-dimensional signal sets is included in Appendix A and the results are summarized in the following paragraphs.

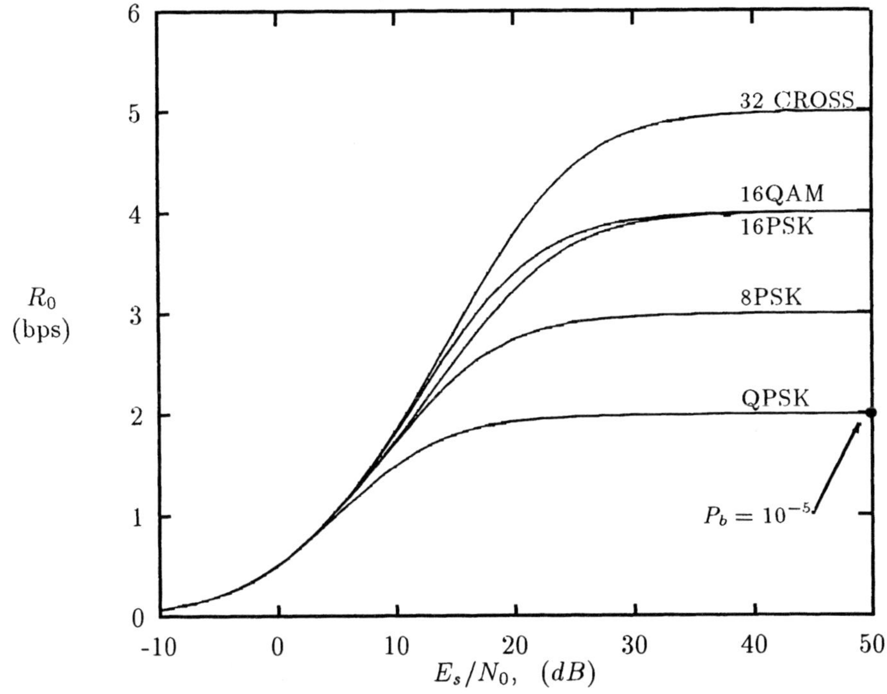


Figure 2.11: Cut-off rates for the Rayleigh fading channel with channel state information.

### Fading Channel with CSI

Assuming that the channel state information is available at the receiver, the Chernoff factor of signals $s$ and $\hat{s}$ for a Rician fading channel is derived as (see Appendix A)

$$\begin{aligned} C(s,\hat{s}) &\equiv C(s,\hat{s},\lambda_{opt}) \\ &= \frac{1+K}{1+K+\frac{1}{4N_0}|s-\hat{s}|^2} \exp\left(-\frac{\frac{K}{4N_0}|s-\hat{s}|^2}{1+K+\frac{1}{4N_0}|s-\hat{s}|^2}\right). \end{aligned} \quad (2.36)$$

The Chernoff factor for a Rayleigh fading channel is obtained from (2.36) by setting $K = 0$,

$$C(s,\hat{s}) = \frac{1}{1+\frac{1}{4N_0}|s-\hat{s}|^2}. \quad (2.37)$$

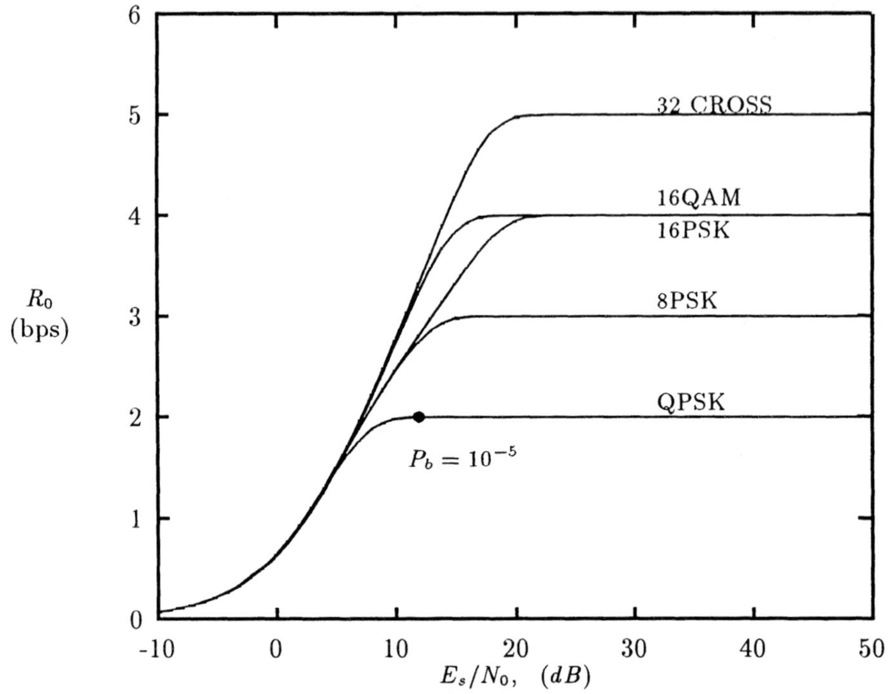


Figure 2.12: Cut-off rates for the AWGN channel.

**Example 2.4** Using (2.35) and(2.37), $R_0$ is plotted for QPSK, 8PSK, 16PSK, 16QAM, and 32 CROSS signal sets on a Rayleigh fading channel in Figure 2.11. The cut-off rate for the AWGN channel is obtained from (2.35) and (2.36) by setting $K = +\infty$. The result is consistent with the findings of [17, page 317] and is shown in Figure 2.12.

From Figures 2.11 and 2.12 it is apparent that at low signal-to-noise ratios the required $E_s/N_0$ to achieve a given $R_0$ in the Rayleigh fading channel is more than that of the AWGN channel. For example, if we consider the required $E_s/N_0$ for rates around 3/4 of the saturated rate, then the Rayleigh fading channel needs 4.5-6 dB more $E_s/N_0$ than the AWGN channel, depending on the selected signal set. However, at high signal-to-noise ratios both channels provide the same rate.

### Fading Channel without CSI

For fading channel without channel state information the Chernoff factor of signals $s$ and $\hat{s}$ is evaluated in Appendix A. For a Rician fading channel this factor is given as

$$C(s,\hat{s},\lambda) = e^{\mathcal{X}'\lambda/N_0}\frac{e^{-K}}{\pi}\int_0^{\pi}(1-\sqrt{\pi}\mathrm{erfc}(\nu)\nu e^{\nu^2})d\theta,$$

where

$$\nu = \frac{\mathcal{X}\lambda/N_0}{\sqrt{1+K}} - \sqrt{K}\cos\theta,$$

$$\mathcal{X} = |s|^2 - |\hat{s}|^2 + |s-\hat{s}|^2.$$

and

$$\mathcal{X}' = |s|^2 - |\hat{s}|^2 + \lambda|s-\hat{s}|^2.$$

The Chernoff factor for a Rayleigh fading channel (K=0) simplifies to

$$C(s,\hat{s},\lambda) \quad = \quad e^{\mathcal{X}'\lambda/N_0}(1-\sqrt{\pi}\mathrm{erfc}(\frac{\lambda}{\mathrm{N}_0}\mathcal{X})e^{(\mathcal{X}\lambda/2\mathrm{N}_0)^2}(\frac{\lambda}{2\mathrm{N}_0}\mathcal{X}). \qquad (2.38)$$

Note that for constant envelope signal sets such as $MPSK$, $|s|^2 = |\hat{s}|^2$ and (2.38) is identical to the results given in [51].

**Example 2.5** Using (2.38) we plot the cut-off rate $R_0$ for QPSK, 8PSK, 16PSK, 16QAM and 32CROSS signal sets on a Rayleigh fading channel without CSI in Figure 2.13. In this case, unlike the case with CSI, the cut-off rate of (2.38) is maximized by optimizing $\lambda$ numerically.

To compare the cut-off rate for Rayleigh fading channels with and without CSI the cut-off rates for these cases are replotted in Figure 2.14. It is observed that, unlike the AWGN and fading channels with CSI, constant envelope signal sets, i.e., MPSK constellations perform much better than rectangular constellations for the fading channels without CSI. Note that the cut-off rate of the 16QAM and 32CROSS signal sets are even inferior to QPSK. This is explained by considering the fact that in an MPSK signal set the boundary of the optimum decision regions are radial and thus invariant to radial scaling of the received signal caused by fading [17]. This makes the MPSK signal sets a suitable choice for designing the modulator part of coded schemes on fading channels. Furthermore, these signals are less sensitive to the nonlinear distortion caused by the nonlinearity of the power amplifiers in the transmitter.

### 2.6.4 Quantized Fading Channels

To complete the discussion of cut-off rate, we now apply the analysis of $R_0$ to the quantized fading channels where either the number of quantized levels, $Q$, is the same as the transmitter alphabet size, $q$ (hard quantized channel), or there is one null zone besides quantized levels (erasure channel). Since the

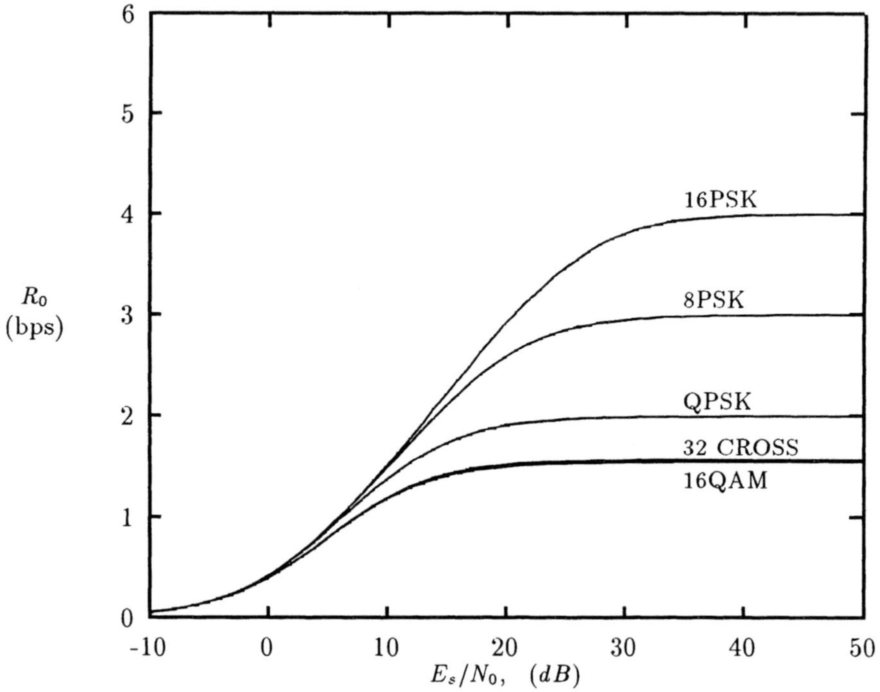


Figure 2.13: Cut-off rates for the Rayleigh fading channel without channel state information.

former is a special case of the latter we begin our discussion with the erasure channel.

### Fading Channel with Erasure Zone

The cut-off rate in (2.21) can be written as

$$R_0 = -\log_2 \left[ \sum_{h=0}^{q-1} [\sum_{l=0}^{q-1} p_l \sqrt{q_{lh}}\ ]^2 \ + \ [\sum_{l=0}^{q-1} p_l \sqrt{q_{le}}\ ]^2 \right], \qquad (2.39)$$

where $q_{le}$ represents the probability of the received signal being in the erasure region provided that the $l$th signal is transmitted. Defining a circular erasure region for MPSK signal sets as in Figure 2.15, $q_{le}$ is independent of the transmitted signal and will be denoted as $q_e$. Using the symmetry of the MPSK signal set and assuming that the signal symbols are equally probable,

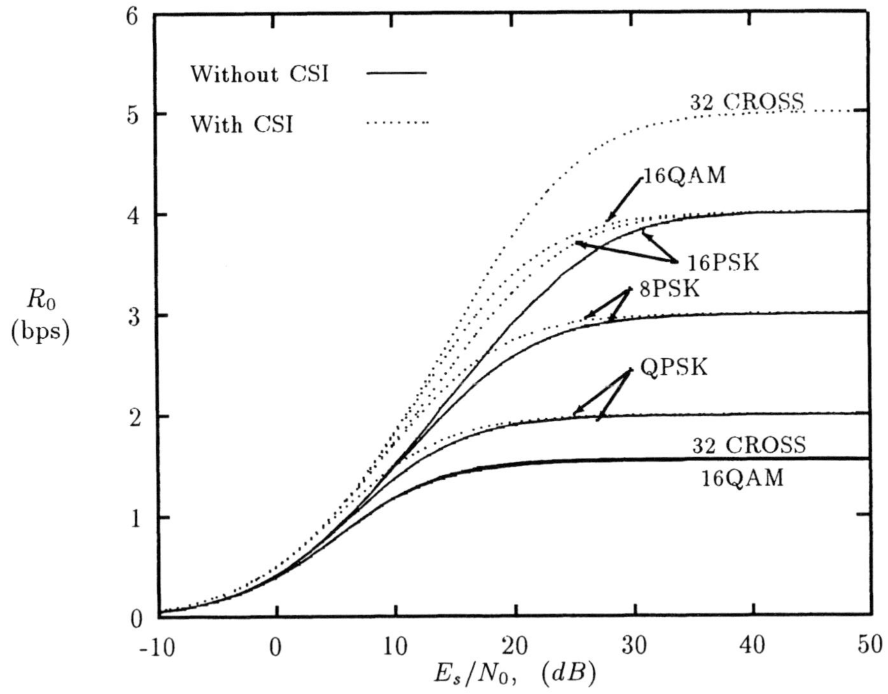


Figure 2.14: Comparison between the cut-off rates of the Rayleigh fading channel with and without channel state information.

i.e., $p_l = 1/q$, then (2.39) simplifies to

$$R_0 = -\log_2 \left[ \frac{1}{q} \left[ \sum_{l=0}^{q-1} \sqrt{q_l} \right]^2 + q_e \right], \tag{2.40}$$

where

$$q_l = E_a[\int_{\Lambda_l} p(r|s_0, a)dr], \quad l = 0, 1, \cdots, q-1, \tag{2.41}$$

and

$$q_e = E_a \left[ \int_{\Lambda_e} p(r \mid s_0, \ a) \ dr \right]. \tag{2.42}$$

In (2.41) and (2.42) $\Lambda_l$ and $\Lambda_e$ represent the decision region of the $l$th signal and the erasure region, respectively.

Assume that the fading amplitude is known at the receiver. In this case a simple erasure zone is defined based on the fading amplitude. If the fading is

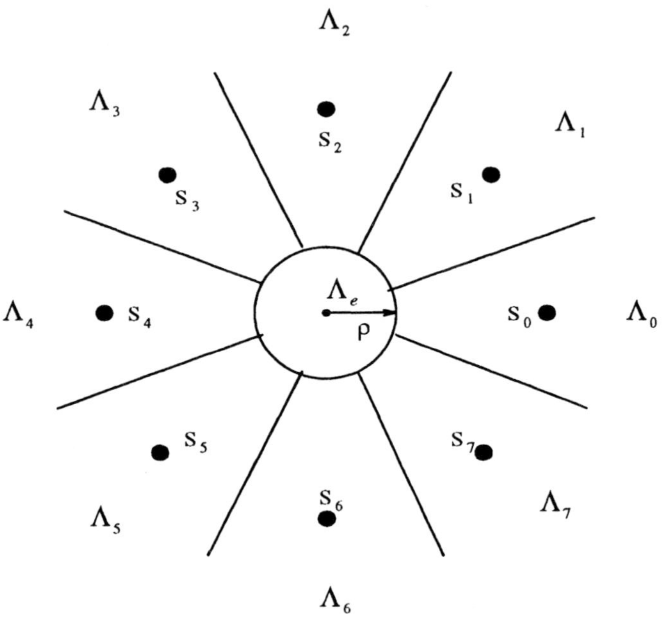


Figure 2.15: Decision regions for an 8PSK signal set with an erasure region.

less than a threshold value, say $a_T$, then the received signal is considered to be in the erasure region. Based on this definition, the probabilities $q_l$ and $q_e$ are evaluated in Appendix B as

$$q_l = \frac{E_s/N_0}{\pi(1+E_s/N_0)} \left[ \frac{\pi}{qE_s/N_0} e^{-a_T^2 \frac{E_s}{N_0}} + \int_{\frac{(2l-1)\pi}{q}}^{\frac{(2l+1)\pi}{q}} d\theta \left( \int_{a_T}^{\infty} \rho^2 \frac{E_s}{N_0} \sqrt{\pi} \cos\theta \right.\right.$$
$$\left.\left. \cdot e^{-\rho^2 E_s/N_0 + \frac{\rho^2 (E_s/N_0)^2 \cos^2\theta}{1+E_s/N_0}} \operatorname{erfc}\left( \frac{-\rho E_s/N_0 \cos\theta}{\sqrt{1+E_s/N_0}} \right) d\rho \right) \right], \quad (2.43)$$

and

$$q_e = 1 - \sum_{l=0}^{q-1} q_l. \quad (2.44)$$

In the above equation we choose $a_T$ in such a way that $R_0$ is maximum.

### Fading Channel with Hard-Decision

By setting the threshold value $a_T = 0$ in (2.43), the transition probability for a fading channel is obtained. In this case $q_e = 0$ and (2.40) simplifies to

$$R_0 = \log_2 q - 2\log_2 \left[ \sum_{l=0}^{q-1} \sqrt{q_l} \right]. \quad (2.45)$$

**Example 2.6** Using (2.40), (2.43), (2.44), and (2.45) the cut-off rates for the hard quantized and erasure channels for some MPSK signal sets are shown in Figure 2.16. In this figure the cut-off rate for soft-decision fading channels with CSI is also included. It is observed that the soft-decision decoding significantly improves the performance of the coded schemes compared to the hard-decision decoding. The reduction in SNR due to soft-decision decoding at rates around 3/4 of the saturated rate is about 7.5 dB compared to hard-decision decoding. Also, the erasure decoding can save about 2 dB at the same rate compared to the hard-decision decoding. It is worth noting that the reduction in SNR due to the soft-decision decoding in fading channels is much more than the one in the AWGN channel. The cut-off rates for the AWGN channel using different decoding strategies are shown in Figure 2.17 for comparison.

### 2.6.5 Discussion

The cut-off rate curves may be interpreted to show the benefit of bandwidth-efficient coded modulation schemes on bandwidth-limited channels. First, we investigate the potential gain of these schemes on the AWGN channel.

Consider a situation in which an uncoded QPSK signal set is used to transmit a message with a rate of 2 bits/symbol over the AWGN channel and with an error probability not to exceed $10^{-5}$. The required $E_s/N_0$ for such an error probability is about 12.9 dB. This point is indicated in Figure 2.12. Now consider a coded system which transmits messages with a rate of 2 bits/symbol by using an 8PSK signal set. Assuming an unlimited coding and decoding effort (very large N), according to Figure 2.12, this coded scheme can provide the same performance of uncoded QPSK with a reduction of 5.2 dB in SNR without bandwidth expansion. It is also apparent that further expansion of the signal set to more than twice the original value may not achieve additional gain.

The same argument can be applied to the Rayleigh fading channel. In such a channel, the uncoded QPSK can provide an error probability of $10^{-5}$ at $E_s/N_0 = 50$ dB. This point is shown in Figure 2.11. It is observed that if the number of channel signal is doubled, e.g., by choosing 8PSK modulation, error-free transmission of 2 bits/symbol is theoretically possible at $E_s/N_0 = 12$ dB. Hence, for Rayleigh fading channel the use of bandwidth-efficient coded-modulation may result in 38 dB reduction in SNR at error probability of $10^{-5}$. This is significantly more than the reduction of SNR in the AWGN channel due to the use of bandwidth-efficient signaling.

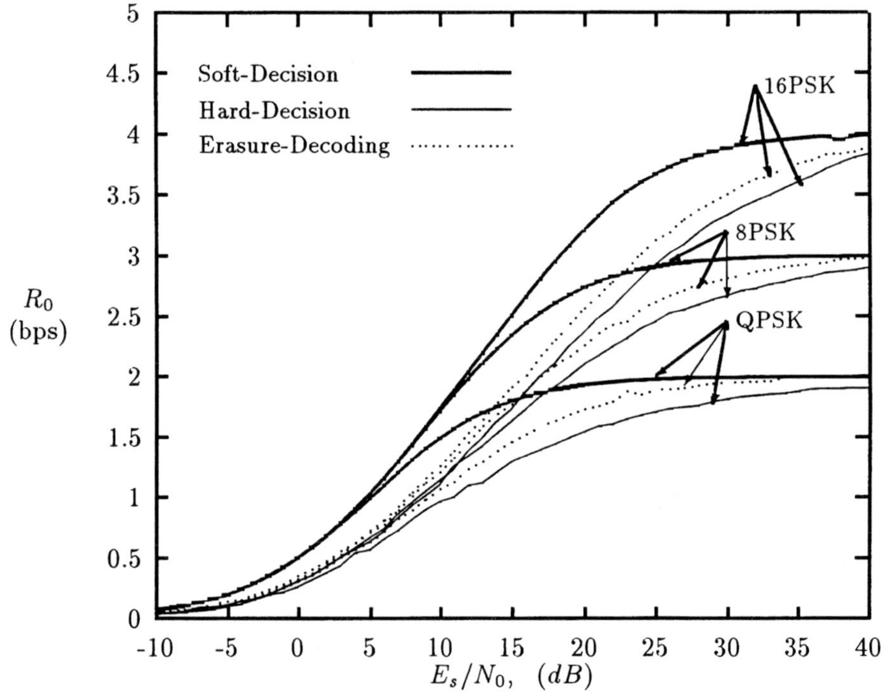


Figure 2.16: Comparison of cut-off rates for the Rayleigh fading channel with channel state information for some MPSK signal sets using different decoding strategies.

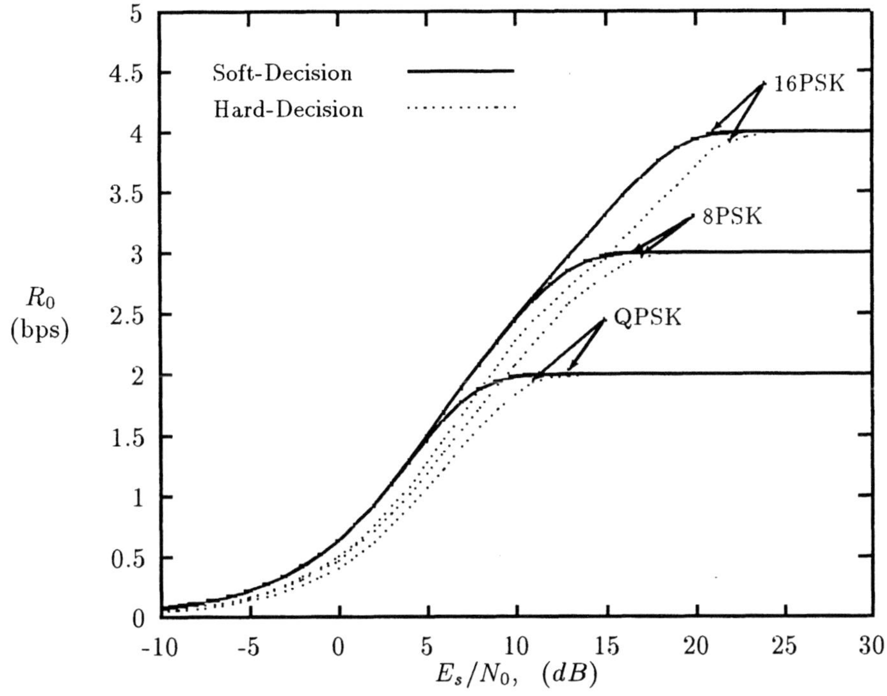


Figure 2.17: Comparison of the cut-off rates for the AWGN channel for some MPSK signal sets using different decoding strategies.

# Problems

**2.1** An orthonormal set is defined as a set of functions $\phi_i(t)$, $i = 1, 2, \cdots, N$, and $0 \le t \le T$ such that

$$\int_0^T \phi_i(t)\phi_j^*(t)dt = \delta_{ij} = \begin{cases} 1 & \text{if } i = j \\ 0 & \text{if } i \neq j \end{cases}$$

**a)** Consider a deterministic, finite-energy signal $x(t)$ defined over $[0,\ T]$ and represented as

$$x(t) = e(t) + \sum_{i=1}^{N} x_i\phi_i(t)$$

where

$$x_i = \int_0^T x(t)\phi_i^*(t)dt$$

and

$$\int_0^T |x(t)|^2 dt < +\infty$$

Show that the mean square error can be represented as

$$E_e(N) \equiv \int_0^T |e(t)|^2 dt = \int_0^T |x(t)|^2 dt - \sum_{i=1}^{N} |x_i|^2$$

and approaches 0 when $N$ approaches infinity and

$$x(t) = \sum_{i=1}^{+\infty} x_i\phi_i(t).$$

If the above statement holds true for all finite-energy $x(t)$, then $\phi_i(t)$'s form a Complete Orthonormal (CON) set specified over $[0,\ T]$ for the class of finite-energy signals.

**b)** Now consider a finite number of deterministic, finite-energy signals $x_j(t)$'s, $j = 1, 2, \cdots, M$, defined over $[0,\ T]$. To find $N$ orthonormal functions $\phi_i(t)$'s such that all $M$ signals $x_j(t)$'s can be represented as

$$x_j(t) = \sum_{i=1}^{N} x_{ji}\phi_i(t)$$

where

$$x_{ji} = \int_0^T x_j(t)\phi_i^*(t)dt$$

we can use the following Gram-Schmidt procedure:

Initialize: $j = 0, i = 0, \phi_0(t) = 0$.

Step 1: $j = j + 1$, Repeat Step 1 until $x_j(t) \neq 0$ or $j > M$.

Step 2: If $j > M$, Stop. Otherwise, go to Step 3.

Step 3: For $k = j$ to $M$ do

Step 3.1: $x_{kl} = \int_0^T x_k(t)\phi_l^*(t)dt$, for $l = 0, 1, \cdots, i$.

Step 3.2: $\Delta x_k(t) = x_k(t) - \sum_{l=0}^{i} x_{kl}\phi_l(t)$.

Step 3.3: If $\Delta x_k(t) \neq 0$, do

$$\begin{aligned} i &= i + 1 \\ E_{\Delta k} &= \int_0^T |\Delta x_k(t)|^2 dt, \\ \phi_i(t) &= \Delta x_k(t)/\sqrt{E_{\Delta k}}. \end{aligned}$$

Prove that the above procedure produces a set of $N$ orthonormal functions $\{\phi_1(t), \phi_2(t), \cdots, \phi_N(t)\}$ where $N \leq M$.

**2.2 Signal-Space Representation of a Random Process:**
Consider a second-order, mean-square continuous random process $X(t)$, for $0 \leq t \leq T$, i.e,

$$E\{|X(t)|^2\} < +\infty \quad \text{(2nd-order process)}$$

and

$$\lim_{\tau \to 0}\{|X(t+\tau) - X(t)|^2\} = 0 \quad \text{(mean-square continuous)}$$

We want to represent $X(t)$ in terms of a complete orthonormal (CON) set $\{\phi_i(t)\}$ specified over $[0,\ T]$ as a Karhunen-Loève expansion,

$$X(t) = \sum_{i=1}^{+\infty} X_i\phi_i(t)$$

where

$$X_i = \int_0^T X(t)\phi_i^*(t)dt \quad \text{(random variable)}$$

and using the mean square sense, i.e.,

$$\lim_{N \to +\infty} E\{|X(t) - \sum_{i=1}^{N} \phi_i(t)|^2\} = 0$$

**a)** Prove that in order to obtain uncorrelated random variables $X_i$'s the CON set $\{\phi_i(t)\}$ has to satisfy the following relation

$$\int_0^T \int_0^T K_{XX}(t,u)\phi_i(t)\phi_j^*(u)dtdu = \lambda_i \delta_{ij}$$

where $K_{XX}(t,u)$ is the autocovariance function

$$K_{XX}(t,u) \equiv E\{(X(t) - m_X(t))(X(u) - m_X(u))^*\}$$

$$m_X(t) \equiv E\{X(t)\}$$

and

$$\delta_{ij} = \begin{cases} 1 & \text{for } i = j \\ 0 & \text{for } i \neq j \end{cases}$$

*Hint*: For uncorrelated random variables $X_i$'s, we obtain

$$E\{(X_i - m_i)(X_j - m_j)^*\} = \lambda_i \delta_{ij}$$

with $m_i \equiv E\{X_i\}$.

**b)** Show that the equation in **a)** can be rewritten as the following homogeneous integral equation

$$\int_0^T K_{XX}(t,u)\phi_i(u)du = \lambda_i \phi_i(t)$$

where $K_{XX}(t,u)$ is called the kernel, $\lambda_i$'s are the eigenvalues and $\phi_i(t)$'s are the corresponding eigenfunctions.

**c)** Show that if $X(t)$ is a white Gaussian process with zero mean, then the above homogeneous integral equation is satisfied for any set of orthonormal functions $\{\phi_i(t)\}$.

**2.3** In a given bit interval, from $nT_b$ to $(n+1)T_b$, a binary frequency shift keying (BFSK) signal can be represented as

$$s(t) = \begin{cases} s_1(t) = A\cos 2\pi f_1 t, & b_n = 1 \\ s_2(t) = A\cos 2\pi f_0 t, & b_n = 0 \end{cases}$$

where $b_n$ is the data bit in the $n$th interval.

**a)** Show that $s_1(t)$ and $s_2(t)$ are orthogonal when

$$f_1 - f_0 = \frac{k}{2T_b},$$

where $k$ is an integer and $k \neq 0$.

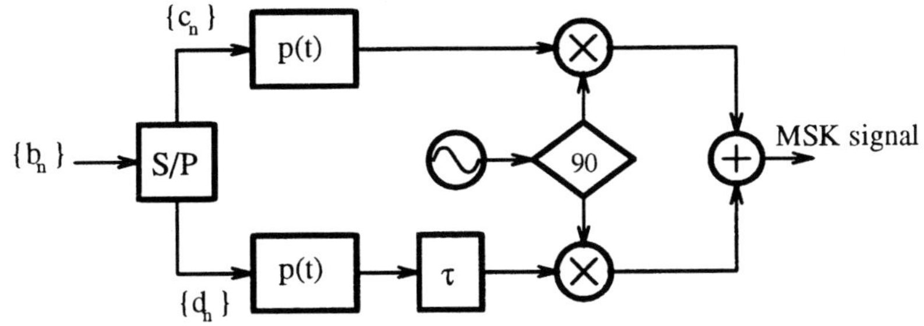

Figure P2.4-1: Block diagram of an MSK modulator.

b) Derive the signal-space representation of BFSK signals, and determine the decision regions for an optimum receiver in an AWGN channel.

c) Prove that the bit-error probability of BFSK signals in an AWGN is

$$P_b = Q\left(\sqrt{\frac{E_b}{N_0}}\right),$$

where $E_b$ is the average energy per bit.

**2.4** Consider the block diagram of a special 4-QAM scheme called Minimum Shift Keying (MSK) in Figure P2.4.1. The input bit serial stream $\{b_m\}$ is split into two parallel symbol streams $\{c_n\}$ and $\{d_n\}$ by the serial-to-parallel (S/P) converter. Note that the time duration of each symbol, $c_n$ or $d_n$, is $T_s = 2T_b$, where $T_b$ is the time duration of an input bit. The MSK signal is represented as

$$\begin{aligned} s_{MSK}(t) &= \sum_{-\infty}^{+\infty} c_n p(t - nT_s) \cos 2\pi f_c t \\ &\quad + \sum_{-\infty}^{+\infty} d_n p(t - nT_s - \tau) \sin 2\pi f_c t, \end{aligned}$$

where $c_n$ and $d_n$ are statistically independent and

$$c_n \text{ (or } d_n) = \begin{cases} 1 & \text{with probability of } 1/2 \\ -1 & \text{with probability of } 1/2, \end{cases}$$

$$p(t) = \begin{cases} \cos(\pi t/2T_b), & |t| \leq T_b \\ 0, & \text{elsewhere} \end{cases}$$

and $\tau = T_b$

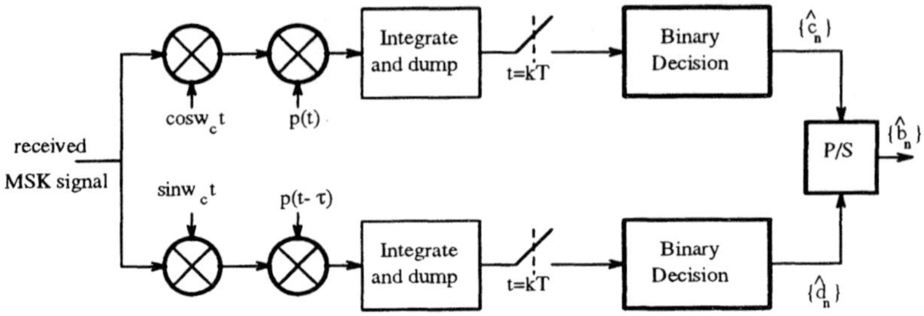


Figure P2.4-2: Block diagram of an MSK demodulator.

**a)** Show that the optimum receiver for the MSK scheme in an AWGN channel is as shown in Figure P2.4.2 and that the bit error probability is

$$P_b = Q\left(\sqrt{\frac{2E_b}{N_0}}\right),$$

where $E_b$ is the average energy per bit.

**b)** Show that the MSK signal, $s_{MSK}(t)$, can be represented as the BFSK signal of Problem 2.3 with $(f_1 - f_0) = 1/2T_b$, i.e.,

$$s_{MSK}(t) = \sum_{-\infty}^{+\infty} A\cos\left[2\pi(f_c + \frac{b_m}{4T_b})(t - mT_b)\right],$$

where $b_m = \pm 1$ with probability of 1/2 and $f_c$ is assumed to be a multiple of $1/T_b$.

**c)** From Problems 2.3 and 2.4, verify that the MSK demodulator of Figure P2.4.2 provides a performance gain of 3 dB relative to the BFSK demodulator of Problem 2.3.

**2.5** A general binary modulator generates one of the two possible waveforms, $s_0(t)$ and $s_1(t)$, according to the value of the input bit. For an equiprobable transmission of $s_0(t)$ and $s_1(t)$, show that the bit error probability of a general binary modulation scheme in an AWGN channel is given by

$$P_b = Q\left(\sqrt{\frac{E_b}{N_0}(1-\gamma)}\right),$$

where $E_b$ is the average energy per bit

$$E_b = \frac{1}{2}\int_0^{T_b} |s_1(t)|^2 dt + \frac{1}{2}\int_0^{T_b} |s_0(t)|^2 dt$$

and

$$\gamma = \frac{1}{E_b} \int_0^{T_b} s_0(t) s_1(t) dt.$$

Verify that the best performance is obtained with an antipodal signaling scheme, i.e., $s_0(t) = -s_1(t)$.

**2.6 Union Bound on the Error Probability of $M$-ary Modulation Schemes in an AWGN Channel**

Consider an $M$-ary modulator using $M$ equiprobably transmitted signaling elements $s_i(t)$, $i = 0, 1, \cdots, (M-1)$, with $P\{s_i(t) \text{ transmitted}\} = 1/M$.

**a)** Show that, in an AWGN, the error probability of an $M$-ary modulation scheme using an ML receiver can be represented as

$$\begin{aligned} P_e &= \frac{1}{M} Pr\{\text{error}|s_i(t) \text{ transmitted}\} \\ &= \frac{1}{M} Pr\left\{ \bigcup_{k=1, k\neq i}^{M} \epsilon_{ik} \right\}, \end{aligned}$$

where $\epsilon_{ik}$ denotes the event that the received vector $\mathbf{r}$ is closer to the signal vector $\mathbf{s}_k$ than to the signal vector $\mathbf{s}_i$ and that $i \neq k$.

**b)** Prove that

**i)**

$$Pr\{\epsilon_{ik}\} = Q\left(\frac{d_{ik}}{\sqrt{2N_0}}\right),$$

where

$$d_{ik} = d_E(\mathbf{s}_i, \mathbf{s}_k), \quad i \neq k.$$

**ii)**

$$Pr\{ \bigcup_{k=1, k\neq i}^{M} \epsilon_{ik}\} \leq \sum_{i=1}^{M} \sum_{k=1}^{M} Q\left(\frac{d_{ik}}{\sqrt{2N_0}}\right)$$

**iii)**

$$P_e \leq (M-1) Q\left(\frac{d_{min}}{\sqrt{2N_0}}\right),$$

where

$$d_{min} = \min_{all\ i\neq k} \{d_{ik}\}.$$

**2.7** Consider an M-ary strict-sense QAM scheme described in Section 2.4.2.

**a)** Show that

$$d_{min} = \min_{all\ i \neq k} \{d_E(\mathbf{s}_i, \mathbf{s}_k)\} = \sqrt{\frac{6E_s}{M-1}}.$$

**b)** Using the results of Problem 2.6, show that $P_e$ of an M-ary strict-sense QAM scheme is upper-bounded as

$$P_e \leq (M-1)Q\left(\sqrt{\frac{3E_s}{(M-1)N_0}}\right).$$

**c)** From (2.19), show that

$$P_e \leq 2P_{e-\sqrt{M}AM} = 4\left(1 - \frac{1}{M}\right) Q\left(\sqrt{\frac{3E_s}{(M-1)N_0}}\right).$$

For large $M$,

$$P_e \leq Q\left(\sqrt{\frac{3E_s}{(M-1)N_0}}\right).$$

**d)** Verify that the upper-bound in c) is tighter than that in b) by a factor of $(M + \sqrt{M})/4$.

**2.8** A QPSK scheme is equivalent to a 4-QAM.

**a)** Show that the error probability of QPSK in an AWGN channel is

$$P_e \leq 2Q\left(\sqrt{\frac{E_s}{N_0}}\right)$$

**b)** Verify that the block diagrams of the QPSK modulator and demodulator are similar to those of Figure P2.4-1 and P2.4-2, respectively, with $\tau = 0$ and

$$p(t) = \begin{cases} 1, & |t| \leq T_b \\ 0, & \text{elsewhere} \end{cases}$$

**c)** An Offset-QPSK (OQPSK) modulator has a similar block diagram as the QPSK with $\tau = T_b$ and $p(t)$ as shown in b). Draw the block diagram of the OQPSK modulator and show that the OQPSK scheme has the same performance as the QPSK.

**2.9 Q-and Error Functions:**
The error function is defined as

$$erf(x) = \frac{2}{\sqrt{\pi}} \int_0^x e^{-t^2} dt, \text{for } x > 0.$$

**a)** Show that $erf(x)$ is an even function and that

$$\lim_{x \to +\infty} erf(x) = 1$$

**b)** Show that the complementary error function

$$\text{erfc(x)} = 1 - \text{erf(x)}$$

can be represented as

$$\begin{aligned} \text{erfc(x)} &= \frac{2}{\sqrt{\pi}} \int_x^{+\infty} e^{-t^2} dt \\ &= \frac{2e^{-x^2}}{\sqrt{\pi}} \int_{-\infty}^0 e^{u(2x-u)} du \end{aligned}$$

*Hint:* let $t = x - u$.

**c)** Show that

$$\text{erfc(x)} = \frac{\text{e}^{-\text{x}^2}}{\text{x}\sqrt{\pi}} \sum_{\text{n}=0}^{+\infty} (-1)^{\text{n}} \frac{(2\text{n}-1)!!}{(2\text{x}^2)^{\text{n}}}$$

where $(2n-1)!! = (2n-1)(2n-3)\cdots 5.3.1$
*Hint:* Use

$$e^{-u^2} = \sum_{n=0}^{+\infty} (-1)^n \frac{u^{2n}}{n!}, \quad v = -2ux,$$

and

$$\int_0^{+\infty} v^n e^{-v} dv = n!$$

for the results of part b).

**d)** Show that

$$\left(1 - \frac{1}{2x^2}\right) \frac{e^{-x^2}}{x\sqrt{\pi}} < \text{erfc(x)} < \frac{\text{e}^{-\text{x}^2}}{\text{x}\sqrt{\pi}}$$

and, for $x \geq 1$, $\quad \text{erfc(x)} < \frac{\text{e}^{-\text{x}^2}}{\sqrt{\pi}}$.
Plot erfc(x) and its upper and lower bounds for x from 0.5 to 4 and observe the tightness of the bounds.

**e)** The Q-function is defined as the tail probability of a normalized Gaussian random variable $X$

$$Q(x) = Pr\{X > x\}, \quad x > 0$$

where

$$p_X(x) = \frac{1}{\sqrt{2\pi}} e^{-x^2/2}$$

Show that

$$Q(x) = \frac{1}{2}\mathrm{erfc}\left(\frac{x}{\sqrt{2}}\right)$$

and

$$\mathrm{erfc}(x) = 2Q(x\sqrt{2}).$$

**f)** Show that

$$\mathrm{erfc}(\sqrt{x+y}) \le \mathrm{erfc}(\sqrt{x})e^{-y}, \quad x \ge 0, y \ge 0.$$

**2.10 Error Probability of M-ary PSK Schemes:**
From Figure 2.10, we can express the error probability of an M-ary PSK scheme as

$$P_e = P_{e|s_0} = Pr\{r \in \overline{\Lambda_0}|s_0\} = 1 - Pr\{r \in \Lambda_0|s_0\}$$

**a)** Verify that

$$\begin{aligned} Pr\{r \in \Lambda_0|s_0\} &= Pr\{|\Theta| \le \frac{\pi}{M}\} \\ &= \int_{-\pi/M}^{+\pi/M} p_\Theta(\theta)d\theta \end{aligned}$$

where

$$\Theta = \tan^{-1}\left(\frac{n_Q}{n_I + \sqrt{E_s}}\right)$$

$n_I, n_Q$ are the inphase and quadrature noise components of the noise vector $\mathbf{n} = \mathbf{r} - \mathbf{s}_0$ and $p_\theta(\theta)$ is the pdf of $\theta$.

**b)** Note that $n_I$ and $n_Q$ are i.i.d. Gaussian random variables with zero-mean and variance $N_0/2$. Using the results of Problem 1.1, show that

$$p_\Theta(\theta) = \frac{1}{2\pi}e^{-E_s/N_0} + \sqrt{\frac{E_s}{\pi N_0}} \cos\theta e^{-\frac{E_s}{N_0}\sin^2\theta}\left[1 - 0.5\mathrm{erfc}\left(\sqrt{\frac{E_s}{N_0}}\cos\theta\right)\right]$$

$$|\theta| \le \pi.$$

c) For high $E_s/N_0$ and $|\theta| < \pi/2$, show that

i)

$$\mathrm{erfc}\left(\frac{\mathrm{E_s}}{\mathrm{N_0}}\cos\theta\right) \approx \sqrt{\frac{\mathrm{N_0}}{\pi \mathrm{E_s}\cos^2\theta}}\mathrm{e}^{-\frac{\mathrm{E_s}}{\mathrm{N_0}}\cos^2\theta}$$

*Hint:* Use the results of Problem 2.9d).

ii)

$$p_\Theta(\theta) \approx \sqrt{\frac{E_s}{N_0}}\cos\theta e^{-\frac{E_s}{N_0}\sin^2\theta}$$

iii)

$$P_e \approx 2Q\left(\sqrt{\frac{2E_s}{N_0}}\sin(\frac{\pi}{M})\right)$$

d) Using the union bound of Problem 2.6, show that

$$P_e \leq (M-1)Q\left(\sqrt{\frac{2E_s}{N_0}}\sin(\frac{\pi}{M})\right)$$

Compare the results obtained in b), c) and in Section 2.4.2.

**2.11 Chebyshev and Chernoff Bounds:**

a) $X$ is a random variable with mean $\mu_x < +\infty$ and variance $\sigma_x^2 < +\infty$. From

$$\sigma_x^2 = \int_{-\infty}^{+\infty}(x-\mu_x)^2 p_X(x)dx$$

where $p_X(x)$ is the pdf of $X$, prove that for a given constant $\gamma > 0$ the probability that $|X-\mu_x| \geq \gamma$ satisfies the Chebyshev inequality

$$Pr\{|X-\mu_x| \geq \gamma\} \leq \frac{\sigma_x^2}{\gamma^2}$$

*Hint:*

$$\sigma_x^2 \geq \int_{|x-\mu_x|\leq\gamma}(x-\mu_x)^2 p_X(x)dx$$

b) Show that the Chebyshev inequality can be represented as

$$E_X[f(X)] \leq E\left[\frac{(X-\mu_x)^2}{\gamma^2}\right] = \frac{\sigma_x^2}{\gamma^2}$$

where

$$f(X) = \begin{cases} 1, & |X - \mu_x| \geq \gamma \\ 0, & \text{elsewhere} \end{cases}$$

**c)** Consider

$$g(X) = \begin{cases} 1, & (X - \mu_x) \geq \gamma \\ 0, & \text{elsewhere} \end{cases}$$

and, for a given $\lambda \geq 0$,

$$g(X) \leq e^{\lambda(X - \mu_x - \gamma)}$$

Prove that the probability that $(X - \mu_x) \geq \gamma$ satisfies the Chernoff bound

$$Pr\{(X - \mu_x) \geq \gamma\} \leq E_X \left[e^{\lambda(X - \mu_x - \gamma)}\right].$$

$\lambda$ is called the Chernoff bound parameter.

**d)** Show that the optimum value of $\lambda$ for the tightest upper bound is the solution to the following equation:

$$E_X \left[(X - \mu_x)e^{\lambda(X - \mu_x)}\right] = \gamma E_X \left[e^{\lambda(X - \mu_x)}\right].$$

**e)** For $p_X(x) = 0.5e^{-|x|}$, derive the Chebyshev and Chernoff bounds. Discuss the tightness of both bounds.

**f)** Repeat e) with $p_X(x) = \frac{1}{\sqrt{2\pi}} e^{-x^2/2}$.

**2.12** From Problem 2.6, the error probability of an $M$-ary signaling scheme in an AWGN channel can be represented as

$$P_e \leq (M - 1)Q\left(\frac{d_{ik}}{\sqrt{2N_0}}\right)$$

For an $M$-ary signaling scheme using $M$ equally probable equal-energy orthogonal signals, show that

**a)** $d_{ik} = \sqrt{2E_s}$, where $E_s$ is the average energy per signal vector.

**b)**

$$P_e < MQ\left(\sqrt{\frac{E_s}{N_0}}\right) < Me^{-E_s/N0}$$

c)

$$P_e < 2^{-T[-\mathcal{R}+0.72P_s/N_0]},$$

where $T$ is the signaling interval, $P_s$ is the average transmitted power and $\mathcal{R}$ is the source rate (in bits/sec). Observe that for rates $\mathcal{R}$ less than $0.72P_s/N_0$ (cut-off rate) the error probability approaches zero exponentially with increasing $T$. However, increasing $T$ implies that the number $M = 2^{\mathcal{R}T}$ of orthogonal signals is exponentially increased.

**2.13** From Problem 2.6, the error probability of an $M$-ary signaling scheme is upper-bounded as

$$P_e \leq (M-1)P_2(\mathbf{s}_i, \mathbf{s}_k) < M P_2(\mathbf{s}_i, \mathbf{s}_k),$$

where $P_2(\mathbf{s}_i, \mathbf{s}_k)$ is the pairwise error probability when $\mathbf{s}_i$ is transmitted and $\mathbf{s}_i \neq \mathbf{s}_k$ is detected instead. We are interested in finding the average of the error probability over the entire ensemble of all $M$-ary signaling schemes

$$\overline{P_e} < M\overline{P_2(\mathbf{s}_i, \mathbf{s}_k)}$$

a) Show that $\overline{P_e}$ can be rewritten as

$$\overline{P_e} < 2^{-T(\mathcal{R}_0-\mathcal{R})} = 2^{-N(R_0-R)}$$

where $T$ is the signaling interval, $N$ is the dimension of the signal vector, $\mathcal{R}$ is the source rate (in bits/sec) and $R$ is the information rate in bits/symbol. $\mathcal{R}_0$ and $R_0$ are the cut-off rates of $\mathcal{R}$ and $R$ respectively

$$R_0 = -\frac{1}{N}\log_2 \overline{P_2(\mathbf{s}_i, \mathbf{s}_k)}$$

b) From (2.11) show that

$$\begin{aligned} P_2(\mathbf{s}_i, \mathbf{s}_k) &= Pr\left\{\ln \frac{p\{\mathbf{r}|\mathbf{s}_k\}}{p\{\mathbf{r}|\mathbf{s}_i\}} \geq 0\right\} \\ &= Pr\{X \geq 0\} \end{aligned}$$

where

$$X = \sum_{j=0}^{N-1} X_j,$$

$X_j$'s are statistically independent, identically distributed random variables taking their values from the set $\{ln\frac{p_{uh}}{p_{lh}}; u, l = 0, 1, \cdots, (q-$

1), $h = 0, 1, \cdots, (Q-1)\}$. $p_{lh}$ represents the channel transition probability

$$p_{lh} = Pr\{r_j = b_h | s_i = a_l\}$$

$a_l$'s and $b_h$'s are the symbols of the transmitter alphabet $\{a_l\}$ of size $q$ and receiver quantizer alphabet $\{b_h\}$ of size $Q$ respectively.

**c)** Using the Chernoff bound (results of Problem 2.11), show that

$$\overline{P_2(\mathbf{s}_i, \mathbf{s}_k)} \leq E\left[e^{\lambda X}\right] = \left(E\left[e^{\lambda X_j}\right]\right)^N, \quad \lambda \geq 0$$

where

$$E\left[e^{\lambda X_j}\right] = \sum_{u=0}^{q-1}\sum_{l=0}^{q-1}\sum_{h=0}^{Q-1} p_u p_l p_{lh} \left(\frac{p_{uh}}{p_{lh}}\right)^{\lambda}$$

**d)** Show that the optimum value of $\lambda$ is 1/2 and that $E[e^{X_j/2}]$ can be rewritten as

$$E\left[e^{X_j/2}\right] = \sum_{h=0}^{Q-1}\left(\sum_{l=0}^{q-1} p_l\sqrt{p_{lh}}\right)^2$$

and that the cut-off rate of $R$ is

$$R_0 = -\log_2 \sum_{h=0}^{Q-1}\left(\sum_{l=0}^{q-1} p_l\sqrt{p_{lh}}\right)^2$$

**2.14** Derive and plot the cut-off rates for BPSK scheme over an AWGN channel and a Rayleigh fading channel with and without CSI. Discuss the results.

**2.15** Repeat Problem 2.14 with the 64-QAM scheme.

**2.16** Repeat Problem 2.14 with the orthogonal BFSK scheme.

**2.17** Derive and plot the cut-off rates for 16-QAM and 16-PSK schemes over a Rician fading channel with and without CSI. Assume $K = 2$. Compare the performance of 16-QAM and 16-PSK.

**2.18** Repeat Problem 2.17 with $K = 4$.

**2.19** Repeat Problem 2.14 with Orthogonal 4FSK scheme.

**2.20** Using the results shown in Figure 2.13, discuss the potential gain of the coded 8PSK relative to uncoded QPSK for $R_0 = 2$ bits/symbol.

**2.21** Derive and plot the cut-off rates for a BPSK scheme using erasure-decoding, soft-decision and hard-decision over a Rayleigh fading channel with CSI.

**2.22** $\frac{\pi}{4}$**-Shift QPSK**

The $\frac{\pi}{4}$-Shift QPSK scheme [133] has been used in digital cellular communications. This modulation scheme uses the signal set of an 8PSK scheme $\mathcal{S}_8 = \{\underline{s}'_k = \sqrt{E_s}e^{j(2k+1)\pi/8}, k = 0, 1, 2, \cdots, 7\}$ as follows. Let $\underline{b}_{n-1} = \sqrt{E_s}e^{j\phi_{n-1}}$ and $\underline{b}_n = \sqrt{E_s}e^{j\phi_n}$ denote the transmitted signal vectors in the $(n-1)$th and $n$th symbol intervals respectively. $\phi_{n-1}$ and $\phi_n$ satisfy the following equation

$$\phi_n = \phi_{n-1} + \Delta\phi_n$$

where

$$\Delta\phi_n = \frac{\pi}{4}(2l + 1)$$

and $l$ is the value of the current 2-bit symbol, i.e., $l = 0, 1, 2, 3$.

a) Verify that the two adjacent signal vectors $\underline{b}_{n-1}$ and $\underline{b}_n$ belong to two disjoint signal subsets $\mathcal{S}_{8\text{even}} = \{\underline{s}'_{2l}\}$ and $\mathcal{S}_{8\text{odd}} = \{\underline{s}'_{2l+1}\}$ where $l = 0, 1, 2, 3$.

b) Consider an unfiltered QPSK signal represented as

$$s_{QPSK}(t) = Re\left\{\sum_{n=-\infty}^{+\infty} \underline{a}_n p(t - nT)e^{-jw_c t}\right\}$$

where $p(t)$ is a rectangular pulse

$$p(t) = \begin{cases} \sqrt{\frac{1}{T}} & \text{for } 0 < t < T \\ 0 & \text{elsewhere} \end{cases}$$

and $\underline{a}_n$ belongs to the QPSK signal set, i.e.,

$$\underline{a}_n \in \mathcal{S}_4 = \{\underline{s}_k = \sqrt{E_s}e^{j(2k+1)\pi/4}, k = 0, 1, 2, 3\}.$$

Verify that the signal subsets $\mathcal{S}_{8\text{even}}$ and $\mathcal{S}_{8\text{odd}}$ can be produced from the signal set $\mathcal{S}_4$ by a phase shift of $-\pi/8$ and $+\pi/8$ respectively, i.e.,

$$\underline{s}'_{2l} = \underline{s}_l e^{-j\pi/8} \text{and} \quad \underline{s}'_{2l+1} = \underline{s}_l e^{+j\pi/8} \quad \text{for } l = 0, 1, 2, 3.$$

Show that the unfiltered $\frac{\pi}{4}$-QPSK signal can be represented as

$$s_{\frac{\pi}{4}-QPSK}(t) = Re\left\{\sum_{n=-\infty}^{+\infty} \underline{b}_n p(t - nT)e^{-j(w_c t - (-1)^n \pi/8)}\right\}$$

Draw the block diagram of the $\frac{\pi}{4}$-QPSK modulator. Sketch the signal space diagram with possible transitions for the $\frac{\pi}{4}$-QPSK and QPSK signals. Observe that the introduced phase shift of $\pm\pi/8$ in even and odd signal vectors greatly reduces the envelope fluctuation in the filtered $\frac{\pi}{4}$-shift QPSK signal. Verify that the filtered QPSK signal exhibits a 100% envelope fluctuation. By reducing the envelope fluctuation, the effect of nonlinearity of the transmit power amplifier on the spectral spreading is reduced (see Chapter 10). This feature makes the $\frac{\pi}{4}$-shift QPSK scheme attractive to mobile communications applications.

**c)** Show that an optimum coherent receiver with a recovered carrier $e^{j(w_c t-(-1)^n\pi/8)}$ and matched filters for a $\frac{\pi}{4}$-shift QPSK scheme in an AWGN has the same bit error probability as the conventional QPSK scheme, i.e.,

$$P_b = \frac{1}{2}\text{erfc}\left(\sqrt{\frac{E_b}{N_0}}\right).$$

**d)** Verify that $\frac{\pi}{4}$-shift QPSK signal can also be generated as follows - Consider a pair of two adjacent signal vectors $(\underline{b}_{n-1}, \underline{b}_n)$. In each two symbol intervals, $2T$, this pair belongs to a signal set of 16 elements

$$\mathcal{S}_{16} = \{(\underline{s}_i', \underline{s}_j'; \underline{s}_i' \in \mathcal{S}_{8\text{even}}, \underline{s}_j' \in \mathcal{S}_{8\text{odd}}\}.$$

In other words, the $\frac{\pi}{4}$-shift QPSK scheme can be viewed as a 16-ary, 4-dimensional modulation scheme which chooses an element of $\mathcal{S}_{16}$ corresponding to the value of the 4-bit input symbol in each $2T$-interval. Show that the minimum Euclidean distance of $\mathcal{S}_{16}$ is $\sqrt{2E_s}$ and that the error probability of the $\frac{\pi}{4}$-shift QPSK scheme is upper-bounded as

$$P_e < \frac{15}{2}\text{erfc}\left(\sqrt{\frac{E_s}{2N_0}}\right).$$

# Chapter 3

# CONVOLUTIONAL CODES

The purpose of this chapter is to give an introduction to convolutional codes as a subclass of linear codes. Our aim in this chapter is to cover the background which is required for the forthcoming subjects in the rest of this book. We start with basic definitions and study the encoder structure of the convolutional codes by presenting different approaches. Then, we consider the distance structure and present the generating functions to find the distance profile for these codes. The Viterbi algorithm, as a maximum likelihood decoding technique for convolutional codes, is explained. Finally, we use the generating functions to obtain an upper bound on the bit error probability of the maximum likelihood decoding of convolutional codes.

## 3.1 BASIC STRUCTURE

A binary[1] rate $k/n$ convolutional encoder with $\nu$ memory elements is a linear finite-state machine, which at any given time unit, accepts $k$ input bits, makes a transition from its state at that time unit to one of the $2^\nu$ possible successor states and outputs $n$ bits. Examples of rate 1/2 and rate 2/3 convolutional codes with $\nu = 2$ are shown in Figure 3.1 and Figure 3.2, respectively. We shall pursue these examples to develop various representations and properties of convolutional codes.

Many approaches can be used in studying convolutional codes. Among them we discuss three approaches, namely, the *generator polynomial matrix* approach, the *state diagram* approach, and the *trellis* approach. These approaches

[1] Throughout this chapter we consider linear binary convolutional codes defined over $GF(2)$. However, the results can be generalized to nonbinary codes.

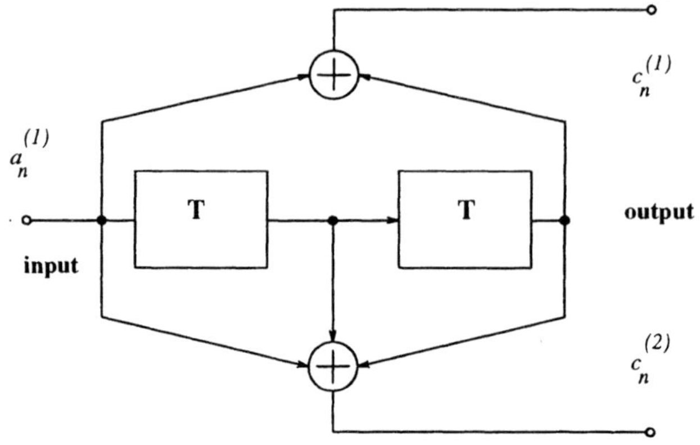


Figure 3.1: Rate 1/2 convolutional code with $\nu = 2$.

will be employed in discussion of the trellis-coded modulation schemes in the rest of this book.

### 3.1.1 Generator Polynomial Matrix

Consider the rate 1/2 convolutional encoder shown in Figure 3.1. It consists of a 2-stage shift register and two modulo-2 adders. The input bits are continuously entered into the left end of the register, and for each new input bit, two output bits are transmitted over the channel. These bits depend not only on the present input bit but also on the two previous input bits, stored in the shift register. For the encoder of Figure 3.1, the output bits can be expressed as

$$\begin{aligned} c_n^{(1)} &= a_n^{(1)} \oplus a_{n-2}^{(1)}, \\ c_n^{(2)} &= a_n^{(1)} \oplus a_{n-1}^{(1)} \oplus a_{n-2}^{(1)}, \end{aligned} \tag{3.1}$$

where subscript $n$ denotes the $n$th time interval and $\oplus$ shows the modulo-2 addition. Equation (3.1) describes the encoding process of the convolutional code using time-domain operations. An alternative description can be obtained using transform-domain involving polynomial arithmetic [74].

A polynomial $G(D)$ of degree $n$ over a field $F$, is expressed as

$$G(D) = g_0 + g_1 D + g_2 D^2 + \cdots + g_n D^n,$$

where the coefficients $\{g_i\}, i = 0, 1, \ldots, n$, are elements of the field $F$ and $D$ is an indeterminate. The indeterminate $D$ can be interpreted as a delay operator. For the polynomials defined over $GF(2)$ the coefficients, $\{g_i\}$, are either zero or one and the addition operation is a modulo-2 addition.

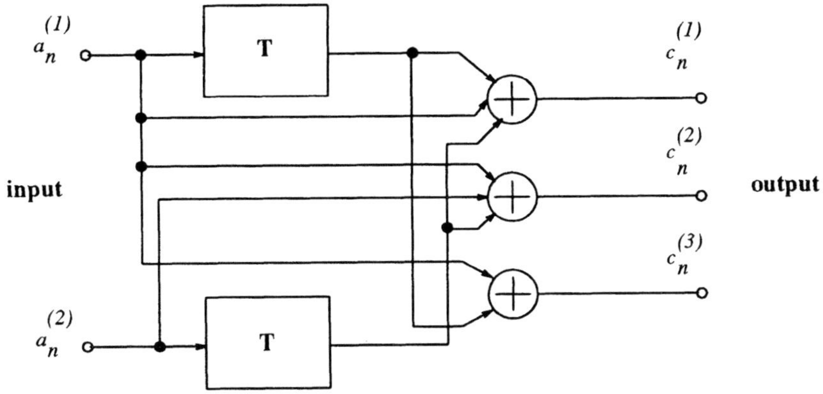


Figure 3.2: Rate 2/3 convolutional code with $\nu = 2$.

Using polynomial notation the input/output relations of (3.1) can be written as

$$\begin{aligned} c^{(1)}(D) &= a^{(1)}(D) \cdot G_1(D) \\ c^{(2)}(D) &= a^{(1)}(D) \cdot G_2(D), \end{aligned}$$

where

$$\begin{aligned} G_1(D) &= 1 + D^2, \\ G_2(D) &= 1 + D + D^2 \end{aligned} \tag{3.2}$$

are called the *generator polynomials* of the convolutional code.

The generator polynomials of a convolutional code can be determined directly from its encoder structure. Considering each shift register stage as a unit-time delay, the coefficients of the generator polynomial corresponding to an output can be obtained by looking at the connections of that output with the sequence of the shift register stages. The coefficient is one if there exists a connection and zero otherwise. This can be seen by examining (3.2) and Figure 3.1.

**Example 3.1** For the rate 2/3 convolutional code of Figure 3.2 the generator polynomials are

$$\begin{array}{lll} G_{11}(D) = 1 + D & G_{12}(D) = 1 & G_{13}(D) = 1 + D \\ G_{21}(D) = D & G_{22}(D) = 1 + D & G_{23}(D) = 0. \end{array}$$

We use the notation $G_{ij}(D)$ to specify the generator polynomial corresponding to the relationship between the $i$th input and the $j$th output.[2]

[2] Since there is only one input, for a rate $1/n$ convolutional encoder we omit the first subscript for simplicity.

The above example reveals that for a rate $k/n$ convolutional code the number of generator polynomials is $nk$. These polynomials can be put together in a $k \times n$ matrix defined as

$$\mathbf{G}(D) = \begin{bmatrix} G_{11}(D) & G_{12}(D) & \cdots & G_{1n}(D) \\ G_{21}(D) & G_{22}(D) & \cdots & G_{2n}(D) \\ \cdots & \cdots & \cdots & \cdots \\ G_{k1}(D) & G_{k2}(D) & \cdots & G_{kn}(D) \end{bmatrix},$$

and called the *generator polynomial matrix*. Representing the $k$ inputs by a row vector $\mathbf{a}(D) = [a^{(1)}(D)\ a^{(2)}(D)\ \cdots\ a^{(k)}(D)]$, and the $n$ outputs by a row vector $\mathbf{c}(D) = [c^{(1)}(D)\ c^{(2)}(D)\ \cdots\ c^{(n)}(D)]$, the encoding process can now be described as a vector-matrix product

$$\mathbf{c}(D) = \mathbf{a}(D) \cdot \mathbf{G}(D). \tag{3.3}$$

**Example 3.2** For the convolutional encoder in Figure 3.2, the generator polynomial matrix is

$$\mathbf{G}(D) = \begin{bmatrix} 1+D & 1 & 1+D \\ D & 1+D & 0 \end{bmatrix},$$

and the encoding operation is expressed as

$$[c^{(1)}(D)\ c^{(2)}(D)\ c^{(3)}(D)] = [a^{(1)}(D)\ a^{(2)}(D)] \begin{bmatrix} 1+D & 1 & 1+D \\ D & 1+D & 0 \end{bmatrix}.$$

Another matrix related to the generator polynomial matrix is the *parity-check polynomial matrix* $\mathbf{H}(\mathbf{D})$. The parity-check polynomial matrix of a rate $k/n$ convolutional code is an $(n-k) \times n$ matrix of polynomials that satisfies

$$\mathbf{G}(D) \cdot \mathbf{H}^T(D) = \mathbf{0}, \tag{3.4}$$

where superscript $T$ represents a matrix transpose and $\mathbf{0}$ is a $k \times (n-k)$ zero matrix. Substituting (3.3) into (3.4) yields

$$\mathbf{c}(D) \cdot \mathbf{H}^T(D) = \mathbf{0}. \tag{3.5}$$

**Example 3.3** The generator polynomial of the rate 2/3 convolutional code in Figure 3.2 is obtained in the previous example as

$$\mathbf{G}(D) = \begin{bmatrix} 1+D & 1 & 1+D \\ D & 1+D & 0 \end{bmatrix}.$$

Let $\mathbf{H}(D) = [H_1(D)\ H_2(D)\ H_3(D)]$ be the parity-check polynomial matrix. Using (3.4), we have

$$\begin{aligned} (1+D)\ H_1(D) + (1)\ H_2(D) + (1+D)\ H_3(D) &= 0 \\ (D)\ H_1(D) + (1+D)\ H_2(D) + (0)\ H_3(D) &= 0. \end{aligned}$$

The desired solution is $H_1(D) = 1 + D^2$, $H_2(D) = D + D^2$, and $H_3(D) = 1 + D + D^2$. Thus,

$$H(D) = [1+D^2 \quad D+D^2 \quad 1+D+D^2].$$

Now we define some important parameters that are associated with a convolutional code. For a rate $k/n$ convolutional encoder there are $n$ generator polynomials for each of the $k$ inputs. The parameter

$$\nu_i = \max_{1 \leq j \leq n} [deg\ G_{ij}(D)]$$

denotes the number of shift-register stages (memories) associated with the $i$th input. The *memory order* of a convolutional encoder is defined as [1]

$$m = \max_{1 \leq i \leq k} \nu_i = \max_{1 \leq i \leq k, 1 \leq j \leq n} [deg\ G_{ij}(D)].$$

The *constraint length*[3] of a convolutional code is defined as

$$\ell_c = n(m+1). \tag{3.6}$$

The constraint length of the convolutional code, defined as (3.6), plays a similar role to that of the block length of a block code [4]. Finally, the *total encoder memory*[4] is defined as the total number of shift register stages in the encoder structure,

$$\nu = \sum_{i=1}^{k} \nu_i.$$

This parameter determines the number of states in the state diagram or trellis representation of the convolutional code.

### 3.1.2 State Diagram

The operation of a convolutional encoder can be described using a state diagram. The state of the encoder is defined as its shift register contents. For a convolutional code with total encoder memory $\nu$ there are $2^\nu$ states in the state diagram. Each state is assigned a number from 0 to $2^{\nu-1}$ which is usually obtained from binary representation of the shift register contents.

The state diagram shows the state transitions and the associated outputs for all possible inputs. For a rate $k/n$ convolutional encoder, there are $2^k$ branches (transitions) leaving, and $2^k$ branches entering each state. Each new sequence of $k$ input bits causes a transition from an initial state to a new one. The encoder output is always uniquely determined by the initial state and current input.

**Example 3.4** Consider the rate 1/2 convolutional encoder shown in Figure 3.1. There are four possible states in the state diagram of this code: $S_0 = 00$, $S_1 = 01$, $S_2 = 10$, and $S_3 = 11$. The state diagram of this encoder is shown in Figure 3.3. A transition from one state to another is shown by a solid line or a dashed line corresponding to an input of "0" or "1", respectively.

---

[3] There are other definitions of constraint length, e.g., see [2], [3]. We use the same definition as that of [1], [4].

[4] This parameter is defined as constraint length in [2].

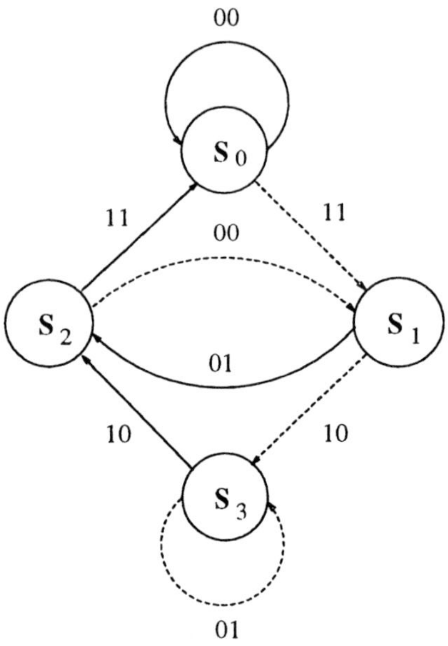


Figure 3.3: State diagram of convolutional encoder shown in Figure 3.1.

The operation of the encoder can be completely described by the state diagram. Knowing the initial state of the encoder, it is easy to calculate the output of the encoder corresponding to any sequence of input bits with the help of this diagram. For example, suppose the encoder starts from initial state $S_0$ and the input sequence is 101100. Based on this input sequence we walk around the state diagram and follow the path $S_0 S_1 S_2 S_1 S_3 S_2 S_0$. In this case the path starts at initial state $S_0$, arrives at final state $S_0$ and the corresponding output sequence is 110100101011.

### 3.1.3 Trellis Representation

It is traditional and instructive to describe a convolutional code by means of a trellis diagram. This diagram which shows the time evolution of the coded sequences, can be obtained from the state diagram by including the dimension of time. Such a diagram for the convolutional encoder shown in Figure 3.1 is represented in Figure 3.4. In drawing the trellis diagram of Figure 3.4 we have used the same convention in specifying the branches introduced by a "0" input bit as a solid line and the branches introduced by a "1" input as a dashed line. Once again the output sequence corresponding to the input sequence 101100 (with initial state $S_0$ ) can be obtained by traveling from state $S_0$ through the trellis diagram according to the input bits. Such a traveling path is shown by a bold line in Figure 3.4. The output sequence corresponding to this path is 110100101011 which agrees with the previous example.

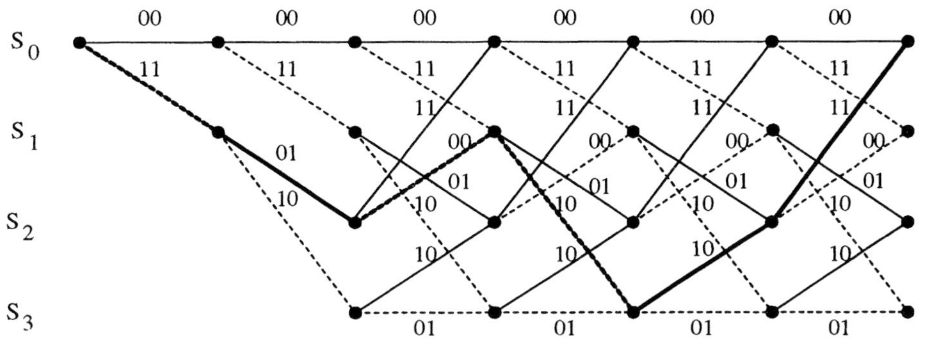


Figure 3.4: Trellis diagram of the encoder shown in Figure 3.1.

### 3.1.4 Systematic Convolutional Codes

The convolutional codes presented in previous discussions are *nonsystematic* codes with a feedforward encoder structure. These codes constitute a subclass of convolutional codes. Another important subclass of convolutional codes are *systematic* codes having the property that the input bits appear unchanged among the output coded bits. In other words, for a rate $k/n$ systematic convolutional code the first $k$ bits of the $n$ output bits are exact replicas of the $k$ input bits. For these codes the generator polynomial matrix can be written as

$$\mathbf{G}(D) = \begin{bmatrix} 1 & 0 & \cdots & 0 & G_{1(k+1)}(D) & \cdots & G_{1n}(D) \\ 0 & 1 & \cdots & 0 & G_{2(k+1)}(D) & \cdots & G_{2n}(D) \\ \cdots & \cdots & \cdots & 0 & \cdots & \cdots & \cdots \\ 0 & 0 & \cdots & 1 & G_{k(k+1)}(D) & \cdots & G_{kn}(D) \end{bmatrix}.$$

Note that the generator polynomial matrix in a systematic convolutional code can be partitioned into two parts: a $k \times k$ identity matrix and a $k \times (n-k)$ matrix of polynomials.

**Example 3.5** Consider the rate 2/3 systematic code with generator polynomial matrix,

$$\mathbf{G}(D) = \begin{bmatrix} 1 & 0 & 1+D^2 \\ 0 & 1 & 1+D+D^2 \end{bmatrix}.$$

Using (3.3), the encoding process for this code can be expressed as

$$[c^{(1)}(D)\ c^{(2)}(D)\ c^{(3)}(D)] = [a^{(1)}(D)\ a^{(2)}(D)] \begin{bmatrix} 1 & 0 & 1+D^2 \\ 0 & 1 & 1+D+D^2 \end{bmatrix}$$

or

$$\begin{aligned} c^{(1)}(D) &= a^{(1)}(D), \\ c^{(2)}(D) &= a^{(2)}(D), \\ c^{(3)}(D) &= a^{(1)}(D) + D^2 a^{(1)}(D) + a^{(2)}(D) + D a^{(2)}(D) + D^2 a^{(2)}(D) \\ &= \{[a^{(1)}(D) + a^{(2)}(D)]D + a^{(2)}(D)\}D + a^{(1)}(D) + a^{(2)}(D). \end{aligned}$$

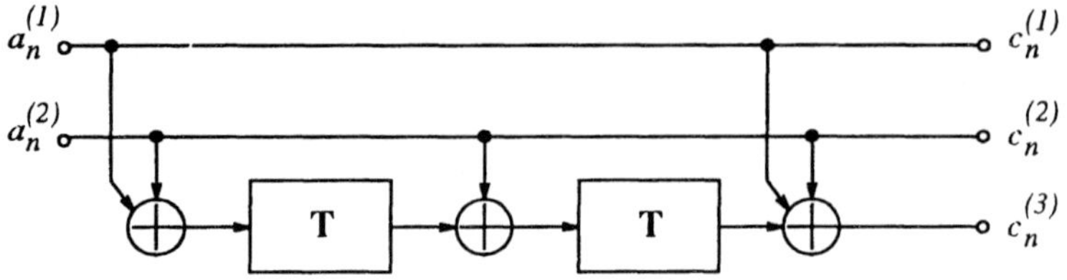


Figure 3.5: Systematic feedforward encoder of Example 3.5.

The two output bits $c^{(1)}$ and $c^{(2)}$ are the exact replicas of the input bits $a^{(1)}$ and $a^{(2)}$, respectively. The third output bit is obtained by a parity-check equation imposed on the current and previous input bits. The realization of the encoder for this systematic code is shown in Figure 3.5.

The systematic feedforward convolutional codes do not perform as well as nonsystematic ones. In [5] it is shown that the error probability of a random ensemble of such codes is greater than the general ensemble of convolutional codes, with the same constraint length, on a memoryless channel with maximum likelihood decoding (see also [3, Chap. 5]).

One advantage of systematic convolutional codes is that they are always *noncatastrophic*. A code is called *catastrophic* when a *finite* number of errors in the received sequence (due to channel noise) produces an *infinite* number of decoded bit errors. Massey and Sain [6] have shown that a necessary and sufficient condition for a rate $k/n$ convolutional code to be noncatastrophic is

$$GCD\left[\Delta_i(D);\ i=1,\cdots,\binom{n}{k}\right]=D^l, \tag{3.7}$$

for some $l \geq 0$, where $\Delta_i(D), i=1,\cdots,\binom{n}{k}$, denote the determinants of all $\binom{n}{k}$ distinct $k \times k$ submatrices of the generator polynomial matrix of the code. In (3.7) $GCD$ shows the *greatest common divisor* operation. This condition implies that for a noncatastrophic code there exists a realizable inverting circuit which recovers the input bits (information sequence) from the coded sequence. For more details about convolutional encoder structure and related topics interested readers should refer to [1], [2], [6].

Forney [2] has shown that for any nonsystematic convolutional code there is an equivalent systematic code with *feedback* which generates the same set of codewords. The following example represents the construction of an equivalent systematic feedback encoder from a nonsystematic feedforward encoder.

**Example 3.6** Consider the nonsystematic encoder of Example 3.2 with generator polynomial matrix

$$\mathbf{G}(D) = \begin{bmatrix} 1+D & 1 & 1+D \\ D & 1+D & 0 \end{bmatrix}.$$

Using some elementary row operations, the generator polynomial matrix can be put into a systematic form $\mathbf{G}^s(D)$

$$\mathbf{G}(D) = \begin{bmatrix} 1+D & 1 & 1+D \\ D & 1+D & 0 \end{bmatrix} \rightarrow \begin{bmatrix} 1+D^2 & 1+D & 1+D^2 \\ D & 1+D & 0 \end{bmatrix}$$

$$\rightarrow \begin{bmatrix} 1+D+D^2 & 0 & 1+D^2 \\ D & 1+D & 0 \end{bmatrix} \rightarrow \begin{bmatrix} 1 & 0 & \frac{1+D^2}{1+D+D^2} \\ 1 & \frac{1+D}{D} & 0 \end{bmatrix}$$

$$\rightarrow \begin{bmatrix} 1 & 0 & \frac{1+D^2}{1+D+D^2} \\ 0 & \frac{1+D}{D} & \frac{1+D^2}{1+D+D^2} \end{bmatrix} \rightarrow \begin{bmatrix} 1 & 0 & \frac{1+D^2}{1+D+D^2} \\ 0 & 1 & \frac{D+D^2}{1+D+D^2} \end{bmatrix} \rightarrow$$

$$\mathbf{G}^s(D) = \begin{bmatrix} 1 & 0 & \frac{1+D^2}{1+D+D^2} \\ 0 & 1 & \frac{D+D^2}{1+D+D^2} \end{bmatrix},$$

where we use the equality $(1+D)^2 = 1+D^2$ when the operations are performed in $GF(2)$. Note that the polynomial division in the third column of the generator polynomial matrix is realizable by a feedback structure [2]. Using (3.3) the output bit sequences of the encoder are expressed in terms of its input bit sequences $a'^{(1)}(D)$ and $a'^{(2)}(D)$ as

$$\begin{aligned} c^{(1)}(D) &= a'^{(1)}(D), \\ c^{(2)}(D) &= a'^{(2)}(D), \\ c^{(3)}(D) &= \tfrac{a'^{(1)}(D)+D^2a'^{(1)}(D)+Da'^{(2)}(D)+D^2a'^{(2)}(D)}{1+D+D^2}. \end{aligned}$$

The last parity-check equation can be put into a more instructive form as

$$c^{(3)}(D)+Dc^{(2)}(D)+D^2c^{(1)}(D) = c^{(1)}(D)+D^2c^{(1)}(D)+Dc^{(2)}(D)+D^2c^{(2)}(D),$$

$$c^{(3)}(D) = a^{(1)}(D) + D\{[c^{(3)}(D) + a^{(2)}(D)] + D[c^{(3)}(D) + a^{(1)}(D) + a^{(2)}(D)]\},$$

which is a difference equation and can be implemented by a feedback structure. A feedback realization of the equivalent systematic code is shown in Figure 3.6. It is easy to verify that the same set of codewords are generated by the encoders shown in Figure 3.2 and Figure 3.6.

The generator polynomial matrix of an equivalent systematic feedback encoder can also be obtained by right-multiplying the latter by a scrambler matrix [2]. That is,

$$\mathbf{G}^s(D) = \mathbf{A} \cdot \mathbf{G}(D), \tag{3.8}$$

where $\mathbf{A}$ is a $k \times k$ matrix and $\mathbf{G}^s(D)$ denotes the generator polynomial matrix of the equivalent systematic encoder.

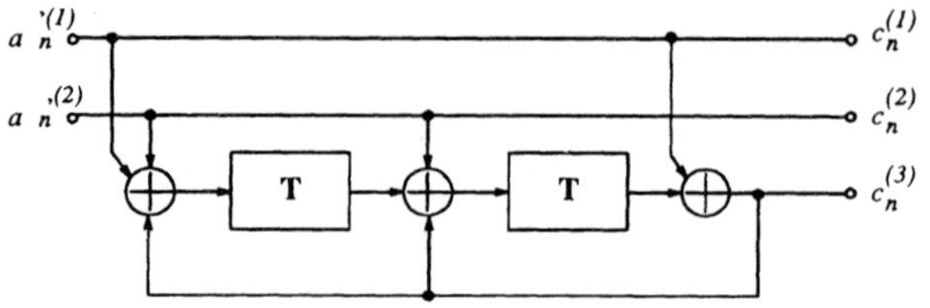


Figure 3.6: Feedback realization of the equivalent systematic encoder of the nonsystematic encoder shown in Figure 3.2.

**Example 3.7** For the rate 2/3 nonsystematic encoder of Figure 3.2 the generator polynomial is

$$\mathbf{G}(D) = \begin{bmatrix} 1+D & 1 & 1+D \\ D & 1+D & 0 \end{bmatrix}.$$

Using (3.8) we have

$$\begin{bmatrix} 1 & 0 & G_{31}^s(D) \\ 0 & 1 & G_{32}^s(D) \end{bmatrix} = \begin{bmatrix} A_{11} & A_{12} \\ A_{21} & A_{22} \end{bmatrix} \begin{bmatrix} 1+D & 1 & 1+D \\ D & 1+D & 0 \end{bmatrix}.$$

Thus

$$\begin{aligned} 1 &= (1+D)A_{11} + (D)A_{12} \\ 0 &= (1)A_{11} + (1+D)A_{12} \\ 0 &= (1+D)A_{21} + (D)A_{22} \\ 1 &= (1)A_{21} + (1+D)A_{22} \\ G_{31}^s(D) &= (1+D)A_{11} \\ G_{32}^s(D) &= (1+D)A_{21}, \end{aligned}$$

which yields

$$\begin{aligned} A_{11} &= \tfrac{1+D}{1+D+D^2}, \\ A_{12} &= \tfrac{1}{1+D+D^2}, \\ A_{21} &= \tfrac{D}{1+D+D^2}, \\ A_{22} &= \tfrac{1+D}{1+D+D^2}, \\ G_{31}^s(D) &= \tfrac{1+D^2}{1+D+D^2}, \\ G_{32}^s(D) &= \tfrac{D+D^2}{1+D+D^2}. \end{aligned}$$

Note that the same result is obtained as in Example 3.6.

## 3.2 DISTANCE PROPERTIES AND GENERATING FUNCTION

The performance of a convolutional code on a binary symmetric channel (BSC) depends on the *Hamming distance*[5] properties of the code. For convolutional

[5] Hamming distance between two codewords $\mathbf{c}_i$ and $\mathbf{c}_j$ is defined as the number of places where they differ.

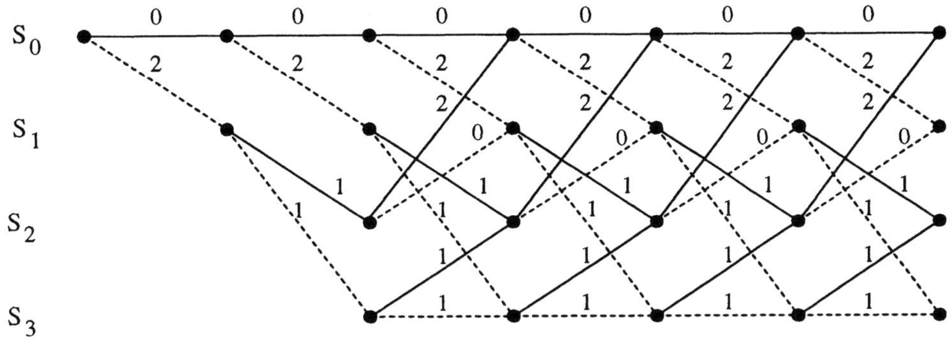


Figure 3.7: Trellis diagram of the encoder shown in Figure 3.1 with branches labeled according to their Hamming distances from all-zero sequence.

codes, which constitute a subclass of linear codes, the set of Hamming distances between coded sequences is the same as the set of distances of the coded sequences from the all-zero sequence. Thus, in the following discussion we consider the distance structure of the convolutional codes with respect to all-zero coded sequence without loss of generality.

The set of distances of a convolutional code can be computed with the aid of the code trellis or state diagram. Consider again the trellis diagram of the rate 1/2 convolutional code of Figure 3.4. This figure is redrawn in Figure 3.7 by labeling the branches according to their *Hamming weights*[6] (Hamming distance from the all-zero code sequence).

To find the set of distances for this convolutional code we consider the paths which diverge from $S_0$ and remerge into it for the first time at some later time. Examples of such paths are: $S_0S_1S_2S_0$, $S_0S_1S_3S_2S_0$, and $S_0S_1S_2S_1S_2S_0$ with Hamming distances of 5, 6, and 6 from the all-zero sequence, respectively. The associated input sequence with these paths are $\cdots 100 \cdots$ (Hamming weight 1), $\cdots 1100 \cdots$ (Hamming weight 2), and $\cdots 10100 \cdots$ (Hamming weight 2), respectively. With enough patience one can determine the distances of all paths from the trellis diagram. Such a trial shows that $S_0S_1S_2S_0$ has the minimum Hamming distance of 5 among all paths of the trellis diagram.

The above-mentioned technique, although instructive, is a time consuming method to evaluate the distance structure of convolutional codes, particularly ones with long length. Using the state diagram one can obtain a closed-form expression, called the *generating function*, whose expansion provides all distance information directly. This method is described for the same rate 1/2 convolutional code as an illustrative example.

Figure 3.8 shows the state diagram of Figure 3.3 in a modified form that

[6] Hamming weight of a sequence is the number of 1's in that sequence, which is equivalent to the Hamming distance of this sequence from the all-zero sequence.

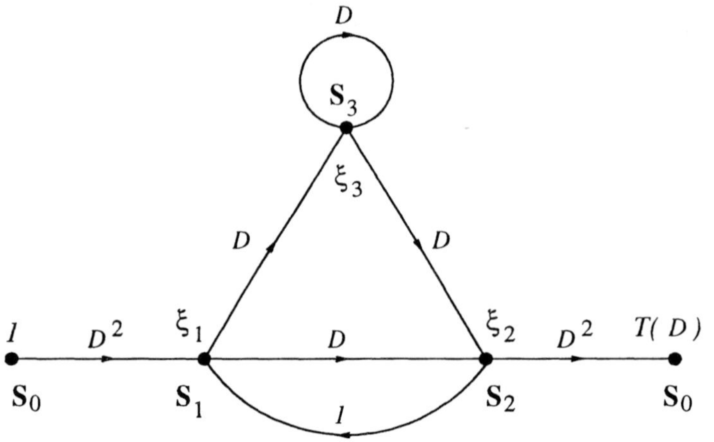


Figure 3.8: The modified state diagram of the convolutional encoder shown in Figure 3.1.

can be convenient for the study of the code distance structure. The branches of this diagram are labeled by *branch gain* $D^i, i = 0, 1, 2$, where the exponent of $D$ denotes the Hamming weight of the branch. Note that in Figure 3.8 the self-loop at $S_0$ has been removed since it has no contribution to the distance properties of the code. Circulating around this loop simply generates the all-zero sequence. Also, $S_0$ has been split into two initial and final states. Now we can trace all of the paths, obtained from trellis digram, on the modified state diagram. As an example, consider the path $S_0S_1S_2S_0$ which starts from initial state $S_0$ and terminates at final state $S_0$. Multiplying the branch gains of the branches of this path yields the *path gain* $D^5$ whose exponent denotes the Hamming distance of the path from the all-zero sequence.

The *distance profile* of the convolutional codes can be determined from its modified state diagram using state equations. For the modified state diagram of Figure 3.8 the state equations are

$$\begin{aligned} \xi_1 &= D^2 + \xi_2, \\ \xi_2 &= D\xi_1 + D\xi_3, \\ \xi_3 &= D\xi_1 + D\xi_3, \end{aligned} \tag{3.9}$$

where $\xi_i$, $i = 1, 2, 3$, are dummy variables which show the values of the path gain from initial state $S_0$ to $S_i$ as influenced by other states. The input to initial state $S_0$ is unity, and the output at final state $S_0$ is the generating function

$$T(D) = D^2\xi_2. \tag{3.10}$$

Solving the state equations given in (3.9) and substituting the value of $\xi_2$ into

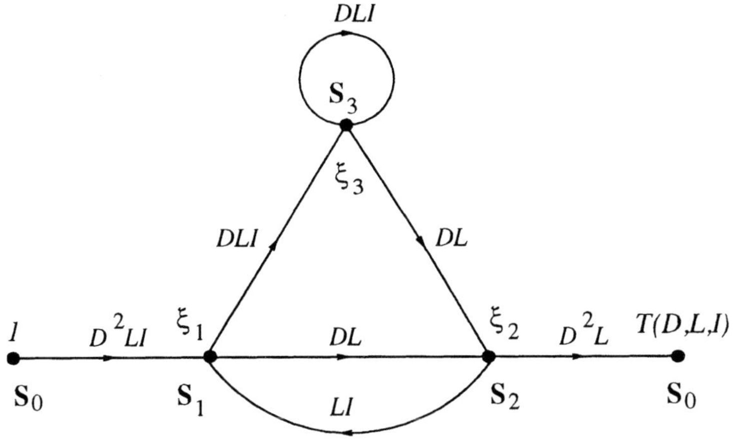


Figure 3.9: The modified state diagram of the convolutional encoder shown in Figure 3.1 with new labels.

(3.10), we obtain

$$\begin{aligned} T(D) &= \frac{D^5}{1-2D} \\ &= D^5 + 2D^6 + 4D^7 + \cdots + 2^k D^{k+5} + \cdots \\ &= \sum_{d=5}^{\infty} a_d D^d. \end{aligned} \tag{3.11}$$

The above equation shows that there is only one path of Hamming distance 5 and two paths of Hamming distance 6 from the all-zero path. In general, the number of paths with Hamming distance $k+5$ from the all-zero sequence is $2^k$. The minimum Hamming distance of this code is 5. The *minimum Hamming distance* of convolutional codes is sometimes called the *free Hamming distance* and denoted as $d^H_{free}$. We use superscript $H$ to distinguish this distance from the free Euclidean distance of trellis-coded modulation schemes which are the subject of the next chapter.

By introducing new labels in the modified state diagram (also called error-state diagram) of the convolutional code more detailed information about the distance structure of these codes can be obtained from the expansion of the augmented generating function. The new labels of the modified state diagram are shown in Figure 3.9. To enumerate the length of a given path the label $L$ is added with exponent 1 for each branch. Thus the exponent of this label will be augmented by one every time a branch is passed through. We also add the label $I^i$ to each branch to enumerate the Hamming weight of the input bits, $i$, associated with each branch of the modified state diagram.

The state equations for the modified state diagram shown in Figure 3.9 are

$$\begin{aligned}\xi_1 &= D^2LI + LI\xi_2,\\ \xi_2 &= DL\xi_1 + DL\xi_3,\\ \xi_3 &= DLI\xi_1 + DLI\xi_3,\end{aligned} \tag{3.12}$$

and the augmented generating function is

$$T(D, L, I) = D^2L\xi_2. \tag{3.13}$$

The desired solution for the augmented generating function is obtained by solving (3.12) and substituting the value of $\xi_3$ into (3.13)

$$\begin{aligned}T(D, L, I) &= \frac{D^5L^3I}{1 - DL(L+1)I}\\ &= D^5L^3I + D^6L^4(1+L)I^2 + D^7L^5(1+L)^2I^3\\ &\quad + \cdots + D^{k+5}L^{3+k}(1+L)^kI^{1+k} + \cdots.\end{aligned}$$

This augmented generating function gives more information about the structure of all paths of the convolutional code. For example, it indicates that the path with minimum Hamming distance of 5 has length 3 and its input sequence weight is one. Of the two paths with Hamming distance 6, one has length 4 while the length of the other path is 5. Both paths have the same input sequence weight, i.e., 2. These results verify our previous observations.

## 3.3 DECODING

There are different algorithms for decoding convolutional codes, e.g., the sequential decoding, the feedback decoding and the Viterbi decoding. Sequential decoding, as the first decoding algorithm for convolutional codes, is introduced by Wozencraft [7]. The other versions of this algorithm, namely, the Fano algorithm and the stack algorithm were proposed by Fano [8] and Zigangirov and Jelenik [9], [10], respectively. The feedback decoding algorithm is an algebraic decoding method based on the threshold decoding developed by Massey [11] (see Problem 3.11). The *Viterbi algorithm* was introduced by Viterbi [12] as a "new probabilistic nonsequential decoding algorithm" for convolutional codes and later [13] was recognized to be in fact a maximum likelihood decoding for convolutional codes. In this section we only consider the Viterbi algorithm, for the other decoding methods we refer the reader to text books [1], [3].

### 3.3.1 Maximum Likelihood Decoding

A maximum likelihood decoder decodes a particular received sequence $\mathbf{r}$ into the code sequence $\mathbf{c}$ which maximizes the likelihood function $P(\mathbf{r}|\mathbf{c})$ or the log-likelihood function $\ln P(\mathbf{r}|\mathbf{c})$. This decoder provides minimum probability of error when all transmitted codewords are equally probable.

Assume that a codeword $\mathbf{c} = (c_1\ c_2\ \cdots c_l)$ of length $l$ is transmitted through a *memoryless* channel and the received sequence is $\mathbf{r} = (r_1\ r_2\ \cdots r_l)$. Since the noise components are independent in a memoryless channel, the likelihood function $P(\mathbf{r}|\mathbf{c})$ can be expressed as

$$P(\mathbf{r}|\mathbf{c}) = \prod_{i=1}^{l} P(r_i|c_i), \tag{3.14}$$

where $P(r_i|c_i)$ is a channel transition probability. We also note that the log-likelihood function, in this case, can be written as

$$\ln P(\mathbf{r}|\mathbf{c}) = \sum_{i=1}^{l} \ln P(r_i|c_i).$$

Next we show that in the case of hard-decision decoding, the maximum likelihood decoding is obtained by choosing the nearest codeword from $\mathbf{r}$ in Hamming distance. Suppose that the codewords are transmitted through a binary symmetric channel (BSC) with a channel transition probability $p$. Assume that the received sequence $\mathbf{r}$ differs from the codeword $\mathbf{c}$ in exactly $d$ positions. In such a case the likelihood function (3.14) can be expressed as

$$P(\mathbf{r}|\mathbf{c}) = p^d(1-p)^{l-d}.$$

Without loss of generality we may assume $p \leq 1/2$. It is then easy to show that maximizing likelihood function $P(\mathbf{r}|\mathbf{c})$ is in fact equivalent to minimizing the Hamming distance $d$ between $\mathbf{r}$ and $\mathbf{c}$.

For soft-decision decoding in an additive white Gaussian noise (AWGN) channel the maximum likelihood decoding is obtained by choosing a codeword which is nearest to the received sequence $\mathbf{r}$ in *Euclidean distance.* For such a case, the likelihood function is written as

$$\begin{aligned} p(\mathbf{r}|\mathbf{c}) &= \prod_{i=1}^{l} \frac{1}{\sqrt{\pi N_0}} e^{-\frac{|r_i - c_i|^2}{N_0}} \\ &= \left(\frac{1}{\pi N_0}\right)^{\frac{l}{2}} \exp\left(-\frac{1}{N_0}\sum_{i=1}^{l} |r_i - c_i|^2\right), \end{aligned} \tag{3.15}$$

where $N_0$ is the single-sided power spectral density of the AWGN and

$$d_E^2(\mathbf{r},\ \mathbf{c}) = \sum_{i=1}^{l} |r_i - c_i|^2$$

is the squared Euclidean distance between $\mathbf{r}$ and $\mathbf{c}$. Considering (3.15) we observe that the maximum likelihood decoding rule for the AWGN channel with soft-decision decoding can be formulated by minimizing the Euclidean distance between the received sequence and the decoded codeword.

### 3.3.2 Viterbi Algorithm

The straightforward implementation of the maximum likelihood decoder for convolutional codes requires examining all code sequences or code paths through the trellis and choosing the one which provides the largest likelihood function (or the smallest distance). Since the number of code paths grows exponentially with sequence length, such a brute force implementation technique would appear absurdly complex. A practical implementation of the maximum likelihood decoding for convolutional codes is the Viterbi algorithm.

The Viterbi algorithm is an efficient method for searching for the maximum likelihood path through the trellis diagram. This process starts from an initial state and attempts to compute the maximum likelihood function for each possible path remerging at a specific state (node). In each state it compares the maximum likelihood (*metric*) of different paths arriving at that state, preserves the path with the largest metric and discards the rest. This path is called *the survivor*. The surviving path is stored at each node level together with its metric. The algorithm proceeds through the trellis in an iterative manner and maintains a relatively small list of paths that are always guaranteed [1, Page 318] to contain the maximum likelihood path. This algorithm can be best explained using the following example.

**Example 3.8** Consider the rate 1/2 convolutional code of Example 3.1. Suppose that the code sequence 110100101011 (the path shown in bold) in Figure 3.4 is transmitted, by using a binary signaling scheme, say BPSK. Assume that the receiver performs hard-decision detection and provides the Viterbi decoder with the received sequence 110110101011 which has one error in the fifth location. In this case the Hamming distance between the received sequence and the code sequence is considered as a metric. The Viterbi decoder chooses the path through the trellis of the code with minimum Hamming distance from the received sequence.

The Viterbi decoding process is shown in Figure 3.10. In this figure the numbers associated with each state represent the metric (Hamming distance) of the surviving path for that state. The surviving paths at different levels of the decoding procedure are shown in Figure 3.11. The Viterbi decoding starts at initial state $S_0$. Initially it sets all the metrics associated with different states to zero. At the first level of the decoding process the decoder computes the Hamming distance (metric) for each of two branches emanating from state $S_0$ and arriving at states $S_0$ and $S_1$. These metrics are stored together with their surviving paths for state $S_0$ and $S_1$ (Figure 3.11(a)). The surviving paths with respected metrics are computed and stored for each state at the second level of the decoding procedure (Figure 3.11(b)). Since there is only one path arriving at each state at this level, no comparison is needed and this single path is stored as the survivor for that state. At the third level of decoding there are two paths merging to each state. In this case the decoder computes the Hamming distance associated with each path, stores the path with minimum Hamming distance and discards the other path. In this way, the decoder proceeds through the

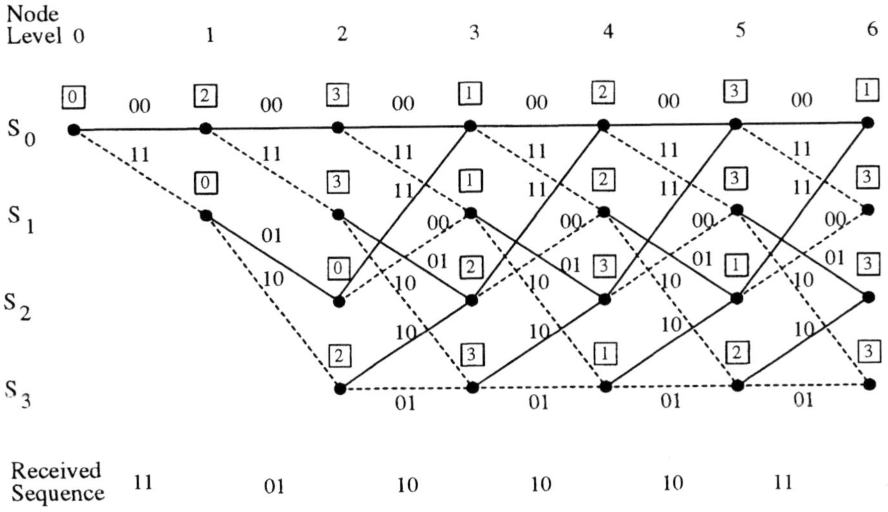


Figure 3.10: Viterbi decoding process of the code shown in Figure 3.1.

trellis, at each level, and preserves only one survivor and its distance from the received signal for each state (see Figure 3.11(c)-(f)).

In decoding a difficulty may arise in a given comparison between merging paths with the same metric values. This tie can be resolved randomly by the flip of a fair coin. The outcome of such a flip does not alter the ultimate performance of the decoder. As an example of a tie consider the paths arriving at state $S_2$ at the fourth level of decoding in Figure 3.10. The distance of both paths from the received signal is 3. Consequently, we choose one of these paths arbitrarily, say the path remerging to $S_2$ from $S_3$, and discard the other one. Note that this choice has no effect on the final decision since it is eliminated at higher levels of decoding.

Note that at the sixth decoding level the code path corresponding to the transmitted sequence has a metric smaller than other paths. If a decision is to be made at this level, this path can be selected as the one nearest to the transmitted code sequence in the Hamming distance sense.

The above example shows that for each state a memory is required to store the entire length of the surviving paths. For most practical applications the convolutionally encoded sequences are very long or even infinite. In such a case, the memory required to preserve the survivors is large and expensive. Moreover, such a large memory length introduces a large decoding delay which is practically undesirable.

In practice, for large coded sequences, it is necessary to truncate the surviving paths to some length $\delta_d$. In this case the decoder at any level $l$ retains only the most recent $\delta_d$ branches back in the trellis. For each new node level

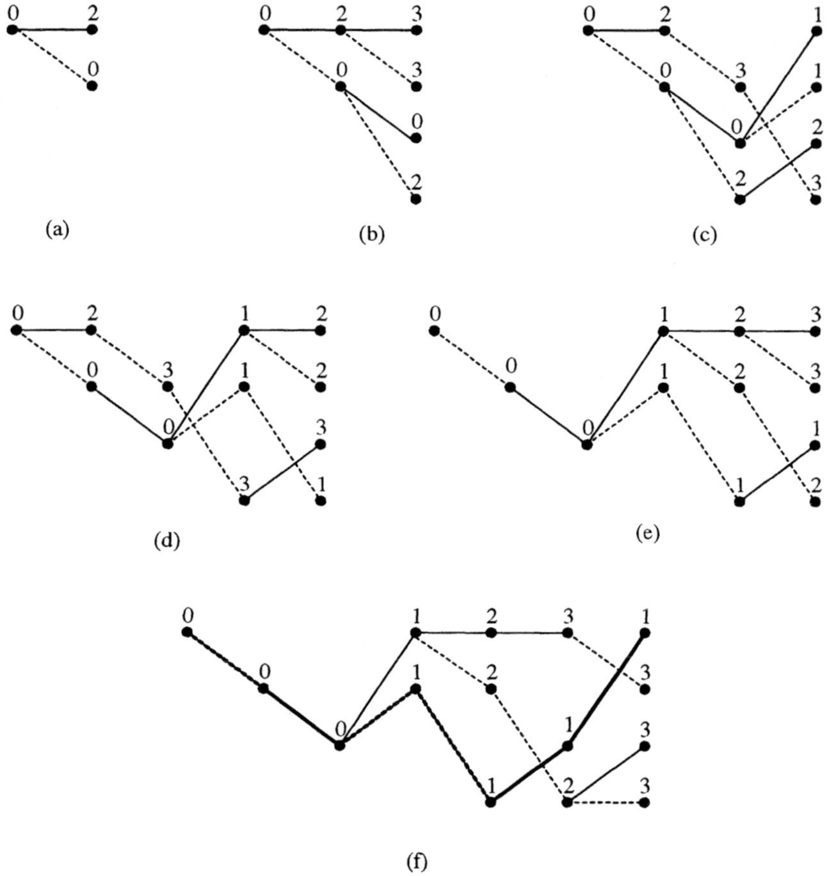

Figure 3.11: Survival paths in the trellis diagram of the encoder shown in Figure 3.1.

the decoder comes to a definite decision on the nodes up to level $l - \delta_d$ at level $l$. If the *truncation depth* $\delta_d$ chosen is large enough, there is a high probability that all surviving sequences at level $l$ contain the same $l - \delta_d$ branches back in time. This can be best seen from Example 3.8 where the first two branches of the surviving paths at the fifth and the sixth level of the decoding procedure coincide. It has been shown [3, Chap. 4] that the performance of the Viterbi algorithm is hardly degraded by a proper truncation of the surviving path memory at a short constraint length.

## 3.4 PERFORMANCE BOUNDS

The most useful performance measure for convolutional codes is *Bit error probability*, $P_b$, which is defined as the expected number of decoded information bit errors per information bits. The bit error probability of a given convolutional code for a given channel is upper-bounded by means of union bounds. We begin our discussion by evaluating this bound for a binary symmetric channel.

Using the linear property of a convolutional code, we assume, without loss of generality, that the all-zero code sequence is transmitted. To find an upper bound on $P_b$, we initially evaluate the *first event error* probability. This is the probability of error for sequences that merge with the all-zero (correct) path *for the first time* at a given node in the trellis. A first error at time $j$ is shown in Figure 3.12 where the correct path $\mathbf{c}$ is eliminated by the incorrect path $\mathbf{c}'$ for the first time.

Suppose the path $\mathbf{c}'$ differs from the correct path $\mathbf{c}$ in $d$ bits. In a binary symmetric channel the maximum likelihood decoder chooses the path $\mathbf{c}'$ in favor of the correct path $\mathbf{c}$ if more than $\lfloor \frac{d}{2} \rfloor$ errors occur, where $\lfloor x \rfloor$ is the largest integer less than or equal to $x$. If the channel transition probability is $p$, then the probability of error in the pairwise comparison of $\mathbf{c}$ and $\mathbf{c}'$ is upper bounded as [1], [14] (see Problem 3.18).

$$P_d \leq \left[2\sqrt{p(1-p)}\right]^d .$$

The above equation shows the first event error probability for a path with distance $d$ from the all-zero (correct) path. There are many paths with different distances that merge with the correct path at a given time $j$ for the first time. Using a union bound an upper bound on the first error probability is obtained by summing the error probabilities of all possible such paths

$$P_e \leq \sum_{d=d_{free}^H}^{\infty} a_d P_d, \tag{3.16}$$

where $a_d$ is the number of codewords of Hamming distance $d$ from the all-zero codeword. Comparing (3.16) with (3.11) it follows that

$$P_e \leq T(D) \Big|_{D=2\sqrt{p(1-p)}} . \tag{3.17}$$

Now we turn to the bit error probability, $P_b$. This probability can be upper-bounded by weighting each pairwise error probability, $P_d$, in (3.16) by the number of incorrectly decoded information bits for the corresponding incorrect path, $n_d$. For a rate $k/n$ convolutional code the result is divided by $k$ to yield

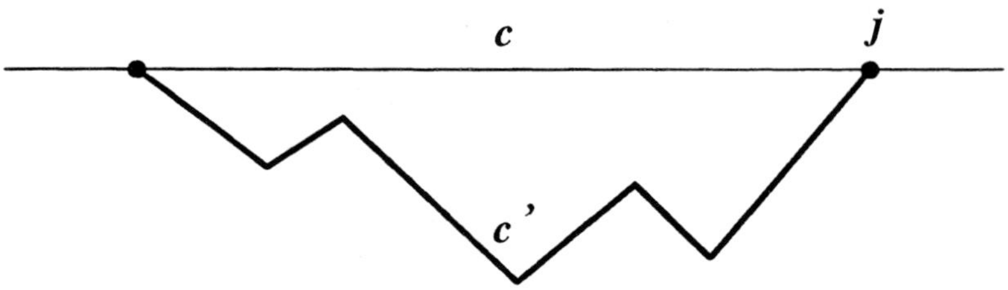


Figure 3.12: First event error at time $j$.

the average bit error probability

$$P_b \le \frac{1}{k} \sum_{d=d^H_{free}}^{\infty} a_d n_d P_d. \tag{3.18}$$

Once again this upper bound can be expressed in terms of the generating function. Recall that the augmented generating function enumerates the number of information bits for each path. In general, this function (with $L = 1$ since we are not interested in path length) can be expressed as

$$T(D,\ I) = \sum_{d=d^H_{free}}^{\infty} a_d D^d I^{n_d},$$

whose derivative at $I = 1$ is

$$\left.\frac{\partial T(D,\ I)}{\partial I}\right|_{I=1} = \sum_{d=d^H_{free}}^{\infty} a_d n_d D^d. \tag{3.19}$$

Comparing (3.19) and (3.18) yields

$$P_b \le \frac{1}{k} \left.\frac{\partial T(D,\ I)}{\partial I}\right|_{I=1,\ D=2\sqrt{p(1-p)}}.$$

Next we consider the performance of convolutional codes when the receiver performs soft-decision detection followed by a maximum likelihood decoder. As an example of such a case consider the binary antipodal transmission in an AWGN channel. Assume again that the correct path **c** and the incorrect path **c**$'$ shown in Figure 3.12, differ in $d$ bit positions. The pairwise error probability for these paths is (see Problem 3.19)

$$P_d = \frac{1}{2}\mathrm{erfc}\left(\sqrt{\frac{\mathrm{d}\mathrm{E_s}}{\mathrm{N_0}}}\right), \tag{3.20}$$

where $E_s$ represents the average energy per transmitted channel symbol. Upper-bounding erfc(x) by $\exp(-x^2)$, the pairwise error probability (3.20) is expressed as

$$P_d \leq \frac{1}{2} e^{-\frac{dE_s}{N_0}}. \tag{3.21}$$

Substituting (3.21) into (3.16) and comparing with (3.11), we obtain

$$P_e \leq \frac{1}{2} T(D) \Big|_{D=\exp(-\frac{E_s}{N_0})}.$$

Similarly, the upper bound on the bit error probability for a binary input AWGN channel is obtained as

$$P_b \leq \frac{1}{2k} \frac{\partial T(D,\ I)}{\partial I} \Big|_{I=1,\ D=\exp(-\frac{E_s}{N_0})}.$$

We may find tighter bounds by using a useful inequality (see Problem 2.9)

$$\mathrm{erfc}(\sqrt{x+y}) \leq \mathrm{erfc}(\sqrt{x}) e^{-y}, \quad x \geq 0,\ y \geq 0. \tag{3.22}$$

Recalling that $d \geq d^H_{free}$, (3.20) may be written as

$$\begin{aligned} P_d &= \frac{1}{2}\mathrm{erfc}\left(\sqrt{\frac{d^H_{free} E_s}{N_0} + \frac{(d - d^H_{free}) E_s}{N_0}}\right) \\ &\leq \frac{1}{2}\mathrm{erfc}\left(\sqrt{\frac{d^H_{free} E_s}{N_0}}\right) e^{-\frac{(d-d^H_{free})E_s}{N_0}}. \end{aligned} \tag{3.23}$$

Substituting (3.23) into (3.16) and comparing with (3.11), we obtain

$$P_e \leq \frac{1}{2}\mathrm{erfc}\left(\sqrt{\frac{d^H_{free} E_s}{N_0}}\right) e^{\frac{d^H_{free} E_s}{N_0}} T(D) \Big|_{D=\exp(-\frac{E_s}{N_0})}.$$

Similarly, for bit error probability we can find

$$P_b \leq \frac{1}{2k}\mathrm{erfc}\left(\sqrt{\frac{d^H_{free} E_s}{N_0}}\right) e^{\frac{d^H_{free} E_s}{N_0}} \frac{\partial T(D,\ I)}{\partial I} \Big|_{I=1,\ D=\exp(-\frac{E_s}{N_0})}. \tag{3.24}$$

For the purpose of comparing coded systems with uncoded systems, we need to evaluate the performance of the coded systems in terms of *energy per information bit*, $E_b$, rather than the signal energy per channel symbol, $E_s$. For a code with rate $k/n$ the energy per information bit, $E_b$, is defined as

$$E_b \equiv \frac{E_s}{(k/n)}. \tag{3.25}$$

Using (3.25) the error bounds of convolutional codes, derived in this section, can be expressed in terms of energy per information bit.

**Example 3.9** Consider the rate 1/2 convolutional code of Figure 3.1. The coded sequence is transmitted through an AWGN channel with single-sided spectral density, $N_0$, using coherent BPSK modulation. The bit error probability of this code with soft-decision decoding is upper-bounded as

$$\begin{aligned} P_b &\leq \frac{1}{2}\frac{D^5}{(1-2D)^2}\Big|_{D=\exp(-E_b/2N_0)} \\ &= \frac{\exp(-5E_b/2N_0)}{2[1-2\exp(-E_b/2N_0)]^2}. \end{aligned} \tag{3.26}$$

A tighter bound is obtained using (3.24)

$$\begin{aligned} P_b &\leq \frac{1}{2}\mathrm{erfc}\left(\sqrt{\frac{5\mathrm{E_b}}{2\mathrm{N_0}}}\right)\mathrm{e}^{\frac{5\mathrm{E_b}}{2\mathrm{N_0}}}\frac{\mathrm{D}^5}{(1-2\mathrm{D})^2}\Big|_{\mathrm{D}=\exp(-\mathrm{E_b}/2\mathrm{N_0})} \\ &= \frac{1}{2}\mathrm{erfc}\left(\sqrt{\frac{5\mathrm{E_b}}{2\mathrm{N_0}}}\right)\frac{1}{[1-2\exp(-\mathrm{E_b}/2\mathrm{N_0})]^2}. \end{aligned} \tag{3.27}$$

If hard-decision decoding is performed prior to decoding, then the upper bound on bit error probability is obtained from (3.17). For a BSC derived from an AWGN channel with coherent BPSK modulation (from (2.14)), the channel transition probability, $p$, is

$$\begin{aligned} p &= \frac{1}{2}\mathrm{erfc}\left(\sqrt{\frac{\mathrm{E_s}}{\mathrm{N_0}}}\right) \\ &= \frac{1}{2}\mathrm{erfc}\left(\sqrt{\frac{\mathrm{E_b}}{2\mathrm{N_0}}}\right). \end{aligned}$$

Thus,

$$P_b \leq \frac{D^5}{(1-2D)^2}\Big|_{D=2\sqrt{p(1-p)}}. \tag{3.28}$$

The upper bounds (3.26), (3.27) and (3.28) are plotted in Figure 3.13 in terms of the energy per information bit-to-noise ratio, $E_b/N_0$. The performance of the uncoded coherent BPSK modulation is also included in this figure for comparison. To examine the tightness of the upper bounds we also include the simulation results for hard-decision decoding and soft-decision decoding of this code from [15].

The simulation results show the tightness of the soft-decision decoding upper bound given by (3.27). We note that the hard-decision decoding bound given by (3.28) is about 1 dB above the simulation results.[7]

[7] A tighter bound for bit error probability of convolutional codes on a BSC can be obtained by evaluating the exact expression of the pairwise error probability, $P_d$, [14] and using (3.16). Another tighter bound for this case is presented in [16].

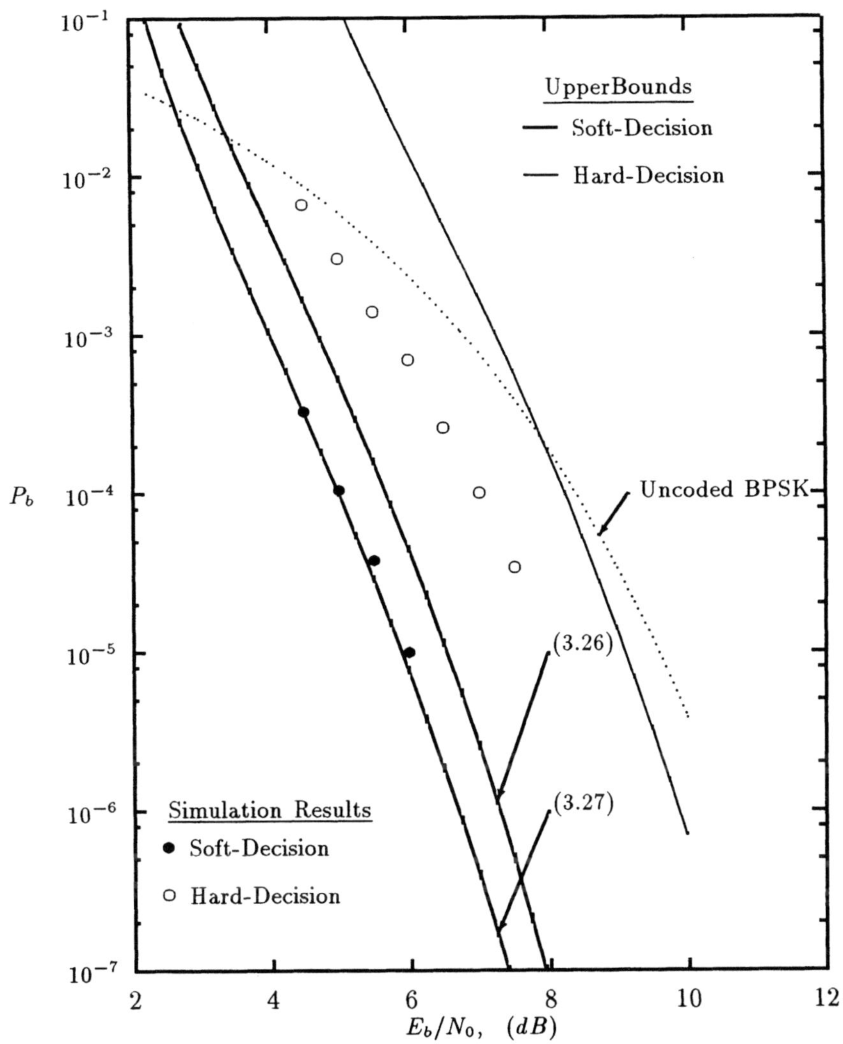


Figure 3.13: Performance of the rate 1/2 convolutional code of Figure 3.1.

The specific example above shows that the performance of convolutional codes with BPSK is superior to the uncoded BPSK modulation at high signal-to-noise ratios. Note that this performance gain is achieved at the expense of lowering the information bit rate per channel bandwidth.

In comparing the performance between soft-decision decoding and hard-decision decoding we note that the difference obtained is about 2.5-3 dB from the upper bounds (about 2 dB from the simulation results) for $10^{-2} \leq P_b \leq 10^{-6}$. Although this comparison is based on performance of a specific code it can be generalized by a random coding analysis. Such analysis shows that [17] about 2-3 dB gain is obtained by the use of soft-decision decoding over the entire range of signal-to-noise ratios.

## Problems

**3.1** The following binary sequence is the input of the encoder shown in Figure 3.1

$$\{a_n^{(1)}\} = 10111000101010$$

**a)** Using (3.1) find the output sequence $\{c_n^{(1)}\}$ and $\{c_n^{(2)}\}$.

**b)** Find the input and output polynomials.

**3.2** Draw the block diagram of the rate 2/3 convolutional encoder which has the following generator polynomial matrix

$$G(D) = \begin{bmatrix} 1+D+D^2 & D^2 & 1 \\ D & 1+D^2 & 1+D+D^2 \end{bmatrix}$$

Find the number of memory elements ($\nu$)

**3.3** The following binary sequences are the inputs of the encoder of Problem 3.2:

$$\{a_n^{(1)}\} = 101100010$$
$$\{a_n^{(2)}\} = 111101011.$$

Find the output sequences $\{c_n^{(1)}\}$, $\{c_n^{(2)}\}$, $\{c_n^{(3)}\}$ using polynomial notation.

**3.4** Find the parity-check polynomial matrix for the convolutional encoder shown in Figure 3.1.

**3.5** Find the parity-check polynomial matrix for the convolutional encoder of Problem 3.2.

**3.6** Draw the state diagram, trellis diagram and modified state diagram of the convolutional encoder shown in Figure 3.2.

Derive the transfer function and determine the free Hamming distance of this convolutional code.

**3.7** Repeat Problem 3.6 for the convolutional encoder of Problem 3.2.

**3.8** Is the code of Problem 3.2 non-catastrophic?

*Hint:* Use (3.7).

**3.9** Construct the equivalent systematic feedback encoder for the code of Problem 3.2.

*Hint:* Follow the procedure given in Example 3.7.

**3.10** Consider a rate 2/3 convolutional code defined by the following generator polynomial matrix:

$$G(D) = \begin{bmatrix} 1+D & D+D^2 & 1+D+D^2 \\ D^2+D^3 & 1 & 1+D+D^2+D^3 \end{bmatrix}$$

**a)** Is this code non-catastrophic? Construct the equivalent systematic feedback encoder.

**b)** Draw the state diagram, trellis diagram, and modified state error state diagram.

**c)** Derive the transfer function and determine the free Hamming distance of the code.

**3.11** Consider a rate 1/2 systematic code defined by the following generator polynomial matrix

$$G(D) = [1 \quad 1+D+D^4+D^6]$$

**a)** Show that the parity check polynomial matrix is

$$H(D) = [1+D+D^4+D^6 \quad 1]$$

**b)** For a transmitted code vector $c(D) = [c^{(1)}(D) \quad c^{(2)}(D)]$, the received vector can be represented as $r(D) = c(D) + e(D)$ where $e(D) = [e^{(1)}(D) \quad e^{(2)}(D)]$ is the error vector. Show that the syndrome vector $s(D) = r(D)H^T(D)$ satisfies the following relation

$$s(D) = e(D)H^T(D)$$

and that its coefficients (called syndrome bits) form the following set of *check-sums*

$$\begin{aligned}
s_0 &= e_0^{(1)} & & & & & & & +e_0^{(2)} \\
s_1 &= e_0^{(1)} & +e_1^{(1)} & & & & & & +e_1^{(2)} \\
s_4 &= e_0^{(1)} & & & +e_3^{(1)} & +e_4^{(1)} & & & +e_4^{(2)} \\
s_6 &= e_0^{(1)} & & +e_2^{(1)} & & & +e_5^{(1)} & +e_6^{(1)} & +e_6^{(2)}
\end{aligned}$$

**c)** The above set of check-sums involves 11 error bits: $e_0^{(1)}, e_1^{(1)}, \cdots, e_5^{(1)}, e_6^{(1)}$, $e_0^{(2)}, e_1^{(2)}, e_4^{(2)}, e_6^{(2)}$. The error bit $e_0^{(1)}$ appears in each check-sum while no other error bit appears more than once. In this case, these check-sums are said to be *orthogonal* on $e_0^{(1)}$. Show that the following *majority* logic decoding rule can be used to correctly decode the value $e_0^{(1)}$ if no more than 2 of these 11 error bits are 1's.

*Majority-logic decoding rule: If more than 2 of 4 check-sums are 1's then decode $e_0^{(1)} = 1$; otherwise decode $e_0^{(1)} = 0$.*

In general, if we can formulate a set of $C$ check-sums orthogonal on an error bit $\epsilon$ and involving $l_\epsilon$ different error bits then $\epsilon$ is decoded as 1 if more than $\lfloor C/2 \rfloor$ check-sums are 1's. This majority-logic decoding is successful if no more than $\lfloor C/2 \rfloor$ of $l_\epsilon$ error bits are 1's. $l_\epsilon$ is called the *effective constraint length* of the code. (Note: $\lfloor C/2 \rfloor$ is the largest integer less than or equal to $C/2$). The estimated value of $\epsilon$ must be fed back to modify the future syndrome calculations in a majority-logic feedback decoder (see [1, Chapter 13]).

**3.12** Consider the rate 1/2 convolutional code defined by the following generator polynomial matrix:

$$G(D) = [1 \quad 1 + D^2 + D^5 + D^6].$$

**a)** Verify that the code is systematic. Find the parity-check polynomial matrix.

**b)** Following the steps given in Problem 3.11, construct the set of check-sum equations orthogonal on $e_0^{(1)}$. Establish the majority-logic decoding rule.

**c)** Discuss the decoding capability and find the effective constraint length of the code.

**3.13** Repeat Problem 3.12 for the rate 1/2 convolutional code defined by

$$G(D) = [1 \quad 1 + D^2 + D^7 + D^{13} + D^{16} + D^{17}].$$

**3.14** Repeat Problem 3.12 for the rate 2/3 convolutional code defined by

$$G(D) = \begin{bmatrix} 1 & 0 & 1+D \\ 0 & 1 & 1+D^2 \end{bmatrix}.$$

**3.15** Repeat Problem 3.12 for the rate 2/3 convolutional code defined by

$$G(D) = \begin{bmatrix} 1 & 0 & 1+D^8+D^9+D^{12} \\ 0 & 1 & 1+D^6+D^{11}+D^{13} \end{bmatrix}.$$

**3.16** Consider the rate 1/2 convolutional code of Example 3.1. Using the Viterbi algorithm, decode the following received sequence:

110101101011.

**3.17** Repeat Problem 3.16 for the following received sequence:

111001101011.

**3.18** Consider the maximum likelihood decoder in a binary symmetric channel with the transition probability $p$. The decoder chooses the path $\mathbf{c}'$ in favor of the correct path $\mathbf{c}$ where the Hamming distance between $\mathbf{c}$ and $\mathbf{c}'$ is $d$.

**a)** Show that the pairwise error probability is

$$P_d \equiv P(\mathbf{c}, \mathbf{c}') = \sum_{i=\frac{d+1}{2}}^{d} \binom{d}{i} p^i (1-p)^{d-i}$$

for $d$ odd.

**b)** Find $P_d$ for $d$ even.

**c)** Show that

i)

$$P_d < p^{d/2}(1-p)^{d/2} \sum_{i=\frac{d+1}{2}}^{d} \binom{d}{i}$$

ii)

$$\sum_{i=\frac{d+1}{2}}^{d} \binom{d}{i} < \sum_{i=0}^{d} \binom{d}{i} = 2^d$$

iii)

$$P_d < \left[2\sqrt{p(1-p)}\right]^d$$

**3.19** Consider the performance of convolutional codes with a soft-decision ML decoder using a binary antipodal transmission scheme in an AWGN channel. The pairwise error probability, $P_d$, denotes the probability that the decoder chooses the path $\mathbf{c}'$ instead of the correct path $\mathbf{c}$ where the Hamming distance between $\mathbf{c}$ and $\mathbf{c}'$ is $d$.

**a)** Show that the Euclidean distance between the paths $\mathbf{c}$ and $\mathbf{c}'$ is

$$d_E(\mathbf{c}, \mathbf{c}') = \sqrt{4dE_s},$$

where $E_s$ is the average energy of the *transmitted* bit.

**b)** Show that the pairwise error probability is

$$P_d = \frac{1}{2}\mathrm{erfc}\sqrt{\frac{\mathrm{d}_{\mathrm{E}}^2(\mathbf{c}, \mathbf{c}')}{4\mathrm{N}_0}} = \frac{1}{2}\mathrm{erfc}\sqrt{\frac{\mathrm{dE_s}}{\mathrm{N}_0}}.$$

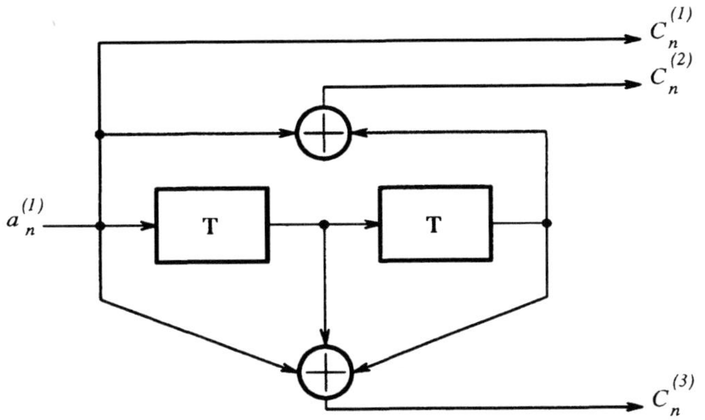


Figure P3.23-1: Block diagram of a convolutional encoder

**3.20** Consider the performance of convolutional codes with a soft-decision ML decoder using a coherent M-ary transmission scheme in an AWGN. Assume that for each branch the transmitter generates one signal vector which is an element of the signal set $\{\mathbf{s}_i, i = 0, 1, \cdots, (M-1)\}$. Show that

**a)** the squared Euclidean distance between two paths $\mathbf{c}$ and $\mathbf{c}'$ is

$$d_E^2(\mathbf{c}, \mathbf{c}') = \sum_{l=1}^{d} \left[ d_E^{(l)}(\mathbf{s}_i, \mathbf{s}_j) \right]^2 ,$$

where $d$ is the total number of different branches.

**b)** The pairwise error probability can be represented as

$$P_d = \frac{1}{2} \mathrm{erfc} \sqrt{\frac{\mathrm{d}_\mathrm{E}^2(\mathbf{c}, \mathbf{c}')}{4\mathrm{N}_0}}.$$

**3.21** Derive and plot the upper bounds on the performance of the rate 2/3 convolutional code of Figure 3.2 for hard-decision and soft-decision ML decoders for a coherent BPSK modulation in an AWGN channel. Compare and discuss the results.

**3.22** Repeat Problem 3.21 with the rate 1/2 convolutional code defined by

$$G(D) = [1 + D + D^3 + D^5 \quad 1 + D^2 + D^3 + D^4 + D^5].$$

**3.23** The encoder of a rate 1/3 convolutional code with $\nu = 2$ is shown in Figure P3.23-1.

a) Determine the generator polynomial matrix $G(D)$ and the parity-check polynomial matrix $H(D)$ of this code.

b) Determine the state diagram and the trellis diagram of this code.

c) The initial state is $S_0 = (00)$ and the input sequence is 1011100. Find the corresponding output sequence.

d) Assume that the transmitted coded sequence is

$$111001100110101010011,$$

and the received sequence is

$$111011100110001010011.$$

Using the Viterbi algorithm find the estimated sequence and compare it with the transmitted sequence.

e) Assume that the coded sequence

$$111001100110101010011$$

is transmitted by using BPSK over an AWGN channel with two-sided noise spectral density of $N_0/2$. The unquantized received signal sequence at the matched filter output is

$$1.1, 0.9, 0.8, -0.7, -0.9, 1, -0.1, 0.2, -0.8, 1.1, 0.9, 0.2, 0.7, -0.9, -0.1, -0.8, 0.9, -0.9, -1, 1.1, 1.2.$$

i) Using the hard-decision Viterbi decoding find the estimated transmitted sequence.

ii) Using soft-decision Viterbi decoding find the estimated transmitted sequence.

iii) Compare the estimated sequences obtained in i) and ii) with the transmitted sequence. Discuss the results.

f) Assuming coherent BPSK demodulation, calculate the probability of bit error for hard-decision and soft-decision Viterbi decoding at $E_s/N_0 = 5$ dB.

**3.24** The encoder of a rate 1/2 convolutional code with $\nu = 3$ is shown in Figure P3.24-1.

a) Determine the generator polynomial matrix $G(D)$ of this code.

b) Find the generator polynomial matrix of the equivalent systematic encoder with feedback, $G^s(D)$.

c) Draw the block diagram of the equivalent systematic encoder.

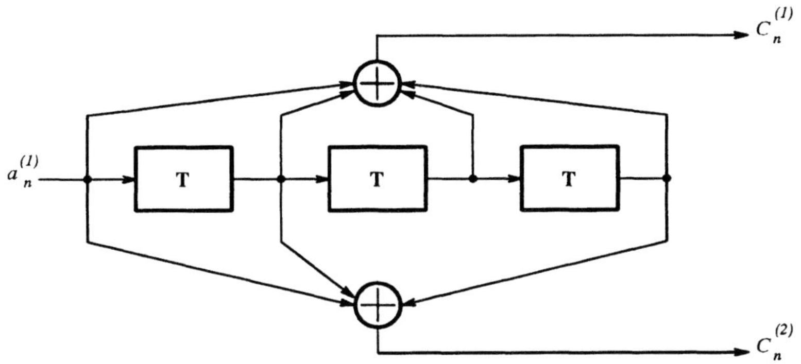


Figure P3.24-1: Block diagram of a convolutional encoder

**d)** Using the modified state diagram, find the minimum Hamming distance, $d_{free}$, of this code. Determine the numbers of paths with Hamming distances of $d_{free}+1$, $d_{free}+2$ and $d_{free}+3$ from the all-zero coded sequence.

**3.25** The state diagram of a rate 1/2 convolutional code with $\nu = 2$ is shown in Figure P3.25-1.

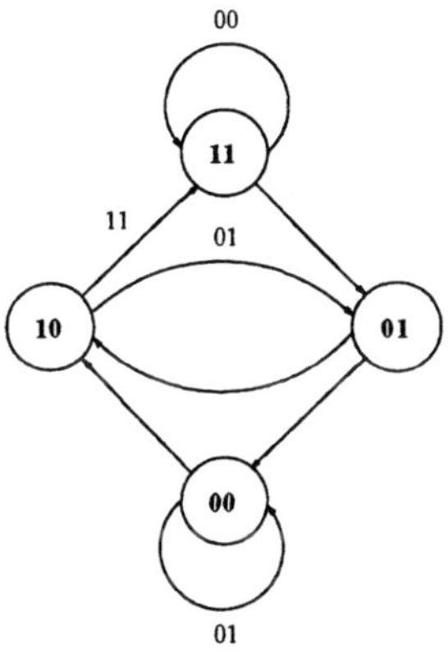


Figure P3.25-1: State diagram of a rate 1/2 convolutional code

**a)** Complete the branch labels of the state diagram.

**b)** Find the generator polynomial matrix $G(D)$ and draw the encoder structure.

**c)** Draw the trellis diagram.

**d)** Is this code catastrophic? Why?

# Chapter 4

# BANDWIDTH-EFFICIENT CODED-MODULATION SCHEMES

In the paper published in 1982 [18], Ungerboeck demonstrated that, by integration of a convolutional code with a bandwidth-efficient modulation scheme, significant coding gain can be achieved, compared to uncoded schemes, without sacrificing data rate or requiring more bandwidth. This technique is called *Trellis-Coded Modulation* (TCM). His idea attracted the interest of many researchers and led to considerable research in searching for new TCM schemes and formalizing the theory.

The bandwidth-efficient coded-modulation schemes are presented in this chapter in two parts. The first part is devoted to the concept and fundamentals of TCM scheme, where a *convolutional code* is combined with a bandwidth-efficient signaling technique. The set partitioning technique as a tool for designing TCM schemes is presented. We address the design of TCM schemes for the AWGN channel. Two representation methods, namely, the *Ungerboeck representation* [18] and the *analytical representation* [19], which are essential in the design and realization of these schemes, are considered. We proceed by evaluating the performance of the Ungerboeck codes on the AWGN channel. The second part of the chapter presents *Block-Coded Modulation* (BCM) techniques, where a *block code* is integrated with a bandwidth-efficient modulation scheme, as a counterpart to TCM schemes.

# 4.1 TRELLIS-CODED MODULATION SCHEMES

The use of error control coding improves the performance of a digital communication system at the expense of bandwidth expansion by an amount equal to the reciprocal of the code rate. For power-limited channels, like deep space communication channels, one may trade the bandwidth expansion for transmitted power in order to achieve a desired system performance. In the past, because of bandwidth expansion the coding has not been popular for the bandwidth-limited channels, such as telephone channels. In fact, in such channels a bandwidth-efficient signaling technique[1] such as Pulse Amplitude Modulation (PAM), Quadrature Amplitude Modulation (QAM), or Multi-Phase Shift Keying (MPSK) modulation is employed to support a high bandwidth efficiency (in bits/s/Hz). In the following we consider a coding technique for bandwidth-limited channels which achieves coding gain without bandwidth expansion.

Traditionally, coding and modulation have been considered as two separate parts of a digital communication system. The modulator and demodulator are usually devised to convert an analog waveform channel to a discrete channel and the encoder and decoder are designed to correct the errors which occurred in the discrete channel. Higher improvement in performance is achieved by lowering the code rate and increasing the number of redundancy in the code, of course, at a cost of decoding complexity and bandwidth expansion.

To avoid bandwidth expansion, one may integrate a code, say a convolutional code, with an expanded bandwidth-efficient signal set and utilize the redundancy resulting from such expansion. To be specific, consider a situation where it is desired to transmit a stream of data through a channel with throughput of 2 bits/s/Hz. One possible solution is to use an uncoded QPSK with throughput of 2 bits/s/Hz. As another solution we may consider a rate 2/3 convolutional code combined with an 8PSK signal set with throughput of 3 bits/s/Hz. This coded 8PSK scheme yields the same throughput as uncoded QPSK, that is, 2 bits/s/Hz. Note that both QPSK and 8PSK schemes require the same bandwidth. If the coded scheme performs superior to the uncoded one at the same SNR we can claim that such an improvement is achieved sacrificing neither data rate nor bandwidth.

Although the expansion of a signal set provides the redundancy required for coding, it shrinks the distance between the signal points if the average signal energy is kept constant (see Figure 4.1). The reduction in the distance between signal points increases the error rate which should be compensated with coding, if the coded scheme is to provide a benefit.

The use of hard-decision demodulation prior to the decoding in a coded scheme causes an irreversible loss of information which translates into a loss of

[1] For these signaling techniques the throughput, defined as the information rate per transmission bandwidth, is more than 1 bit/s/Hz.

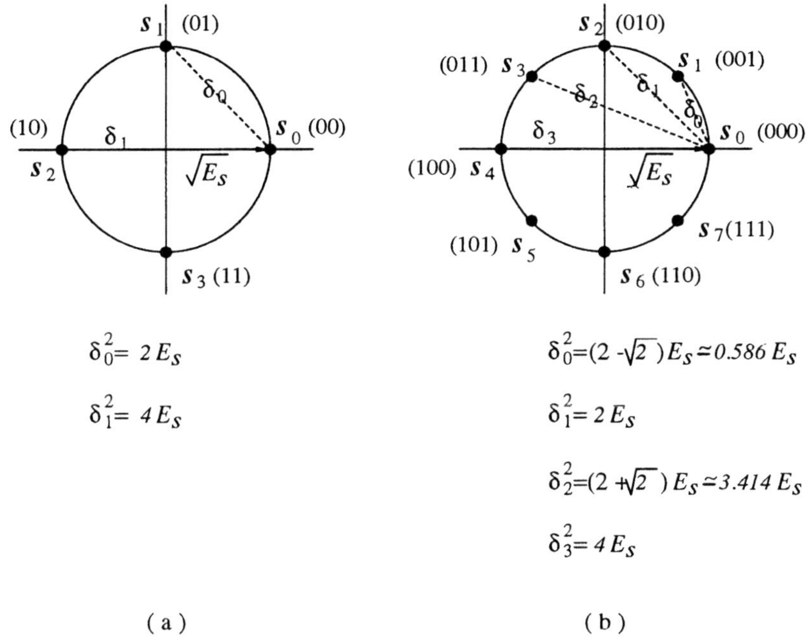


Figure 4.1: Two examples of MPSK signal set: a) QPSK, b) 8PSK.

SNR. Recall from Chapter 3 that for a convolutional code with binary transmission this loss is about 2-3 dB. In a bandwidth-efficient coded scheme where, due to the expansion of the signal set, a compensation of power penalty is required, it is necessary to avoid such a hard-decision loss by employing *soft-decision* instead. As a consequence, demodulation and decoding should be combined in a single step and the decoder should operate directly on the "soft" output samples of the channel.

If the maximum likelihood criterion is applied in soft-decision decoding on the AWGN channel, the decision rule of the optimum sequence decoder will depend on the *Euclidean distance* (see Section 3.3.1). In other words, the optimum decoder chooses the code sequence which is nearest to the received sequence in terms of Euclidean distance. In this case, the performance of the coded scheme is determined mainly by the minimum Euclidean distance between coded sequences. Consequently, an efficient coding scheme *should be designed based on maximizing the minimum Euclidean distance between coded sequences rather than Hamming distance.* This is clarified by considering the two following examples.

**Example 4.1** Let us consider the situation where a stream of data is to be transmitted with 2 bits/s/Hz over an AWGN channel. As a coded solution, we employ a rate 2/3 convolutional code with an 8PSK signal set. For this purpose we

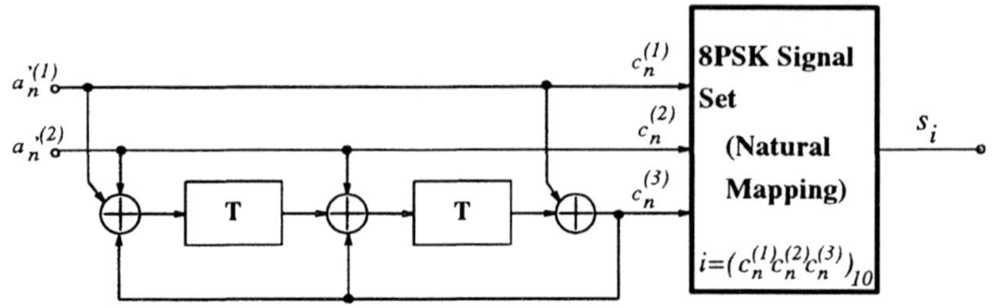


Figure 4.2: Encoder structure of a rate 2/3 convolutional coded 8PSK scheme.

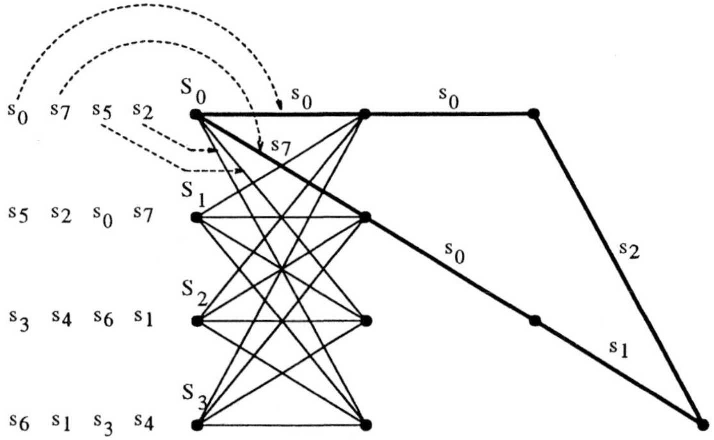

Figure 4.3: Trellis diagram of the coded scheme shown in Figure 4.2.

choose the rate 2/3 convolutional code of Example 3.1 which is designed based on maximizing *free Hamming distances.* The mapping of 3 output bits of the convolutional encoder to the 8PSK signal points is done using a natural mapping, in which each 3 bits are assigned to a signal point of Figure 4.1(b) with an equivalent decimal subscript. The encoder of this coded scheme is illustrated in Figure 4.2 using a systematic feedback realization. The trellis diagram of this coded scheme is shown in Figure 4.3 with branches labeled with associated signal points.

Soft-decision decoding of this coded scheme is performed using the Viterbi algorithm. The Viterbi algorithm searches the most likely path through the trellis diagram and chooses the path closest to the received sequence in Euclidean distance sense. Later we will show that the performance of this decoding algorithm depends on the minimum Euclidean distance between a pair of paths forming an error event. This minimum Euclidean distance is sometimes called *free Euclidean distance* and is denoted by $d_{free}$. We can find the squared free

Euclidean distance of this coded scheme by inspecting its trellis diagram

$$\begin{aligned} d^2_{free} &= d^2_E(s_0,\ s_7) + d^2_E(s_0,\ s_0) + d^2_E(s_2,\ s_1) \\ &= \delta_0^2 + 0 + \delta_0^2 = 2\delta_0^2 \\ &= 1.172E_s. \end{aligned}$$

The error event corresponding to $d_{free}$ is shown in Figure 4.3 by bold lines. In general, coded schemes resulting from integration of a convolutional code with a nonbinary modulation scheme are *nonlinear* and the transmitted all-zero sequence cannot be utilized in the evaluation of the distance structure or error bounds. Instead, every possible codeword should be considered as a transmitted one.

To compare the coded and uncoded schemes it is common to use the *coding gain* parameter, defined as the difference between the values of coded and uncoded signal-to-noise ratios (in dB) required to achieve the same error event probability:

$$g \equiv \mathrm{SNR}\ |_{\mathrm{uncoded}} - \mathrm{SNR}\ |_{\mathrm{coded}}\ .$$

At high signal-to-noise ratios this gain is expressed as

$$g_\infty \equiv g|_{\mathrm{SNR}\to\infty} = 10\log\frac{(d^2_{free}/E_s)_{\mathrm{coded}}}{(d^2_{free}/E_s)_{\mathrm{uncoded}}},$$

where $g_\infty$ represents the *asymptotic coding gain* and $E_s$ is the average signal energy. Note that for uncoded schemes $d_{free}$ is simply the minimum Euclidean distance between signal points.

For a bandwidth-efficient coded scheme the asymptotic coding gain is evaluated in comparison to the uncoded scheme with the same throughput. For the coded scheme, presented in Example 4.1, the asymptotic coding gain is computed as

$$g_\infty = 10\log\frac{1.172}{2} = -2.3\ \mathrm{dB},$$

where 2 is the normalized minimum Euclidean distance of the uncoded QPSK (see Figure 4.1(a)). This result which shows the performance degradation of the coded scheme compared to the uncoded one is quite disappointing. In the following we consider another example which shows the importance of designing the coded schemes based on the free Euclidean distance rather than on the free Hamming distance.

**Example 4.2** As an alternative coded 8PSK scheme, we may use the TCM scheme of Figure 4.4 introduced by Ungerboeck [18]. This scheme, consisting of a rate 2/3 convolutional code and an 8PSK signal set mapping, is devised based on maximizing the free Euclidean distance between coded sequences. The trellis diagram of this TCM scheme is shown in Figure 4.5. This four-state trellis diagram contains parallel paths between its states.

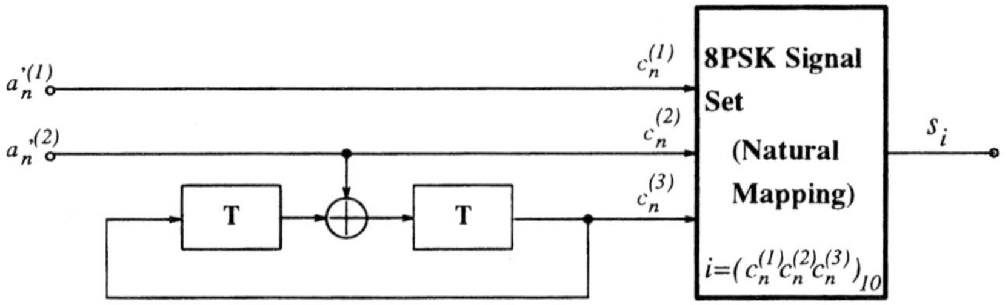


Figure 4.4: Encoder structure of the rate 2/3 8PSK TCM scheme [18].

The free Euclidean distance of this scheme can be computed by inspecting its trellis diagram. We observe that the branches of this trellis are labeled in such a way that the squared Euclidean distance between signal points associated with those branches diverging from or remerging in one state is at least $\delta_1^2$. Furthermore, the parallel paths are associated with signal points with squared Euclidean distance $\delta_3^2$.

Now, consider the error events associated with parallel paths. The squared Euclidean distance between two signal points associated with each of these error events is $\delta_3^2 = 4E_s$. Next, consider the error events associated with paths that diverge in one state and remerge at the same or another state after more than one transition. The minimum Euclidean distance associated with such error events can be expressed as

$$d_E^2(\text{diverging pair of paths}) + \cdots + d_E^2(\text{remerging pair of paths}) = \delta_1^2 + \cdots + \delta_1^2,$$

which is larger than the minimum squared Euclidean distance between parallel paths. One such error event is shown by the dashed line in Figure 4.5. The squared Euclidean distance between two paths of this error event is

$$\begin{aligned} d_E^2(s_0,\ s_2) + d_E^2(s_0,\ s_1) + d_E^2(s_0,\ s_2) &= \delta_1^2 + \delta_0^2 + \delta_1^2 = \delta_3^2 + \delta_0^2 \\ &= 4.586E_s. \end{aligned}$$

Thus, the error events associated with parallel paths have the minimum squared Euclidean distance among the entire error events, and hence, determine the free Euclidean distance of this coded scheme.

The asymptotic coding gain for this TCM scheme can be computed as

$$g_\infty = 10 \log \frac{4}{2} = 3\text{dB},$$

which shows a 3 dB asymptotic improvement compared to the uncoded QPSK.

The preceding example shows that if the integrated coding and modulation scheme is designed based on increasing the minimum Euclidean distance between coded sequences, the loss from the expansion of the signal set can be overcome and a significant coding gain can be achieved. In the following we consider a technique which is essential in the design of the trellis-coded schemes.

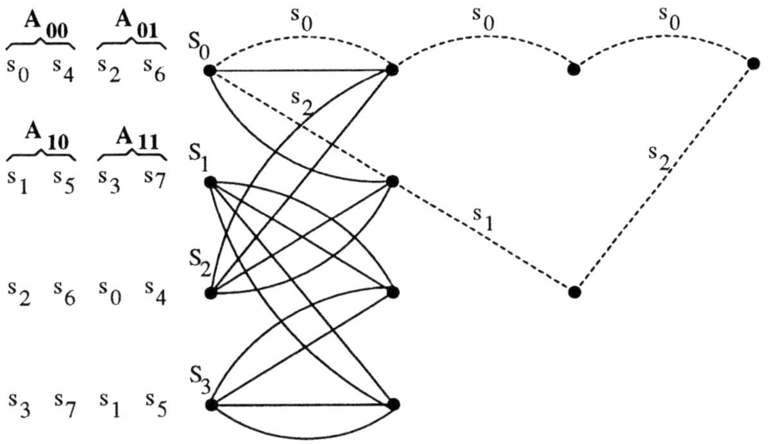


Figure 4.5: Trellis diagram of the coded scheme shown in Figure 4.4.

# 4.2 DESIGN OF TCM SCHEMES

In TCM schemes, a redundant $2^{m+1}$-ary signal set is used in order to transmit $m$ bits in each signaling interval. The $m$ input bits are encoded by a rate $m/(m+1)$ trellis encoder and $m+1$ encoded bits are mapped to the signal points of the $2^{m+1}$-ary signal set in such a way that the minimum Euclidean distance between channel-signal sequences is maximized. This is done by following a rule called *mapping by set partitioning*.

## 4.2.1 Mapping by Set Partitioning

The mapping by set partitioning rule is based on successive partitioning of the expanded $2^{m+1}$-ary signal set into *subsets* with increasing minimum Euclidean distances, $\Delta_0 < \Delta_1 < \Delta_2 < \cdots$, between the signal points of these subsets. As an illustrative example, consider the set partitioning of an 8PSK signal set, shown in Figure 4.6. In the first step of this procedure the 8PSK signal set is partitioned into two subsets $\mathbf{A}_0$ and $\mathbf{A}_1$ of four points each. The minimum Euclidean distance between signals of these subsets is $\Delta_1 = \delta_1$ which is larger than the minimum Euclidean distance between signals of the original 8PSK set, i.e., $\Delta_0 = \delta_0$. The second step of partitioning divides each of the subsets $\mathbf{A}_0$ and $\mathbf{A}_1$ into two subsets with two signal points. The minimum Euclidean distance between signal points of the subsets in this level of partitioning is $\Delta_2 = \delta_3$. Further partitioning of these subsets leads to eight subsets with a single signal point in each.

In general, it is not necessary to continue partitioning to the level that each subset contains only a single signal point. In fact, the partitioning is repeated $\tilde{m}+1$ times ($\tilde{m} \leq m$) to the degree that the minimum Euclidean distance

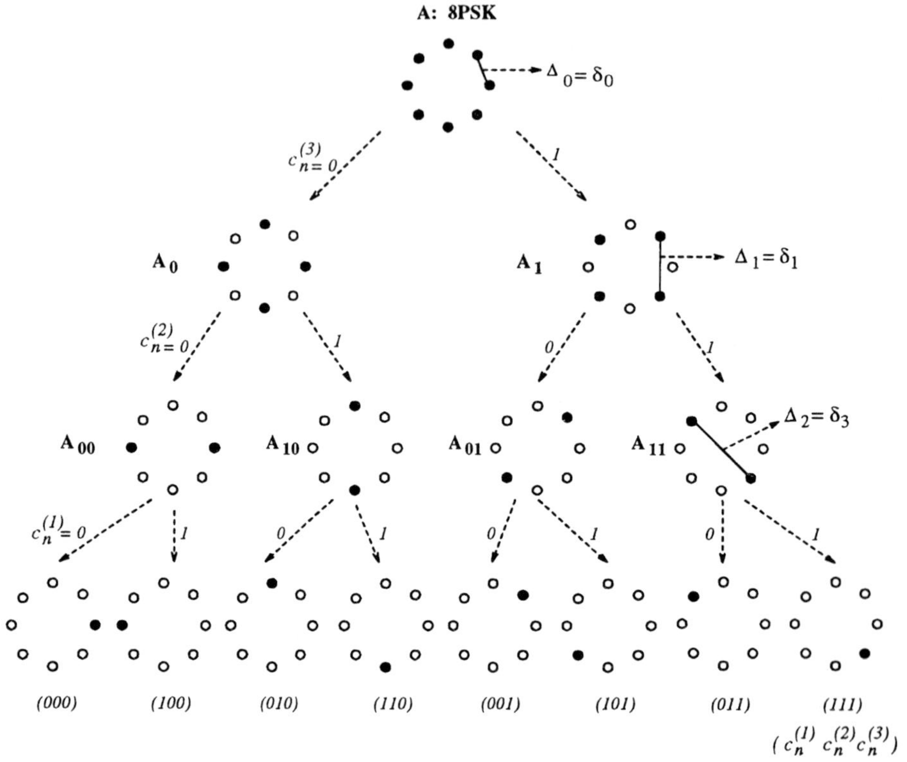


Figure 4.6: Set partitioning of an 8PSK signal set.

between the signal points at the final level of partitioning, $\Delta_{\tilde{m}+1}$, is equal or larger than the desired minimum Euclidean distance of the TCM scheme to be designed.

A general structure of a TCM encoder is shown in Figure 4.7. In this figure, at each time instant $n$, a block of $m$ information bits $(a_n^{(1)}, a_n^{(2)}, \cdots, a_n^{(m)})$ enters the TCM encoder. From these $m$ information bits, $\tilde{m} \leq m$ bits are encoded by a rate $\tilde{m}/(\tilde{m}+1)$ binary convolutional encoder into $\tilde{m}+1$ coded bits while the remaining $m - \tilde{m}$ bits are left uncoded. The $(\tilde{m}+1)$ output bits of the convolutional encoder are used to select one of the $2^{\tilde{m}+1}$ possible subsets of the expanded signal set while the remaining $m-\tilde{m}$ uncoded bits are used to select one of the $2^{m-\tilde{m}}$ signals in this subset.

The number of states in the trellis diagram of a TCM scheme depends on the number of memories in the encoder. For an encoder with $\nu$ memories the total number of states is $2^{\nu}$. The number of paths (transitions) between two consecutive states in the trellis diagram depends upon the number of uncoded bits, i.e., $m - \tilde{m}$. For a TCM encoder with $m = \tilde{m}$ the states are joined by

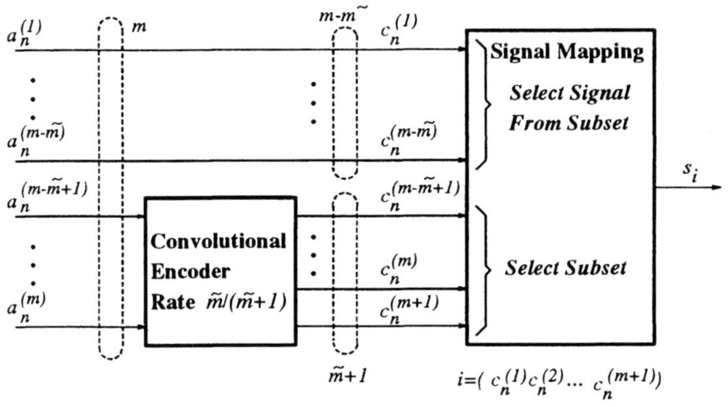


Figure 4.7: General structure of a TCM encoder.

*single transitions* while for an encoder with $m > \tilde{m}$, $2^{m-\tilde{m}}$ *parallel transitions* exist between states. These parallel transitions are associated with the $2^{m-\tilde{m}}$ signals of the subsets in the lowest layer of the set partitioning tree.

Denoting the minimum Euclidean distance between parallel transitions by $\Delta_{\tilde{m}+1}$ and the minimum Euclidean distance between non-parallel paths in the TCM trellis diagram by $d_{free}(\tilde{m})$, the free Euclidean distance of a TCM code can be expressed as

$$d_{free} = \min[\Delta_{\tilde{m}+1}, d_{free}(\tilde{m})], \tag{4.1}$$

which was previously used for the specific TCM scheme of Example 4.2. In the special case of $m = \tilde{m}$, the subsets in the lowest layer of the set partitioning contain only one signal and the free Euclidean distance of the TCM code is determined solely by the code trellis diagram.

### 4.2.2 Ungerboeck TCM Design Rules

To construct TCM schemes based on maximizing the free Euclidean distance Ungerboeck [18] introduced the following heuristic design rules.

- **R1** Parallel transitions, if present, are associated with signals of the subsets in the lowest layer of set partitioning tree. These signals have minimum Euclidean distance $\Delta_{\tilde{m}+1}$.

- **R2** The transitions originating from or merging into one state are associated with signals of the first step of set partitioning. The Euclidean distance between these signals is at least $\Delta_1$.

- **R3** All signals are used with equal frequency in the trellis diagram.

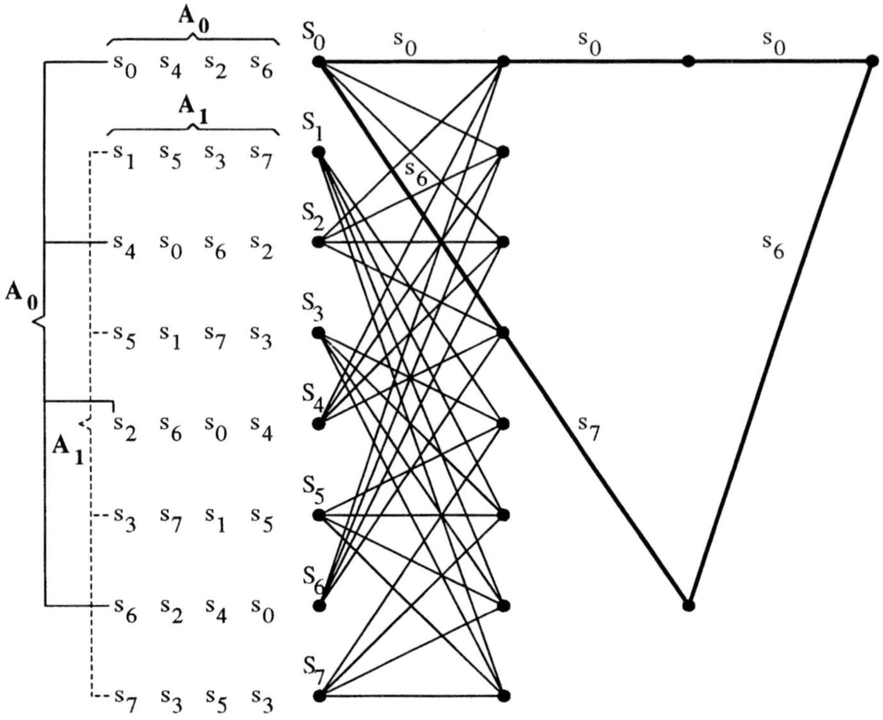


Figure 4.8: Trellis diagram of the 8-state 8PSK TCM scheme [18].

As an example consider the 4-state 8PSK TCM of Figure 4.5. The parallel paths are associated with signals from one of the four subsets, $\mathbf{A}_{00}$, $\mathbf{A}_{01}$, $\mathbf{A}_{10}$ and $\mathbf{A}_{11}$, with minimum Euclidean distance $\Delta_2$. The signals associated with transitions diverging from or arriving at one state are from subsets $\mathbf{A}_0$ and $\mathbf{A}_1$, with minimum Euclidean distance $\Delta_1$.

**Example 4.3** As an illustrative example of heuristic TCM code design, consider the design of an 8-state PSK TCM scheme [18]. This scheme with throughput of 2 bits/s/Hz is comparable to uncoded QPSK. Design procedure consists of selecting a suitable trellis diagram for this scheme and assigning the signals from a partitioned 8PSK signal set to the transitions of this diagram. For obvious reasons the parallel transitions should be avoided in the trellis diagram of this code, since they limit the free Euclidean distance to $\Delta_2$ which is already achieved by a 4-state 8PSK TCM. A suitable trellis diagram is shown in Figure 4.8. The transitions of the trellis diagram are associated with a partitioned 8PSK signal set (see Figure 4.6) according to the rules R1, R2, and R3. Observe that the signals from subsets $\mathbf{A}_0$ and $\mathbf{A}_1$, at the first level of set partitioning, are assigned to the transitions originating from the states with even and odd subscripts respectively in Figure 4.8. This alternative assignment guarantees

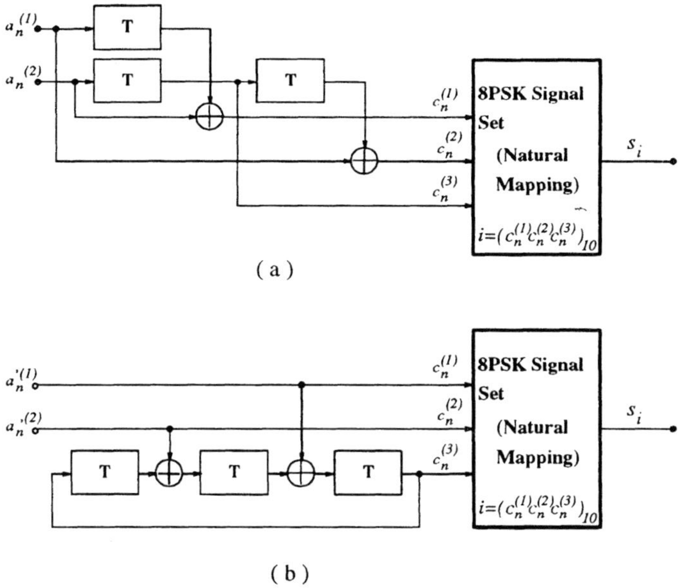


Figure 4.9: Two encoder realizations of the 8-state 8PSK TCM scheme [18]: a) equivalent feedforward encoder, b) equivalent systematic encoder with feedback.

that the signals merging into one state are selected from either $\mathbf{A}_0$ or $\mathbf{A}_1$. This is shown in Figure 4.8.

The squared free Euclidean distance of this scheme is obtained as

$$\begin{aligned} d_{free}^2 &= \Delta_1^2 + \Delta_0^2 + \Delta_1^2 \\ &= 4.586 E_s, \end{aligned}$$

which, when compared with an uncoded QPSK with $\delta_0^2 = 2E_s$, provides

$$g_\infty = 10 \log \frac{4.586}{2} = 3.6 \text{ dB}$$

asymptotic coding gain. An error event corresponding to the squared free Euclidean distance is denoted in Figure 4.8 by bold lines. Note that the 8-state 8PSK scheme provides 0.6 dB more asymptotic coding gain than 4-state 8PSK TCM at the cost of more decoding complexity.

We may design other 8-state 8PSK schemes with the same free Euclidean distance using the same design rules. The 8-state TCM scheme of this example is selected among equivalent codes because of its encoder structure. With some effort we can show that the encoder of this scheme coincides with the general structure of Figure 4.7, i.e., a binary convolutional code followed by a mapping. A realization of the encoder of this TCM scheme using feedforward

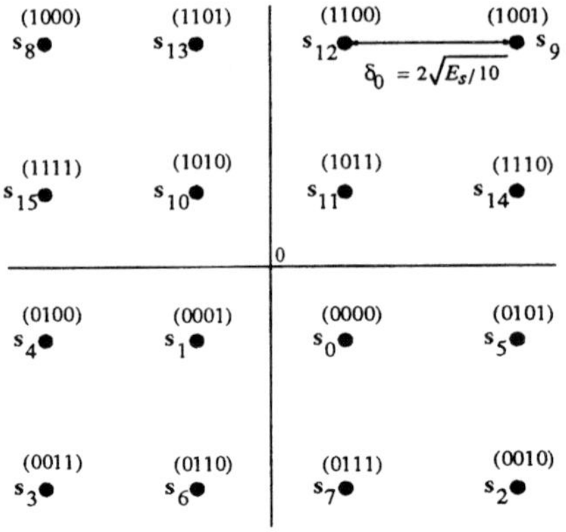


Figure 4.10: 16QAM signal constellation.

convolutional encoder is shown in Figure 4.9(a). An equivalent systematic convolutional encoder with feedback (see Section 3.1.4) can be obtained for this TCM scheme as shown in Figure 4.9(b).

So far we have described the concept and design of TCM schemes based on examples of MPSK signal constellations. Other trellis-coded schemes are developed for one- or two-dimensional modulation schemes such as ASK and QAM schemes. As an illustrative example we consider another 8-state trellis-coded scheme owing to Ungerboeck [18] which uses a 16QAM expanded signal set.

**Example 4.4** A 16QAM signal constellation is shown in Figure 4.10, with the Euclidean distance $\delta_0$ between adjacent signal points. This constellation is of lattice type and is called a *lattice of type* $\mathcal{Z}_2$. The set partitioning of the lattice structure of type $\mathcal{Z}_2$ results in subsets with doubling minimum squared Euclidean distance in each step. The average signal energy of a 16QAM signal set is simply obtained as[2]

$$\begin{aligned} E_s &= \frac{1}{16}(2\delta_0^2 + 10\delta_0^2 + 10\delta_0^2 + 18\delta_0^2) \\ &= \frac{10}{4}\delta_0^2. \end{aligned}$$

Thus,

$$\delta_0 = 2\sqrt{\frac{E_s}{10}}.$$

The throughput of 16QAM is 4 bits/s/Hz. For the 8-state 16QAM trellis code we have $m = 3$. This trellis-coded scheme has throughput of 3 bits/s/Hz and can be compared with an uncoded 8PSK modulation. We choose $\tilde{m} = 2$ and

[2]It is assumed that the signals are equiprobable.

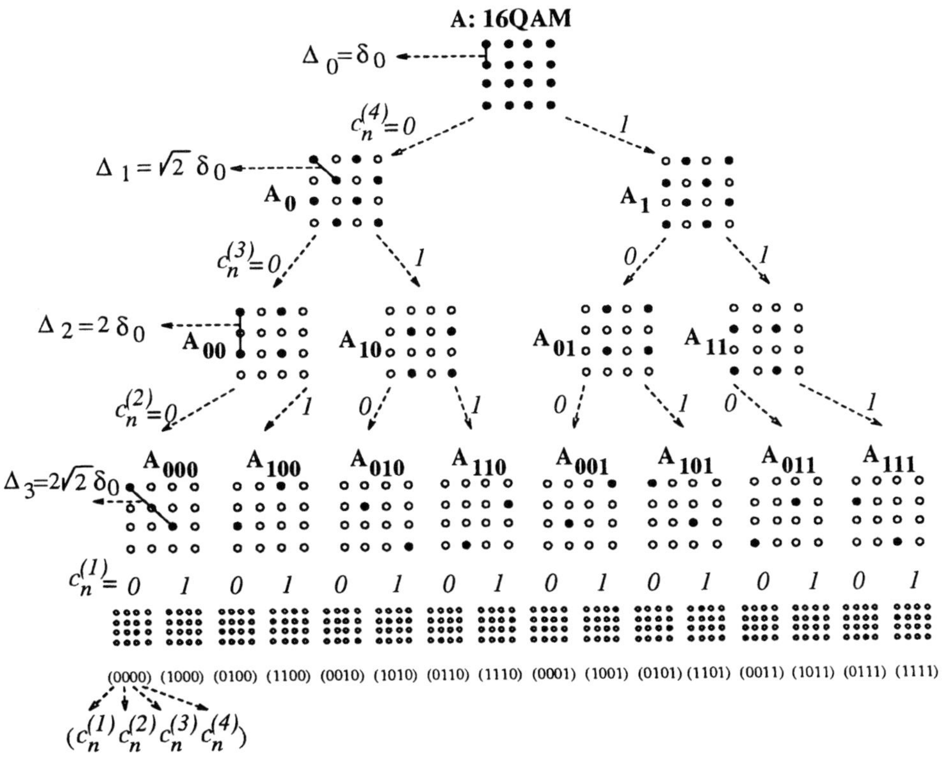


Figure 4.11: Set partitioning of a 16QAM signal set.

carry out set partitioning $\tilde{m} + 1 = 3$ times. This set partitioning is illustrated in Figure 4.11.

The trellis diagram of 8-state 16QAM TCM is shown in Figure 4.12 with associated signals according to the heuristic design rules. The paths originating from or remerging at the same state are associated with signals of subsets $\mathbf{A}_0$ or $\mathbf{A}_1$. The parallel paths are associated with the two signals within each of the subsets at the last set partitioning step, i.e., $\mathbf{A}_{000}$, $\mathbf{A}_{001}$, $\cdots$ and $\mathbf{A}_{111}$.

The squared free Euclidean distance of this scheme is obtained by inspection and using (4.1) as

$$\begin{aligned} d_{free}^2 &= \min\,(\Delta_3^2,\ \Delta_1^2 + \Delta_0^2 + \Delta_1^2) \\ &= \min\,(8\Delta_0^2,\ 5\Delta_0^2) \\ &= 5\Delta_0^2 = 2E_s, \end{aligned}$$

which translates to an asymptotic coding gain

$$g_\infty = 10 \log \frac{2}{2-\sqrt{2}} = 5.3 \text{ dB}$$

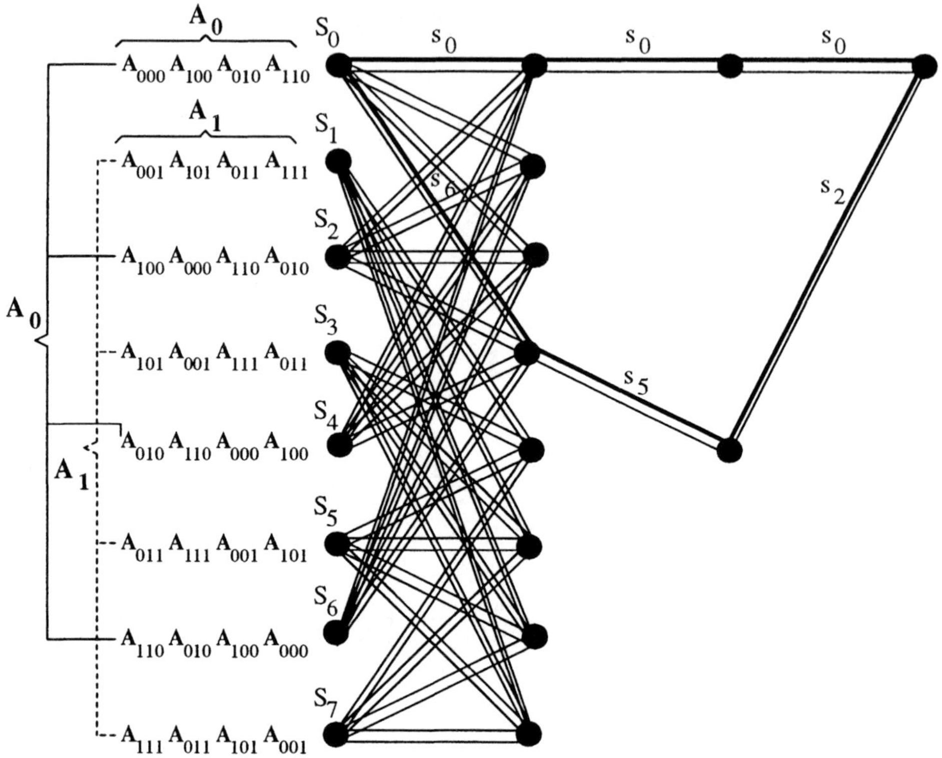


Figure 4.12: Trellis diagram of the 8-state 16QAM TCM scheme [18].

compared to uncoded 8PSK[3]. Note that for this TCM scheme the free Euclidean distance is limited by the code trellis diagram rather than by the parallel paths. Two encoder realizations of the 8-state 16QAM TCM scheme using feedforward convolutional encoder and equivalent systematic encoder with feedback are shown in Figure 4.13.

TCM schemes with a small number of states can be constructed using the heuristic design rules. To construct a larger TCM scheme, however, a computer code-search program [18] is necessary. In the following the results of such a code-search program for some trellis-coded schemes are provided. These results are reported in [20].

Table 4.1 shows the asymptotic coding gains of good TCM codes constructed for the 8PSK signal constellation. The coding gain is given compared to uncoded QPSK. The TCM schemes in this table are denoted by the coefficients of the parity-check polynomials in *octal form*. For example for the

[3]The coding gain of this TCM scheme compared to another uncoded two-dimensional signaling with 3 bits/s/Hz throughput, namely, 8AMPM is only 4 dB (see Problem 4.5).

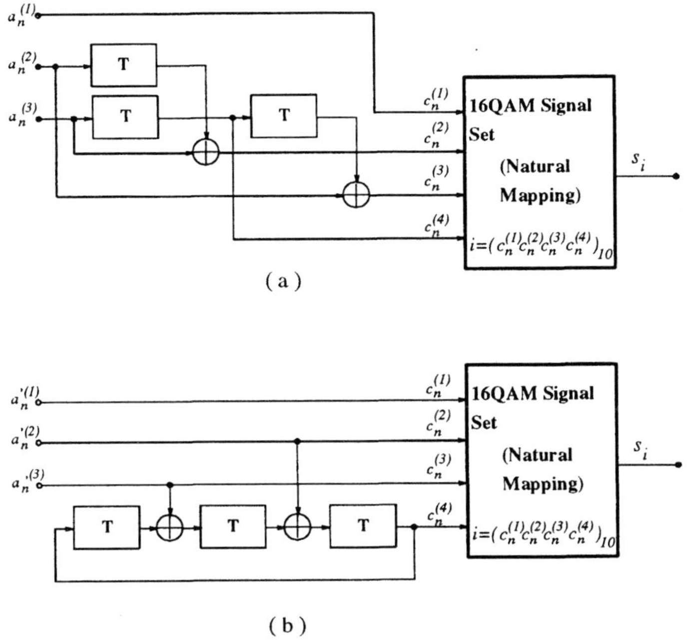


Figure 4.13: Two encoder realizations of the 8-state 16QAM TCM scheme [18]: a) feedforward encoder, b) equivalent systematic encoder with feedback.

8-state 8PSK TCM scheme ($\nu = 3$), $\mathbf{H}(D) = [H_1(D)\ H_2(D)\ H_3(D)]$ and

$$\begin{array}{rclclcl} H_1(D) & = & D^2 & = & (000\,100)_2 & = & (04)_8, \\ H_2(D) & = & D & = & (000\,010)_2 & = & (02)_8, \\ H_3(D) & = & D^3+1 & = & (001\,001)_2 & = & (11)_8. \end{array} \tag{4.2}$$

The TCM schemes constructed for the 16PSK signal constellation are listed in Table 4.2. Throughput of these schemes is 3 bits/s/Hz, and hence, they are compared with uncoded 8PSK.

The results of these tables show that about 3-4 dB coding gains can be obtained by using 4- and 8-state TCM codes. More improvement in the coding gain can be achieved by increasing the number of encoder states at the expense of decoding complexity.

### 4.2.3 Two TCM Encoder Realizations

The encoder of Ungerboeck codes may be realized either with a feedforward convolutional encoder or an equivalent systematic encoder with feedback (see Section 3.1.4). Such encoder realizations were previously presented as examples of TCM codes. In the following we show how to obtain a feedforward

Table 4.1: Trellis-Coded 8PSK Schemes [20]

| Number of States $2^\nu$ | Number of Coded Bits $\tilde{m}$ | $H_1$ | $H_2$ | $H_3$ | Normalized Free Euclidean Distance $d^2_{free}/E_s$ | Asymptotic Coding Gain $g_\infty$ (dB) |
|---|---|---|---|---|---|---|
| 4 | 1 | - | 2 | 5 | 4.000* | 3.01 |
| 8 | 2 | 04 | 02 | 11 | 4.586 | 3.60 |
| 16 | 2 | 16 | 04 | 23 | 5.172 | 4.13 |
| 32 | 2 | 34 | 16 | 45 | 5.758 | 4.59 |
| 64 | 2 | 066 | 030 | 103 | 6.343 | 5.01 |
| 128 | 2 | 122 | 054 | 277 | 6.586 | 5.17 |
| 256 | 2 | 130 | 072 | 435 | 7.515 | 5.75 |

* The asterisk indicates that free Euclidean distance is determined by parallel paths.

or feedback realization if the coefficients of the parity-check polynomials are given.

Consider a rate $m/(m+1)$ TCM code. An equivalent systematic realization of the encoder of this code contains the exact replicas of $m$ input bits at the output and an extra output bit is obtained by a parity-check equation imposed on the current and the previous input bits. This bit can be found recalling (3.5)

$$\mathbf{c}(D) \cdot \mathbf{H}^T(D) = \mathbf{0},$$

where

$$\mathbf{c}(D) = [c^{(1)}(D)\ c^{(2)}(D)\ \cdots\ c^{(m+1)}(D)]$$

and

$$\mathbf{H}(D) = [H_1(D)\ H_2(D)\ \cdots\ H_{m+1}(D)]$$

are the output codeword and the parity-check polynomial matrices, respectively.

**Example 4.5** Consider the 8-state 8PSK TCM code with parity-check polynomials given by (4.2).[4] The parity-check equation for this scheme can be obtained as

$$\begin{aligned}\mathbf{c}(D) \cdot \mathbf{H}^T(D) &= [c^{(1)}(D)\ c^{(2)}(D)\ c^{(3)}(D)] \cdot \begin{bmatrix} D^2 \\ D \\ D^3+1 \end{bmatrix} \\ &= D^2 c^{(1)}(D) + D c^{(2)}(D) + D^3 c^{(3)}(D) + c^{(3)}(D) = 0,\end{aligned}$$

[4] This TCM scheme is equivalent to the 8-state TCM scheme of Example 4.3, except that the latter uses a different set of parity-check polynomials.

Table 4.2: Trellis-Coded 16PSK Schemes [20]

| Number of States $2^\nu$ | Number of Coded Bits $\tilde{m}$ | $H_1$ | $H_2$ | $H_3$ | Normalized Free Euclidean Distance $d^2_{free}/E_s$ | Asymptotic Coding Gain $g_\infty$ (dB) |
|---|---|---|---|---|---|---|
| 4 | 1 | - | 2 | 5 | 1.324 | 3.54 |
| 8 | 1 | - | 04 | 13 | 1.476 | 4.01 |
| 16 | 1 | - | 04 | 23 | 1.628 | 4.44 |
| 32 | 1 | - | 10 | 45 | 1.910 | 5.13 |
| 64 | 1 | - | 024 | 103 | 2.000* | 5.33 |
| 128 | 1 | - | 024 | 277 | 2.000* | 5.33 |
| 256 | 2 | 374 | 176 | 435 | 2.085 | 5.51 |

* The asterisk indicates that free Euclidean distance is determined by parallel paths.

which yields the encoder input/output relations

$$\begin{aligned} c^{(1)}(D) &= a'^{(1)}(D), \\ c^{(2)}(D) &= a'^{(2)}(D), \\ c^{(3)}(D) &= \frac{D^2 a'^{(1)}(D) + D a'^{(2)}(D)}{1 + D^3}, \end{aligned}$$

where $a'^{(1)}(D)$ and $a'^{(2)}(D)$ are the input bits. The realization of the above equation as a systematic encoder with feedback is shown in Figure 4.14(a).

It is also possible to realize a feedforward encoder for the Ungerboeck codes. Consider again a rate $m/(m+1)$ TCM code with the total number of encoder memory elements, $\nu$. To find a feedforward encoder structure for such a TCM code two steps are performed. First, we specify how many of the memory elements are associated with each of the input bits. Next, we determine a way to connect each output modulo-2 adder to the memory cells. These two steps should be carried out in such a way that the feedforward encoder produces the same set of codewords as the systematic encoder with feedback. In general, this can be done by an algorithm proposed for this purpose in [21]. In the following we consider a simple example.

**Example 4.6** Consider again the 8-state 8PSK TCM of Example 4.5. The generator polynomial matrix of this code is a $2 \times 3$ matrix

$$\begin{bmatrix} G_{11}(D) & G_{12}(D) & G_{13}(D) \\ G_{21}(D) & G_{22}(D) & G_{23}(D) \end{bmatrix},$$

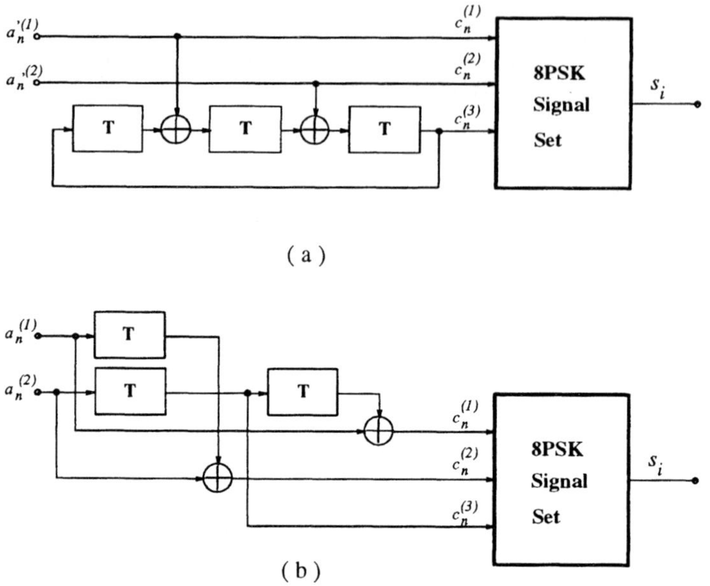


Figure 4.14: Two encoder realizations of the 8-state 8PSK TCM scheme with parity-check polynomials given in (4.2): a) systematic encoder with feedback, b) equivalent feedforward encoder.

where $G_{ij}(D)$ is a polynomial which determines the $i$th input/$j$th output relation (see Section 3.1.1).

The total encoder memory of this code is 3. Let us assign one memory to the first input bit, $a'^{(1)}(D)$, and two memories to the second input bit, $a'^{(2)}(D)$. It is obvious that for such a case the degree of polynomials $G_{1j}(D)$, $j = 1, 2, 3$, can be at most one while the degree of polynomials $G_{2j}(D)$, $j = 1, 2, 3$, can be at most two.

To determine $G_{ij}(D)$ we recall (3.4)

$$\begin{aligned} \mathbf{G}(D) \cdot \mathbf{H}^T(D) &= \begin{bmatrix} G_{11}(D) & G_{12}(D) & G_{13}(D) \\ G_{21}(D) & G_{22}(D) & G_{23}(D) \end{bmatrix} \cdot \begin{bmatrix} D^2 \\ D \\ D^3+1 \end{bmatrix} \\ &= \mathbf{0}, \end{aligned}$$

which yields

$$\begin{aligned} D^2 G_{11}(D) + D G_{12}(D) + (D^3+1) G_{13}(D) &= 0, \\ D^2 G_{21}(D) + D G_{22}(D) + (D^3+1) G_{23}(D) &= 0. \end{aligned}$$

Imposing the constraint on the degree of $G_{ij}(D)$ the desired solution is obtained as

$$\begin{array}{lll} G_{11}(D) = 1 & G_{12}(D) = D & G_{13}(D) = 0 \\ G_{21}(D) = D^2 & G_{22}(D) = 1 & G_{23}(D) = D, \end{array}$$

or

$$\mathbf{G}(D) = \begin{bmatrix} 1 & D & 0 \\ D^2 & 1 & D \end{bmatrix}.$$

Recall that (see Section 3.1.1) the generator polynomial $G_{ij}(D)$ determines the connections of the $j$th output with the sequence of the shift register stages associated with the $i$th input. Thus, using the above generator polynomial matrix the feedforward encoder of the 8-state 8PSK TCM scheme is realized as Figure 4.14(b).

# 4.3 REPRESENTATION OF TCM SCHEMES

In the following we consider two representation approaches for trellis-coded modulation schemes. The first approach, introduced by Ungerboeck [20], specifies these schemes as a linear convolutional code and a mapping rule. The second approach, introduced by Calderbank and Mazo [19], combines the two steps of the Ungerboeck representation into one and describes TCM schemes with an analytical relation between TCM signals and input bits.

## 4.3.1 Ungerboeck Representation

We have used this representation explicitly in describing the concept of TCM earlier. Here we consider this representation in more detail. Consider a rate $m/(m+1)$ TCM scheme designed based on Ungerboeck design rules. This scheme can be represented by a rate $m/(m+1)$ linear convolutional code followed by a memoryless mapper as in Figure 4.15. The TCM output signal at time $n$, say $s_n$, is related to the output of the convolutional encoder $\mathbf{c}_n = (c_n^{(1)}, c_n^{(2)}, \cdots, c_n^{(m+1)})$ by the mapping rule

$$s_n = \mathcal{M}(\mathbf{c}_n), \tag{4.3}$$

where $\mathcal{M}(.)$ is a one-to-one mapping. The binary $(m+1)$-tuple $\mathbf{c}_n$ is called the label of signal $\mathbf{s}_n$ [22].

### Symmetrical Structure of Ungerboeck Codes

In the following we consider a lemma due to Zehavi and Wolf [23] which results from the linearity of convolutional codes.

**Lemma 4.1** *For a rate $m/(m+1)$ convolutional code the set of $2^m$ vectors $\mathbf{c}$ from the all-zero state $S_0$ forms a commutative group $\mathcal{G}_0$. Furthermore, the set of $2^m$ vectors from any other state is either $\mathcal{G}_0$ or its coset $\mathcal{G}_0 + \tilde{\mathbf{c}}$ (only one coset exists).*

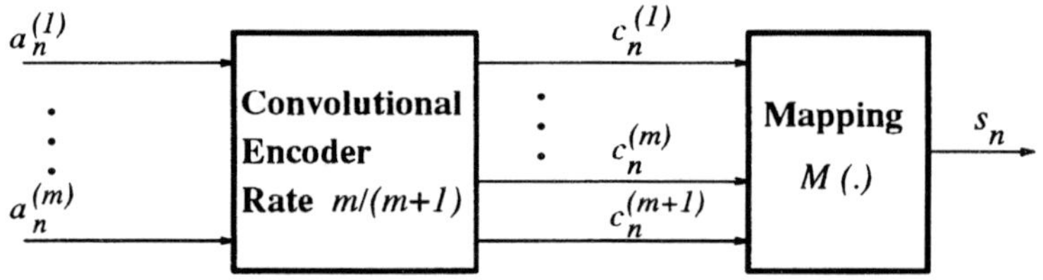


Figure 4.15: Ungerboeck representation of TCM codes.

**Proof** This lemma can be shown by using the linear property of convolutional codes. Suppose the encoder is in the all-zero state. In this case each of the $(m+1)$ outputs is the modulo-2 sum of some subset of the $m$ inputs. Because of the linearity of convolutional codes the sum of $m$-tuple inputs generates the corresponding sum of the $(m+1)$-tuple outputs, so we have closure. Since each $(m+1)$-tuple is modulo-2 inverse of its own, the $(m+1)$-tuples $\mathbf{c}$ form a subgroup $\mathcal{G}_0$. Next assume that the encoder is in a state other than the all-zero state. The $2^m$ outputs in this state can be found by adding a fixed $(m+1)$-tuple, say $\tilde{\mathbf{c}}$, to each of $(m+1)$-tuples in $\mathcal{G}_0$. The resulting set is either $\mathcal{G}_0$ or a coset of $\mathcal{G}_0$. Only one such coset exists, since the intersection of $\mathcal{G}_0$ and $\mathcal{G}_0 + \tilde{\mathbf{c}}$ is an empty set and their union has cardinality $2^{m+1}$.

□

Using the foregoing lemma we can specify two types of states. The first type produces output $\mathbf{c}$ from the set $\mathcal{G}_0$, while the second type generates outputs from $\mathcal{G}_0 + \tilde{\mathbf{c}}$. Since the outputs of the convolutional encoder are mapped to the signal points by a one-to-one mapping rule, the encoder partitions the signal set into two subsets. These are the subsets resulting from the first level of the set partitioning procedure, considered earlier.

**Example 4.7** Consider the 4-state 8PSK TCM scheme of Example 4.2. The outputs of the convolutional code, $\mathbf{c}$, corresponding to different states are

$$\begin{array}{lrcllcl} S_0: & \mathcal{G}_0 & = & \{000,\ 100,\ 010,\ 110\} & \rightarrow & \mathbf{A}_0 \\ S_1: & \mathcal{G}_0 + (001) & = & \{001,\ 101,\ 011,\ 111\} & \rightarrow & \mathbf{A}_1 \\ S_2: & \mathcal{G}_0 & = & \{010,\ 110,\ 000,\ 100\} & \rightarrow & \mathbf{A}_0 \\ S_3: & \mathcal{G}_0 + (001) & = & \{011,\ 111,\ 001,\ 101\} & \rightarrow & \mathbf{A}_1, \end{array}$$

which belong to the set $\mathcal{G}_0 = \{000,\ 010,\ 100,\ 110\}$ or its only coset $\mathcal{G}_0 + (001) = \{001,\ 011,\ 101,\ 111\}$. These outputs partition the signal set $\mathbf{A}$ into two subsets $\mathbf{A}_0 = \{s_0,\ s_2,\ s_4,\ s_6\}$ and $\mathbf{A}_1 = \{s_1,\ s_3,\ s_5,\ s_7\}$.

The above example shows that the first level of partitioning corresponds to the $(m+1)$th component of the label $\mathbf{c}$ and the value of this component is chosen from one of the two subsets. To be specific, consider the 8PSK signal

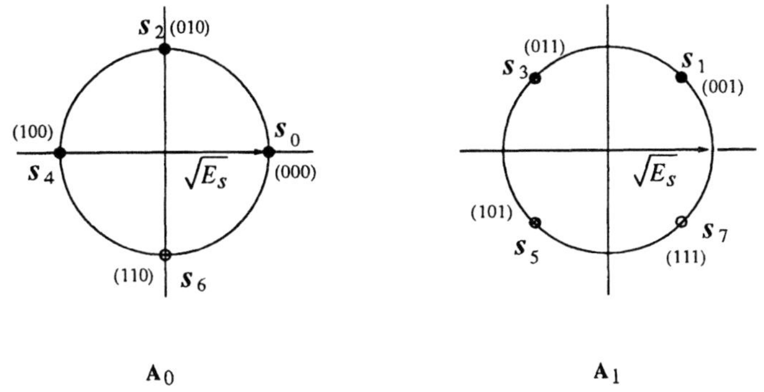


Figure 4.16: The isometry relation between the two subsets of the 8PSK signal set at the first level of set partitioning.

set in the foregoing example which is partitioned into two subsets $\mathbf{A}_0$ and $\mathbf{A}_1$ based on the value $c_n^{(3)}$. For labels with $c_n^{(3)} = 0$ the signal points are selected from $\mathbf{A}_0$ while the labels with $c_n^{(3)} = 1$ choose the signals from $\mathbf{A}_1$. The subsets $\mathbf{A}_0$ and $\mathbf{A}_1$ are related by an *isometry*, that is, a one-to-one mapping that preserves the distances. This can be seen from Figure 4.16 where the subset $\mathbf{A}_1$ can be obtained from $\mathbf{A}_0$ by a $\pi/4$ rotation.

Next we generalize the foregoing discussion by considering a signal set which is partitioned at the first level of the partitioning procedure by the value of $\kappa$th component, $c_n^{(\kappa)}$, into two subsets $\mathbf{A}_0^\kappa$ and $\mathbf{A}_1^\kappa$. Now we introduce the $\kappa$-*isometric* TCM scheme by the following definition.

**Definition 4.1** [22] *A TCM scheme is called $\kappa$-isometric if*

*1) the subsets $\mathbf{A}_0^\kappa$ and $\mathbf{A}_1^\kappa$ are related by an isometry, and*

*2) the labels of two corresponding signals $\mathcal{M}(\mathbf{c})$ and $\mathcal{M}(\hat{\mathbf{c}})$ in the isometry, i.e., $\mathbf{c}$ and $\hat{\mathbf{c}}$, differ only in their $\kappa$th component.*

**Example 4.8** Consider the subsets $\mathbf{A}_0$ and $\mathbf{A}_1$ of the 8PSK signal set with *natural labeling*shown in Figure 4.16. These two subsets are related by an isometry ($\pi/4$ rotation). The corresponding signals in the isometry, that is, (**000**, **001**),(**010**, **011**), (**100**, **101**) and (**110**, **111**) have labels that differ in the 3rd component.

**Definition 4.2** [22] *The TCM scheme represented as Figure 4.15 is called an* Ungerboeck code *if it is $\kappa$-isometric for some $\kappa$.*

Note that it is customary to choose $\kappa = m + 1$ in actual designs. This can be observed from the examples of TCM schemes considered earlier.

Now we state a theorem, due to Biglieri and McLane [22], which will be used in the performance evaluation of Ungerboeck codes later on.

**Theorem 4.1** *Ungerboeck codes (a $\kappa$-isometric code) have the following properties.*

*1) The encoded m-tuples formed from the labels $\mathbf{c}_n$ by deleting the $\kappa$th component, that is*

$$\mathbf{c}_n^\star = (c_n^{(1)}, \cdots, c_n^{(\kappa-1)}, c_n^{(\kappa+1)}, \cdots, c_n^{(m+1)}),$$

*are independent sequences.*

*2) The sequence of $(m+1)$-tuples $\mathbf{c}_n$ are uniquely determined from the sequence of m-tuples $\mathbf{c}_n^\star$.*

A $\kappa$-isometric TCM code having the above properties is called *uniform* [22].

**Proof** To prove these properties we observe that in a rate $m/(m+1)$ convolutional code there exists a dependency between encoded vectors $\mathbf{c}_n$ which involves only one of their components. Specifically, any encoded sequence $(c_n^{(1)}, \cdots, c_n^{(\kappa-1)}, c_n^{(\kappa+1)}, \cdots, c_n^{(m+1)})$ can occur at time $n$ and the sequence $c_n^\kappa$ is determined by the parity-check equations imposed by the convolutional code.

□

**Example 4.9** Consider the 8-state 8PSK TCM scheme of Figure 4.14. Looking at the equivalent systematic encoder with feedback of this TCM scheme (Figure 4.14(a)) we observe that (see Example 4.5) the component $c_n^{(3)}$ is determined by the parity-check equation

$$c^{(3)}(D) = \frac{D^2 c^{(1)}(D) + Dc^{(2)}(D)}{1 + D^3}$$

from the components $c_n^{(1)}$ and $c_n^{(2)}$. The components $c_n^{(1)}$ and $c_n^{(2)}$ are determined by input bits and form an independent sequence $\mathbf{c}_n^\star = (c_n^{(1)}, c_n^{(2)})$ which determines uniquely the sequence $\mathbf{c}_n = (c_n^{(1)}, c_n^{(2)}, c_n^{(3)})$. Recalling the 3-isometric property of this TCM scheme, we see that the 8-state 8PSK TCM scheme is uniform.

### 4.3.2 Analytical Representation

In the previous section TCM schemes were introduced as a convolutional code and a rule (mapping by set partitioning) that maps the output of this code onto a fixed signal constellation. Calderbank and Mazo in [19] combine these two steps into one and introduce a new description of these schemes, analytical rather than graphical, which expresses the TCM output as a series of expansion of products (of all orders) of the encoder bits. In the following their approach is briefly described.

Assume a rate $m/(m+1)$ TCM scheme with $\nu$ memories in its encoder. The output of this scheme (each channel symbol) depends not only on the most recent $m$ bits, but also on the $\nu$ previous bits, i.e.,

$$s_n = f(a_n^{(1)},\ a_n^{(2)},\ \cdots,\ a_n^{(m)},\ a_n^{(m+1)},\ \cdots,\ a_n^{(m+\nu)}), \tag{4.4}$$

where $s_n$ is the transmitted channel symbol at time $n$, and $a_n^{(i)}$, $i = 1, 2, \cdots, m$ are the most recent $m$ bits and $a_n^{(i)}$, $i = m+1, m+2, \cdots, m+\nu$ are the contents of the encoder memories ($\nu$ previous bits). In analytical representation it is customary to assume that the input bits take the bipolar values $\pm 1$ [19]. Now we use the linear transformation

$$a = \frac{1-b}{2} \tag{4.5}$$

to express $s_n$ in terms of bipolar variables $b_i$. Dropping the time index $n$ for convenience, (4.4) can be written as a sum of products of $b_i$'s [19]

$$\begin{aligned} s &= f(b_1,\ \cdots,\ b_m,\ b_{m+1},\ \cdots, b_{m+\nu}) \\ &= d_0 + \sum_{i=1}^{m+\nu} b_i d_i + \sum_{\substack{i,j=1 \\ j>i}}^{m+\nu} b_i b_j d_{ij} + \cdots + b_1 \cdots b_{m+\nu} d_{1\ldots(m+\nu)}, \end{aligned} \tag{4.6}$$

where $d_i$'s are constants that can be determined based on $2^{m+\nu}$ values of function $f$.

The $d_i$'s can be determined as follows. A simple argument shows that the number of $d_i$'s is $2^{m+\nu}$, and hence, they can be shown as a column vector $\mathbf{d}$ of dimension $2^{m+\nu}$. Similarly, the values taken by $f(b_1,\ \cdots,\ b_m,\ b_{m+1},\ \cdots, b_{m+\nu})$ can be regarded as a column vector $\mathbf{f}$ of dimension $2^{m+\nu}$. Now (4.6) can be written as

$$\mathbf{f} = \mathbf{B}_h \cdot \mathbf{d},$$

where

$$\mathbf{B}_h = [1\ b_1\ b_2\ \cdots\ b_{m+\nu}\ b_1 b_2\ b_1 b_3\ \cdots\ b_1 b_2 \cdots b_{m+\nu}]$$

is a $2^{m+\nu} \times 2^{m+\nu}$ matrix. Note that the vectors corresponding to rows (columns) of $\mathbf{B}_h$ are orthogonal and have norm-squares equal to $2^{m+\nu}$. In fact, $\mathbf{B}_h$ is a Hadamard matrix and its inverse can be obtained as

$$\mathbf{B}_h^{-1} = \frac{1}{2^{m+\nu}} \mathbf{B}_h^T,$$

where the superscript $T$ represents a matrix transpose. Thus, the coefficients of products in (4.6) can be obtained as

$$\mathbf{d} = \frac{1}{2^{m+\nu}} \mathbf{B}_h^T \cdot \mathbf{f}. \tag{4.7}$$

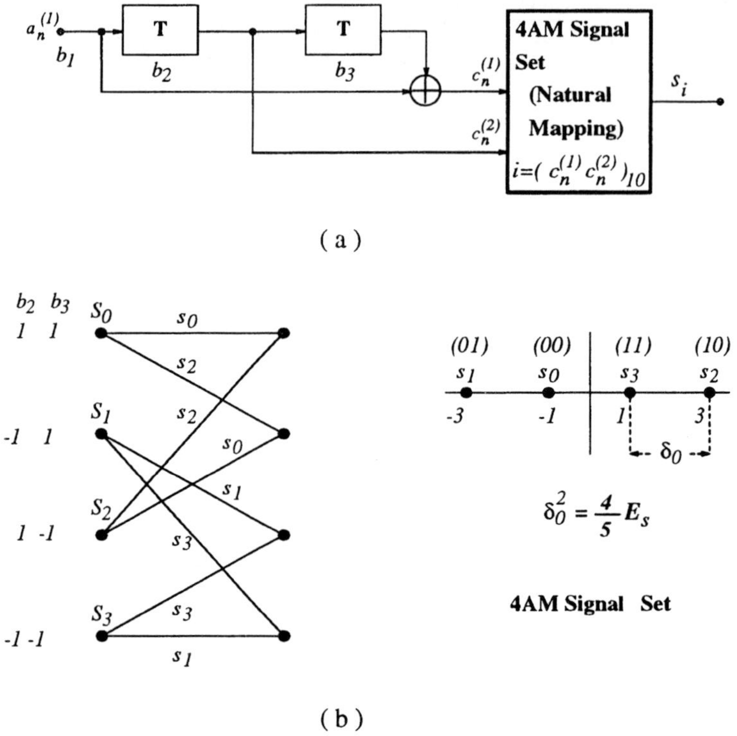


Figure 4.17: Four-state 4AM TCM scheme of Example 4.10: a) encoder (Ungerboeck representation), b) trellis diagram.

**Example 4.10** As an illustrative example we consider the analytical description of a one-dimensional TCM code. Consider a rate 1/2 ($m = 1$) 4-state ($\nu = 2$) 4AM TCM scheme [18]. The encoder and the trellis diagram of this coded scheme are shown in Figure 4.17(a) and Figure 4.17(b), respectively. This scheme provides 2.5 dB asymptotic coding gain with respect to uncoded 2AM.

The analytical representation of 4-state 4AM TCM can be expressed as

$$\begin{aligned} f(b_1\ b_2\ b_3) &= d_0 + d_1 b_1 + d_2 b_2 + d_3 b_3 \\ &+ d_{12} b_1 b_2 + d_{13} b_1 b_3 + d_{23} b_2 b_3 \\ &+ d_{123} b_1 b_2 b_3, \end{aligned}$$

where $b_1$ corresponds to the only input bit and $b_2$ and $b_3$ correspond to the contents of the two memories. Using the linear transformation of (4.5) the following relations are obtained from the code trellis of Figure 4.17(b)

$$\begin{aligned} f(\ 1\ \ \ 1\ \ \ 1) &= s_0 = -1 & \qquad f(\ 1\ \ \ 1\ -1) &= s_2 = \ \ 3 \\ f(-1\ \ \ 1\ \ \ 1) &= s_2 = \ \ 3 & \qquad f(-1\ \ \ 1\ -1) &= s_0 = -1 \\ f(\ 1\ -1\ \ \ 1) &= s_1 = -3 & \qquad f(\ 1\ -1\ -1) &= s_3 = \ \ 1 \\ f(-1\ -1\ \ \ 1) &= s_3 = \ \ 1 & \qquad f(-1\ -1\ -1) &= s_1 = -3 \end{aligned}$$

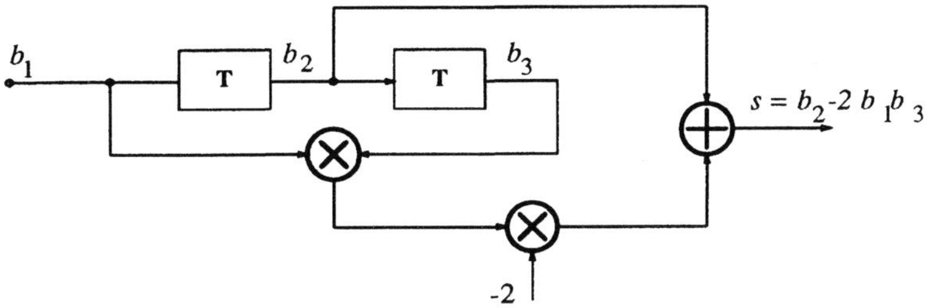


Figure 4.18: The encoder of the rate 1/2 4AM TCM scheme based on analytical representation.

where, for simplicity, we assume that $\delta_0 = 2$, that is, $E_s = 5$. The column vectors **f** and **d** are

$$\mathbf{f} = \begin{bmatrix} -1 \\ 3 \\ -3 \\ 1 \\ 3 \\ -1 \\ 1 \\ -3 \end{bmatrix}, \qquad \mathbf{d} = \begin{bmatrix} d_0 \\ d_1 \\ d_2 \\ d_3 \\ d_{12} \\ d_{13} \\ d_{23} \\ d_{123} \end{bmatrix}.$$

The matrix $\mathbf{B}_h$ in this case can be written as

$$\mathbf{B}_h = \left[\begin{array}{cccccccc} 1 & b_1 & b_2 & b_3 & b_1 b_2 & b_1 b_3 & b_2 b_3 & b_1 b_2 b_3 \\ \hline 1 & 1 & 1 & 1 & 1 & 1 & 1 & 1 \\ 1 & -1 & 1 & 1 & -1 & -1 & 1 & -1 \\ 1 & 1 & -1 & 1 & -1 & 1 & -1 & -1 \\ 1 & -1 & -1 & 1 & 1 & -1 & -1 & 1 \\ 1 & 1 & 1 & -1 & 1 & -1 & -1 & -1 \\ 1 & -1 & 1 & -1 & -1 & 1 & -1 & 1 \\ 1 & 1 & -1 & -1 & -1 & -1 & 1 & 1 \\ 1 & -1 & -1 & -1 & 1 & 1 & 1 & -1 \end{array}\right].$$

Replacing **f**, **d** and $\mathbf{B}_h$ in (4.7) and solving for $d$'s yields

$$\begin{aligned} d_0 &= d_1 = d_3 = d_{12} = d_{23} = d_{123} = 0 \\ d_2 &= 1 \\ d_{13} &= -2. \end{aligned}$$

Thus, the analytical description of this TCM scheme is

$$s = f(b_1\ b_2\ b_3) = b_2 - 2b_1 b_3.$$

An alternative encoder structure for the 4-state 4AM TCM based on the foregoing expression is shown in Figure 4.18.

# 4.4 TCM DECODER

Like convolutional codes the TCM schemes are described by using a trellis diagram. Each TCM transmitted sequence is represented by a trellis path. The task of TCM decoder, based on maximum likelihood criterion, is to observe the received sequence and to choose the most likely transmitted sequence based on this observation. An efficient search method for finding the most likely path, corresponding to the transmitted sequence through the trellis diagram of the TCM scheme, is accomplished with a well-known algorithm, namely, the Viterbi algorithm (see Section 3.3.2).

In soft-decision decoding of TCM schemes using the Viterbi algorithm each trellis branch is labeled by a number based on the observed received sequence, called the branch metric. Using the maximum likelihood decoder on the AWGN channel results in a metric which is defined as the Euclidean distance between the coded sequence and the received sequence (see Section 3.3.1). The branch metric for a TCM scheme on the AWGN channel, hence, is defined as the Euclidean distance between the signal associated with the branch and the respective received signal. The Viterbi algorithm finds a path through the TCM trellis which is closest to the received sequence in the Euclidean distance sense.

Decoding of TCM schemes with parallel paths can be performed in two steps. In the first step a branch among the parallel paths, merging to a state which has the minimum distance from the received signal, is selected. This branch and its metric are stored as the representative of the parallel paths. In the second step the Viterbi algorithm searches the code trellis for the most likely transmitted path considering only the representative branches instead of parallel paths.

# 4.5 PERFORMANCE EVALUATION

The performance of a TCM scheme on the AWGN channel can be evaluated by considering different performance measures. Two of these performance measures, namely, the free Euclidean distance and the (asymptotic) coding gain have already been presented in previous discussions. In the following we consider some other parameters which are used to characterize a given TCM code.

**Average Number of Nearest Neighbors at Free Distance $N(d_{free})$**

This parameter gives the average number of signal paths with free Euclidean distance $d_{free}$ from a transmitted sequence. This number is used, together with $d_{free}$, in the asymptotic evaluation of the *error event* probability defined in the following section.

**Error Event Probability $P_e$**

Two paths of finite length through the trellis diagram of the code form an error event if they start from the same state, merge in the same state, and do not simultaneously occupy the same state in between. Mathematically, an error event of length $l$ is defined by two coded sequences $\mathbf{s}_n$ and $\hat{\mathbf{s}}_n$

$$\begin{aligned} \mathbf{s}_n &= (s_n,\ s_{n+1},\ \cdots,\ s_{n+1+l}) \\ \hat{\mathbf{s}}_n &= (\hat{s}_n,\ \hat{s}_{n+1},\ \cdots,\ \hat{s}_{n+1+l}), \end{aligned}$$

such that

$$\begin{aligned} S_n &= \hat{S}_n & \qquad S_{n+l} &= \hat{S}_{n+l} \\ S_i &\neq \hat{S}_i & i &= n+1,\ \cdots,\ n+l-1, \end{aligned}$$

where $S_i$ and $\hat{S}_i$ denote the state of the code sequences $\mathbf{s}_n$ and $\hat{\mathbf{s}}_n$ at time $i$, respectively. The probability of an error event starting at time $n$, given that the decoder has estimated the correct transmitter state at that time, is called error event probability or $P_e$.

**Bit Error Probability $P_b$**

This parameter is defined as the expected number of decoded information bit errors per information bit.

## 4.5.1 Performance Evaluation of TCM Codes on the AWGN Channel

Performance of the TCM schemes is evaluated by means of upper and lower bounds on the error event probability or the bit error probability. It is based on the *generating function approach* which has been extensively used in the performance evaluation of convolutional codes. This approach involves finding a generating function for the code which, when combined with a union bound, gives the upper bound on the probability of error attainable by the code.

The generating function is derived from the state diagram of the code and expresses the weight distribution of the code in closed form. This function enumerates the distance, length, and the number of errors on any incorrect path, compared to the correct one, at any step in the Viterbi decoding process. For linear codes, without loss of generality, it is assumed that the correct path is the all-zero codeword. In such a case, the generating function is obtained as a transfer function of a state diagram containing $2^{\nu}$ states, where $\nu$ is the total memory of the encoder.

In general, the TCM schemes are nonlinear. The Euclidean distance between any incorrect path and the correct one depends upon the correct path. In such a case the error probability bound can still be derived by the generating function approach. However, the generating function in this case is obtained as a transfer function of an expanded pair state diagram containing $2^{2\nu}$ states [3], [24], [25]. This approach is extremely general, but has a computational complexity which increases with the squared number of states in the trellis

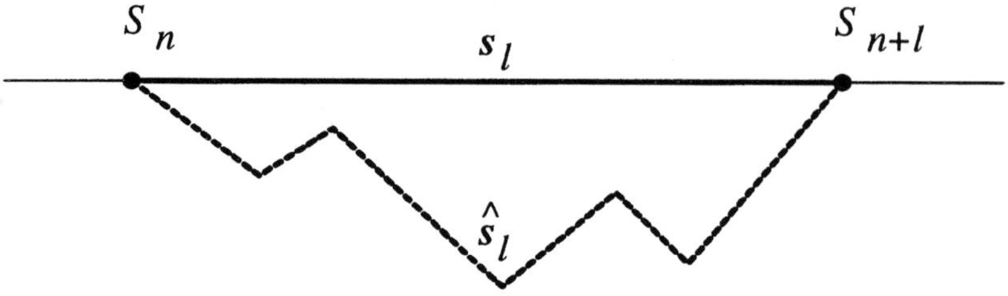


Figure 4.19: Example of an error event with length $l$.

diagram of the code. A rather low complex approach in performance analysis of TCM schemes uses a generating function of an error-state diagram with $2^\nu$ states whose branches are labeled by $2^\nu \times 2^\nu$ matrices [26].

Among the TCM schemes, some have the *Uniform Error Property* (UEP). According to [22] a TCM scheme is said to have UEP if it is possible to evaluate its error probability bound using the generating function of a state diagram containing $2^\nu$ states whose branch labels are scalars. In the preceding section we showed that Ungerboeck codes have uniform structure. In the following analysis we will show that UEP follows from such uniformity.

**Error Event Probability**

We start by deriving the upper bound on the error event probability. Consider two paths through the code trellis that originate from the same state $l$ times later without occupying the same state in between (see Figure 4.19). As stated earlier, we refer to these paths as an error event with length $l$. Let the coded sequences corresponding to these paths be $\mathbf{s}_l$ and $\hat{\mathbf{s}}_l$. Without loss of generality we take $\mathbf{s}_l$ to be the transmitted sequence and $\hat{\mathbf{s}}_l$ to be the chosen one by the maximum likelihood decoder. The probability of choosing $\hat{\mathbf{s}}_l$ instead of the transmitted sequence $\mathbf{s}_l$ is called the *pairwise error probability* and is denoted by $P_2(\mathbf{s}_l, \hat{\mathbf{s}}_l)$. Using the union bound an upper bound to error event probability is achieved by summing over $l$, $l = 1, \cdots, \infty$, and considering all possible transmitted sequences as

$$P_e \leq \sum_{l=1}^{\infty} \sum_{\mathbf{s}_l} \sum_{\hat{\mathbf{s}}_l \neq \mathbf{s}_l} P(\mathbf{s}_l) P_2(\mathbf{s}_l,\ \hat{\mathbf{s}}_l), \tag{4.8}$$

where $P(\mathbf{s}_l)$ is the *a priori* probability of transmitting $\mathbf{s}_l$.

Considering the one-to-one mapping between the TCM signals and their labels, the upper bound of (4.8) can be expressed in terms of labels rather than TCM signals. Let $\mathbf{C}_l = (\mathbf{c}_n,\ \mathbf{c}_{n+1},\ \cdots,\ \mathbf{c}_{n+l-1})$ and $\hat{\mathbf{C}}_l = (\hat{\mathbf{c}}_n,\ \hat{\mathbf{c}}_{n+1},\ \cdots,\ \hat{\mathbf{c}}_{n+l-1})$ be two label sequences corresponding to signal sequences $\mathbf{s}_l$ and $\hat{\mathbf{s}}_l$, respectively. The two label sequences $\mathbf{C}_l$ and $\hat{\mathbf{C}}_l$ can be related by introducing a sequence of binary error vectors $\mathbf{E}_l = (\mathbf{e}_n, \mathbf{e}_{n+1}, \cdots, \mathbf{e}_{n+l-1})$ as

$$\hat{\mathbf{c}}_i = \mathbf{c}_i \oplus \mathbf{e}_i, \qquad i = n, n+1, \cdots, n+l-1, \tag{4.9}$$

where $\oplus$ shows modulo-2 (binary) addition. Replacing $\mathbf{s}_l$ and $\hat{\mathbf{s}}_l$ by their labels in (4.8) and using (4.9) the upper bound of error event probability can be written as

$$\begin{aligned} P_e &\leq \sum_{l=1}^{\infty} \sum_{\mathbf{C}_l} \sum_{\hat{\mathbf{C}}_l \neq \mathbf{C}_l} P(\mathbf{C}_l) P_2(\mathbf{C}_l,\ \hat{\mathbf{C}}_l) \\ &= \sum_{l=1}^{\infty} \sum_{\mathbf{C}_l} \sum_{\mathbf{E}_l \neq \mathbf{0}} P(\mathbf{C}_l) P_2(\mathbf{C}_l,\ \mathbf{C}_l \oplus \mathbf{E}_l). \end{aligned} \tag{4.10}$$

The pairwise error probability $P_2(\mathbf{C}_l,\ \mathbf{C}_l \oplus \mathbf{E}_l)$ can be upper-bounded by using the Bhattacharyya bound (see Problem 4.13)

$$P_2(\mathbf{C}_l, \mathbf{C}_l \oplus \mathbf{E}_l) \leq D^{d_E^2[\mathcal{M}(\mathbf{C}_l), \mathcal{M}(\mathbf{C}_l \oplus \mathbf{E}_l)]}, \tag{4.11}$$

where $D = e^{-\frac{1}{4N_0}}$ for the additive white Gaussian noise channel with single-sided power spectral density $N_0$. Substituting (4.11) into (4.10) yields

$$P_e \leq \sum_{l=1}^{\infty} \sum_{\mathbf{E}_l \neq \mathbf{0}} \mathcal{F}(\mathbf{E}_l, D), \tag{4.12}$$

where

$$\mathcal{F}(\mathbf{E}_l,\ D) = \sum_{\mathbf{C}_l} P(\mathbf{C}_l) D^{d_E^2[\mathcal{M}(\mathbf{C}_l),\ \mathcal{M}(\mathbf{C}_l \oplus \mathbf{E}_l)]}. \tag{4.13}$$

The error probability bound (4.12) can be evaluated by using the generating function method if (4.13) can be expressed in a product form. In general, (4.13) can be written in a product form by introducing an error-state diagram whose branches have $2^\nu \times 2^\nu$ matrix labels [26]. We refer the reader to [26], [27, Chap. 4] for this general case. In the following we evaluate the error probability bound for uniform TCM codes [22].

To write (4.13) as a form of product we observe that

$$d_E^2[\mathcal{M}(\mathbf{C}_l),\ \mathcal{M}(\mathbf{C}_l \oplus \mathbf{E}_l)] = \sum_{i=1}^{l} d_E^2[\mathcal{M}(\mathbf{c}_i),\ \mathcal{M}(\mathbf{c}_i \oplus \mathbf{e}_i)].$$

Since the components of $\mathbf{C}_l$ are dependent, in general

$$\sum_{\mathbf{C}_l} P(\mathbf{C}_l) \neq \prod_{i=1}^{l} \sum_{\mathbf{c}_i} P(\mathbf{c}_i),$$

which hinders the writing of (4.13) in a product form. This hindrance can be overcome by using the properties of uniform TCM schemes stated in Theorem 4.1.

Now consider a $\kappa$-isometric TCM code (Ungerboeck code). Form a sequence $\mathbf{C}_l^{(\kappa)}$ from the sequence $\mathbf{C}_l$ with values of $\kappa$th components $c_i^{(\kappa)}$, $i = 1, 2, \cdots, l$. (4.13) can then be written as

$$\begin{aligned} \mathcal{F}(\mathbf{E}_l,\ D) &= \sum_{\mathbf{C}_l=\mathbf{C}_l^{(\kappa)}} P(\mathbf{C}_l) D^{d_E^2[\mathcal{M}(\mathbf{C}_l),\ \mathcal{M}(\mathbf{C}_l \oplus \mathbf{E}_l)]} \\ &= 2^{-ml} \sum_{\mathbf{C}_l=\mathbf{C}_l^{(\kappa)}} D^{d_E^2[\mathcal{M}(\mathbf{C}_l),\ \mathcal{M}(\mathbf{C}_l \oplus \mathbf{E}_l)]}, \end{aligned} \tag{4.14}$$

where the second row results from the fact that the labels in the sequence $\mathbf{C}_l^{(\kappa)}$ are independent (see Theorem 4.1). Note that because of $\kappa$-isometry, $\mathcal{F}(\mathbf{E}_l,\ D)$ in (4.14) does not depend on the values of the chosen sequence since the Euclidean distances are invariant with respect to this sequence. Substituting (4.14) into (4.12) we have

$$P_e \leq \sum_{l=1}^{\infty} \sum_{\mathbf{E}_l \neq \mathbf{0}} \prod_{i=1}^{l} \mathcal{F}(\mathbf{e}_i, D), \tag{4.15}$$

where

$$\mathcal{F}(\mathbf{e}_i,\ D) = \frac{1}{2^m} \sum_{\mathbf{c}_i=\mathbf{c}_i^{(\kappa)}} D^{d_E^2[\mathcal{M}(\mathbf{c}_i),\ \mathcal{M}(\mathbf{c}_i \oplus \mathbf{e}_i)]}$$

is called the *error weight profile*.

**Example 4.11** Consider a rate 2/3 ($m = 2$) 8PSK Ungerboeck TCM scheme. By setting $c^{(3)} = 0$ (labels corresponding to signals of the subset $\mathbf{A}_0$) the error weight profile can be obtained as

$$\mathcal{F}(\mathbf{e}_i,\ D) = \frac{1}{4} \sum_{(c^{(1)},c^{(2)},0)} D^{d_E^2[\mathcal{M}(c^{(1)},c^{(2)},0),\ \mathcal{M}(\ (c^{(1)},c^{(2)},0) \oplus (e^{(1)},e^{(2)},e^{(3)})\ )]}.$$

Using Figure 4.6 the error weight profiles for all possible error vectors are obtained as

$$\begin{aligned} \mathcal{F}(000,\ D) &= 1 & \mathcal{F}(100,\ D) &= D^{\delta_3^2} \\ \mathcal{F}(001,\ D) &= D^{\delta_0^2} & \mathcal{F}(101,\ D) &= D^{\delta_2^2} \\ \mathcal{F}(010,\ D) &= D^{\delta_1^2} & \mathcal{F}(110,\ D) &= D^{\delta_1^2} \\ \mathcal{F}(011,\ D) &= \tfrac{1}{2}(D^{\delta_0^2} + D^{\delta_2^2}) & \mathcal{F}(111,\ D) &= \tfrac{1}{2}(D^{\delta_0^2} + D^{\delta_2^2}) \end{aligned}$$

where $\delta_0^2 = 0.586E_s$, $\delta_1^2 = 2E_s$, $\delta_2^2 = 3.414E_s$, and $\delta_3^2 = 4E_s$. Note that by setting $c^{(3)} = 1$ the same error weight profile is achieved for every $\mathbf{e}_i$.

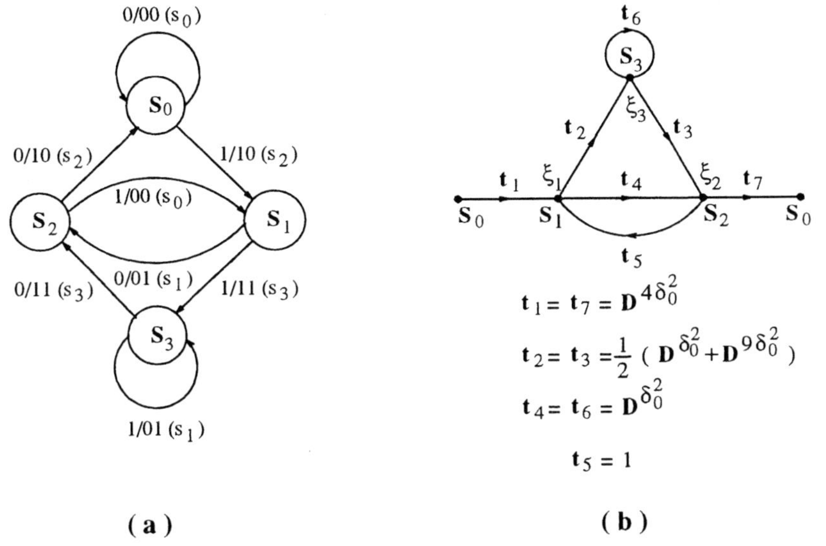


Figure 4.20: a) The state diagram and b) the error-state diagram of 4-state 4AM TCM.

The error event upper bound of (4.15) can be expressed as

$$P_e \leq T(D)\Big|_{D=e^{-\frac{1}{4N_0}}},$$

where

$$T(D) = \sum_{l=1}^{\infty} \sum_{\mathbf{E}_l \neq \mathbf{0}} \prod_{i=1}^{l} \mathcal{F}(\mathbf{e}_i, D)$$

is the generation function of an error-state diagram with $2^\nu$ states which enumerates all possible error sequences. The error-state diagram is derived from the code (modified) state diagram of the TCM scheme by labeling its branches with error weight profiles.

**Example 4.12** The state diagram of the encoder of the 4-state 4AM TCM scheme of Example 4.10 is shown in Figure 4.20(a). Using the 4AM signal set of Figure 4.17(b) the error weight profiles are obtained as (set $c^{(2)} = 0$ or $c^{(2)} = 1$)

$$\begin{aligned} \mathcal{F}(00,\ D) &= 1 & \mathcal{F}(10,\ D) &= D^{4\delta_0^2} \\ \mathcal{F}(01,\ D) &= D^{\delta_0^2} & \mathcal{F}(11,\ D) &= \tfrac{1}{2}(D^{\delta_0^2} + D^{9\delta_0^2}) \end{aligned}$$

where $\delta_0^2 = \frac{4}{5}E_s$ and $E_s$ is the average energy of the 4AM signal set. The error-state diagram of this TCM scheme is obtained from the state diagram of Figure 4.20(a) by labeling its branches by corresponding error weight profiles and splitting the all-zero state into two initial and final states as Figure 4.20(b).

### A Tighter Upper Bound

A tighter upper bound on error event probability can be found by using the exact expression for the pairwise error probability (see Problem 4.16)

$$P_2(\mathbf{C}_l,\ \hat{\mathbf{C}}_l) = \frac{1}{2}\mathrm{erfc}\left(\sqrt{\frac{\mathrm{d}_\mathrm{E}^2}{4\mathrm{N}_0}}\right), \tag{4.16}$$

where $d_E^2 \equiv d_E^2[\mathcal{M}(\mathbf{C}_l),\ \mathcal{M}(\hat{\mathbf{C}}_l)]$ is the squared Euclidean distance between the channel signal sequences corresponding to $\mathbf{C}_l$ and $\hat{\mathbf{C}}_l$. Using $d_E^2 = d_{free}^2 + (d_E^2 - d_{free}^2)$ and recalling the inequality of (3.22) we can upper-bound (4.16) as

$$P_2(\mathbf{C}_l,\ \hat{\mathbf{C}}_l) = \frac{1}{2}\mathrm{erfc}\left(\sqrt{\frac{\mathrm{d}_\mathrm{free}^2}{4\mathrm{N}_0}}\right) \cdot \mathrm{e}^{\frac{\mathrm{d}_\mathrm{free}^2}{4\mathrm{N}_0}} \cdot \mathrm{e}^{-\frac{\mathrm{d}_\mathrm{E}^2}{4\mathrm{N}_0}}. \tag{4.17}$$

Using (4.17) instead of the Bhattacharyya bound in (4.11) leads to the following tighter upper bound on error event probability

$$P_e \leq \frac{1}{2}\mathrm{erfc}\left(\sqrt{\frac{\mathrm{d}_\mathrm{free}^2}{4\mathrm{N}_0}}\right) \cdot \mathrm{e}^{\frac{\mathrm{d}_\mathrm{free}^2}{4\mathrm{N}_0}} \cdot \mathrm{T(D)}\Big|_{\mathrm{D}=\mathrm{e}^{-\frac{1}{4\mathrm{N}_0}}}. \tag{4.18}$$

### Asymptotic Estimate on $P_e$

An asymptotic estimate on the error event probability can be obtained from (4.18) by considering only the error events with free Euclidean distance

$$P_e \approx \frac{1}{2}N(d_{free})\mathrm{erfc}\left(\sqrt{\frac{\mathrm{d}_\mathrm{free}^2}{4\mathrm{N}_0}}\right),$$

where $N(d_{free})$ is the average number of nearest neighbors at free Euclidean distance from the transmitted sequence.

### Bit Error Probability

The bit error probability can be upper-bounded by weighting the pairwise error probabilities in (4.10) by the number of incorrect input bits associated with each error vector and dividing the result by $m$. To enumerate this number, the branch labels of the error-state diagram are modified as $I^r\mathcal{F}(\mathbf{e}_i, D)$, where $r$ represents the number of incorrect input bits associated with $\mathbf{e}_i$. Using the modified error-state diagram the bit error probability is upper-bounded as

$$P_b \leq \frac{1}{m}\frac{\partial T(D,\ I)}{\partial I}\Big|_{I=1,\ D=e^{-\frac{1}{4N_0}}}, \tag{4.19}$$

where $T(D,\ I)$ is the generating function of the modified error-state diagram. Similar to $P_e$, a tighter upper bound is obtained by using (4.17) instead of the

Bhattacharyya bound

$$P_b \leq \frac{1}{2m}\mathrm{erfc}\left(\sqrt{\frac{\mathrm{d}^2_{\mathrm{free}}}{4\mathrm{N}_0}}\right) \cdot \mathrm{e}^{\frac{\mathrm{d}^2_{\mathrm{free}}}{4\mathrm{N}_0}} \cdot \left.\frac{\partial \mathrm{T(D,\ I)}}{\partial \mathrm{I}}\right|_{\mathrm{I}=1,\ \mathrm{D}=\mathrm{e}^{-\frac{1}{4\mathrm{N}_0}}} . \tag{4.20}$$

**Example 4.13** Now we consider an illustrative example of the computation of error bounds for the 4-state 4AM TCM code of Example 4.10. The error-state diagram of this code is shown in Figure 4.20(b). To evaluate the upper bound of $P_b$, the branch labels of this diagram are modified as

$$\begin{aligned} t_1 &= ID^{4\delta_0^2} & t_5 &= I \\ t_2 &= \tfrac{1}{2}I(D^{\delta_0^2} + D^{9\delta_0^2}) & t_6 &= ID^{\delta_0^2} \\ t_3 &= \tfrac{1}{2}(D^{\delta_0^2} + D^{9\delta_0^2}) & t_7 &= D^{4\delta_0^2}. \\ t_4 &= D^{\delta_0^2} \end{aligned}$$

The state equations for the error-state diagram shown in Figure 4.20(b) are

$$\begin{aligned} \xi_1 &= t_1 + t_5\xi_2, \\ \xi_2 &= t_4\xi_1 + t_3\xi_3, \\ \xi_3 &= t_2\xi_1 + t_6\xi_3, \end{aligned} \tag{4.21}$$

and the generating function is

$$T(D,\ I) = t_7\xi_2. \tag{4.22}$$

Solving the system of equations (4.21) yields

$$\xi_2 = \frac{t_1[t_2t_3 + t_4(1-t_6)]}{(1-t_6)(1-t_4t_5) - t_2t_3t_5}.$$

Substituting the solution of $\xi_2$ into (4.22), we have

$$T(D,\ I) = \frac{t_1t_7[t_2t_3 + t_4(1-t_6)]}{(1-t_6)(1-t_4t_5) - t_2t_3t_5}. \tag{4.23}$$

Next, we substitute the values of $t_i, i = 1, 2, \cdots, 7$, into (4.23). We obtain

$$T(D,\ I) = \frac{\alpha_1 I^2 + \beta_1 I}{\alpha_2 I^2 + \beta_2 I + \gamma_2},$$

where

$$\begin{aligned} \alpha_1 &= -D^{10\delta_0^2}(3 - 2D^{8\delta_0^2} - D^{16\delta_0^2}), \\ \beta_1 &= 4D^{9\delta_0^2}, \\ \alpha_2 &= D^{2\delta_0^2}(3 - 2D^{8\delta_0^2} - D^{16\delta_0^2}), \\ \beta_2 &= -8D^{\delta_0^2}, \\ \gamma_2 &= 4, \end{aligned}$$

and $\delta_0^2 = \frac{4E_s}{5}$. Taking the derivative of $T(D, I)$ with respect to $I$ and evaluating the result at $I = 1$ we obtain

$$\left.\frac{\partial T(D,\ I)}{\partial I}\right|_{I=1} = \frac{\alpha_1\beta_2 - \alpha_2\beta_1 + 2\alpha_1\gamma_2 + \beta_1\gamma_2}{(\alpha_2 + \beta_2 + \gamma_2)^2}.$$

Using (4.19) and (4.20) the upper bounds on $P_b$ versus $E_b/N_0$ are shown in Figure 4.21. Note that the throughput of the rate 1/2 4AM scheme is 1 bit/s/Hz, and hence, $E_b = E_s$.

**Example 4.14** As another example we consider the 4-state 8PSK TCM scheme. The encoder and the trellis diagram of this coded scheme are shown in Figures 4.4 and 4.5, respectively. The error-state diagram is found to be the same as that of Figure 4.20(b) except for the branch labels. Using the results of Example 4.12 the branch labels of the error-state diagram are expressed as

$$\begin{aligned}
t_1 &= I\mathcal{F}(010,\ D) + I^2\mathcal{F}(110,\ D) &&= (I + I^2)D^{\delta_1^2} \\
t_2 &= I\mathcal{F}(011,\ D) + I^2\mathcal{F}(111,\ D) &&= \tfrac{1}{2}(I + I^2)(D^{\delta_0^2} + D^{\delta_2^2}) \\
t_3 &= \mathcal{F}(011,\ D) + I\mathcal{F}(111,\ D) &&= \tfrac{1}{2}(1 + I)(D^{\delta_0^2} + D^{\delta_2^2}) \\
t_4 &= \mathcal{F}(001,\ D) + I\mathcal{F}(101,\ D) &&= D^{\delta_0^2} + ID^{\delta_2^2} \\
t_5 &= I\mathcal{F}(000,\ D) + I^2\mathcal{F}(100,\ D) &&= I + I^2D^{\delta_3^2} \\
t_6 &= I\mathcal{F}(001,\ D) + I^2\mathcal{F}(101,\ D) &&= ID^{\delta_0^2} + I^2D^{\delta_2^2} \\
t_7 &= \mathcal{F}(010,\ D) + I\mathcal{F}(110,\ D) &&= (1 + I)D^{\delta_1^2}.
\end{aligned}$$

Note that the two error weight profiles in each of the $t_i$'s correspond to the parallel transitions between states. Substituting the above values of $t_i, i = 1, 2, \cdots, 7$, into (4.23) we can obtain the generating function of the error-state diagram. The derivative of this function with respect to $I$ at $I = 1$ is found to be [22]

$$\left.\frac{\partial T(D,\ I)}{\partial I}\right|_{I=1} = 4D^{4.586E_s}\frac{\hat{\alpha}(2 - \alpha - \alpha^2) + (1 - 2\alpha + 2\alpha^2)D^{2.828E_s}}{(1 - 2\alpha - \alpha D^{4E_s})^2},$$

where $\alpha = D^{0.586E_s} + D^{3.414E_s}$ and $\hat{\alpha} = \alpha \cdot D^{-0.586E_s}$.

Note that the error events associated with parallel transitions are not involved in the above generating function. Since the distance between signals associated with parallel transitions is $d_{free} = 2\sqrt{E_s}$, the bit error probability for these paths is found to be $\frac{1}{4}\mathrm{erfc}(\sqrt{E_s/N_0})$. Adding this term to (4.19) or (4.20) gives the upper bound on $P_b$ as

$$P_b \leq \frac{1}{4}\mathrm{erfc}\left(\sqrt{\frac{E_s}{N_0}}\right) + \frac{1}{2}\left.\frac{\partial T(D,\ I)}{\partial I}\right|_{I=1,\ D=e^{-\frac{1}{4N_0}}},$$

or

$$P_b \leq \frac{1}{4}\mathrm{erfc}\left(\sqrt{\frac{E_s}{N_0}}\right) + \frac{1}{4}\mathrm{erfc}\left(\sqrt{\frac{4.586E_s}{4N_0}}\right)e^{\frac{4.586E_s}{4N_0}}\left.\frac{\partial T(D,\ I)}{\partial I}\right|_{I=1,\ D=e^{-\frac{1}{4N_0}}}.$$

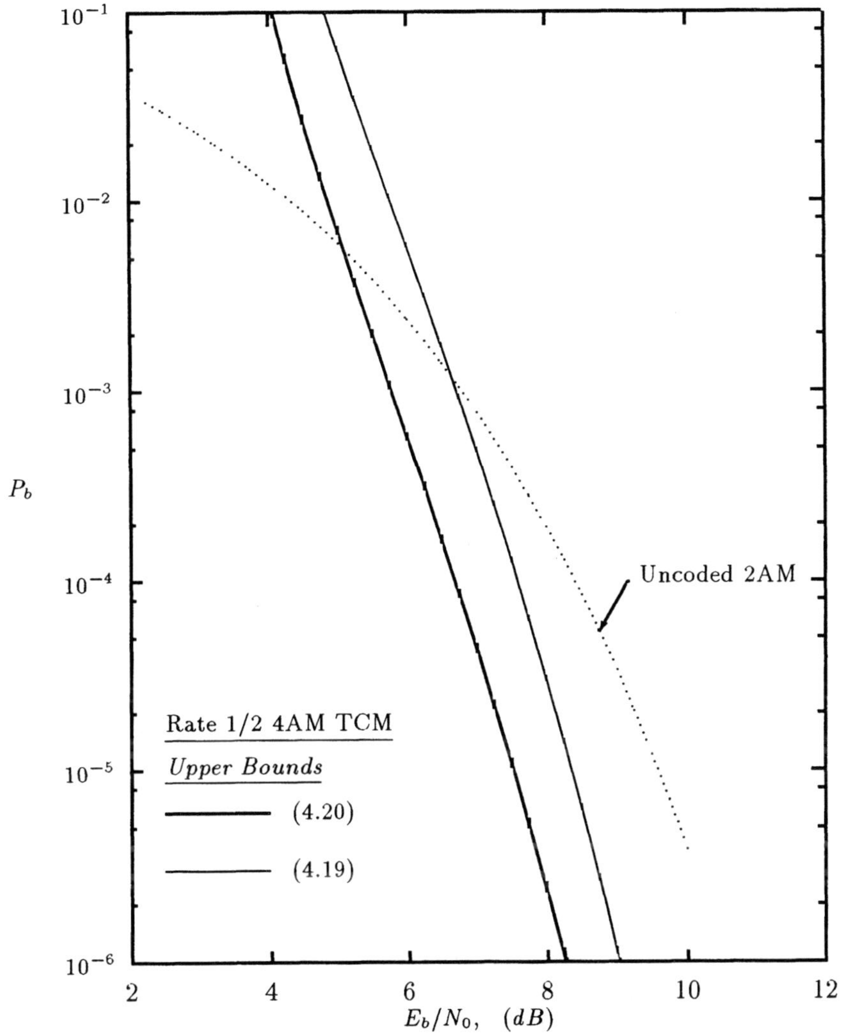


Figure 4.21: Performance of the rate 1/2 4-state 4AM TCM scheme.

The above upper bounds can be expressed in terms of the average bit energy $E_b$. Since the throughput of the rate 2/3 8PSK scheme is 2 bits/s/Hz then the average bit energy $E_b$ can be related to the average symbol energy $E_s$ as $E_b = \frac{1}{2}E_s$. The upper bounds are shown in Figure 4.22. The simulation results are also included to show the tightness of the upper bounds.

## 4.6 BLOCK-CODED MODULATION SCHEMES

Block-coded modulation (BCM) schemes as a counterpart to TCM schemes have been studied by several authors. Imai and Hirakawa [28] introduced a multilevel coding technique in which different error-correcting codes with different error-correcting capabilities are used with multilevel modulation schemes. Based on the set partitioning method, Cusack [29] constructed block-codes by using Reed-Muller codes and a QAM signal set. Later, Sayegh [30] generalized Cusack's work to other signal constellations by using known binary codes. More BCM schemes are described and constructed in [31], [32], [33], [34], and [35]. In the following we briefly describe the BCM schemes based on the work of Sayegh. For more details on the other schemes refer to the above-mentioned references.

A block diagram of a BCM scheme using a $2^m$-ary signal constellation is shown in Figure 4.23. The BCM encoder consists of $m$ binary codes (called binary component codes) with length $n$ shown as $(n, k_i, d_i), i = 1, 2, \cdots, m$, where $k_i$ and $d_i$ are the information length and the minimum Hamming distance of the $i$th code, respectively. The $k$ input bits are partitioned into $m$ blocks with $k_i$ bits in the $i$th block. The $i$th block is encoded by the $i$th binary encoder $(n, k_i, d_i)$ which generates a codeword with $n$ bits. The outputs of $m$ encoders can be expressed as a binary array with $n$ columns and $m$ rows

$$\begin{array}{lcccc} \mathbf{c}^{(1)}: & c_1^{(1)} & c_2^{(1)} & \cdots & c_n^{(1)} \\ \mathbf{c}^{(2)}: & c_1^{(2)} & c_2^{(2)} & \cdots & c_n^{(2)} \\ \cdots & \cdots & \cdots & \cdots & \cdots \\ \mathbf{c}^{(m)}: & c_1^{(m)} & c_2^{(m)} & \cdots & c_n^{(m)}. \end{array} \tag{4.24}$$

Each column of this array corresponds to a signal point in the $2^m$-ary signal space. We use natural labeling in mapping the columns of the binary array to the signal points of the signal set with a convention that the bit in the first row, $c_i^{(1)}$, shows the most significant bit and the bit in the last row, $c_i^{(m)}$ shows the least significant bit. The $n$ signals corresponding to $n$ columns of the binary array are transmitted one at a time.

The array in (4.24) contains $mn$ bits of which $mR_c$ bits are information bits. Denoting the number of information bits in the $i$th row by $k_i$, we can

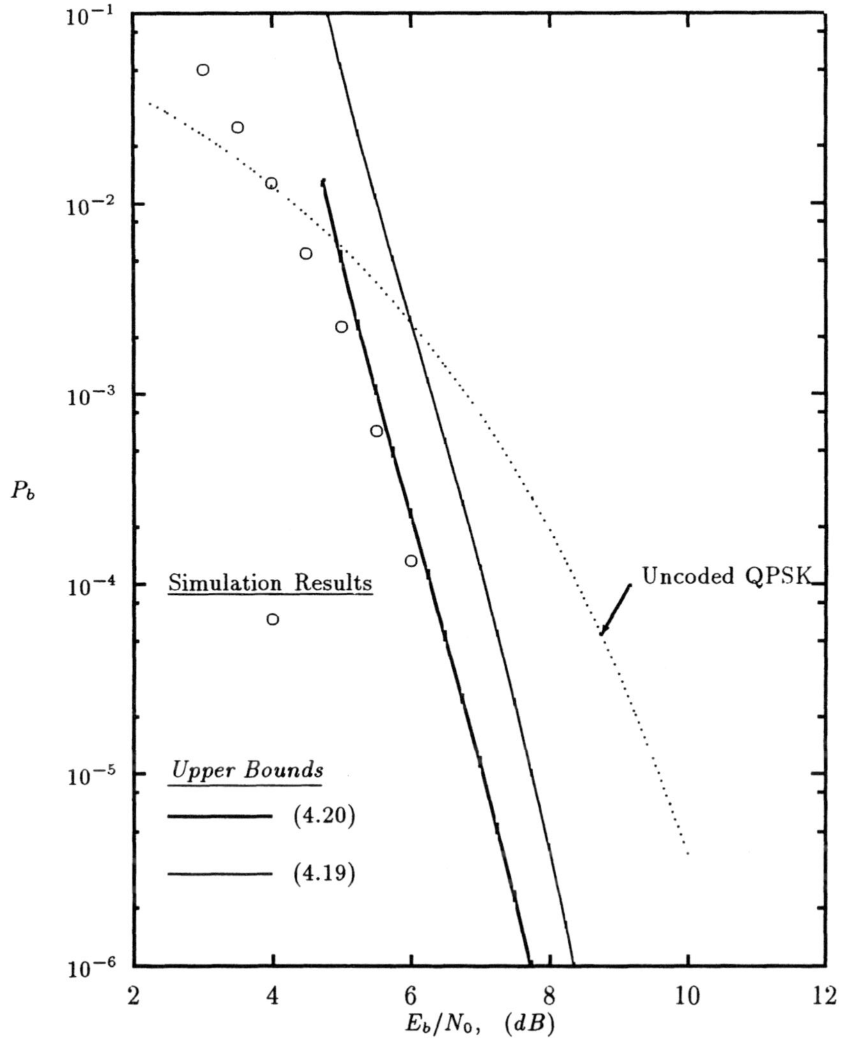


Figure 4.22: Performance of the rate 2/3 4-state 8PSK TCM scheme.

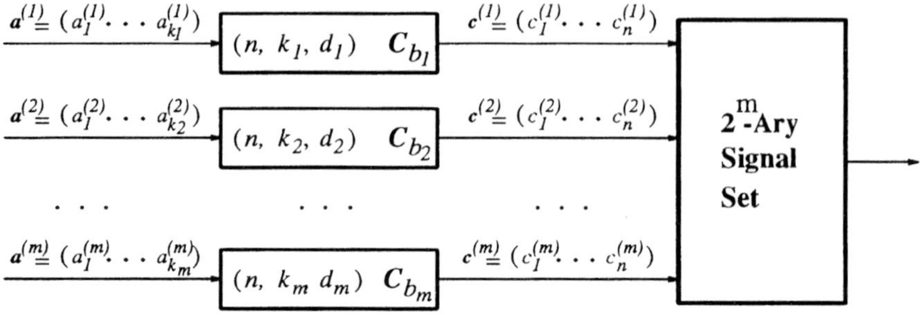


Figure 4.23: Block diagram of a BCM encoder.

write

$$k_1 + k_2 + \cdots + k_m = mR_c n, \tag{4.25}$$

where $R_c$ is the rate of the block-coded scheme. For a given rate $R_c$, the values of $k_i$'s are chosen subject to the above condition in such a way that the minimum Euclidean distance between the BCM codewords is maximized.

**Example 4.15** Consider the 8PSK signal set of Figure 4.24.[5] Using this signal set a block-coded scheme with three binary component codes is constructed. The first component code, $C_{b_1}$, is the binary repetition (8, 1, 8)[6] code which consists of the all-zero and the all-one codewords. The second component code, $C_{b_2}$, is the binary parity check (8, 7, 2) code consisting of the all-even weight codewords. The third component code, $C_{b_3}$, is the trivial (8, 8, 1) binary code consisting of all possible binary 8-tuples. Using (4.25) the rate of this block-coded scheme can be obtained as

$$R_c = \frac{k_1 + k_2 + k_3}{3n} = \frac{1+7+8}{3 \times 8} = \frac{2}{3},$$

which translates to 2 bits/s/Hz throughput (comparable with uncoded QPSK). An example of the binary array follows:

$$\begin{matrix} 0 & 0 & 0 & 0 & 0 & 0 & 0 & 0 \\ 1 & 0 & 1 & 1 & 0 & 1 & 0 & 0 \\ 1 & 1 & 0 & 1 & 0 & 1 & 1 & 0, \end{matrix}$$

which corresponds to the transmitted BCM codeword

$$s_6\ s_4\ s_2\ s_6\ s_0\ s_6\ s_4\ s_0.$$

[5]Note that in natural labeling of signal points of the 8PSK signal set in Figure 4.24 the order of bits is different from that in Figure 4.1. We use this convention in order to be consistent with the literature, e.g., [30], [35].

[6]The $(n, k, d_{min})$ code has a block length of $n$, $k$ information bits and a minimum Hamming distance of $d_{min}$.

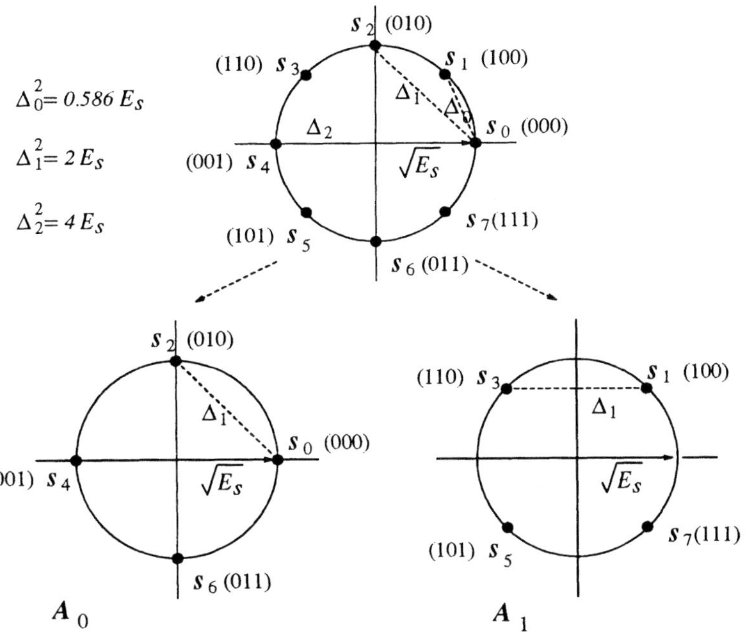


Figure 4.24: 8PSK signal set of Example 4.15.

### 4.6.1 Distance Considerations

We explain the distance property of the Block-Coded Modulation (BCM) schemes by considering a specific signal constellation, say, 8PSK. This discussion can be extended to other signal sets.

Consider a block-coded 8PSK scheme consisting of three binary component codes $C_{b_1}$, $C_{b_2}$, and $C_{b_3}$ with minimum Hamming distances $d_1$, $d_2$, and $d_3$, respectively. The minimum Euclidean distance between two distinct codewords of this block-coded scheme can be evaluated by considering their binary arrays.

Consider two BCM codewords. Three cases are possible.

**(i)** The binary arrays of these two codewords are different in the first row. Since the first rows are codewords of the binary component code $C_{b_1}$, they can be different in at least $d_1$ positions. For each one of the positions where two first rows differ (regardless of the values of the other rows) the corresponding points in the signal space will be at a Euclidean distance of at least $\Delta_0$ (see Figure 4.24). Thus,

$$d_E^2 \geq \Delta_0^2 d_1. \tag{4.26}$$

**(ii)** The binary arrays of the two codewords have the same first rows. The same argument shows that the squared Euclidean distance between such codewords is lower-bounded as

$$d_E^2 \geq \Delta_1^2 d_2. \tag{4.27}$$

**(iii)** Finally for two BCM codewords with binary arrays which only differ in the third rows the lower bound on the squared Euclidean distance is

$$d_E^2 \geq \Delta_2^2 d_3. \tag{4.28}$$

Considering (4.26), (4.27) and (4.28) the minimum squared Euclidean distance, $d_e^2$, of the block-coded 8PSK scheme is obtained as

$$d_e^2 \geq \min\left(\Delta_0^2 d_1,\ \Delta_1^2 d_2,\ \Delta_2^2 d_3\right). \tag{4.29}$$

We can easily find pairs of codewords for which the inequality in (4.29) holds with the equality. Thus,

$$d_e^2 = \min\left(\Delta_0^2 d_1,\ \Delta_1^2 d_2,\ \Delta_2^2 d_3\right).$$

**Example 4.16** For the BCM scheme of Example 4.15, we have

$$\begin{aligned} d_e^2 &= \min\left(8\Delta_0^2,\ 2\Delta_1^2,\ \Delta_2^2\right) \\ &= \min\left(8 \times 0.586E_s,\ 2 \times 2E_s,\ 4E_s\right) \\ &= \min\left(4.688E_s,\ 4E_s,\ 4E_s\right) \\ &= 4E_s, \end{aligned}$$

which translates to an asymptotic coding gain

$$g_\infty = 10 \log \frac{4}{2} = 3 \text{ dB}$$

compared to uncoded QPSK.

### 4.6.2 Decoding of BCM Schemes

The maximum likelihood decoding of BCM schemes, in general, requires the comparison of the received signal with each of $2^k$, $k = \sum_{i=1}^{m} k_i$, codewords and choosing the most likely codeword. For codes with large numbers of codewords this method becomes prohibitively complex. A suboptimum decoding technique is presented in [30].

Some BCM schemes have a trellis structure, and hence can be decoded by a soft-decision Viterbi decoding algorithm. As an illustrative example we consider the trellis construction for the BCM scheme of Example 4.15 in the following example. This method is taken from [35].

**Example 4.17** Consider the BCM scheme of Example 4.15. The encoder of this scheme consists of three binary component codes $C_{b_1}$, $C_{b_2}$, and $C_{b_3}$ which generate the $c_i^{(1)}$, $c_i^{(2)}$, and $c_i^{(3)}$ components of the BCM signal label at time $1 \leq i \leq n$. Since $C_{b_1}$ is a repetition code with two all-zero and all-one codewords, the value of $c_i^{(1)}$ determines that all of the signals of the BCM codeword are either from the subset $\mathbf{A}_0$ or the subset $\mathbf{A}_1$ (see Figure 4.24). Therefore, the value of $c_i^{(1)}$ is one of the parameters which determines the state of the code.

To construct the trellis diagram we need to define the state of the BCM encoder. Denoting the Hamming weight of the sequence $c_1^{(2)}$, $c_2^{(2)}$, $\cdots$, $c_l^{(2)}$ as $W_H^{(2)}(l)$, the following parameter is defined

$$\sigma_l = \begin{cases} 0, & \text{if } W_H^{(2)}(l) \text{ is even,} \\ 1, & \text{if } W_H^{(2)}(l) \text{ is odd.} \end{cases}$$

Now, we define the state of the BCM scheme at time $l$ as

$$S(l) = (c_l^{(1)},\ \sigma_l).$$

Four possible states exist

$$\begin{array}{ll} S_0 = (0,\ 0) & \qquad S_2 = (1,\ 0) \\ S_1 = (0,\ 1) & \qquad S_3 = (1,\ 1). \end{array}$$

The trellis diagram of the BCM scheme of Example 4.15 is shown in Figure 4.25. The trellis starts from the initial state $S_I$ and terminates at the final state $S_F$. The parallel transitions between states at time $l$ correspond to $c_l^{(3)} = 0$ and $c_l^{(3)} = 1$, respectively. Each valid BCM codeword corresponds to one of the paths in the trellis diagram starting at $S_I$ and terminating at $S_F$. The path corresponding to the codeword $(s_6\ s_4\ s_2\ s_4\ s_0\ s_4\ s_0)$ of Example 4.15 is shown by bold lines.

## 4.7 BIBLIOGRAPHICAL NOTES

The idea of improving the performance of the digital communication systems by looking at modulation and coding as a single entity was originally suggested by Massey in 1974 [36]. Following his idea, trellis-coded modulation schemes were developed by Ungerboeck in 1976 [18]. At the same time similar ideas were also developed independently by Imai and Hirakawa [28]. Following the idea of TCM many reseachers contributed in this area by searching for new TCM schemes and formalizing the theory. Calderbank and Mazo [19] gave a new description of the trellis-coded modulation schemes and showed that asymmetric one-dimensional TCM schemes provide more coding gain than symmetric ones. Extension of Ungerboeck's two-dimensional TCM schemes to multiple-dimentional TCM schemes was performed by Calderbank and Sloane [37], [38], [39] and Wei [40]. Some trellis codes were constructed in [41], [42] by Divsalar *et al.* using two-dimensional as well as one-dimensional asymmetric signal

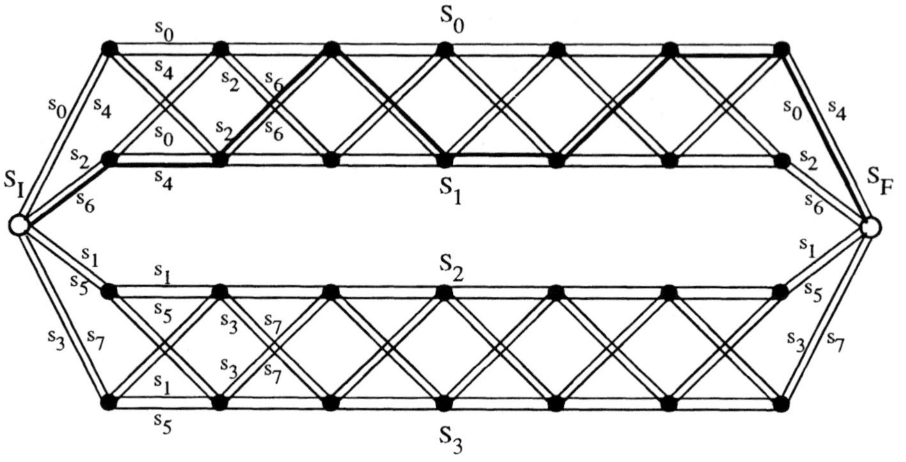


Figure 4.25: Trellis diagram of the BCM scheme of Example 4.15.

sets. Multiple trellis-coded modulation (MTCM) schemes were introduced by Divsalar and Simon [43] as an extension of multidimensional schemes using MPSK signaling. The rotationally TCM schemes were presented by Wei [44], [45] and one of these schemes with a coding gain of 4 dB was adopted by CCITT [46] for use in new high-speed voiceband modems in 1984.

# Problems

**4.1** Consider a rate 2/3 convolutional code defined by

$$G(D) = \begin{bmatrix} 1+D+D^2 & D^2 & 1 \\ D & 1+D^2 & 1+D+D^2 \end{bmatrix}$$

This code is used with an 8PSK signal set (Figure 4.1) to provide a throughput of 2 bits/s/Hz.

**a)** Find the free Euclidean distance.

**b)** Find the asymptotic coding gain compared to the uncoded QPSK scheme.

**c)** Compare your results with those of Examples 4.1, 4.2 and 4.3.

**4.2** Design an 8-state 16-PSK TCM scheme with $\tilde{m} = 1$ using the mapping rule by set partitioning described in Section 4.2.1. Compute the squared free Euclidean distance of the scheme and its asymptotic coding gain compared to the uncoded 8PSK.

**4.3** Repeat Problem 4.2 for a 4-state 16-PSK TCM with $\tilde{m} = 1$.

**4.4** Compare the performance of the 8-state 16-PSK TCM of Problem 4.2 with that of the 16-QAM TCM of Example 4.4.

**4.5** **a)** Design the signal set for an optimum uncoded two-dimensional signaling scheme with a throughput of 3 bits/s/Hz. Derive its performance in an AWGN channel.

**b)** Compare the performance of the following schemes: scheme obtained in a), 16-QAM TCM scheme of Example 4.4 and 16-PSK TCM of Problem 4.2.

**4.6** Using the signal set of the octal two-dimensional signaling scheme of Problem 4.5a), design the 4-state TCM scheme with a throughput of 2 bits/s/Hz. Compare the performance of the resulting TCM scheme with that of uncoded QPSK.

**4.7** Design a 16-state 8PSK TCM scheme with $\tilde{m} = 2$ using the set partitioning approach (Section 4.2.1). Compute the squared Euclidean distance and asymptotic coding gain of this scheme compared to the uncoded QPSK.

**4.8** Compare the performance of the TCM scheme of Problem 4.6 to that of the 4-state 8PSK TCM scheme of Example 4.2.

**4.9** Using Definitions 4.1 and 4.2 and Theorem 4.1 verify the uniformity of the 8-state 16-PSK TCM scheme of Problem 4.2.

**4.10** Is the 8-state 8PSK TCM scheme of Example 4.5 uniform ?

**4.11** Is the 8-state 16-QAM TCM scheme of Example 4.4 uniform ?

**4.12 a)** Construct a 4-state 8PSK TCM code with $\tilde{m} = 1$ using set partitioning.

**b)** Compute the free Euclidean distance and asymptotic coding gain of this code compared to the uncoded QPSK.

**c)** Provide the analytical description of the code in a). Derive $\mathbf{f}$, $\mathbf{B}_n$ and $\mathbf{d}$.

**d)** Find the analytical description of 8-state 8PSK TCM scheme of Example 4.5.

**4.13** Consider the pairwise error probability $P_2(\mathbf{s}_l, \hat{\mathbf{s}}_l)$.

**a)** For a maximum likelihood decoder, prove that

$$P_2(\mathbf{s}_l, \hat{\mathbf{s}}_l) = \int_{\mathbf{r}} f(\mathbf{r}) p_{R|S}(\mathbf{r}|\mathbf{s}_l) d\mathbf{r}$$

where $\mathbf{r}$ is the received vector, $p_{R|S}(\mathbf{r}|\mathbf{s}_l)$ is the channel transition probability density function, and

$$f(\mathbf{r}) = \begin{cases} 1 & \text{if } \frac{p_{R|S}(\mathbf{r}|\hat{\mathbf{s}}_l)}{p_{R|S}(\mathbf{r}|\mathbf{s}_l)} \geq 1 \\ 0 & \text{elsewhere} \end{cases}$$

**b)** Show that i)

$$f(\mathbf{r}) \leq \frac{p_{R|S}(\mathbf{r}|\hat{\mathbf{s}}_l)}{p_{R|S}(\mathbf{r}|\mathbf{s}_l)}$$

ii)

$$P_2(\mathbf{s}_l, \hat{\mathbf{s}}_l) \leq \int_{\mathbf{r}} \sqrt{p_{R|S}(\mathbf{r}|\hat{\mathbf{s}}_l) p_{R|S}(\mathbf{r}|\mathbf{s}_l)} d\mathbf{r}$$

Note that the above expression is called the Bhattacharyya bound (a special case of Chernoff bound).

**c)** For an AWGN channel with variance $N_0/2$, show that
i)

$$p_{R|S}(\mathbf{r}, \mathbf{s}) = \frac{1}{(\pi N_0)^{N/2}} exp\left(-\frac{d_E^2(\mathbf{r}, \mathbf{s})}{N_0}\right),$$

where $N$ is the dimension of $\mathbf{r}$ and $\mathbf{s}$.
ii)

$$\begin{aligned} P_2(\mathbf{s}_l, \hat{\mathbf{s}}_l) &\leq \frac{1}{(\pi N_0)^{N/2}} \int_{-\infty}^{+\infty} exp\left(-\frac{d_E^2(\mathbf{r}, \mathbf{s}) + d_E^2(\mathbf{r}, \hat{\mathbf{s}}_l)}{2N_0}\right) d\mathbf{r} \\ &= exp\left(-\frac{d_E^2(\mathbf{s}_l, \hat{\mathbf{s}}_l)}{4N_0}\right) \end{aligned}$$

**4.14** Find the error weight profiles for the 4-state 8PSK TCM code of Problem 4.12 and for a rate 3/4 16PSK Ungerboeck TCM scheme.

**4.15** Find the error weight profiles for the 8-state 8PSK TCM code of Example 4.3 and for a rate 2/3 8AM Ungerboeck TCM scheme.

**4.16** Consider the pairwise error probability $P_e(\mathbf{s}_l, \hat{\mathbf{s}}_l)$ of two paths $\mathbf{s}_l$, and $\hat{\mathbf{s}}_l$ as shown in Figure 4.19.

**a)** Show that the squared Euclidean distance between two paths $d_E^2(\mathbf{s}_l, \hat{\mathbf{s}}_l)$ is the sum of the squared Euclidean distances between different branches.

**b)** For a maximum likelihood decoder in an AWGN channel with a coherent receiver, prove that the pairwise error probability is

$$P_2(\mathbf{s}_l, \hat{\mathbf{s}}_l) = \frac{1}{2}\mathrm{erfc}\left(\frac{\mathrm{d_E}(\mathrm{s_l}, \hat{\mathrm{s}}_\mathrm{l})}{2\sqrt{\mathrm{N_0}}}\right).$$

**4.17** Derive the generating function and upper bound on the bit error probability of the 8-state 8PSK TCM of Example 4.5 in an AWGN channel.

**4.18** Repeat Problem 4.17 for the 8-state 16-PSK TCM scheme of Problem 4.2.

**4.19** Repeat Problem 4.17 for the 4-state 16-PSK TCM scheme of Problem 4.3.

**4.20** Consider an 8PSK BCM scheme using the signal set of Figure 4.24 with three binary component codes. $C_{b_1}$ is a BCH(15,5,7) code, $C_{b_2}$ is a Hamming (15,11,3) code and $C_{b_3}$ is the trivial (15,15,1) code.

**a)** Find the rate of this 8PSK BCM scheme.

**b)** Compute the minimum distance of this BCM scheme and its asymptotic coding gain compared to the uncoded QPSK.

**4.21** Consider a 16-QAM BCM scheme using the signal set of Figure 4.10 with four binary component codes. $C_{b1}$ is a repetition (15,1,15) code, $C_{b2}$ is a BCH(15,5,7) code, $C_{b3}$ is a Hamming (15,11,3) code, and $C_{b4}$ is a parity-check (15,14,2) code.

**a)** Find the rate of this 16-QAM BCM scheme.

**b)** Compute the minimum distance of this BCM scheme and its asymptotic coding gain compared to uncoded QPSK.

**4.22** Consider a rate 1/2 QPSK BCM scheme using the signal set of Figure 4.1(a) with two following binary component codes:

$C_{b_1}$: repetition (4,1,3) code
$C_{b_2}$: parity-check (4,3,2) code

**a)** Compute the minimum distance and asymptotic coding gain of this BCM scheme compared to the uncoded BPSK.

**b)** Draw the trellis diagram of this BCM scheme.

**4.23 Coded $\frac{\pi}{4}$-shift QPSK:**
Consider the $\frac{\pi}{4}$-shift QPSK scheme discussed in Problem 2.22b).

**a)** Using the Ungerboeck TCM design rule, design a 2-state rate 2/3 trellis-coded $\frac{\pi}{4}$-shift QPSK scheme.

**b)** Construct the trellis diagram and the error-state diagram and derive the tight upper-bound on the bit error probability of the 2-state coded scheme in an AWGN channel.

**4.24** Repeat Problem 4.23 for a 4-state, rate 2/3 trellis-coded $\frac{\pi}{4}$-shift QPSK scheme.

**4.25** Repeat Problem 4.23 for an 8-state, rate 2/3 trellis-coded $\frac{\pi}{4}$-shift QPSK scheme.

**4.26** A TCM is designed by using a rate-1/2 convolutional encoder with $\nu = 1$. The generator polynomial matrix of the convolutional encoder is

$$\mathbf{G}(D) = [1 + D \quad 1].$$

The output bits of the convolutional encoder are mapped to one signal element of a 4-AM signal set.

**a)** Draw the encoder of the convolutional code. Is this encoder systematic?

**b)** Draw the trellis diagram of the convolutional code.

**c)** Determine the mapping rule so that the minimum Euclidean distance between the TCM coded sequences is maximized. What is the minimum Euclidean distance?

**d)** Find the asymptotic coding gain of this TCM scheme compared to the uncoded BPSK.

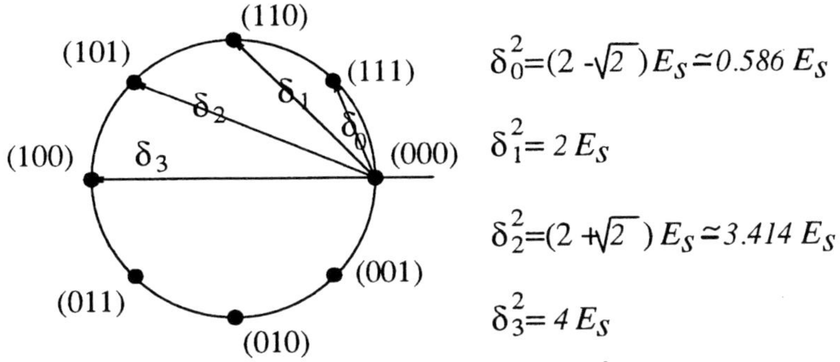

Figure P4.28-1: An 8PSK signal set

**4.27** A TCM scheme is designed by using a rate 1/2 convolutional code with $\nu = 3$. The generator polynomial matrix of the convolutional code is

$$\mathbf{G}(D) = [1 + D + D^3 \quad 1].$$

The output bits of the convolutional encoder are mapped to one signal element of a 4-AM signal set.

**a)** Draw the encoder of the convolutional code.

**b)** Draw the trellis diagram of the convolutional code.

**c)** Determine the mapping rule so that the minimum Euclidean distance of the TCM scheme is maximized. Find the minimum Euclidean distance.

**d)** Find the asymptotic coding gain of this TCM scheme compared to BPSK.

**e)** Using the Ungerboeck method (i.e., mapping by set partitioning) design a rate 1/2 4-AM TCM scheme with $\nu = 3$. Draw its trellis diagram and determine its asymptotic coding gain compared to uncoded BPSK. Compare this coding gain with that obtained in **d)**.

**4.28** Consider an 8PSK signal set with bit assignements to the signal points is shown in Figure P3.28-1. A 16-state rate 2/3 TCM is designed by using this signal set with the following parity-check polynomials

$$H_3(D) = D^4 + D + 1 \quad H_2(D) = D^2$$
$$H_1(D) = D^3 + D^2 + D.$$

**a)** Determine the feedforward encoder structure of the TCM scheme

**b)** Determine the feedback realization of the equivalent systematic encoder.

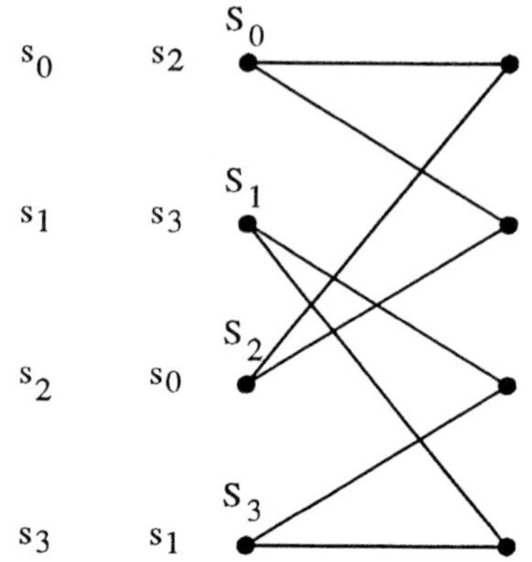

Figure P4.29-1: Trellis diagram

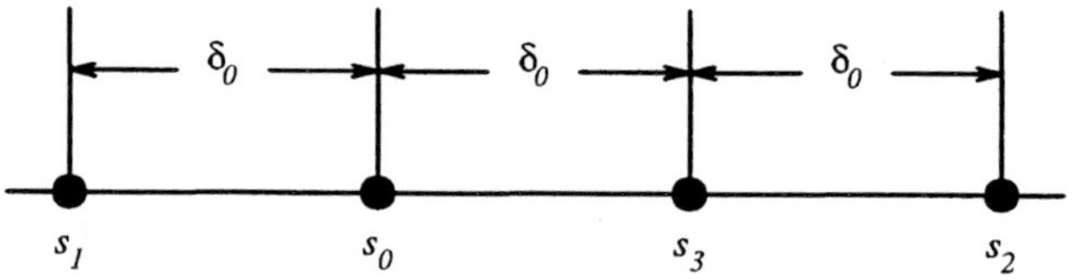

Figure P4.29-2: 4AM signal set

c) Is this TCM scheme $\kappa$-isometric? Why?

**4.29** The trellis diagram of a rate 1/2 4-state 4-AM TCM is shown in Figure P4.29-1. The 4AM signal set is illustrated in Figure 4.29-2.

**a)** Determine the minimum Euclidean distance of this TCM scheme and its asymptotic coding gain compared to the uncoded 2AM.

**b)** Using this code, design a rate 2/3 4-state 8AM and determine its asymptotic coding gain compared to the uncoded 4AM.

**c)** Design a rate $m/(m+1)$ 4-state $2^{m+1}$AM TCM scheme and determine its asymptotic coding gain compared to the uncoded $2^m$AM. Determine the asymptotic coding gain when $m \to \infty$.

**4.30** Consider a rate 2/3 4-state 8PSK TCM scheme with the following parity-check polynomial matrix

$$\mathbf{H}(D) = [0 \quad 1 + D^2 \quad D].$$

**a)** Determine the minimum Euclidean distance of this TCM scheme and its asymptotic coding gain compared to the uncoded QPSK.

**b)** Adding an uncoded bit to the encoder of the above TCM, design a rate 3/4 4-state 16PSK TCM scheme. Find the minimum Euclidean distance of this code and determine its asymptotic coding gain compared to the uncoded 8PSK.

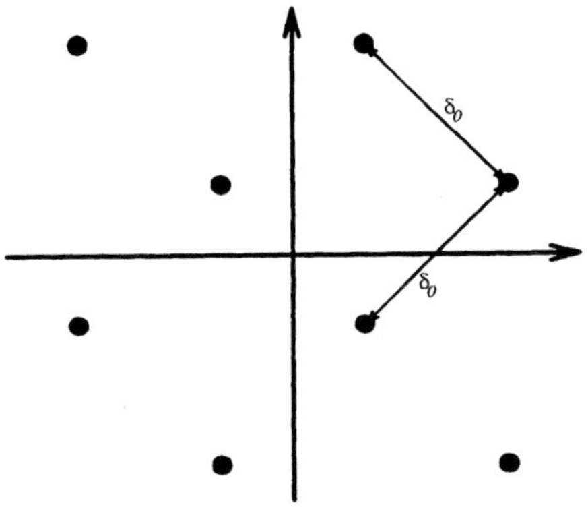


Figure P4.31-1: An 8AMPM signal set

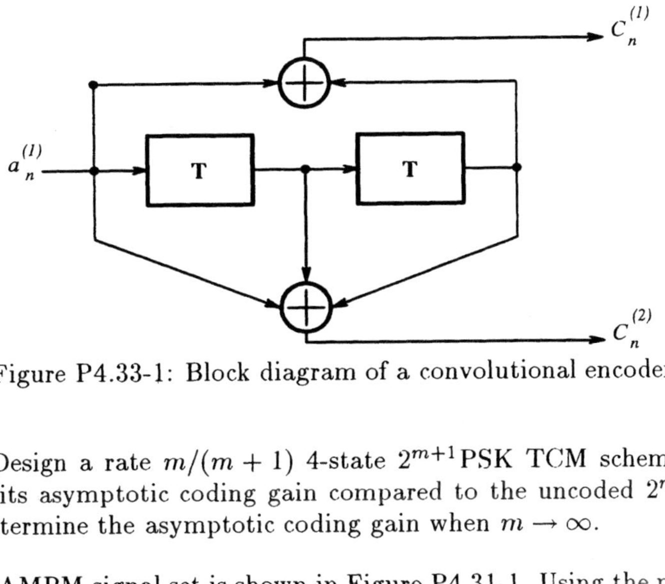


Figure P4.33-1: Block diagram of a convolutional encoder

c) Design a rate $m/(m+1)$ 4-state $2^{m+1}$PSK TCM scheme and find its asymptotic coding gain compared to the uncoded $2^m$PSK. Determine the asymptotic coding gain when $m \to \infty$.

**4.31** An 8AMPM signal set is shown in Figure P4.31-1. Using the mapping by set partitioning rule, design a rate 2/3 4-state TCM scheme. Show that this TCM scheme is $\kappa$-isometric. Find the asymptotic coding gain of this TCM scheme compared to the uncoded QPSK.

**4.32** Using the mapping by set partitioning rule, design a rate 2/3 2-state 8AMPM TCM scheme. Find the upper bound on the bit error probability for this TCM scheme in an AWGN channel.

**4.33** The encoder block diagram and trellis diagram of a rate 1/2 convolutional code are shown in Figures P4.33-1 and P4.33-2, respectively.

**a)** Find the minimum Hamming distance of this code.

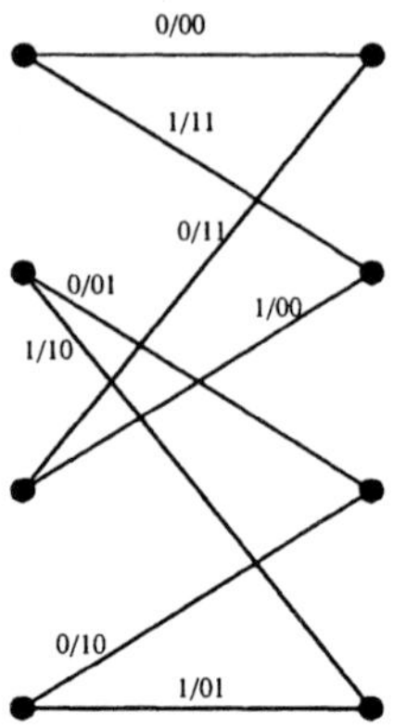


Figure P4.33-2: Trellis diagram

**b)** Using the Gray coding the output bits of the convolutional encoder are mapped to one signal element of a QPSK signal set. Draw the trellis diagram of the resulting TCM scheme and find its minimum Euclidean distance.

**c)** Determine the asymptotic coding gain of this TCM scheme compared to the uncoded BPSK.

**d)** Is the squared Euclidean distance of the TCM scheme in b) proportional to the Hamming distance of the convolutional code? Why?

**e)** Assume that the TCM coded sequences are transmitted over an AWGN channel with the two-sided noise spectral density of $N_0/2$. For a coherent detection and $E_b/N_0 = 5$ dB find the probability of bit error for hard-decision and soft-decision Viterbi decoding schemes.

# Chapter 5

# TRELLIS-CODED MODULATION SCHEMES FOR FADING CHANNELS

In Chapter 4, we presented a background on the aspects of trellis-coded modulation schemes and evaluated the performance of these schemes on the AWGN channel. In this chapter we present the design and performance of the TCM schemes for a fading channel. We begin with the performance evaluation of trellis codes over a Rician fading channel. The parameters which dominate the performance of TCM schemes on fading channels are extracted from this evaluation. This leads to the observation that on channels with a small value of the Rician parameter $K$, the *length of the shortest error event path* (the effective length of the code) and the *product of branch distances* along this path are main design criteria [47], [48]. The free Euclidean distance, in such a case, is considered as a secondary design parameter.

Next we consider the construction of TCM schemes based on fading channel design criteria. In this context we consider both conventional and Multiple Trellis-Coded Modulations (MTCM). The performance analysis of the *coherent* detection of these schemes on a slow-fading channel is then presented by using different approaches, i.e., the upper bound analysis and the exact solution [49].

# 5.1 DESIGN CRITERIA FOR TCM ON FADING CHANNELS

TCM schemes can be used for the transmission of reliable digital information over bandwidth-limited fading channels. When combined with interleaving and deinterleaving of sufficient depth, these schemes can provide significant coding gains compared to uncoded ones, provided that some new design criteria are utilized in designing the code. To obtain these design criteria we first evaluate the performance of the trellis-coded MPSK schemes over a Rician fading channel; an illustrative example follows.

## 5.1.1 Performance of a Trellis-Coded MPSK on a Rician Fading Channel

A general block diagram of a TCM scheme on a fading channel is shown in Figure 5.1. Input bits are encoded by a trellis encoder to produce a sequence of signals $\mathbf{s}_l = (s_1, s_2, \cdots, s_l)$ where each signal $s_i$ is a two-dimensional vector chosen from an MPSK signal set. Using complex notation we can represent each of the signals $s_i$ by a point in a complex plane. The coded signals are interleaved to spread the bursts of errors caused by slowly varying fading process. The sequence of interleaved signals is denoted by $\mathbf{s}'_l = (s'_1, s'_2, \cdots, s'_l)$. The in-phase and quadrature components of the interleaved coded signals are pulse-shaped for no intersymbol interference and then are translated to radio frequencies (RF) for transmission over the channel (see Problem 5.1). The channel corrupts the transmitted signal by introducing a fading gain and an additive white Gaussian noise term.

At the receiver the in-phase and quadrature components of the received signal are demodulated and quantized for soft-decision decoding. The channel estimator also provides an estimate of the channel gain which can be used in the decoding process to improve the performance of the coded system. Using the terminology of [50] and [51] the estimate of channel gain is called *Channel State Information* (CSI).

The output sequence of modulator $\mathbf{r}'_l = (r'_1, r'_2, \cdots, r'_l)$ and the channel state information sequence $\hat{\mathbf{g}}'_l = (\hat{g}'_1, \hat{g}'_2, \cdots, \hat{g}'_l)$ are deinterleaved to produce the sequences $\mathbf{r}_l = (r_1, r_2, \cdots, r_l)$ and $\hat{\mathbf{g}}_l = (\hat{g}_1, \hat{g}_2, \cdots, \hat{g}_l)$, respectively. The two sequences $\mathbf{r}_l$ and $\hat{\mathbf{g}}_l$ are the inputs to TCM decoder which performs maximum likelihood (ML) decoding.

A discrete time model for the system of Figure 5.1 is shown in Figure 5.2. Using this model the received signal at time $i$ can be written as

$$r_i = g_i \cdot s_i + n_i, \tag{5.1}$$

where $n_i$ is a sample of a zero-mean complex Gaussian noise process with variance $\sigma_n^2 = N_0/2$ and the complex channel gain $g_i$ is a sample of a complex

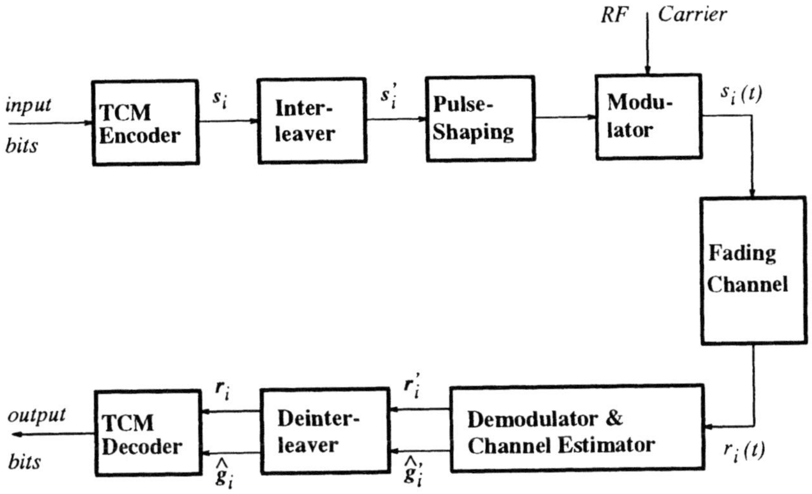


Figure 5.1: Block diagram of the trellis coded system on fading channels.

Gaussian process with variance $\sigma_g^2$. Using phasor notation, the complex channel gain $g_i$ is expressed as

$$g_i = a_i \cdot e^{j\phi_i},$$

where $a_i$ and $\phi_i$ are the amplitude and the phase processes, respectively.

To analyze the performance of the system under consideration we make some assumptions. First we assume that the receiver performs *coherent detection*, and hence the channel phase shift is compensated by the receiver. In this case (5.1) can be simplified as

$$r_i = a_i s_i + n_i, \tag{5.2}$$

where $a_i$ is the amplitude of the fading process. For a channel with only a diffused multipath component the fading amplitude is *Rayleigh distributed* with a probability density function (pdf)

$$p_A(a) = 2ae^{-a^2}, \qquad a \geq 0. \tag{5.3}$$

When there is a single dominant, nonfading component in the received signal along with a multipath fading, the amplitude $a_i$ is *Rician distributed* with a pdf (see Chapter 1)

$$p_A(a) = 2a(1+K)e^{-(K+a^2(1+K))} I_0(2a\sqrt{K(1+K)}\ ), \qquad a \geq 0 \tag{5.4}$$

where $K$ is the *Rician parameter* defined as the ratio of the energy of the direct component to the energy of the diffused multipath component and $I_0(\cdot)$ is the

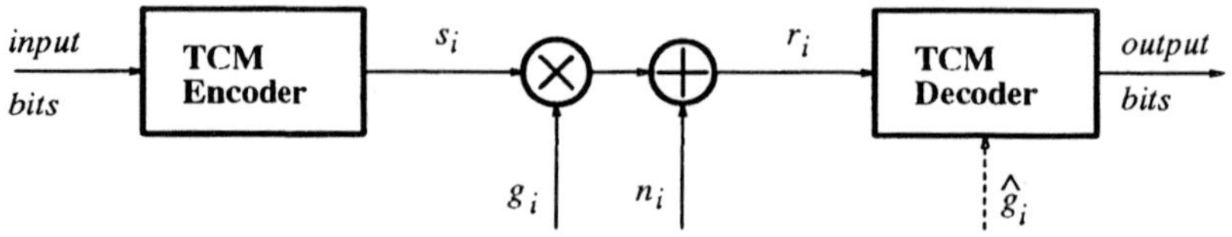


Figure 5.2: The baseband system model of Figure 5.1.

zero-order modified Bessel function of the first kind[1]. Note that both pdf's (5.3) and (5.4) are *normalized*, that is, the fading amplitude $a$ has a mean-squared value of unity. We choose a normalized probability density function so that the measured signal energy at the receiver represents the average signal energy per channel symbol, $E_s$.

Next we assume that the interleaving is ideal, that is, an interleaver with infinite depth is used. This assumption implies that the fading amplitudes are statistically independent and provides a memoryless channel model for the performance analysis.

### Upper Bound on Pairwise Error Probability

In this section we present an upper bound on pairwise error probability [51] which reveals the effect of some dominant factors in the performance of TCM schemes on fading channels. This leads to specifying the design criteria for constructing trellis codes suitable for transmission over fading channels. The results of this section will be used to derive the upper bounds on bit error probability later.

Let $\mathbf{r}_l = (r_1, r_2, \cdots, r_l)$ be the received signal where its $i$th component $r_i$ is given by (5.2). Maximum likelihood decoder (usually implemented by a Viterbi algorithm) chooses the coded sequence that most likely corresponds to the sequence of received signals. This is achieved by computing a decoding metric between the sequence of received signals $\mathbf{r}_l$ and a possible sequence of transmitted signals $\mathbf{s}_l$. This metric is related to the conditional channel probabilities (see Chapter 2)

$$m(\mathbf{r}_l, \mathbf{s}_l) = \ln p_{\mathbf{N}}(\mathbf{r}_l|\mathbf{s}_l), \tag{5.5}$$

when the channel state information is not used and

$$m(\mathbf{r}_l, \mathbf{s}_l; \hat{\mathbf{a}}_l) = \ln p_{\mathbf{N}}(\mathbf{r}_l|\mathbf{s}_l, \hat{\mathbf{a}}_l), \tag{5.6}$$

when CSI is used in decoding process. In (5.6) $\hat{\mathbf{a}}_l = (\hat{a}_1, \hat{a}_2, \cdots, \hat{a}_l)$ is the sequence of fading amplitude estimates. For an ideal CSI, the estimate $\hat{\mathbf{a}}$ is replaced by $\mathbf{a}_l = (a_1, a_2, \cdots, a_l)$ where $a_i$ is the fading amplitude at time

[1] $I_0(x) = \frac{1}{2\pi}\int_0^{2\pi} e^{x\cos t}\,dt$

$i$. Since the channel under consideration is memoryless (ideal interleaving) the metrics in (5.5) and (5.6) have an additive property and can be written as

$$m(\mathbf{r}_l,\ \mathbf{s}_l) = \sum_{i=1}^{l} \ln p_N(r_i|s_i),$$

and

$$m(\mathbf{r}_l,\ \mathbf{s}_l;\ \hat{\mathbf{a}}_l) = \sum_{i=1}^{l} \ln p_N(r_i|s_i,\ \hat{a}_i), \tag{5.7}$$

respectively.

The ML decoder chooses the coded sequence for which the defined metric is maximized. Let $\mathbf{s}_l = (s_1,\ s_2,\ \cdots,\ s_l)$ be the transmitted sequence. The decoder makes an error if it decodes the coded sequence $\hat{\mathbf{s}}_l = (\hat{s}_1,\ \hat{s}_2,\ \cdots,\ \hat{s}_l)$ instead. This is called a *pairwise error probability* and is denoted by $P_2(\mathbf{s}_l,\ \hat{\mathbf{s}}_l)$. We derive the upper-bound of the pairwise error probability with or without CSI as follows[2].

### Ideal Channel State Information

In this case it is assumed that the fading amplitude is perfectly estimated so that $\hat{a}_i = a_i$. Considering that $n_i$ in (5.2) is a complex Gaussian random variable, the metric $m(r_i,\ s_i;\ a_i)$ can be expressed as

$$m(r_i,\ s_i;\ a_i) = -|r_i - a_i s_i|^2, \tag{5.8}$$

where the variance factor $N_0/2$ is discarded since it is independent of the transmitted signal and has no effect on the decoder decision. In (5.8), $|(\cdot)|^2$ represents the squared Euclidean norm of $(\cdot)$.

Between the two valid code sequences $\mathbf{s}_l$ and $\hat{\mathbf{s}}_l$ the optimum decoder incorrectly decides $\hat{\mathbf{s}}_l$ if

$$m(\mathbf{r}_l,\ \hat{\mathbf{s}}_l;\ \mathbf{a}_l) \geq m(\mathbf{r}_l,\ \mathbf{s}_l;\ \mathbf{a}_l).$$

Using (5.7), the above inequality can be written as

$$\sum_{i=1}^{l} m(r_i,\ \hat{s}_i;\ a_i) \geq \sum_{i=1}^{l} m(r_i,\ s_i;\ a_i).$$

Next the pairwise error probability is given by

$$P_2(\mathbf{s}_l,\ \hat{\mathbf{s}}_l) = E_{\mathbf{a}_l}[P_2(\mathbf{s}_l,\ \hat{\mathbf{s}}_l|\mathbf{a}_l)],$$

where

$$P_2(\mathbf{s}_l,\ \hat{\mathbf{s}}_l|\mathbf{a}_l) = Pr[m(\mathbf{r}_l,\ \hat{\mathbf{s}}_l;\ \mathbf{a}_l) \geq m(\mathbf{r}_l,\ \mathbf{s}_l;\ \mathbf{a}_l)\ |\ \mathbf{a}_l], \tag{5.9}$$

[2] Part of this discussion was presented earlier in Chapter 2 . We repeat it here so that the reader can proceed without referring to Chapter 2.

is the conditional pairwise error probability, conditioned on the fading amplitude sequence $\mathbf{a}_l$, and $E$ shows the statistical expectation operation. The conditional pairwise error probability (5.9) is upper-bounded by using the Chernoff bound [17], [4] (see Problem 2.11)

$$\begin{aligned} P_2(\mathbf{s}_l, \hat{\mathbf{s}}_l|\mathbf{a}_l) &\leq E_{\mathbf{n}}[\exp(\lambda\{m(\mathbf{r}_l, \hat{\mathbf{s}}_l; \mathbf{a}_l) - m(\mathbf{r}_l, \mathbf{s}_l; \mathbf{a}_l)\})|\mathbf{a}_l] \\ &= \prod_{i=1}^{l} E_{n_i}[\exp(\lambda\{|r_i - a_i s_i|^2 - |r_i - a_i \hat{s}_i|^2\})], \end{aligned} \quad (5.10)$$

where the second row is obtained by considering the additive property of the metric and using (5.8). Using the results of Appendix A (see (A.3)), the conditional pairwise error probability given in (5.10) is upper-bounded as

$$P_2(\mathbf{s}_l, \hat{\mathbf{s}}_l|\mathbf{a}_l) \leq \prod_{i=1}^{l} \exp(-\lambda(1 - N_0\lambda)a_i^2|s_i - \hat{s}_i|^2).$$

This upper bound can be tightened by optimizing the Chernoff parameter $\lambda$. By setting $\lambda_{\text{opt}} = \frac{1}{2N_0}$ (see Appendix A, (A.5)), we obtain

$$P_2(\mathbf{s}_l, \hat{\mathbf{s}}_l|\mathbf{a}_l) \leq \prod_{i=1}^{l} \exp(-\frac{1}{4N_0} a_i^2|s_i - \hat{s}_i|^2). \quad (5.11)$$

To find the pairwise error probability on a Rician fading channel, we average (5.11) over the pdf of (5.4). Using (A.7) and (A.8) (see Appendix A), we obtain

$$P_2(\mathbf{s}_l, \hat{\mathbf{s}}_l) \leq \prod_{i=1}^{l} \frac{1+K}{1+K+\frac{1}{4N_0}|s_i - \hat{s}_i|^2} \exp[-\frac{K\frac{1}{4N_0}|s_i - \hat{s}_i|^2}{1+K+\frac{1}{4N_0}|s_i - \hat{s}_i|^2}], \quad (5.12)$$

which simplifies at high signal-to-noise ratios to

$$P_2(\mathbf{s}_l, \hat{\mathbf{s}}_l) \leq \prod_{i\in\eta} \frac{(1+K)e^{-K}}{\frac{1}{4N_0}|s_i - \hat{s}_i|^2}, \quad (5.13)$$

where $\eta$ is the set of all $i$ for which $s_i \neq \hat{s}_i$. Denoting the number of elements in $\eta$ by $l_\eta$, (5.13) can be put in an instructive form

$$P_2(\mathbf{s}_l, \hat{\mathbf{s}}_l) \leq \frac{(\,(1+K)e^{-K})^{l_\eta}}{(\frac{1}{4N_0})^{l_\eta} d_P^2(l_\eta)}, \quad (5.14)$$

where

$$d_P^2(l_\eta) = \prod_{i\in\eta} |s_i - \hat{s}_i|^2.$$

is the squared product distance of signals $s_i \neq \hat{s}_i$ along the error event $(\mathbf{s}_l, \hat{\mathbf{s}}_l)$ path. Using the terminology of [52], $l_\eta$ is called the *effective length* of the error event $(\mathbf{s}_l, \hat{\mathbf{s}}_l)$.

**No Channel State Information**

For a fading channel without the channel state information, i.e., $\hat{a}_i = 1$, the metric $m(r_i, s_i)$ is

$$m(r_i, s_i) = -|r_i - s_i|^2.$$

Similar to the case with CSI the conditional pairwise error probability is upper-bounded as

$$P_2(\mathbf{s}_l, \hat{\mathbf{s}}_l|\mathbf{a}_l) \leq \prod_{i=1}^{l} E_{n_i}[\exp(\lambda\{|r_i - s_i|^2 - |r_i - \hat{s}_i|^2\})].$$

The right-hand side of the above equation can be simplified by using the results of Appendix A. Setting $|s_i| = |\hat{s}_i|$ in (A.13), the conditional error probability for MPSK signals can be expressed as

$$P_2(\mathbf{s}_l, \hat{\mathbf{s}}_l|\mathbf{a}_l) \leq \prod_{i=1}^{l} \exp(-\lambda(a_i - N_0\lambda)|s_i - \hat{s}_i|^2). \tag{5.15}$$

Denormalizing the Chernoff parameter $\lambda$, that is, replacing $\lambda$ by $\frac{\lambda}{N_0}$ in (5.15) we obtain

$$P_2(\mathbf{s}_l, \hat{\mathbf{s}}_l|\mathbf{a}_l) \leq \prod_{i=1}^{l} \exp(-\frac{\lambda}{N_0}(a_i - \lambda)|s_i - \hat{s}_i|^2). \tag{5.16}$$

Note that unlike the case with CSI, (5.16) cannot be optimized over $\lambda$ to yield a constant value for this parameter. In this case we defer the optimization procedure after averaging over the fading amplitude distribution.

Averaging (5.16) over the pdf of (5.4), the pairwise error probability is upper-bounded as

$$P_2(\mathbf{s}_l, \hat{\mathbf{s}}_l) \leq \min_{\lambda} \prod_{i=1}^{l} \exp(\frac{\lambda^2}{N_0}|s_i - \hat{s}_i|^2)\frac{e^{-K}}{\pi}\int_0^{\pi}(1 - \sqrt{\pi}\text{erfc}(\nu)\text{e}^{\nu^2}\nu\,)d\theta, \tag{5.17}$$

where $\nu$ is defined by (A.15) as

$$\nu = \frac{\frac{\lambda}{2N_0}|s_i - \hat{s}_i|^2}{\sqrt{1+K}} - \sqrt{K}\cos(\theta). \tag{5.18}$$

The pairwise error probability of (5.17) can be simplified at high signal-to-noise ratios by noticing that the first term of (5.18) is dominant and $\nu$ becomes independent of $\theta$. In this case, the evaluation of the integral becomes trivial and after some mathematical manipulations, (5.18) is simplified to [47] (see Problem 5.2)

$$P_2(\mathbf{s}_l,\ \hat{\mathbf{s}}_l) \leq \frac{\exp\left(\frac{\lambda^2}{N_0}\sum_{i\in\eta}|s_i-\hat{s}_i|^2\right)\ (1+K)^{l_\eta}e^{-l_\eta K}}{2^{l_\eta}\lambda^{2l_\eta}(\frac{1}{2N_0})^{2l_\eta}d_P^4(l_\eta)}.$$

Optimizing the Chernoff parameter $\lambda$ provides a tighter upper bound for $P_2(\mathbf{s}_l,\ \hat{\mathbf{s}}_l)$ [47]

$$P_2(\mathbf{s}_l,\ \hat{\mathbf{s}}_l) \leq \frac{(2e/l_\eta)^{l_\eta}}{(1/N_0)^{l_\eta}} \cdot \frac{\left(\sum_{i\in\eta}|s_i-\hat{s}_i|^2\right)^{l_\eta}}{d_P^4(l_\eta)} \cdot (1+K)^{l_\eta}e^{-l_\eta K},$$

with the optimum value of $\lambda$ such that

$$\lambda_{\mathrm{opt}}^2 = \frac{l_\eta}{(\frac{1}{N_0})\sum_{i\in\eta}|s_i-\hat{s}_i|^2}.$$

### 5.1.2 Code Design Criteria

The foregoing analysis leads to a set of design parameters for good trellis codes over fading channels. To characterize these parameters we consider the error event probability of TCM on fading channels as a performance measure. For this purpose we choose the case when the perfect channel state information is available at the receiver. This discussion can also be extended to the other cases.

A union bound on the error event probability $P_e$ is presented in Chapter 4 (see (4.8)). To evaluate this upper bound on a Rician fading channel the pairwise error probability $P_2(\mathbf{s}_l,\ \hat{\mathbf{s}}_l)$ is replaced by (5.12) or at high signal-to-noise ratios (SNR's) by (5.14). At high SNR's the upper bound on $P_e$ can be put in the following form

$$P_e \leq \sum_{l_\eta}\sum_{d_P^2(l_\eta)} \alpha(l_\eta,\ d_P^2(l_\eta))\frac{(\,(1+K)e^{-K})^{l_\eta}}{(\frac{1}{4N_0})^{l_\eta}d_P^2(l_\eta)}, \tag{5.19}$$

where $\alpha(l_\eta,\ d_P^2(l_\eta))$ is the average number of code sequences having the effective length $l_\eta$ and the squared product distance $d_P^2(l_\eta)$. Among all terms in (5.19) the one with the smallest effective length $l_\eta$ and the smallest product distance $d_P^2(l_\eta)$ dominates the error event probability at high signal-to-noise ratios. Denoting $\min(l_\eta)$ by $L$ and the corresponding squared product distance by $d_P^2(L)$, the error event probability is asymptotically approximated as

$$P_e \simeq \alpha(L,\ d_P^2(L))\frac{(\,(1+K)e^{-K})^L}{(\frac{1}{4N_0})^L\ d_P^2(L)}. \tag{5.20}$$

The parameter $L$ is called the *effective length* of the code [52].

We observe that the error event probability asymptotically varies with $L$th power of SNR. This effect is similar to what is normally achieved with a *time diversity* technique[3]. Because of this similarity the parameter $L$ is also called the *time diversity* of the TCM code.

It is clear from (5.20) that primary design parameters of good TCM codes for fading channels are the effective length of the code (time diversity) $L$ and the minimum squared product distance $d_P^2(L)$. This is in contrast with the AWGN channel where the free Euclidean distance is considered as a main design parameter. Thus, codes designed based on the AWGN channel design criterion can be quite poor for fading channels and conversely suboptimal designs for the AWGN channel can be suitable choices for fading channels. For example, (5.20) shows that parallel transitions are most harmful in a fading channel, because they limit the effective length of the code to one. On the other hand, some of the good TCM codes, designed for the AWGN channel, have parallel transitions (see Tables 4.1 and 4.2). The design of good TCM schemes for fading channels will be addressed in the sequel, but we pause here to state some points.

The design criteria, i.e., maximizing the effective length of the code and the minimum product distance, is of primary importance at high signal-to-noise ratios for Rician channels with a small value of $K$ or Rayleigh fading channels[4]. For large values of $K$ the effect of the free Euclidean distance on the performance of TCM becomes significant. For instance, the upper bound on the pairwise error probability, given in (5.12), for large values of $K$, that is $K \gg \frac{1}{4N_0}|s_i - \hat{s}_i|^2$, becomes[5]

$$
\begin{aligned}
P_2(\mathbf{s}_l, \hat{\mathbf{s}}_l) &\leq \prod_{i=1}^{l} \exp(-\frac{1}{4N_0}|s_i - \hat{s}_i|^2) \\
&= \exp(-\frac{1}{4N_0}\sum_{i=1}^{l}|s_i - \hat{s}_i|^2) \\
&= \exp(-\frac{1}{4N_0} d_E^2(\mathbf{s}_l, \hat{\mathbf{s}}_l)\ ), \qquad (5.21)
\end{aligned}
$$

where $d_E(\mathbf{s}_l, \hat{\mathbf{s}}_l)$ is the Euclidean distance between code sequences $\mathbf{s}_l$ and $\hat{\mathbf{s}}_l$. Note that (5.21) is the same as the pairwise error probability obtained for the AWGN channel by using the Bhattacharyya bound (see (4.11)).

A similar expression for the pairwise error probability can be obtained at low signal-to-noise ratios by neglecting $\frac{1}{4N_0}|s_i - \hat{s}_i|^2$ with respect to $K+1$, showing that the dominant factor in the performance of TCM at low SNR's is the free

[3] This technique involves sending a symbol several times over the channel, in the hope that not all of the transmissions will be subjected to deep fades [17] (see Section 2.5).

[4] Note that a Rayleigh fading channel is a special case of a Rician channel with $K = 0$.

[5] Note that $K = \infty$ corresponds to the no fading case or to the AWGN channel.

Euclidean distance rather than $L$ (see Problem 5.3). This is not surprising and can be explained as follows. Between the fading process and the AWGN, the latter is more effective at low signal-to-noise ratios, and hence dominates the bit error probability. To combat the additive noise component, we have to increase the Euclidean distance between the coded sequences. However, at high signal-to-noise ratios the effect of the channel fading amplitude is dominant. In this case, increasing the time diversity (effective length) of the code has long been known [17] as a remedy.

The foregoing discussion leads to the conclusion that at high signal-to-noise ratios and for small values of $K$ maximizing the effective length of the code $L$ and the minimum squared product distance $d_P^2(L)$ is the main objective for the code design for fading channels. A simple argument shows that maximizing the minimum product distance for codes with small effective lengths is more effective than for codes with large effective lengths.

Consider a TCM scheme with an effective length $L$ and a minimum product distance $d_{P_1}^2(L)$. Suppose that this code is redesigned to provide a minimum product distance $d_{P_2}^2(L)$ with the same $L$. Using (5.20), the increase in coding gain due to increasing the minimum product distance from $d_{P_1}^2(L)$ to $d_{P_2}^2(L)$ is obtained as

$$\Delta g_\infty \equiv \mathrm{SNR}_1 - \mathrm{SNR}_2|_{P_{e_1}=P_{e_2}} = \frac{10}{L}\log\frac{d_{P_2}^2(L)}{d_{P_1}^2(L)}\cdot\frac{\alpha_1}{\alpha_2},$$

where the coefficient in (5.20) is simply shown by $\alpha_1$ and $\alpha_2$ for schemes 1 and 2, respectively.

It is observed that the increase in the coding gain due to the increase in the minimum product distance is inversely proportional to the effective length $L$. For instance, doubling the minimum product distance of a code with a time diversity of 2 provides 1.5 dB more coding gain (assuming the same $\alpha$ for both schemes), while the same increment in the minimum product distance yields only 0.75 dB more coding gain for an effective length 4.

### 5.1.3 Maximum Effective Length of TCM Codes

In the following we show that the maximum effective length that a conventional TCM code[6] can achieve is limited by the number of input bits $m$ and the number of encoder memories $\nu$.

As stated earlier, the effective length of a TCM code, $L$, is the minimum number of distinct symbols between any two codewords. This parameter is limited by the error event which has the smallest length among all possible error events. The maximum $L$ is achieved when the branches of such error events corresponding to two different paths are labeled by distinct symbols.

[6] The term "conventional" is used for codes whose trellis branches have one symbol. The TCM codes with more than one symbol per trellis branch, i.e., MTCM schemes, will be considered separately in the sequel.

Consider a rate $m/(m+1)$ conventional TCM scheme with $\nu$ memories in its encoder. The trellis diagram of such a code has $2^\nu$ states and $2^m$ branches leaving each state during each symbol interval. Parallel transitions occur if $\nu$ is less than $m$, or when the trellis encoder has $(m - \tilde{m}) > 0$ uncoded bits (see Figure 4.7). Parallel transitions limit the effective length of the code to one and must be avoided on fading channels.

Two code sequences constitute an error event of length $l$ if they start from the same state and rejoin at the same state for the first time after $l$ intervals. The smallest value of $l$ is $\lfloor \frac{\nu}{m} \rfloor$, where $\lfloor (\cdot) \rfloor$ denotes the integer part of the number $(\cdot)$. To see this, we consider the two code sequences diverging from a common state at time $n$ because of different input sequences. Suppose that the input bits differ only in one position at time $n$ and remain the same for other time intervals. Since $m$ input bits are shifted into $\nu$ memories at each time interval, this differing bit can remain in the encoder memory for *at least* $\lfloor \frac{\nu}{m} \rfloor$ time intervals and can cause the two code sequences to remerge at a common state after $\lfloor \frac{\nu}{m} \rfloor + 1$ times.

The maximum effective length is achieved if the branches of the error events are labeled by different symbols. Thus,

$$L_{\max} = \lfloor \frac{\nu}{m} \rfloor + 1. \tag{5.22}$$

Note that $L$ will always be equal to $L_{\max}$ if the cardinality of the signal set is $2^{m+\nu}$. In this case, each branch of the trellis diagram can be associated with a distinct symbol. However, if the cardinality of the signal set is less than $2^{m+\nu}$ then the maximum effective length might not be achieved.

In Tables 5.1 and 5.2 we list the effective length for the set of rate 2/3 8PSK and rate 3/4 16PSK trellis codes designed by Ungerboeck (see Tables 4.1 and 4.2) for the AWGN channel. It is observed that some of these codes have parallel transitions, and hence, provide an effective length of one. The 4-state rate 2/3 8PSK TCM scheme and all of rate 3/4 16PSK TCM schemes can be given as examples. Furthermore, most of the codes with no-parallel transitions do not achieve the maximum effective length.

## 5.2 DESIGN OF TCM FOR FADING CHANNELS

In this section we present the TCM schemes constructed based on the fading channel design criteria. We start with the design of a 4-state 8PSK TCM. The construction of the code is performed by introducing some design rules [53] rather than by using an exhaustive computer search.

Table 5.1: Effective Length of the Rate 2/3 Trellis-Coded 8PSK Schemes Shown in Table 4.1.

| Number of Memories $\nu$ | Number of States $2^\nu$ | Number of Coded Bits $\tilde{m}$ | Number of Parallel Paths $2^{m-\tilde{m}}$ | Effective Length $L$ | Maximum Effective Length $L_{\max}$ |
|---|---|---|---|---|---|
| 2 | 4 | 1 | 2 | 1 | 2 |
| 3 | 8 | 2 | 1 | 2 | 2 |
| 4 | 16 | 2 | 1 | 3 | 3 |
| 5 | 32 | 2 | 1 | 2 | 3 |
| 6 | 64 | 2 | 1 | 3 | 4 |
| 7 | 128 | 2 | 1 | 4 | 4 |
| 8 | 256 | 2 | 1 | 3 | 5 |

Table 5.2: Effective Length of the Rate 3/4 Trellis-Coded 16PSK Schemes Shown in Table 4.2.

| Number of Memories $\nu$ | Number of States $2^\nu$ | Number of Coded Bits $\tilde{m}$ | Number of Parallel Paths $2^{m-\tilde{m}}$ | Effective Length $L$ | Maximum Effective Length $L_{\max}$ |
|---|---|---|---|---|---|
| 2 | 4 | 1 | 4 | 1 | 1 |
| 3 | 8 | 1 | 4 | 1 | 2 |
| 4 | 16 | 1 | 4 | 1 | 2 |
| 5 | 32 | 1 | 4 | 1 | 2 |
| 6 | 64 | 1 | 4 | 1 | 3 |
| 7 | 128 | 1 | 4 | 1 | 3 |
| 8 | 256 | 2 | 2 | 1 | 3 |

### 5.2.1 4-State Rate 2/3 8PSK TCM Schemes

Ungerboeck's 4-state 8PSK TCM with 3 dB gain over uncoded QPSK on the AWGN channel was considered as an example in Chapter 4 ( see Figure 4.4). This code has been designed for the AWGN channel [18] and its trellis diagram contains parallel transitions (see Figure 4.5) implying that a one-step error event can occur. This limits the achievable effective length to one. Thus, if this code is used for fading channels with interleaving, the asymptotic steepest rate of descent of bit (event) error probability will be inversely proportional to $E_s/N_0$.

To increase the effective length of the code, parallel transitions must be avoided. For Rayleigh fading channels, Wilson and Leung [48] suggest a 4-state 8PSK TCM without parallel transitions. The trellis diagram of this code is shown in Figure 5.3. We refer to this code as the Wilson-Leung code. As is shown, the trellis diagram of the Wilson-Leung code is fully connected and results in shortest error event paths of length two, i.e., asymptotically the rate of descent of $P_b$ or $P_e$ with $E_s/N_0$ follows the inverse square law. This is the maximum achievable time diversity with 4-state 8PSK TCM schemes (see (5.22)). The coding gain of this code in the AWGN channel is 1.1 dB which is 1.9 dB inferior to the Ungerboeck code (see Problem 5.5).

This code has been designed to increase the shortest error event path length. However, the product of the squared branch distances along this path has not been optimized. Figure 5.3 indicates that this parameter is limited by $d_P^2(s_0 s_0 s_0,\ s_2 s_0 s_1) = \delta_0^2 \cdot \delta_1^2 = 1.172 E_s^2$ which is the product of the squared distances between unlike pairs of the paths $(s_0, s_0, s_0)$ and $(s_2, s_0, s_1)$. Note that this is an error event of actual length three but effective length two. In the following we shall introduce new rules [53] to avoid such paths and we shall design a new 4-state TCM based on maximizing $d_P^2$.

#### Code Design Rules

The trellis diagram of the 4-state 8PSK TCM scheme without parallel transitions is fully connected. This allows us to represent the signal associated with transitions between states of consecutive stages by a $4 \times 4$ matrix $\mathbf{B}$, whose $ij$th element is denoted by $b_{ij}$, which represents the signal associated with the path from state $i$ at stage $k$ to state $j$ at stage $k+1$ of the trellis diagram. Also, note that the elements of the $i$th row indicate the associated signals with paths diverging from state $i$ and the elements of the $j$th column show the associated signals with paths remerging to state $j$.

Using set partitioning, the 8PSK signal set can be partitioned into two subsets $\mathbf{A}_0 = (s_0, s_2, s_4, s_6)$ and $\mathbf{A}_1 = (s_1, s_3, s_5, s_7)$ with intra-set distances $\delta_1$ and $\delta_3$ (see Figure 4.6). Signal points from either of the subsets $\mathbf{A}_0$ and $\mathbf{A}_1$ are assigned to the alternative rows of matrix $\mathbf{B}$. To assign signal points to the elements of matrix $\mathbf{B}$ the following rules are utilized. To express these rules we define the "state difference" as the *number of bits in which two states differ.*

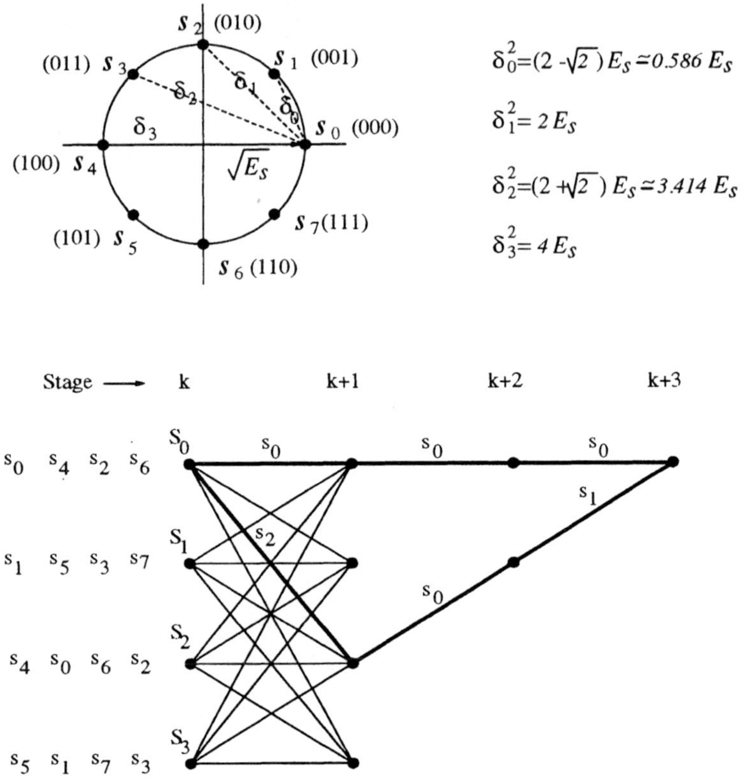


Figure 5.3: Trellis diagram of a 4-state 8PSK TCM scheme without parallel transitions (Wilson-Leung code).

For instance, the "state difference" 01 and 10 can be considered to be two.

- **Rule 1** Elements of each row of matrix **B** are associated with signals from subsets $\mathbf{A}_0$ or $\mathbf{A}_1$ such that the distance between a pair of branches diverging from one state to two adjacent states with "state differences" of two or one is $\delta_1$ or $\delta_3$, respectively. This rule is shown schematically in Figure 5.4.

This rule guarantees that the distance between each pair of diverging paths from state $i$ is at least $\delta_1$. The same rule cannot be applied for assigning the signal points to the elements in one column, since there will be at least two elements in one column whose distances differ by only $\delta_0$. In other words, among the remerging branches at state $j$ there will be at least one pair with distance $\delta_1$. If we assign the signals with distance $\delta_0$ to the pair of branches remerging at state $j$ from the two states with a "state difference" of one, then the minimum product of squared branch distances of the error events of actual

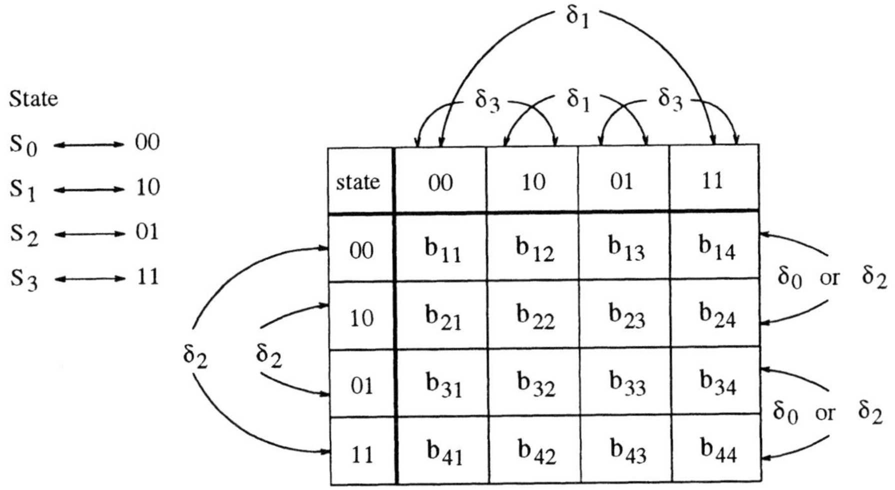


Figure 5.4: The schematic representation of Rule 1 and Rule 2.

length two (diverging at stage $k$ and remerging at stage $k+2$) will be limited to $\delta_0^2 \cdot \delta_3^2 = 2.344E_s^2$ which is greater than $\delta_0^2 \cdot \delta_1^2 = 1.172E_s^2$. Thus,

- **Rule 2** In assigning signal points to each column of matrix **B**, the signals with distances $\delta_2$ are associated with branches remerging at state $j$ (column $j$) from two adjacent states with "state difference" of two. The pair of paths remerging to state $j$ from two states with "state difference" of one may be associated with signals with distances $\delta_0$ or $\delta_2$.

Using Rule 1 and Rule 2 the minimum product of the squared distances of the error events of actual length two will be

$$d_P^2 = \min\left(\delta_0^2 \cdot \delta_3^2,\ \delta_1^2 \cdot \delta_2^2\right) = \delta_0^2 \cdot \delta_3^2 = 2.344E_s^2.$$

To limit $d_P^2$ to the error event path with actual length two the third rule is introduced as

- **Rule 3** The same signals can be used in two different columns whose associated states have "state difference" of two. In other words, if a signal appears in column 1, it can only be used in column 4, and the signals of column 2 can only be repeated in column 3.

It can be seen that each pair of branches with like signals between stage $k+1$ and stage $k+2$ are connected to two adjacent states with "state difference" two. Furthermore, according to the second rule, remerging branches from these two adjacent states at stage $k+2$ to a state at stage $k+3$ have intra-set distance $\delta_2$. Therefore, Rule 3 guarantees that the product of the squared branch distances

| Step 1 | Step 2 | Step 3 | Step 4 |
|---|---|---|---|
| $s_0$ $s_4$ $s_2$ $s_6$ | $s_0$ $s_4$ $s_2$ $s_6$ | $s_0$ $s_4$ $s_2$ $s_6$ | $s_0$ $s_4$ $s_2$ $s_6$ |
| X X X X | X X X X | $s_3$ $s_7$ $s_1$ $s_5$ | $s_3$ $s_7$ $s_1$ $s_5$ |
| X X X X | $s_6$ $s_2$ $s_4$ $s_0$ | $s_6$ $s_2$ $s_4$ $s_0$ | $s_6$ $s_2$ $s_4$ $s_0$ |
| X X X X | X X X X | X X X X | $s_5$ $s_1$ $s_7$ $s_3$ |

Figure 5.5: The 4-state 8PSK TCM code design procedure of Example 5.1.

for the shortest error events of length two and actual length three are always $\delta_1^2 \cdot \delta_2^2 > \delta_0^2 \cdot \delta_3^2$.

### Code Design Examples

Using the above code design rules we consider the design of two 4-state TCM schemes in the following examples.

**Example 5.1** Figure 5.5 shows an example of the code design procedure in 4 steps. In the first step the signals from subset $\mathbf{A}_0$ are assigned to the first row of matrix $\mathbf{B}$ according to Rule 1. Using the third rule the elements of the third row can be obtained in step 2. For the signals of the second and the fourth rows the subset $\mathbf{A}_1$ must be used. At this stage we may use signal points $s_1$ or $s_3$ as the first element of the second row according to Rule 2. Choosing $s_3$, the second and the third elements of this row can be found as $s_7$ and $s_1$, respectively, according to Rule 1 and Rule 1 + Rule 2. Finally, the fourth row of matrix $\mathbf{B}$ can be obtained from the second row by using the third rule. This completes the code design. The trellis diagram of this code is shown in Figure 5.6.

The free Euclidean distance of this code can be obtained simply by using the design rules

$$d_{\text{free}}^2 = \min\,(\delta_0^2 + \delta_3^2,\ \delta_1^2 + \delta_2^2,\ \delta_0^2 + \delta_0^2 + \delta_1^2) = 3.172E_s,$$

which implies that the coding gain of this code over uncoded QPSK on the AWGN channel is 2 dB which is only 1dB inferior to the 4-state Ungerboeck code. However, this new scheme provides 0.9 dB more coding gain compared to the Wilson-Leung scheme over the AWGN channel. The new scheme outperforms the Ungerboeck code over fading channels. Also, the coding gain of the new scheme is more than that of the Wilson-Leung code over these channels. This can be justified by noting that $d_P^2(L)$ of the new scheme is *doubled.*

**Example 5.2** (see Problem 5.6). Consider the code design procedure of Example 5.1 (see Figure 5.5). Choosing the signal points $s_1$ in step 2 (instead of $s_3$) and

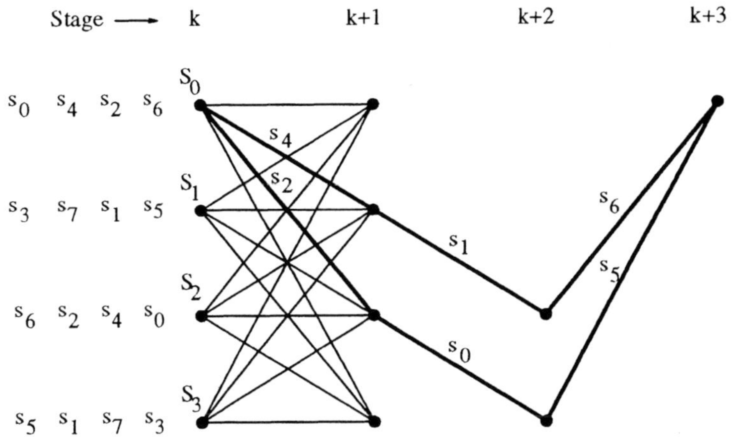


Figure 5.6: Trellis diagram of the 4-state 8PSK TCM scheme of Example 5.1.

repeating the same procedure as Example 5.1 another 4-state TCM code will be obtained. The trellis diagram of this code is shown in Figure 5.7. This code has the same $d^2_{\text{free}}$ and $d^2_P(L)$ as the code of Example 5.1.

## Encoder Realization

The encoder of the new schemes may be realized by using the analytical description of the TCM codes (see Chapter 4). The following example illustrates the analytical description of the 4-state TCM scheme, designed in Example 5.1.

**Example 5.3** Using the trellis diagram of Figure 5.6 the following relations can be obtained between the phase of trellis-coded 8PSK signals and the input and memory bits:

$$\begin{array}{ll}
\theta_0 = f(\ 1,\ \ 1,\ \ 1,\ \ 1) = \frac{\pi}{8} & \theta_8 = f(\ 1,\ \ 1,\ \ 1,\ -1) = -\frac{7\pi}{8} \\
\theta_1 = f(-1,\ \ 1,\ \ 1,\ \ 1) = -\frac{7\pi}{8} & \theta_9 = f(-1,\ \ 1,\ \ 1,\ -1) = \frac{\pi}{8} \\
\theta_2 = f(\ 1,\ -1,\ \ 1,\ \ 1) = \frac{5\pi}{8} & \theta_{10} = f(\ 1,\ -1,\ \ 1,\ -1) = -\frac{3\pi}{8} \\
\theta_3 = f(-1,\ -1,\ \ 1,\ \ 1) = -\frac{3\pi}{8} & \theta_{11} = f(-1,\ -1,\ \ 1,\ -1) = \frac{5\pi}{8} \\
\theta_4 = f(\ 1,\ \ 1,\ -1,\ \ 1) = \frac{3\pi}{8} & \theta_{12} = f(\ 1,\ \ 1,\ -1,\ -1) = -\frac{5\pi}{8} \\
\theta_5 = f(-1,\ \ 1,\ -1,\ \ 1) = -\frac{5\pi}{8} & \theta_{13} = f(-1,\ \ 1,\ -1,\ -1) = \frac{3\pi}{8} \\
\theta_6 = f(\ 1,\ -1,\ -1,\ \ 1) = \frac{7\pi}{8} & \theta_{14} = f(\ 1,\ -1,\ -1,\ -1) = -\frac{\pi}{8} \\
\theta_7 = f(-1,\ -1,\ -1,\ \ 1) = -\frac{\pi}{8} & \theta_{15} = f(-1,\ -1,\ -1,\ -1) = \frac{7\pi}{8}.
\end{array} \tag{5.23}$$

Replacing (5.23) as a vector form ($\theta$ instead of $\mathbf{f}$) in (4.7) and solving for $\mathbf{d}$, the desired result is obtained as

$$\theta = (-b_3 + 4b_1b_4 - 2b_2b_3b_4)\frac{\pi}{8}. \tag{5.24}$$

The realization of the TCM encoder based on (5.24) is shown in Figure 5.8(a).

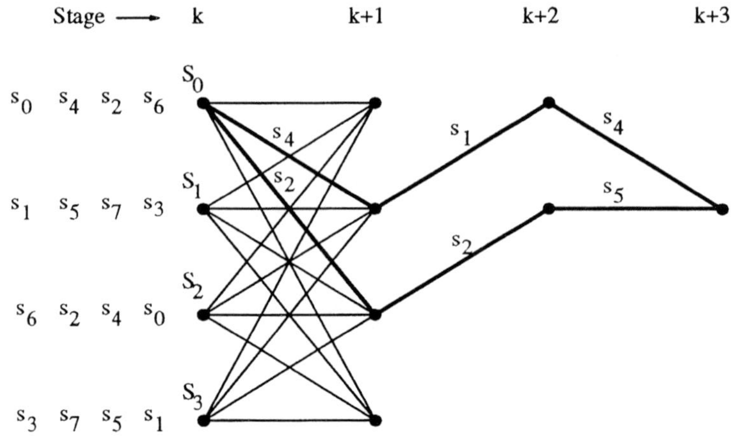


Figure 5.7: Trellis diagram of the 4-state 8PSK TCM scheme of Example 5.2.

The encoder of the TCM of Example 5.1 can also be realized by using the Ungerboeck representation (a linear convolutional encoder followed by a mapping rule). Indeed, it is shown in [54], that if the analytical expression of a rate $m/(m+1)$ TCM scheme has only $m+1$ terms (codes with minimal transmitter complexity) then a convolutional encoder realization can be found by using a simple procedure.

**Example 5.4** The scheme described by (5.24) is of minimal complexity, and hence, can be converted to the Ungerboeck representation. Assume that natural mapping is used in associating the outputs of the convolutional encoder to the signal points. Denoting the input and output bits of the convolutional encoder at time $n$ by $a_n^{(i)}$, $(i = 1,2)$ and $c_n^{(j)}$, $(j = 1,2,3)$, respectively, the following associations are found:

$$\begin{aligned} c_n^{(1)} &= a_{n-1}^{(1)} \oplus a_n^{(2)}, \\ c_n^{(2)} &= a_n^{(1)} \oplus a_{n-1}^{(1)} \oplus a_{n-1}^{(2)}, \\ c_n^{(3)} &= a_{n-1}^{(2)}, \end{aligned} \tag{5.25}$$

where $\oplus$ shows the modulo-2 addition. Equation (5.25) defines the underlying convolutional code. It is realized in Figure 5.8(b).

Using (5.25) the generator polynomial matrix of this code is found to be

$$\mathbf{G}(D) = \begin{bmatrix} D & D+1 & 0 \\ 1 & D & D \end{bmatrix}.$$

With a similar procedure, presented in Example 3.3, the parity-check polynomial matrix is obtained as

$$\mathbf{H}(D) = [D^2 + D,\ D^2,\ D^2 + D + 1],$$

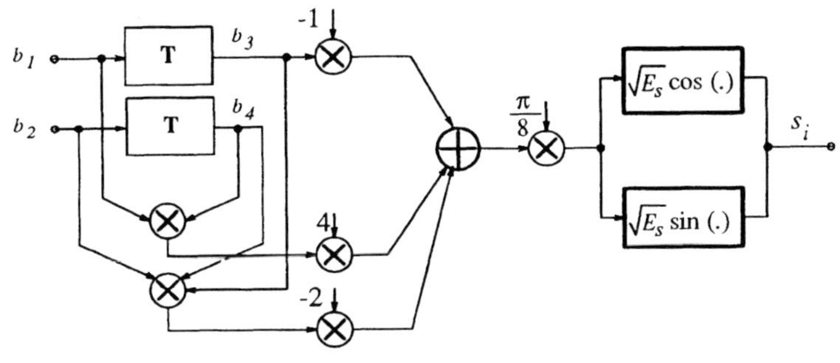


( a )

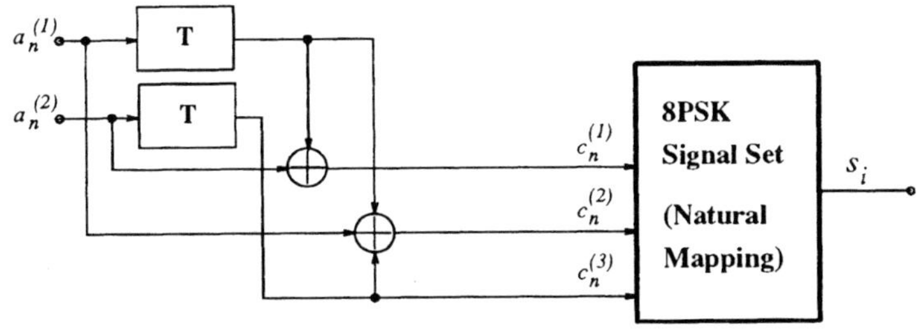


( b )

Figure 5.8: Two different encoder realizations of the TCM scheme of Example 5.1: a) analytical representation, b) Ungerboeck representation.

which can be put in octal form

$$\begin{array}{lllllll} H_1(D) & = & D^2 + D & = & (110)_2 & = & (6)_8 \\ H_2(D) & = & D^2 & = & (100)_2 & = & (4)_8 \\ H_3(D) & = & D^2 + D + 1 & = & (111)_2 & = & (7)_8. \end{array} \tag{5.26}$$

Using (5.26) an equivalent systematic encoder with feedback for this TCM scheme can also be realized.

### 5.2.2 Trellis Codes with Good Effective Lengths

Using an exhaustive search, Schlegel and Costello [52], [55] constructed a class of rate 2/3 8PSK trellis codes with a maximal effective length for $\nu \geq 3$. Like Ungerboeck [18] their search is restricted to systematic convolutional codes followed by natural mapping to an 8PSK signal set. The search is performed by imposing a set of rejection rules based on the fading channel code design criteria. We list their findings in Table 5.3 and for details about the construction method refer the reader to [55, Chap. 7]. The minimum squared product

Table 5.3: Rate 2/3 8PSK Trellis Codes Designed for Fading Channels [52]

| Number of States $2^\nu$ | $H_1$ | $H_2$ | $H_3$ | Normalized Free Euclidean Distance $d^2_{free}/E_s$ | Effective Length $L$ | Normalized minimum Product Distance $d^2_P(L)/(E_s)^L$ |
|---|---|---|---|---|---|---|
| 8 | 04 | 02 | 11 | 4.59(4.59)* | 2(2)* | 8.00(8.00)* |
| 16 | 16 | 04 | 23 | 5.17(5.17) | 3(3) | 4.68(4.68) |
| 32 | 36 | 14 | 43 | 5.17(5.76) | 3(2) | 16.00(8.00) |
| 64 | 154 | 036 | 103 | 4.34(6.34) | 4(3) | 8.00(16.00) |
| 128 | 314 | 076 | 223 | 4.93(6.59) | 4(4) | 8.00(2.75) |
| 256 | 164 | 336 | 673 | 6.34(7.52) | 5(3) | 5.49(16.00) |
| 512 | 244 | 756 | 1413 | 6.34(7.52) | 5(3) | 18.75(16.00) |
| 1024 | 504 | 1676 | 3303 | 6.34(7.52) | 5(4) | 128.00(32.00) |
| 2048 | 1264 | 3436 | 6403 | 7.52 | 6 | 10.98 |

* The numbers inside parentheses are the respective parameters corresponding to the Ungerboeck codes partly given in Table 4.1.

distances tabulated in Table 5.3 are given in a normalized form by dividing the squared Euclidean distance corresponding to each error event branch by $E_s$. The parity-check polynomials are represented in octal form.

Also included in this table are the effective length and the normalized squared free Euclidean and minimum product distances of the Ungerboeck codes. These parameters are included inside parentheses. The codes obtained for $\nu = 3$ and $\nu = 4$ are the same as the Ungerboeck codes, hence these codes are good for both the AWGN and fading channels. However, trellis codes with good effective lengths for $\nu \geq 5$ are suboptimal on the AWGN channel because of their smaller free Euclidean distances compared to the corresponding Ungerboeck codes.

A set of good rate 2/3 8PSK and rate 3/4 16PSK trellis codes was constructed by Du and Vucetic in [56], based on fading channel code design criteria. They adopt *Gray mapping* in assigning the output of the underlying convolutional code to the signal points in order to improve the bit error probability. The results of their exhaustive search for 8PSK TCM are listed in Table 5.4. The parameters of the 8PSK codes constructed by Schlegel and Costello are also included for comparison in this table.

Table 5.4: Rate 2/3 8PSK Trellis Codes† Designed for Fading Channels [56]

| Number of States $2^\nu$ | $H_1$ | $H_2$ | $H_3$ | Normalized Free Euclidean Distance $d^2_{free}/E_s$ | Effective Length $L$ | Normalized minimum Product Distance $d^2_P(L)/(E_s)^L$ |
|---|---|---|---|---|---|---|
| 4 | 05 | 03 | 07 | 3.17 | 2 | 2.34 |
| 32 | 73 | 67 | 45 | 5.17(5.17)* | 3(3)* | 32.00(16.00)* |
| 128 | 235 | 201 | 333 | 5.76(4.93) | 4(4) | 64.00(8.00) |
| 256 | 621 | 573 | 417 | 6.93(6.34) | 5(5) | 4.69(5.49) |
| 512 | 1767 | 1503 | 1065 | 7.52(6.34) | 5(5) | 16.00(18.75) |

† The Gray mapping is used in assigning the outputs of underlying convolutional codes of this table to the 8PSK signal set.

* The numbers inside parentheses are the respective parameters corresponding to the Schlegel-Costello codes given in Table 5.3.

### 5.2.3 Multiple Trellis-Coded Modulation (MTCM) for Fading Channels

In the preceding sections we considered the conventional trellis codes constructed for fading channels. In these schemes one symbol is assigned to each branch of the code trellis. In constructing conventional TCM for fading channels the effective length of the code increased by preventing parallel transitions and increasing the number of encoder states. In this section we address an alternative approach for increasing the effective length of the code by assigning more than one symbol per trellis branch. This technique was proposed by Divsalar and Simon [43] and is referred to as *multiple trellis-coded modulation* (MTCM).

A general block diagram of an MTCM encoder is shown in Figure 5.9. A group of $n_a$ binary input bits is encoded by a trellis encoder. The $n_c$ binary output bits of the encoder are grouped into $k^*$ $m_i$-tuples and each $m_i$-tuple is mapped to a signal point from a $2^{m_i}$-ary signal set. The parameter $k^*$ is called the *multiplicity* of the code and represents the number of coded symbols assigned to each branch in the trellis diagram. The throughput of such a scheme is $n_a/k^*$ which, in general, may not be an integer-valued number unless $n_a$ is chosen as an integer multiple of the multiplicity $k^*$. With such a choice the MTCM scheme can be compared to an uncoded scheme with the same throughput.

To explain the aspects of the MTCM schemes we focus mainly on a special

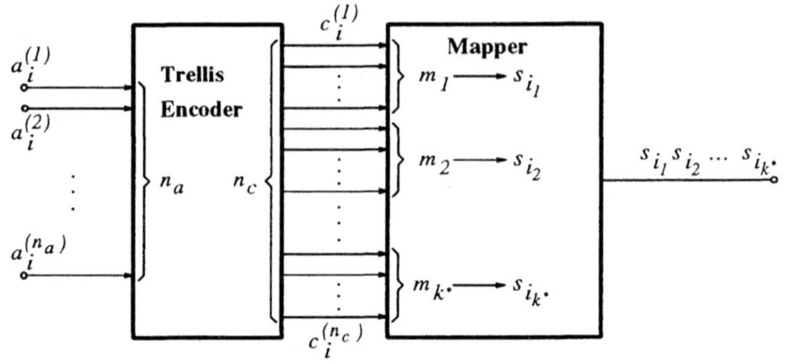


Figure 5.9: A general block diagram of an MTCM scheme.

case where $n_a = mk^*$ and $n_c = (m+1)k^*$. In this case $mk^*$ input bits are encoded to $(m+1)k^*$ bits and $k^*$ groups of $(m+1)$ bits are mapped to signals of a signal set with cardinality $2^{m+1}$. The throughput of such a scheme is $m$ (integer). Before addressing the design of the MTCM schemes for fading channels we consider the following example.

**Example 5.5** The trellis diagram of a two-state rate 4/6 8PSK MTCM scheme with multiplicity $k^* = 2$ is shown in Figure 5.10. For this MTCM scheme, the number of input bits is 4 implying that $2^4 = 16$ branches leave each state. Since the number of states is 2, there are 16/2=8 parallel transitions between trellis states. The parallel transitions are associated with pairs of 8PSK signals in a way that all of them differ from each other in both signal positions. This implies that the effective length of the error events, constrained to the parallel transitions, is 2.

Next we consider the transitions originating from or arriving at a given state. The signal assignment of such transitions is done in such a way that every two possible transitions differ in at least one signal position. This implies that the effective length of error events other than parallel transitions is at least two. Thus, this MTCM scheme achieves an effective length of $L = 2$ which is determined by its multiplicity, i.e., $k^* = 2$.

The foregoing example shows that the multiple trellis-code schemes, in spite of the parallel transitions in their trellis, can provide effective lengths of more than one. The effective length of the MTCM schemes can be improved by increasing the multiplicity and the number of encoder states provided that appropriate signal allocation is done in the code design procedure. In the following we consider a set partitioning method [57] for the design of these scheme based on the fading channel code design criteria.

## Set Partitioning for MTCM

Like conventional TCM schemes the main goal in designing an MTCM

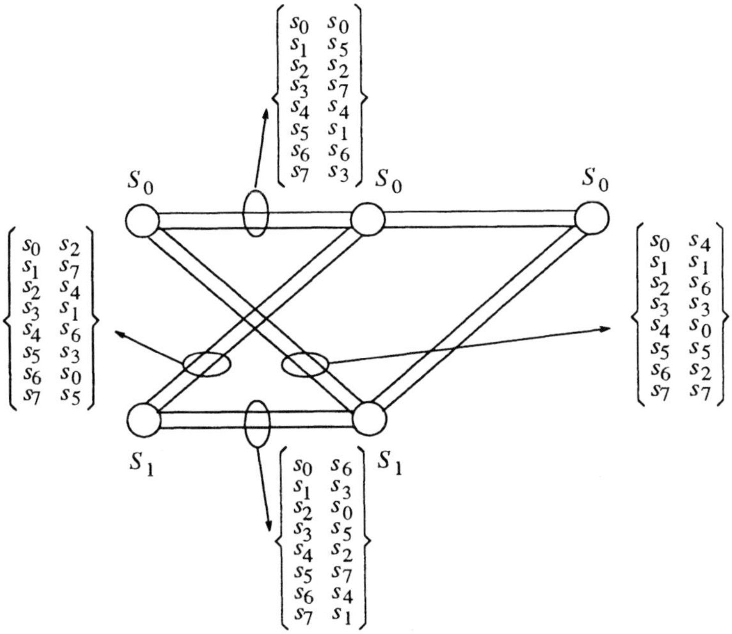


Figure 5.10: Trellis diagram of the two-state rate 4/6 8PSK MTCM with $k^* = 2$ of Example 5.5.

scheme for fading channels is to maximize 1) the effective length of the code $L$, and 2) the minimum product distance $d_P^2(L)$. This goal is achieved by employing a set partitioning technique [57] as explained in the following. For the sake of simplicity we describe the set partitioning procedure for multiplicity of two, i.e., $k^* = 2$ and then we generalize this method for higher multiplicities.

Let $\mathbf{A}$ denote an MPSK signal set with elements $s_k = \sqrt{E_s}e^{(j\frac{2k\pi}{M})}$, $k = 0, 1, \cdots, M-1$ (see Figure 4.1(b)). The set of all possible 2-tuples $(s_k,\ s_l)$ where $s_k$ and $s_l$ are elements of $\mathbf{A}$ is denoted by $\mathbf{A} \times \mathbf{A}$ and is called the *Cartesian product* of $\mathbf{A}$ by itself. The number of elements (cardinality) of $\mathbf{A} \times \mathbf{A}$ is $M^2$. For an MTCM scheme with $k^* = 2$ the associated signals with branches of the trellis diagram are chosen from subsets of $\mathbf{A} \times \mathbf{A}$ which are obtained in the following set partitioning procedure.

At the first level of set partitioning the set $\mathbf{A} \times \mathbf{A}$ is partitioned into $M$ distinct subsets $\mathbf{A} \times \mathbf{B}_k$, $k = 0, 1, \cdots, M-1$, with $M$ elements in each. The elements of $\mathbf{A} \times \mathbf{B}_k$ are ordered 2-tuples with the first and second signals chosen from $\mathbf{A}$ and $\mathbf{B}_k$, respectively. Based on the earlier discussions, this partitioning is carried out in such a way that 1) each of the 2-tuples of $\mathbf{A} \times \mathbf{B}_k$ has distinct signals in the two signal positions and 2) the minimum product of the distances between signals of all pairs of 2-tuples of $\mathbf{A} \times \mathbf{B}_k$ is maximized.

To meet the first of the above objectives the subsets $\mathbf{A}\times\mathbf{B}_k$, $k = 0, 1, \cdots, M-1$, are chosen such that the $l$th 2-tuple of the subset $\mathbf{A}\times\mathbf{B}_k$ is the ordered pair $(s_l,\ s_{\mu l\oplus k})$ where $\mu$ is an odd integer number and $\oplus$ denotes the modulo-$M$ addition. It can be easily shown that for a pair of 2-tuples $(s_{l_1},\ s_{\mu l_1\oplus k})$ and $(s_{l_2},\ s_{\mu l_2\oplus k})$ if the first signals $s_{l_1}$ and $s_{l_2}$ are distinct, i.e., $l_1 \neq l_2$, then the second signals $s_{\mu l_1\oplus k}$ and $s_{\mu l_2\oplus k}$ will also be distinct.

Next we determine the integer $\mu$. The value(s) of $\mu$ is (are) chosen such that the second objective, i.e., maximizing the minimum product distance, is accomplished. Before determining the value(s) of $\mu$ we observe the following property.

**Property 5.1** *The intradistance structure*[7] *of all of the partitions* $\mathbf{A}\times\mathbf{B}_k$, $k = 0, 1, \cdots, M-1$, *is identical.*

**Proof** To prove this property consider a pair of 2-tuples $(s_{l_1},\ s_{\mu l_1\oplus k})$ and $(s_{l_2},\ s_{\mu l_2\oplus k})$ from the subset $\mathbf{A}\times\mathbf{B}_k$. The squared Euclidean distances between the signals of these 2-tuples are obtained as

$$\begin{aligned} d_E^2(s_{l_1},\ s_{l_2}) &= 4E_s\sin^2(\ (l_1-l_2)\frac{\pi}{M}\ ), \\ d_E^2(s_{\mu l_1\oplus k},\ s_{\mu l_2\oplus k}) &= 4E_s\sin^2(\ \mu(l_1-l_2)\frac{\pi}{M}\ ), \end{aligned} \tag{5.27}$$

which are independent of $k$. This can also be seen from observing that each of the $M$ sets $\mathbf{B}_k$, $k = 0, 1, \cdots, M-1$ in the Cartesian products $\mathbf{A}\times\mathbf{B}_k$ can be obtained by an appropriate cyclic shift of the other sets.

□

Based on the above property, it is sufficient to consider only one of the subsets, say, $\mathbf{A}\times\mathbf{B}_0$, in the study of the intradistance structure of the subsets $\mathbf{A}\times\mathbf{B}_k$.

To determine the value(s) of $\mu$ we consider the product of the squared distances between signals of the pair of 2-tuples $(s_{l_1},\ s_{\mu l_1\oplus k})$ and $(s_{l_2},\ s_{\mu l_2\oplus k})$. Using (5.27) this product can be written as

$$\begin{aligned} d_P^2(\mu) &= d_E^2(s_{l_1},\ s_{l_2})\cdot d_E^2(s_{\mu l_1\oplus k},\ s_{\mu l_2\oplus k}) \\ &= 16E_s^2\sin^2(\ (l_1-l_2)\frac{\pi}{M}\ )\cdot\sin^2(\ \mu(l_1-l_2)\frac{\pi}{M}\ ) \\ &= 16E_s^2\sin^2(l\frac{\pi}{M})\cdot\sin^2(\ \mu l\frac{\pi}{M}), \end{aligned} \tag{5.28}$$

where $l_1 - l_2 = l$. Note that, because of the symmetry of the MPSK signal set around the circle, $s_l = s_{l_1-l_2}$ is also an MPSK signal. The desired value(s) of

[7] Intradistance is defined as the minimum distance among the elements belonging to the same subset.

Table 5.5: Solutions of (5.29) for various values of $M$

| $M$ | $\mu^*$ |
|---|---|
| 4 | 1, 3 |
| 8 | 3, 5 |
| 16 | 7, 9 |
| 32 | 7, 9, 23, 25 |
| 64 | 19, 27, 37, 45 |

$\mu$ is (are) obtained by maximizing (5.28). Denoting the desired value(s) of $\mu$ by $\mu^*$, the maximin solution of $\mu^*$ is [57]

$$d_P^2(\mu^*) = \max_{\mu^*=1,3,\cdots,M-1} \left\{ \min_{l=1,3,\cdots,\frac{M}{2}-1} \left[ 16E_s^2 \sin^2(l\frac{\pi}{M}) \cdot \sin^2(\mu^* l\frac{\pi}{M}) \right] \right\}. \tag{5.29}$$

Note that if $\mu^*$ is a solution of (5.29) then its additive inverse, that is, $M - \mu^*$ is also a valid solution. We list the solutions of (5.29) in Table 5.5 for $M = 4$, 8, 16, 32, and 64 [57] (see Problem 5.13).

**Example 5.6** Consider an 8PSK signal set $\mathbf{A} = \{s_0, s_1, \cdots, s_7\}$. The Cartesian product of $\mathbf{A} \times \mathbf{A}$ can be partitioned at the first level of the set partitioning procedure into 8 partitions $\mathbf{A} \times \mathbf{B}_k$, $k = 0, 1, \cdots, 7$, by choosing $\mu^* = 3$ or 5. Specifically, if we choose $\mu^* = 5$ these partitions are obtained as

$$\mathbf{A} \times \mathbf{B}_0 = \begin{bmatrix} s_0 & s_0 \\ s_1 & s_5 \\ s_2 & s_2 \\ s_3 & s_7 \\ s_4 & s_4 \\ s_5 & s_1 \\ s_6 & s_6 \\ s_7 & s_3 \end{bmatrix} \quad \mathbf{A} \times \mathbf{B}_1 = \begin{bmatrix} s_0 & s_1 \\ s_1 & s_6 \\ s_2 & s_3 \\ s_3 & s_0 \\ s_4 & s_5 \\ s_5 & s_2 \\ s_6 & s_7 \\ s_7 & s_4 \end{bmatrix}$$

$$\mathbf{A} \times \mathbf{B}_2 = \begin{bmatrix} s_0 & s_2 \\ s_1 & s_7 \\ s_2 & s_4 \\ s_3 & s_1 \\ s_4 & s_6 \\ s_5 & s_3 \\ s_6 & s_0 \\ s_7 & s_5 \end{bmatrix} \quad \mathbf{A} \times \mathbf{B}_3 = \begin{bmatrix} s_0 & s_3 \\ s_1 & s_0 \\ s_2 & s_5 \\ s_3 & s_2 \\ s_4 & s_7 \\ s_5 & s_4 \\ s_6 & s_1 \\ s_7 & s_6 \end{bmatrix}$$

$$\mathbf{A}\times\mathbf{B}_4 = \begin{bmatrix} s_0 & s_4 \\ s_1 & s_1 \\ s_2 & s_6 \\ s_3 & s_3 \\ s_4 & s_0 \\ s_5 & s_5 \\ s_6 & s_2 \\ s_7 & s_7 \end{bmatrix} \quad \mathbf{A}\times\mathbf{B}_5 = \begin{bmatrix} s_0 & s_5 \\ s_1 & s_2 \\ s_2 & s_7 \\ s_3 & s_4 \\ s_4 & s_1 \\ s_5 & s_6 \\ s_6 & s_3 \\ s_7 & s_0 \end{bmatrix}$$

$$\mathbf{A}\times\mathbf{B}_6 = \begin{bmatrix} s_0 & s_6 \\ s_1 & s_3 \\ s_2 & s_0 \\ s_3 & s_5 \\ s_4 & s_2 \\ s_5 & s_7 \\ s_6 & s_4 \\ s_7 & s_1 \end{bmatrix} \quad \mathbf{A}\times\mathbf{B}_7 = \begin{bmatrix} s_0 & s_7 \\ s_1 & s_4 \\ s_2 & s_1 \\ s_3 & s_6 \\ s_4 & s_3 \\ s_5 & s_0 \\ s_6 & s_5 \\ s_7 & s_2 \end{bmatrix} . \tag{5.30}$$

The minimum squared product distance of each of the partitions in (5.30) is $d_P^2 = \delta_0^2 \cdot \delta_2^2 = 2E_s^2$.

In general, the second level of partitioning for MTCM schemes does not follow the tree structure of the traditional set partitioning. Following the traditional approach, one may partition each of the $M$ subsets $\mathbf{A}\times\mathbf{B}_k$, $k = 0, 1, \cdots, M-1$, into two subsets with the first subset containing the 2-tuples of the even rows of $\mathbf{A}\times\mathbf{B}_k$ and the second subset containing the 2-tuples of the odd rows. By using this method, although the 2-tuples of each subset at the second level of partitioning are distinct in the two symbol positions compared to the other 2-tuples of the same subset, it is not always true that the minimum product distance of each subset at this level has the maximum possible value.

To perform the second level of partitioning with maximized intradistance product structure for each of the resulting subsets (of dimensionality $\frac{M}{2}$), we form the subsets[8] $\mathbf{A}\times\mathbf{B}_k^*$, $k = 0, 1, \cdots, M-1$, according to (5.29) with $M$ replaced by $\frac{M}{2}$. Then the subsets of the second partitioning level are achieved by an odd-even split of each of the subsets $\mathbf{A}\times\mathbf{B}_k^*$.

Note that if one of the solutions of (5.29) (including the additive inverse) is equal to one of the solutions of (5.29) with $M$ replaced by $\frac{M}{2}$, then the subsets $\mathbf{A}\times\mathbf{B}_i$ and $\mathbf{A}\times\mathbf{B}_i^*$ are the same. In this case the first two levels of the set partitioning follow a *tree* structure. It is shown that [57] for $M = 4$, 8, and 32 there always exists a solution of $\mu^*$ which satisfies (5.29) with both values of $M$ and $\frac{M}{2}$. Thus, for $M = 4$, 8, and 32 the two levels of set partitioning form a tree whereas for M=16, they do not. The higher levels of set partitioning are performed similarly to the second level by choosing the alternate rows of the sets determined by solving (5.29) with $M$ replaced by $\frac{M}{2^p}$ where $p$ shows the partitioning level.

[8] We use asterisk superscript for $\mathbf{B}_k^*$ to emphasize that, in general, the subsets $\mathbf{A}\times\mathbf{B}_k^*$ may not be the same as the subsets $\mathbf{A}\times\mathbf{B}_k$.

**Example 5.7** Consider an 8PSK signal set with the first level of set partitioning for $k^* = 2$ demonstrated, in Example 5.6. Since for $M = 8$ the two first levels of set partitioning follow a tree structure, the subsets of the second partitioning level can be obtained by an odd-even split of the subsets of the first partitioning level as

$$\mathbf{A}_0 \times \mathbf{B}_{00} = \begin{bmatrix} s_0 & s_0 \\ s_2 & s_2 \\ s_4 & s_4 \\ s_6 & s_6 \end{bmatrix} \quad \mathbf{A}_1 \times \mathbf{B}_{01} = \begin{bmatrix} s_1 & s_5 \\ s_3 & s_7 \\ s_5 & s_1 \\ s_7 & s_3 \end{bmatrix}$$

$$\mathbf{A}_0 \times \mathbf{B}_{10} = \begin{bmatrix} s_0 & s_1 \\ s_2 & s_3 \\ s_4 & s_5 \\ s_6 & s_7 \end{bmatrix} \quad \mathbf{A}_1 \times \mathbf{B}_{11} = \begin{bmatrix} s_1 & s_6 \\ s_3 & s_0 \\ s_5 & s_2 \\ s_7 & s_4 \end{bmatrix}$$

$$\mathbf{A}_0 \times \mathbf{B}_{20} = \begin{bmatrix} s_0 & s_2 \\ s_2 & s_4 \\ s_4 & s_6 \\ s_6 & s_0 \end{bmatrix} \quad \mathbf{A}_1 \times \mathbf{B}_{21} = \begin{bmatrix} s_1 & s_7 \\ s_3 & s_1 \\ s_5 & s_3 \\ s_7 & s_5 \end{bmatrix}$$

$$\mathbf{A}_0 \times \mathbf{B}_{30} = \begin{bmatrix} s_0 & s_3 \\ s_2 & s_5 \\ s_4 & s_7 \\ s_6 & s_1 \end{bmatrix} \quad \mathbf{A}_1 \times \mathbf{B}_{31} = \begin{bmatrix} s_1 & s_0 \\ s_3 & s_2 \\ s_5 & s_4 \\ s_7 & s_6 \end{bmatrix}$$

$$\mathbf{A}_0 \times \mathbf{B}_{40} = \begin{bmatrix} s_0 & s_4 \\ s_2 & s_6 \\ s_4 & s_0 \\ s_6 & s_2 \end{bmatrix} \quad \mathbf{A}_1 \times \mathbf{B}_{41} = \begin{bmatrix} s_1 & s_1 \\ s_3 & s_3 \\ s_5 & s_5 \\ s_7 & s_7 \end{bmatrix}$$

$$\mathbf{A}_0 \times \mathbf{B}_{50} = \begin{bmatrix} s_0 & s_5 \\ s_2 & s_7 \\ s_4 & s_1 \\ s_6 & s_3 \end{bmatrix} \quad \mathbf{A}_1 \times \mathbf{B}_{51} = \begin{bmatrix} s_1 & s_2 \\ s_3 & s_4 \\ s_5 & s_6 \\ s_7 & s_0 \end{bmatrix}$$

$$\mathbf{A}_0 \times \mathbf{B}_{60} = \begin{bmatrix} s_0 & s_6 \\ s_2 & s_0 \\ s_4 & s_2 \\ s_6 & s_4 \end{bmatrix} \quad \mathbf{A}_1 \times \mathbf{B}_{61} = \begin{bmatrix} s_1 & s_3 \\ s_3 & s_5 \\ s_5 & s_7 \\ s_7 & s_1 \end{bmatrix}$$

$$\mathbf{A}_0 \times \mathbf{B}_{70} = \begin{bmatrix} s_0 & s_7 \\ s_2 & s_1 \\ s_4 & s_3 \\ s_6 & s_5 \end{bmatrix} \quad \mathbf{A}_1 \times \mathbf{B}_{71} = \begin{bmatrix} s_1 & s_4 \\ s_3 & s_6 \\ s_5 & s_0 \\ s_7 & s_2 \end{bmatrix} .$$

Note that the minimum squared product distance of each of the subsets in the second level of the set partitioning is $d_P^2 = \delta_1^2 \cdot \delta_1^2 = 4E_s^2$.

**Example 5.8** Now, we consider a 4-state rate 4/6 multiple trellis-coded 8PSK scheme with $k^* = 2$ [57]. For such a code there are $2^4 = 16$ branches leaving each of the states. The trellis diagram of this code is shown as Figure 5.11. Since each

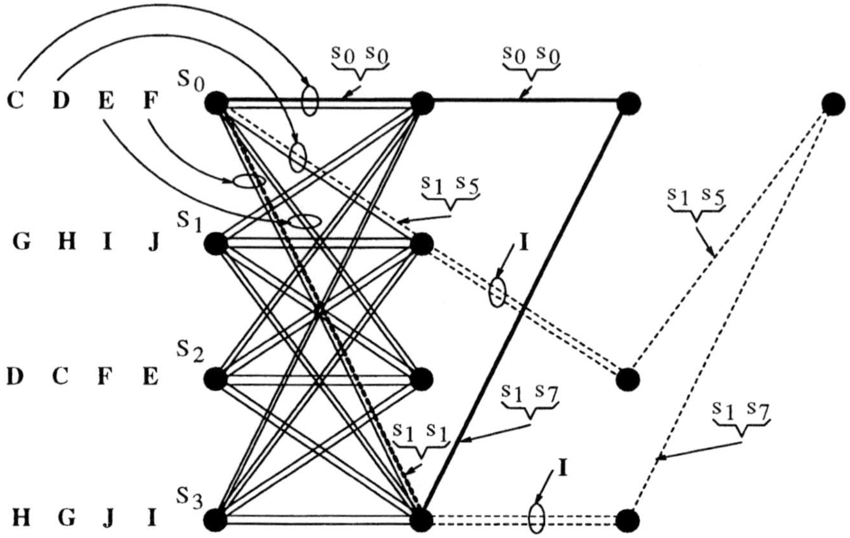


Figure 5.11: Trellis diagram of the 4-state rate 4/6 multiple trellis-coded 8PSK scheme with $k^* = 2$.

of the states of this diagram is connected to all possible four states, then the number of parallel transitions between two states is $\frac{16}{4} = 4$. These transitions are associated with the 2-tuples of the subsets chosen from the partitions of the second level of the set partitioning. These subsets are

$$\mathbf{C} = \begin{bmatrix} s_0 & s_0 \\ s_2 & s_2 \\ s_4 & s_4 \\ s_6 & s_6 \end{bmatrix} \quad \mathbf{D} = \begin{bmatrix} s_1 & s_5 \\ s_3 & s_7 \\ s_5 & s_1 \\ s_7 & s_3 \end{bmatrix} \quad \mathbf{E} = \begin{bmatrix} s_0 & s_4 \\ s_2 & s_6 \\ s_4 & s_0 \\ s_6 & s_2 \end{bmatrix} \quad \mathbf{F} = \begin{bmatrix} s_1 & s_1 \\ s_3 & s_3 \\ s_5 & s_5 \\ s_7 & s_7 \end{bmatrix}$$

$$\mathbf{G} = \begin{bmatrix} s_0 & s_2 \\ s_2 & s_4 \\ s_4 & s_6 \\ s_6 & s_0 \end{bmatrix} \quad \mathbf{H} = \begin{bmatrix} s_1 & s_7 \\ s_3 & s_1 \\ s_5 & s_3 \\ s_7 & s_5 \end{bmatrix} \quad \mathbf{I} = \begin{bmatrix} s_0 & s_6 \\ s_2 & s_0 \\ s_4 & s_2 \\ s_6 & s_4 \end{bmatrix} \quad \mathbf{J} = \begin{bmatrix} s_1 & s_3 \\ s_3 & s_5 \\ s_5 & s_7 \\ s_7 & s_1 \end{bmatrix}.$$

The effective length of this MTCM scheme is $L = 2$ which is determined by parallel transitions. However, the parallel transitions are not the only error events having effective length of 2. By inspection, examples of error events with more than one trellis branch whose effective lengths are 2 can be found. One example of such an error event with three trellis branches is shown in Figure 5.11 by dashed lines. Among the signals associated with the two paths of this error event only two are distinct.

The error events associated with parallel transitions have minimum squared product distances among the error events with the effective length 2. Thus, the minimum squared product distance of this MTCM scheme is

$$d_P^2(L) = \delta_1^2 \cdot \delta_1^2 = 4E_s^2.$$

Comparing this parameter with that of the 2-state rate 4/6 MTCM scheme of Example 5.5 (with $d_P^2(L) = \delta_0^2 \cdot \delta_2^2 = 2E_s^2$), we observe that the doubling of the number of states in this case gains 1.5 dB in SNR.

Finally, the free Euclidean distance of the 4-state rate 4/6 MTCM scheme of this example can be found as

$$d_{free}^2 = \delta_0^2 + \delta_0^2 + \delta_0^2 + \delta_0^2 = 2.344E_s.$$

An error event corresponding to the free Euclidean distance is shown by bold lines in Figure 5.11. Considering the free Euclidean distance we observe that this MTCM scheme only provides a 0.7 dB coding gain with respect to uncoded QPSK in an AWGN channel.

The set partitioning procedure for $k^* = 2$ can be extended for MTCM schemes with higher multiplicities. An appropriate method for this purpose is to form $k^*/2$-fold Cartesian products of the subsets in a given level of set partitioning, created for $k^* = 2$. As a specific example, consider the subsets of the first level of set partitioning of a multiple trellis-coded 8PSK with $k^* = 2$, given in (5.30). The subsets of the first partitioning level for $k^* = 4$ are obtained from the subsets of (5.30) as $\mathbf{A} \times \mathbf{B}_i \times \mathbf{A} \times \mathbf{B}_j$ where $i$ and $j$ run from 0 to 7. There are $8^2 = 64$ such subsets. As an illustrative example, we can specify one of these subsets as

$$\mathbf{A} \times \mathbf{B}_0 \times \mathbf{A} \times \mathbf{B}_2 = \begin{bmatrix} s_0 & s_0 & s_0 & s_2 \\ s_1 & s_5 & s_1 & s_7 \\ s_2 & s_2 & s_2 & s_4 \\ s_3 & s_7 & s_3 & s_1 \\ s_4 & s_4 & s_4 & s_6 \\ s_5 & s_1 & s_5 & s_3 \\ s_6 & s_6 & s_6 & s_0 \\ s_7 & s_3 & s_7 & s_5 \end{bmatrix}.$$

It is obvious that the $k^*$-tuples of the subsets formed by the above procedure are distinct in all of the $k^*$ signal positions.

The number of subsets obtained from the set partitioning of MTCM signals is usually large. In designing an MTCM scheme with a given number of states, if the number of partitioned subsets in a particular level of set partitioning is more than the number of subsets required by the trellis, we should choose those subsets having the largest *interdistances*[9]. Such a choice was earlier shown in Example 5.5 explicitly. Note that in assigning the 2-tuples to the branches of the two-state trellis diagram of Figure 5.10 the subsets $\mathbf{A} \times \mathbf{B}_0$, $\mathbf{A} \times \mathbf{B}_2$, $\mathbf{A} \times \mathbf{B}_4$, and $\mathbf{A} \times \mathbf{B}_6$ are used among the 8 possible subsets given in (5.30). The interdistance between the subsets $\mathbf{A} \times \mathbf{B}_0$, $\mathbf{A} \times \mathbf{B}_2$, $\mathbf{A} \times \mathbf{B}_4$, and $\mathbf{A} \times \mathbf{B}_6$ is

[9]The interdistance is defined as the minimum distance between elements in two different subsets of the same partitioning level.

$\delta_0^2 + \delta_0^2 = 1.172E_s$ while the interdistance between the subsets given in (5.30) is only $\delta_0^2 = 0.586E_s$.

Another partitioning technique for multiplicities greater than two, i.e., $k^* > 2$, is obtained by generalizing (5.29) as [57]

$$d_P^2(\mu_1^*, \mu_2^*, \cdots, \mu_{k^*-1}^*) = \max_{\mu_1^*, \mu_2^*, \cdots, \mu_{k^*-1}^*} \left\{ \min_{l=1,3,\cdots,\frac{M}{2}-1} \left[ (4E_s)^{k^*} \sin^{k^*}\left(\tfrac{l\pi}{M}\right) \cdot \prod_{i=1}^{k^*-1} \sin^2\left(\mu_i^* \tfrac{l\pi}{M}\right) \right] \right\},$$

where $\mu_i^*, i = 1, 2, \cdots, k^* - 1$, takes on values from the set of odd integers $1, 3, \cdots, M/2 - 1$. The set of solutions $\mu_i^*, i = 1, 2, \cdots, k^* - 1$, can be used to generate all of the subsets in any level of the set partitioning procedure. More examples of MTCM schemes for fading channels can be found in [27, Chapter 10] and [57].

## 5.3 PERFORMANCE ANALYSIS OF COHERENT DETECTION OF TCM ON FADING CHANNELS

In this section we evaluate the performance of the trellis-coded schemes on fading channels by deriving the upper bounds on error event or bit error probabilities. First we use the upper bound on the pairwise error probability, derived earlier in Section 5.2, and the generating function approach to derive the desired upper bounds for both Rician and Rayleigh fading channels. Next an exact solution to the pairwise error probability on a Rayleigh fading channel [49] will be employed to estimate an accurate bit error probability.

In the following discussions we again assume that the receiver performs coherent detection. Also, the assumption of the perfect interleaving on a slow fading channel is used. In this case the samples of the fading amplitude are independent and result in a memoryless discrete channel model.

### 5.3.1 The Generating Function Bounds on $P_e$ and $P_b$

To find an upper bound on error event or bit error probabilities the conditional pairwise error probabilities given in (5.11) and (5.16) corresponding to the fading channels with CSI and without CSI, respectively, can be put in the form

$$P_2(\mathbf{s}_l, \ \hat{\mathbf{s}}_l|\mathbf{a}_l) \leq \prod_{i=1}^{l} D^{d^2(s_i, \ \hat{s}_i)}, \tag{5.31}$$

where[10]

$$d^2(s_i,\ \hat{s}_i) = \begin{cases} a_i^2|s_i - \hat{s}_i|^2, & \text{with CSI} \\ 4\lambda(a_i - \lambda)|s_i - \hat{s}_i|^2, & \text{without CSI} \end{cases} \tag{5.32}$$

and $D = \exp(-\frac{1}{4N_0})$. Averaging (5.31) over the fading amplitude and recalling that the samples of fading are independent we obtain

$$P_2(\mathbf{s}_l,\ \hat{\mathbf{s}}_l) \leq \prod_{i=1}^{l} E_{a_i}[D^{d^2(s_i,\ \hat{s}_i)}]. \tag{5.33}$$

Representing $\mathbf{s}_l = (s_1, s_2, \cdots, s_l)$ and $\hat{\mathbf{s}}_l = (\hat{s}_1, \hat{s}_2, \cdots, \hat{s}_l)$ by their labels $\mathbf{C}_l = (\mathbf{c}_1, \mathbf{c}_2, \cdots, \mathbf{c}_l)$ and $\hat{\mathbf{C}}_l = (\hat{\mathbf{c}}_1, \hat{\mathbf{c}}_2, \cdots, \hat{\mathbf{c}}_l)$, respectively, and introducing the sequence of the binary error vectors $\mathbf{E}_l = (\mathbf{e}_1,\ \mathbf{e}_2,\ \cdots,\ \mathbf{e}_l)$ (see Section 4.5) the pairwise error probability of (5.33) is expressed as

$$P_2(\mathbf{C}_l,\ \hat{\mathbf{C}}_l) \leq \prod_{i=1}^{l} E_{a_i}[D^{d^2[\mathcal{M}(\mathbf{c}_i),\ \mathcal{M}(\mathbf{c}_i \oplus \mathbf{e}_i)]}], \tag{5.34}$$

where $\mathcal{M}$ is a one-to-one mapping as defined in (4.3). Substituting (5.34) into (4.10) yields the upper bound on error event probability as[11]

$$P_e \leq \frac{1}{2} \sum_{l=1}^{\infty} \sum_{\mathbf{C}_l} \sum_{\hat{\mathbf{E}}_l \neq \mathbf{0}} P(\mathbf{C}_l) \prod_{i=1}^{l} E_{a_i}[D^{d^2[\mathcal{M}(\mathbf{c}_i),\ \mathcal{M}(\mathbf{c}_i \oplus \mathbf{e}_i)]}]. \tag{5.35}$$

Using the same argument as Section 4.5 for the *uniform* TCM schemes (Ungerboeck type codes) the upper bound of (5.35) can be expressed as

$$\begin{aligned} P_e &\leq \frac{1}{2} \sum_{l=1}^{\infty} \sum_{\mathbf{E}_l \neq \mathbf{0}} \prod_{i=1}^{l} \overline{\mathcal{F}}(\mathbf{e}_i,\ D) \\ &= \frac{1}{2} T(I, D), \end{aligned}$$

where

$$\overline{\mathcal{F}}(\mathbf{e}_i, D) = \frac{1}{2^m} \sum_{\mathbf{c}_i = \mathbf{c}_i^\kappa} E_{a_i}[D^{d^2[\mathcal{M}(\mathbf{c}_i),\ \mathcal{M}(\mathbf{c}_i \oplus \mathbf{e}_i)]}]. \tag{5.36}$$

The foregoing analysis shows that the upper bound on $P_e$ can be evaluated by deriving the generating function of an error-state diagram. The same error-state diagram which is used for evaluating the performance of the uniform TCM

---

[10] Note that the definition of $d^2(s_i,\ \hat{s}_i)$ for the fading channel without CSI given in (5.32) is only for trellis-coded MPSK schemes, where $|s_i| = |\hat{s}_i|$. The extension of the general case of non-constant envelope signals, i.e., $|s_i| \neq |\hat{s}_i|$ can be done by using the results of Appendix A.

[11] We introduce a factor of 1/2 in the upper bound given in (5.35). This is due to the fact that it is possible to introduce a factor of 1/2 in the Chernoff bound as discussed in [58, Appendix 4B].

schemes on the AWGN channel can be utilized for this purpose provided that its branches are labeled by the error weight profiles given in (5.36).

**Example 5.9** Consider a rate 2/3 ($m = 2$) uniform (Ungerboeck code) 8PSK TCM scheme. By setting $c^{(3)} = 0$ (labels corresponding to signals of the subset $\mathbf{A}_0$) the error weight profiles on a fading channel can be obtained as

$$\overline{\mathcal{F}}(\mathbf{e}_i,\ D) = \frac{1}{4} \sum_{(c^{(1)},c^{(2)},0)} E_{a_i}[D^{d_E^2[\mathcal{M}(c^{(1)},c^{(2)},0),\ \mathcal{M}(\ (c^{(1)},c^{(2)},0)\oplus(e^{(1)},e^{(2)},e^{(3)})\ )]}].$$

Using the results of Example 4.11 and Appendix A the error weight profiles for a Rayleigh fading channel are computed as

$$\begin{array}{lllllll}
\overline{\mathcal{F}}(000,\ D) & = & 1 & \quad & \overline{\mathcal{F}}(100,\ D) & = & f(\delta_3^2) \\
\overline{\mathcal{F}}(001,\ D) & = & f(\delta_0^2) & & \overline{\mathcal{F}}(101,\ D) & = & f(\delta_2^2) \\
\overline{\mathcal{F}}(010,\ D) & = & f(\delta_1^2) & & \overline{\mathcal{F}}(110,\ D) & = & f(\delta_1^2) \\
\overline{\mathcal{F}}(011,\ D) & = & \frac{1}{2}(f(\delta_0^2) + f(\delta_2^2)) & & \overline{\mathcal{F}}(111,\ D) & = & \frac{1}{2}(f(\delta_0^2) + f(\delta_2^2)),
\end{array} \tag{5.37}$$

where $\delta_0^2 = 0.586E_s$, $\delta_1^2 = 2E_s$, $\delta_2^2 = 3.414E_s$, and $\delta_3^2 = 4E_s$ and

$$f(\delta_i^2) = \begin{cases} \frac{1}{1+\delta_i^2/4N_0}, & \text{with CSI} \\ e^{\frac{\lambda^2\delta_i^2}{N_0}}[1 - \sqrt{\pi}\mathrm{erfc}(\frac{\lambda\delta_i^2}{2N_0})e^{(\frac{\lambda\delta_i^2}{2N_0})^2}(\frac{\delta_i^2}{2N_0})]. & \text{without CSI} \end{cases} \tag{5.38}$$

As with the case of no fading, the upper bound on the bit error probability $P_b$ is obtained as

$$P_b \leq \frac{1}{2m} \frac{\partial T(D,\ I)}{\partial I} |_{I=1}\ , \tag{5.39}$$

where for uniform TCM schemes $T(D,\ I)$ is the generating function of the modified error-state diagram whose branches are modified as $I^r\overline{\mathcal{F}}(\mathbf{e}_i,\ D)$ and $r$ shows the number of incorrect input bits associated with $\mathbf{e}_i$.

In the following we evaluate the upper bounds on $P_b$ for three different 4-state rate 2/3 8PSK TCM schemes, namely, the Ungerboeck code with parallel transitions (Figure 4.4), the Wilson-Leung code (Figure 5.3), and the 4-state TCM scheme designed for fading channels in Section 5.2.1 (Figure 5.6). Hereafter we denote these codes by 4U, 4WL, and 4N, respectively. The channel under consideration is a Rayleigh fading channel with CSI.

**Example 5.10** The error-state diagram of the 4-state Ungerboeck code (4U) is the same as Figure 4.20(b), with branch labels

$$\begin{array}{lllll}
t_1 & = & I\overline{\mathcal{F}}(010,\ D) + I^2\overline{\mathcal{F}}(110,\ D) & = & (I + I^2)f(\delta_1^2) \\
t_2 & = & I\overline{\mathcal{F}}(011,\ D) + I^2\overline{\mathcal{F}}(111,\ D) & = & \frac{1}{2}(I + I^2)(f(\delta_0^2) + f(\delta_2^2)) \\
t_3 & = & \overline{\mathcal{F}}(011,\ D) + I\overline{\mathcal{F}}(111,\ D) & = & \frac{1}{2}(1 + I)(f(\delta_0^2) + f(\delta_2^2)) \\
t_4 & = & \overline{\mathcal{F}}(001,\ D) + I\overline{\mathcal{F}}(101,\ D) & = & f(\delta_0^2) + If(\delta_2^2) \\
t_5 & = & I\overline{\mathcal{F}}(000,\ D) + I^2\overline{\mathcal{F}}(100,\ D) & = & I + I^2 f(\delta_3^2) \\
t_6 & = & I\overline{\mathcal{F}}(001,\ D) + I^2\overline{\mathcal{F}}(101,\ D) & = & If(\delta_0^2) + I^2 f(\delta_2^2) \\
t_7 & = & \overline{\mathcal{F}}(010,\ D) + I\overline{\mathcal{F}}(110,\ D) & = & (1 + I)f(\delta_1^2),
\end{array}$$

where the values of $f(\delta_i^2)$ are given as in (5.38). Using the modified error-state diagram, the generating function $T(D, I)$ is obtained as (see Example 4.13)

$$T(D,\ I) = \frac{t_1 t_7[t_2 t_3 + t_4(1 - t_6)]}{(1 - t_6)(1 - t_4 t_5) - t_2 t_3 t_5}.$$

After some mathematical manipulation we have

$$\left.\frac{\partial T(D,\ I)}{\partial I}\right|_{I=1} = 4f^2(\delta_1^2)\frac{\alpha(2 - \alpha - \alpha^2) + f(\delta_2^2)(1 - 2\alpha + 2\alpha^2)}{(1 - 2\alpha - \alpha f(\delta_3^2)\ )^2},$$

where $\alpha = f(\delta_0^2) + f(\delta_2^2)$.

Note that the error associated with parallel transitions is not involved in the above generating function. Recalling that the squared Euclidean distance between parallel transitions is $\delta_3^2 = 4E_s$ the upper bound on the bit error probability, after including the term corresponding to the parallel transitions, can be expressed as

$$P_b \le \frac{1}{4}f(\delta_3^2) + \frac{1}{4}\left.\frac{\partial T(D,\ I)}{\partial I}\right|_{I=1}.$$

Using the above equation the upper bound on $P_b$ can be computed in terms of the signal energy-to-noise ratio $E_s/N_0$ or the bit energy-to-noise ratio $E_b/N_0$. Figure 5.12 shows the upper bound on $P_b$ versus $E_b/N_0$ for a Rayleigh fading channel with CSI.

**Example 5.11** The error-state diagram of the 4-state Wilson-Leung code (4WL) is shown in Figure 5.13. Using the trellis diagram of this code, shown in Figure 5.3, the branch labels of the error-state diagram are obtained as

$$\begin{aligned}
t_1 &= I\overline{\mathcal{F}}(100,\ D) &&= If(\delta_3^2)\\
t_2 &= I\overline{\mathcal{F}}(010,\ D) &&= If(\delta_1^2)\\
t_3 &= I^2\overline{\mathcal{F}}(110,\ D) &&= I^2 f(\delta_1^2)\\
t_4 &= I\overline{\mathcal{F}}(101,\ D) &&= If(\delta_2^2)\\
t_5 &= I\overline{\mathcal{F}}(110,\ D) &&= If(\delta_1^2)\\
t_6 &= I^2\overline{\mathcal{F}}(011,\ D) &&= \tfrac{1}{2}I^2(f(\delta_0^2) + f(\delta_2^2))\\
t_7 &= I^2\overline{\mathcal{F}}(111,\ D) &&= \tfrac{1}{2}I^2(f(\delta_0^2) + f(\delta_2^2))\\
t_8 &= I\overline{\mathcal{F}}(001,\ D) &&= If(\delta_0^2)\\
t_9 &= I\overline{\mathcal{F}}(011,\ D) &&= \tfrac{1}{2}I(f(\delta_0^2) + f(\delta_2^2))\\
t_{10} &= I\overline{\mathcal{F}}(000,\ D) &&= I\\
t_{11} &= I^2\overline{\mathcal{F}}(010,\ D) &&= I^2 f(\delta_1^2)\\
t_{12} &= I\overline{\mathcal{F}}(111,\ D) &&= \tfrac{1}{2}I(f(\delta_0^2) + f(\delta_2^2))\\
t_{13} &= \overline{\mathcal{F}}(001,\ D) &&= f(\delta_0^2)\\
t_{14} &= \overline{\mathcal{F}}(100,\ D) &&= f(\delta_3^2)\\
t_{15} &= \overline{\mathcal{F}}(101,\ D) &&= f(\delta_2^2).
\end{aligned}$$

The generating function is now obtained from the error-state diagram of Figure 5.13 by solving the nodal equations of this graph

$$T(D,\ I) = \xi_1 t_{13} + \xi_2 t_{14} + \xi_3 t_{15},$$

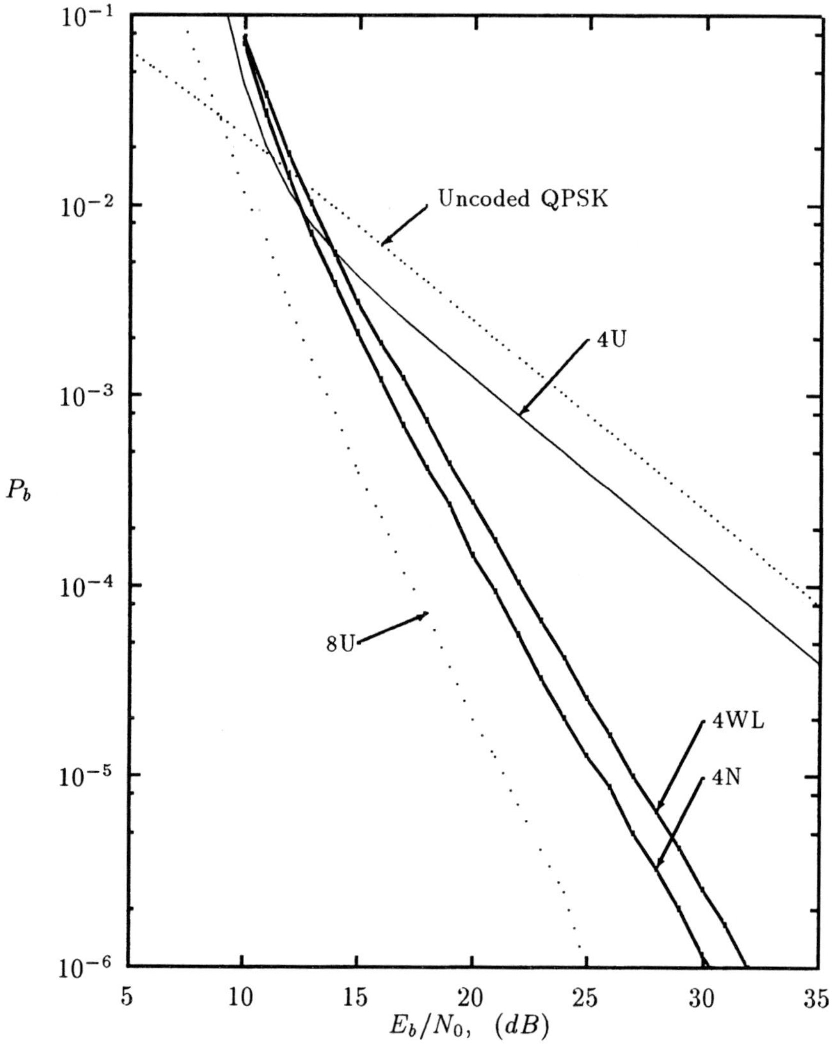


Figure 5.12: Performance of some rate 2/3 8PSK TCM schemes on a Rayleigh fading channel with CSI.

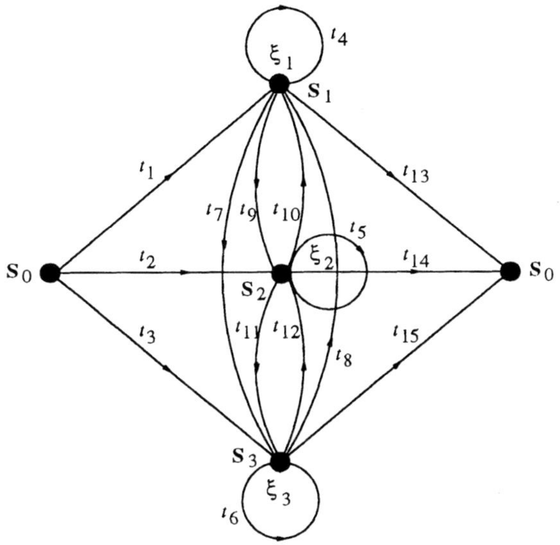


Figure 5.13: Error-state diagram of the TCM schemes shown in Figure 5.3 and Figure 5.6.

where

$$\begin{bmatrix} \xi_1 \\ \xi_2 \\ \xi_3 \end{bmatrix} = \begin{bmatrix} 1-t_4 & -t_{10} & -t_8 \\ -t_9 & 1-t_5 & -t_{12} \\ -t_7 & -t_{11} & 1-t_6 \end{bmatrix}^{-1} \cdot \begin{bmatrix} t_1 \\ t_2 \\ t_3 \end{bmatrix}.$$

After some mathematical manipulation, the generating function $T(D, I)$ is simplified to

$$T(D,I) = \frac{A_1 I + A_2 I^2 + A_3 I^3 + A_4 I^4}{1 + B_1 I + B_2 I^2 + B_3 I^3 + B_4 I^4}, \tag{5.40}$$

where

$$\begin{aligned} A_1 &= f(\delta_3^2)(f(\delta_0^2) + f(\delta_1^2)) \\ A_2 &= \frac{1}{2}(f(\delta_0^2) + f(\delta_2^2))(2f(\delta_1^2) + f^2(\delta_3^2) - 2f(\delta_1^2)f(\delta_3^2)) \\ A_3 &= (f^2(\delta_2^2) - f^2(\delta_0^2))(\frac{1}{2}f(\delta_3^2) - f(\delta_1^2)) \\ A_4 &= 0 \\ B_1 &= -f(\delta_1^2) - f(\delta_2^2) \\ B_2 &= f(\delta_1^2)f(\delta_2^2) - f(\delta_0^2) - f(\delta_2^2) \\ B_3 &= \frac{1}{2}(f(\delta_2^2) - f(\delta_0^2)) \\ B_4 &= 0, \end{aligned} \tag{5.41}$$

and $f(\delta_i^2)$, $i = 0, 1, 2, 3$, are given in(5.38). Using (5.40), (5.41), and (5.39) the upper bound on the bit error probability of 4WL versus $E_b/N_0$ for a Rayleigh fading channel with CSI is plotted in Figure 5.12.

**Example 5.12** For the 4-state TCM scheme of Figure 5.6 the error-state diagram is the same as Figure 5.13 with branch labels

$$
\begin{aligned}
t_1 &= I\overline{\mathcal{F}}(100,\ D) &&= If(\delta_3^2)\\
t_2 &= I\overline{\mathcal{F}}(010,\ D) &&= If(\delta_1^2)\\
t_3 &= I^2\overline{\mathcal{F}}(110,\ D) &&= I^2f(\delta_1^2)\\
t_4 &= I\overline{\mathcal{F}}(111,\ D) &&= \tfrac{1}{2}I(f(\delta_0^2)+f(\delta_2^2))\\
t_5 &= I\overline{\mathcal{F}}(100,\ D) &&= If(\delta_3^2)\\
t_6 &= I^2\overline{\mathcal{F}}(011,\ D) &&= \tfrac{1}{2}I^2(f(\delta_0^2)+f(\delta_2^2))\\
t_7 &= I^2\overline{\mathcal{F}}(101,\ D) &&= I^2f(\delta_2^2)\\
t_8 &= I\overline{\mathcal{F}}(001,\ D) &&= If(\delta_0^2)\\
t_9 &= I\overline{\mathcal{F}}(001,\ D) &&= If(\delta_0^2)\\
t_{10} &= I\overline{\mathcal{F}}(010,\ D) &&= If(\delta_1^2)\\
t_{11} &= I^2\overline{\mathcal{F}}(000,\ D) &&= I^2\\
t_{12} &= I\overline{\mathcal{F}}(111,\ D) &&= \tfrac{1}{2}I(f(\delta_0^2)+f(\delta_2^2))\\
t_{13} &= \overline{\mathcal{F}}(011,\ D) &&= \tfrac{1}{2}(f(\delta_0^2)+f(\delta_2^2))\\
t_{14} &= \overline{\mathcal{F}}(110,\ D) &&= f(\delta_1^2)\\
t_{15} &= \overline{\mathcal{F}}(101,\ D) &&= f(\delta_2^2).
\end{aligned}
$$

The generating function for this TCM scheme is obtained as (5.40) with

$$
\begin{aligned}
A_1 &= f^2(\delta_1^2)+\frac{1}{2}f(\delta_0^2)f(\delta_3^2)+\frac{1}{2}f(\delta_2^2)f(\delta_3^2)\\
A_2 &= f(\delta_0^2)f(\delta_1^2)f(\delta_3^2)-\frac{1}{2}f(\delta_0^2)f^2(\delta_3^2)-\frac{1}{2}f(\delta_2^2)f^2(\delta_3^2)\\
&\quad +f(\delta_1^2)f(\delta_2^2)\\
A_3 &= \frac{1}{2}f^2(\delta_0^2)f(\delta_1^2)-\frac{1}{4}f^2(\delta_0^2)f(\delta_3^2)-\frac{1}{2}f(\delta_0^2)f(\delta_2^2)f(\delta_3^2)\\
&\quad \frac{3}{4}f^2(\delta_2^2)f(\delta_3^2)-\frac{1}{2}f(\delta_1^2)f^2(\delta_2^2)-f(\delta_1^2)f(\delta_2^2)f(\delta_3^2)\\
&\quad +f(\delta_1^2)f(\delta_2^2)\\
A_4 &= f^2(\delta_0^2)f^2(\delta_1^2)-f^2(\delta_0^2)f(\delta_1^2)f(\delta_3^2)+f(\delta_1^2)f^2(\delta_2^2)f(\delta_3^2)\\
&\quad -2f(\delta_0^2)f^2(\delta_1^2)f(\delta_2^2)+\frac{1}{4}f^2(\delta_0^2)f^2(\delta_3^2)+\frac{1}{2}f(\delta_0^2)f(\delta_2^2)f^2(\delta_3^2)\\
&\quad +\frac{1}{2}f^2(\delta_0^2)f(\delta_1^2)-\frac{1}{4}f^2(\delta_0^2)f(\delta_3^2)+\frac{1}{2}f(\delta_0^2)f(\delta_2^2)f(\delta_3^2)\\
&\quad -\frac{3}{4}f^2(\delta_2^2)f^2(\delta_3^2)-\frac{1}{4}f^2(\delta_2^2)f(\delta_3^2)+f^2(\delta_1^2)f^2(\delta_2^2)\\
&\quad -\frac{1}{2}f(\delta_1^2)f^2(\delta_2^2)\\
B_1 &= -\frac{1}{2}(f(\delta_0^2)+f(\delta_2^2))-f(\delta_3^2)\\
B_2 &= -\frac{1}{2}(f(\delta_0^2)+f(\delta_2^2))+\frac{1}{2}f(\delta_0^2)f(\delta_3^2)+\frac{1}{2}f(\delta_2^2)f(\delta_3^2)\\
&\quad -f(\delta_0^2)f(\delta_1^2)\\
B_3 &= \frac{1}{2}f(\delta_0^2)f(\delta_3^2)+\frac{1}{2}f(\delta_2^2)f(\delta_3^2)+\frac{1}{4}f^2(\delta_0^2)+\frac{1}{4}f^2(\delta_2^2)
\end{aligned}
$$

$$
\begin{aligned}
& -\frac{1}{2}f(\delta_0^2)f(\delta_2^2) - \frac{1}{2}f(\delta_0^2) - \frac{1}{2}f(\delta_2^2) \\
B_4 \quad = \quad & -\frac{1}{4}f^2(\delta_0^2)f(\delta_3^2) - \frac{1}{4}f^2(\delta_2^2)f(\delta_3^2) + \frac{1}{2}f(\delta_0^2)f(\delta_2^2) \\
& +\frac{1}{2}f(\delta_0^2)f(\delta_2^2)f(\delta_3^2) + \frac{1}{2}f^2(\delta_0^2)f(\delta_1^2) - \frac{1}{2}f(\delta_1^2)f^2(\delta_2^2) \\
& +\frac{1}{4}f^2(\delta_2^2) - \frac{3}{4}f^2(\delta_0^2). \qquad (5.42)
\end{aligned}
$$

Using (5.40), (5.42), and (5.39) the upper bound on the bit error probability of 4N versus $E_b/N_0$ for a Rayleigh fading channel with CSI is plotted in Figure 5.12.

The results shown in Figure 5.12 compare the performance of the three 4-state TCM schemes on a Rayleigh fading channel with CSI. For bit error probabilities of $10^{-2}$ or less, the 4N scheme has better performance than the other two schemes. The coding gain of this code for bit error probabilities in the area of $10^{-3}$, which is important in digital speech transmission, is about 1.1 dB and 4.5 dB compared to the 4WL and 4U schemes, respectively. For the lower bit error probabilities, both the 4N and 4WL codes have much more coding gain than the 4U code because of their doubled effective lengths. However, the performance of the 4N code is about 1.5 dB superior to that of the 4WL code at higher signal-to-noise ratios. This is because the minimum squared product distance of the former scheme is twice that of the latter.

The foregoing examples show that the evaluation of $T(D,\ I)$ involves the analytical computation of a $(2^\nu - 1) \times (2^\nu - 1)$ matrix inversion where $2^\nu$ is the number of trellis states. This computation becomes tedious for codes with large numbers of trellis states. To avoid such tedious computations a numerical approach is applied [3, Chap. 4]. The numerical computation is performed by upper-bounding the derivative of the generating function $T(D,\ I)$ with respect to $I$ as [3, Page 248]

$$
\left.\frac{\partial T(D,\ I)}{\partial I}\right|_{I=1} < \frac{T(D,\ 1+\epsilon) - T(D,\ 1)}{\epsilon}, \qquad (5.43)
$$

where $\epsilon \ll 1$ can be chosen small enough, e.g., $10^{-6}$, to approximate the derivative. Now, for a fixed value of $D$ the values of $T(D,\ 1+\epsilon)$ and $T(D,\ 1)$ can be computed numerically.

**Example 5.13** Consider the 8-state rate 2/3 Ungerboeck TCM scheme shown in Figure 4.8. Recall that this code is good for both the AWGN and fading channels. The error-state diagram of the 8-state TCM scheme is shown in

Figure 5.14. The branch labels of the error-state diagram are

$$
\begin{aligned}
t_1 &= I\overline{\mathcal{F}}(100,\ D) &&= If(\delta_3^2)\\
t_2 &= I\overline{\mathcal{F}}(010,\ D) &&= If(\delta_1^2)\\
t_3 &= I^2\overline{\mathcal{F}}(110,\ D) &&= I^2 f(\delta_1^2)\\
t_4 &= \overline{\mathcal{F}}(100,\ D) &&= f(\delta_3^2)\\
t_5 &= I\overline{\mathcal{F}}(000,\ D) &&= I\\
t_6 &= I\overline{\mathcal{F}}(110,\ D) &&= If(\delta_1^2)\\
t_7 &= I^2\overline{\mathcal{F}}(010,\ D) &&= I^2 f(\delta_1^2)\\
t_8 &= \overline{\mathcal{F}}(001,\ D) &&= f(\delta_0^2)\\
t_9 &= I\overline{\mathcal{F}}(101,\ D) &&= If(\delta_2^2)\\
t_{10} &= I\overline{\mathcal{F}}(011,\ D) &&= \tfrac{1}{2}I(f(\delta_0^2)+f(\delta_2^2))\\
t_{11} &= I^2\overline{\mathcal{F}}(111,\ D) &&= \tfrac{1}{2}I^2(f(\delta_0^2)+f(\delta_2^2))\\
t_{12} &= \overline{\mathcal{F}}(101,\ D) &&= f(\delta_2^2)\\
t_{13} &= I\overline{\mathcal{F}}(001,\ D) &&= If(\delta_0^2)\\
t_{14} &= I\overline{\mathcal{F}}(111,\ D) &&= \tfrac{1}{2}I(f(\delta_0^2)+f(\delta_2^2))\\
t_{15} &= I^2\overline{\mathcal{F}}(011,\ D) &&= \tfrac{1}{2}I^2(f(\delta_0^2)+f(\delta_2^2))\\
t_{16} &= \overline{\mathcal{F}}(010,\ D) &&= f(\delta_1^2)\\
t_{17} &= I\overline{\mathcal{F}}(110,\ D) &&= If(\delta_1^2)\\
t_{18} &= I\overline{\mathcal{F}}(000,\ D) &&= I\\
t_{19} &= I^2\overline{\mathcal{F}}(100,\ D) &&= I^2 f(\delta_3^2)\\
t_{20} &= \overline{\mathcal{F}}(011,\ D) &&= \tfrac{1}{2}(f(\delta_0^2)+f(\delta_2^2))\\
t_{21} &= I\overline{\mathcal{F}}(111,\ D) &&= \tfrac{1}{2}I(f(\delta_0^2)+f(\delta_2^2))\\
t_{22} &= I\overline{\mathcal{F}}(001,\ D) &&= If(\delta_0^2)\\
t_{23} &= I^2\overline{\mathcal{F}}(101,\ D) &&= I^2 f(\delta_2^2)\\
t_{24} &= \overline{\mathcal{F}}(110,\ D) &&= f(\delta_1^2)\\
t_{25} &= I\overline{\mathcal{F}}(010,\ D) &&= If(\delta_1^2)\\
t_{26} &= I\overline{\mathcal{F}}(100,\ D) &&= If(\delta_3^2)\\
t_{27} &= I^2\overline{\mathcal{F}}(000,\ D) &&= I^2\\
t_{28} &= \overline{\mathcal{F}}(111,\ D) &&= \tfrac{1}{2}(f(\delta_0^2)+f(\delta_2^2))\\
t_{29} &= I\overline{\mathcal{F}}(011,\ D) &&= \tfrac{1}{2}I(f(\delta_0^2)+f(\delta_2^2))\\
t_{30} &= I\overline{\mathcal{F}}(101,\ D) &&= If(\delta_2^2)\\
t_{31} &= I^2\overline{\mathcal{F}}(001,\ D) &&= I^2 f(\delta_0^2).
\end{aligned}
$$

The generating function for the 8-state TCM scheme is obtained by simultaneous solution of the state equations of the error-state diagram shown in Figure 5.14 as

$$T(D,\ I) = \xi_2 t_4 + \xi_4 t_{16} + \xi_6 t_{24},$$

where

$$
\begin{bmatrix} \xi_1\\ \xi_2\\ \xi_3\\ \xi_4\\ \xi_5\\ \xi_6\\ \xi_7 \end{bmatrix}
=
\begin{bmatrix}
1 & -t_5 & 0 & -t_{17} & 0 & -t_{25} & 0\\
0 & 1-t_6 & 0 & -t_{18} & 0 & -t_{26} & 0\\
0 & -t_7 & 1 & -t_{19} & 0 & -t_{27} & 0\\
-t_8 & 0 & -t_{12} & 1 & -t_{20} & 0 & -t_{28}\\
-t_9 & 0 & -t_{13} & 0 & 1-t_{21} & 0 & -t_{29}\\
-t_{10} & 0 & -t_{14} & 0 & -t_{22} & 1 & -t_{30}\\
-t_{11} & 0 & -t_{15} & 0 & -t_{23} & 0 & 1-t_{30}
\end{bmatrix}^{-1}
\cdot
\begin{bmatrix} t_1\\ t_2\\ t_3\\ 0\\ 0\\ 0\\ 0 \end{bmatrix}.
$$

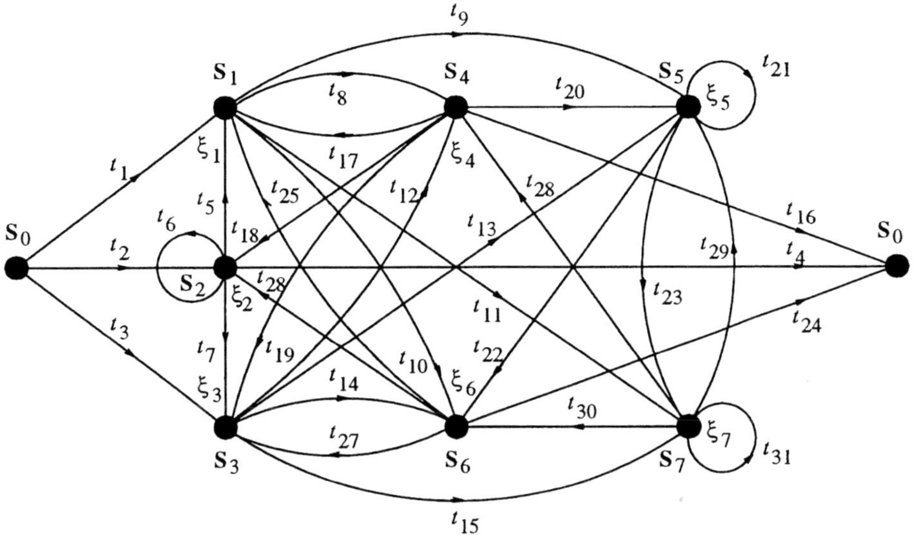


Figure 5.14: Error-state diagram of the TCM schemes shown in Figure 4.8.

Using (5.43) and (5.39) the upper bound on the bit error probability of the 8-state 8PSK TCM scheme (8U) versus $E_b/N_0$ for a Rayleigh fading channel with CSI is included in Figure 5.12.

To examine the tightness of the upper bound on $P_b$, given in (5.39), we evaluate the system performance through computer simulation. Although this technique is straightforward and relatively easy to implement, it requires an unacceptably large computer time when smaller values of error probability are needed.

The computer simulation is based on the baseband system model shown in Figure 5.2. The channel is modeled as an additive white Gaussian noise $n_i$ with a time varying fading process $g_i$, representing the complex signal gain. The in-phase and quadrature components of the channel complex gain are generated as two independent Gaussian random variables. In such a case, the fading amplitude $a_i$ is Rayleigh distributed. Samples of the fading process are generated independently in order to simulate the ideal interleaving.

The simulation results along with the upper bounds on $P_b$ for the 4-state 8PSK (4N) (see Figure 5.6) and the 8-state 8PSK (8U) (see Figure 4.8) TCM schemes are shown in Figure 5.15. It is observed that at high signal-to-noise ratios the true performance of the 4N scheme and the 8U scheme deviate about 2.5 dB and 2 dB, respectively, from the analytical upper bounds.

So far we have considered some examples of the evaluation of the upper bound on $P_b$ for a Rayleigh fading channel with CSI. In the following we evaluate the performance of the 4-state TCM schemes, namely, 4N, 4W, and 4U, on a Rician fading channel with CSI.

**Example 5.14** Consider a Rician fading channel with the Rician parameter $K$. On such a channel the error weight profiles for a rate 2/3 ($m = 2$) uniform 8PSK TCM scheme can be obtained as (5.37) with (see Appendix A)

$$f(\delta_i^2) = \frac{1+K}{1+K+\delta_i^2/4N_0} \exp(-\frac{K\delta_i^2/4N_0}{1+K+\delta_i^2/4N_0}), \quad \text{with CSI} \tag{5.44}$$

and

$$f(\delta_i^2) = \frac{e^{-K+\lambda^2\delta_i^2/N_0}}{\pi} \int_0^{\pi} (1-\sqrt{\pi}\text{erfc}(\nu)e^{\nu^2}\nu)d\theta, \quad \text{without CSI} \tag{5.45}$$

where

$$\nu = \frac{\lambda\delta_i^2/2N_0}{\sqrt{1+K}} - \sqrt{K}\cos\theta. \tag{5.46}$$

Using (5.44) and the results of Examples 5.10-5.12, the upper bounds on $P_b$ for the three different 4-state rate 2/3 8PSK TCM schemes are plotted in Figure 5.16 for different values of the Rician parameter $K$. Note that the results for $K = +\infty$ correspond to the AWGN channel.

It is interesting to compare the performance of the three 4-state TCM schemes once again based on the results of Figure 5.16. Observe that for small values of $K$ the 4N TCM scheme outperforms the other two TCM schemes. However, as the Rician parameter $K$ increases, that is as the channel approaches the AWGN channel, the performance of the 4-state Ungerboeck code (4U), which has the largest Euclidean distance, becomes better than the 4N and 4WL schemes. For instance, note that the intersection of the upper bound curve of the 4U scheme with the other upper bound curves moves towards the lower bit error probabilities as $K$ approaches the larger values. This is in agreement with our earlier discussion about the TCM design criteria for fading channels.

As a final example of upper bound analysis we consider the case of a Rayleigh fading channel without CSI.

**Example 5.15** In this example we evaluate the upper bound on $P_b$ for two 4-state TCM schemes, namely the 4U and 4N TCM schemes, on a Rayleigh fading channel without CSI. For the Ungerboeck 4-state TCM scheme (4U) we use the upper bound on $P_b$, given in Example 5.10 with the values of $f(\delta_i^2)$, $i = 0, 1, 2, 3$, defined for a Rayleigh fading channel without CSI in (5.38). For the 4-state TCM scheme without parallel transitions (4N) the generating function $T(D,\ I)$ is obtained by substituting (5.42) into (5.40) and using the values of $f(\delta_i^2)$, $i = 0, 1, 2, 3$, given in (5.38) for a Rayleigh fading channel without CSI.

The upper bounds on $P_b$ versus $E_b/N_0$ for the 4U and 4N schemes are plotted in Figure 5.17. In evaluating these upper bounds the Chernoff parameter $\lambda$ is optimized numerically, for each value of $E_b/N_0$, to tighten the bound. The corresponding upper bounds for a Rayleigh fading channel with CSI are also included in Figure 5.17 for comparison.

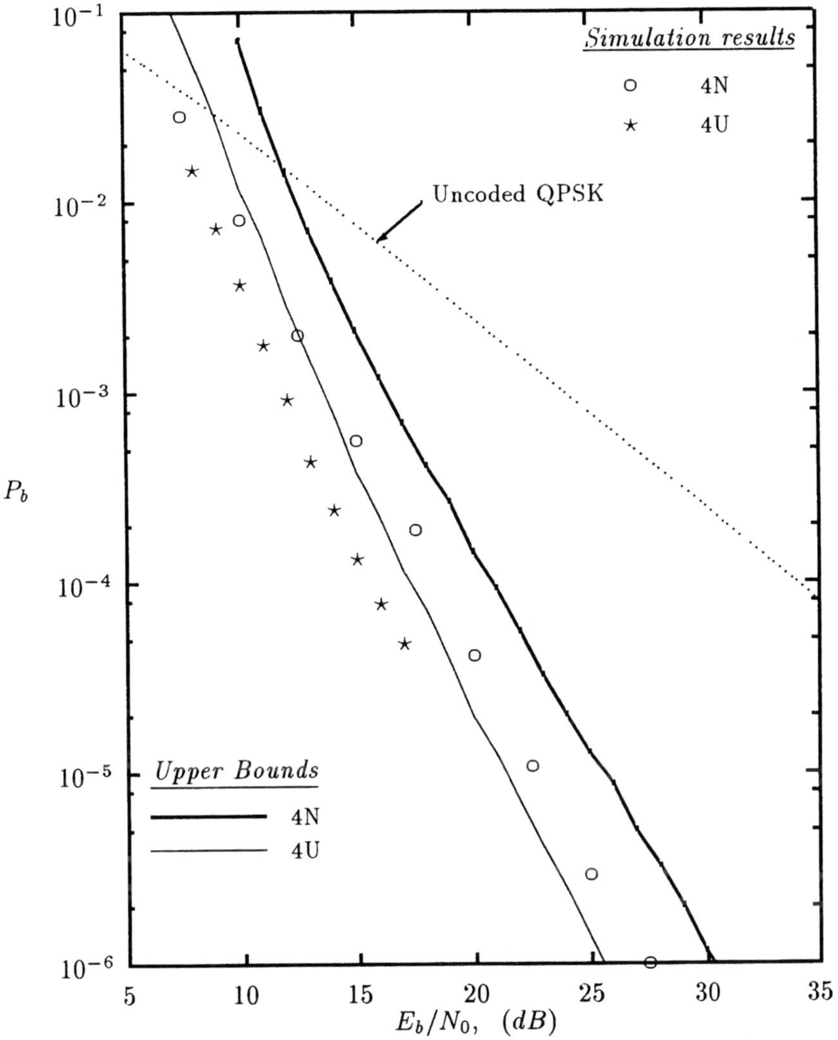


Figure 5.15: Comparison of the simulation results and upper bounds for 4-state and 8-state TCM schemes on a Rayleigh fading channel with CSI.

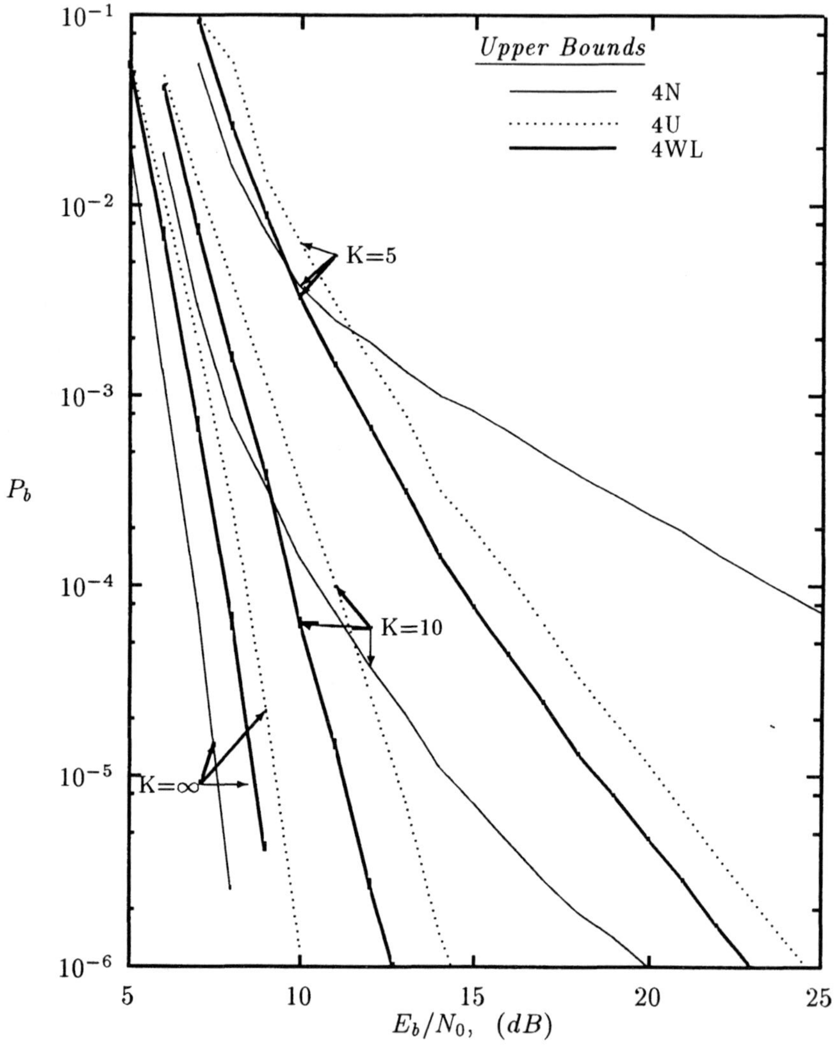


Figure 5.16: Performance of three different rate 2/3 4-state 8PSK TCM schemes on Rician fading channels with CSI.

The results shown in Figure 5.17 illustrate the benefit of using the channel state information (CSI) in the decoding process. It is interesting to note that the use of CSI in the decoding process improves the performance of the 4-state TCM scheme without parallel transitions (4N) more than that of the 4-state TCM scheme with parallel transitions (4U). This can be also denoted by observing the simulation results performed for these schemes on a Rayleigh fading channel with/without CSI. These results (Figure 5.18) show that, for the 4-state Ungerboeck code (4U), the use of the CSI slightly improves the performance of the scheme at low signal-to-noise ratios. However, its effect at high signal-to noise ratios is negligible. This may be explained by considering the contribution of different error events with different lengths at low signal-to-noise ratios and the effect of the dominant error event with the shortest length at high SNR's. For the 4U TCM scheme error events with length one are dominant at high signal-to-noise ratios. This means that for high SNR's the decision can be made by observing the received signal in one signaling interval. Thus, in such a case, the detection of the coded signals are similar to the detection of uncoded MPSK signals which is independent of the CSI because of their angular decision regions [17]. Based on this fact it seems that the use of CSI is more effective for codes with higher time diversities. Figure 5.18 shows that the improvement for the 4N TCM scheme due to the use of CSI is about 1.5 dB.

### 5.3.2 An Exact Expression for the Pairwise Error Probability of TCM Schemes on a Rayleigh Fading Channel

In the preceding sections we used the Chernoff bound to evaluate the pairwise error probabilities for Rician (Rayleigh) fading channels. Then, using the nature of the product form of the pairwise error probability, we developed an upper bound on bit error probability which can be efficiently evaluated by using the generating function method.

While the Chernoff bound technique is often used as an upper bound for the pairwise error probability because of its product form, it could be loose for a wide range of signal-to-noise ratios of interest. This is particularly true for the channels where the line-of-sight component is weak, i.e., the channel approaches the Rayleigh fading channel. In the following we present an *exact* analytical expression for the pairwise error probability of TCM schemes on a *Rayleigh* fading channel based on the work of Cavers and Ho [49]. This analysis is not limited to coherent detection and so the results can be applied to differential coherent detection as well. In the following analysis the assumption of ideal interleaving is used.

We begin by considering the baseband system model shown in Figure 5.2. Based on this model the received signal at time $i$ is expressed as (5.1). As stated

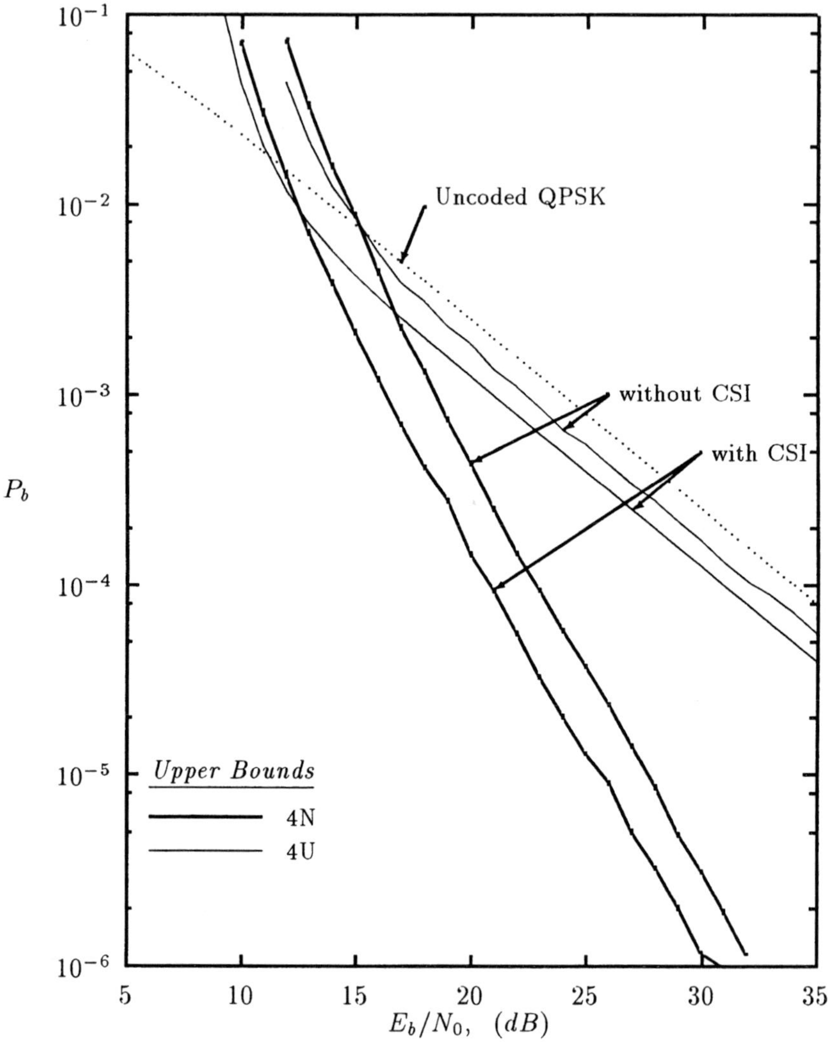


Figure 5.17: Performance of the 4U and 4N TCM schemes on a Rayleigh fading channel with/without CSI.

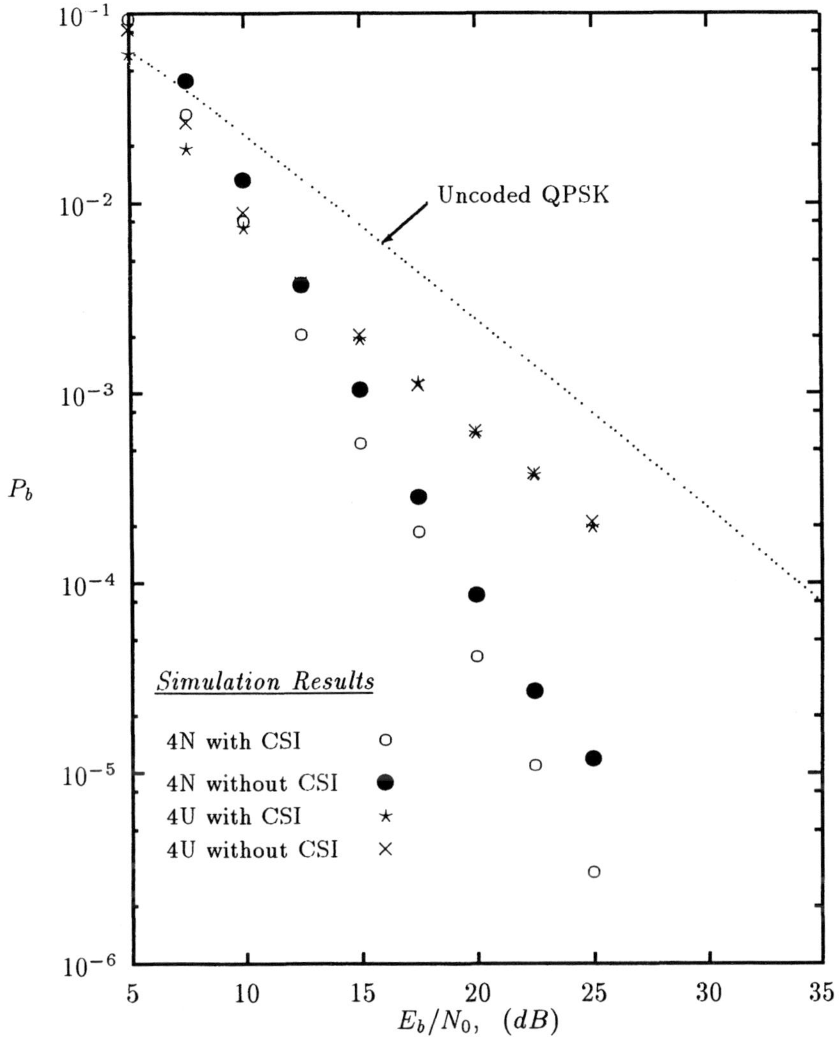


Figure 5.18: Simulation results for the 4U and 4N TCM schemes on a Rayleigh fading channel with/without CSI.

earlier, in this equation $n_i$ shows a sample of a zero-mean complex Gaussian noise process with a variance of $\sigma_n^2 = N_0/2$ and $g_i$ shows the complex channel gain. For a normalized Rayleigh fading channel $g_i$ is a sample of a zero-mean complex Gaussian process with a variance of $\sigma_g^2 = 1/2$. Using (5.1) it is easily shown that for a given transmitted signal $s_i$ the received signal $r_i$ is a complex Gaussian random variable with a zero mean and a variance of

$$\begin{aligned}\sigma_{r_i|s_i}^2 &= \sigma_g^2|s_i|^2 + \sigma_n^2 \\ &= \frac{1}{2}(|s_i|^2 + N_0).\end{aligned} \tag{5.47}$$

Assume that the receiver is provided with an estimate of the complex gain, say $\hat{g}_i$. The complex gain can be estimated by using different methods, such as a pilot tone or an embedded pilot sequence, or by differential detection. In this analysis we assume that the estimate of the complex channel gain $\hat{g}_i$ is a sample of a zero-mean Gaussian random process with a variance of $\sigma_{\hat{g}}^2$, which is correlated to $g_i$ by a correlation coefficient of

$$\mu = \frac{\sigma_{g\hat{g}}^2}{\sigma_g \sigma_{\hat{g}}}, \tag{5.48}$$

where $\sigma_{g\hat{g}}^2 = \frac{1}{2}E[g_i\hat{g}_i^*]$ is the covariance of $g_i$ and $\hat{g}_i$. Now, assuming that $\hat{g}_i$ is given, then $g_i$ is a complex Gaussian random variable with a mean of $m_{g|\hat{g}} = \mu\hat{g}_i\sigma_g/\sigma_{\hat{g}}$ and a variance of

$$\sigma_{g|\hat{g}}^2 = (1 - |\mu|^2)\sigma_g^2. \tag{5.49}$$

Using (5.1) and considering the fact that $g_i$ and $\hat{g}_i$ are two correlated complex Gaussian random variables we can show that for a given transmitted signal $s_i$ and a given $\hat{g}_i$, the received signal $r_i$ is a Gaussian random variable with a mean of

$$m_{r_i|\hat{g}_i,\, s_i} = \mu s_i \hat{g}_i \frac{\sigma_g}{\sigma_{\hat{g}}}, \tag{5.50}$$

and a variance of

$$\begin{aligned}\sigma_{r_i|\hat{g}_i,\, s_i}^2 &= \sigma_{g|\hat{g}}^2|s_i|^2 + \sigma_n^2 \\ &= \frac{1}{2}(|s_i|^2(1 - |\mu|^2) + N_0).\end{aligned} \tag{5.51}$$

Finally, (5.1) implies that for a given transmitted signal $s_i$ the received signal $r_i$ and the channel gain estimate $\hat{g}_i$ are two correlated complex Gaussian random variables with a covariance of

$$\sigma_{r_i,\, \hat{g}_i|s_i}^2 = \frac{1}{2}E[r_i\hat{g}_i^*] = s_i\sigma_{g\hat{g}}^2. \tag{5.52}$$

The foregoing discussion provides us with enough information to evaluate an exact expression for the pairwise error probability.

**Pairwise Error Probability**

As stated earlier in this chapter, given that the sequence of the coded symbols $\mathbf{s}_l = (s_1, s_2, \cdots, s_l)$ is transmitted a maximum likelihood (ML) decoder picks the erroneous coded sequence $\hat{\mathbf{s}} = (\hat{s_1}, \hat{s_2}, \cdots, \hat{s}_l)$ if

$$m(\mathbf{r}_l, \hat{\mathbf{s}}_l; \hat{\mathbf{g}}_l) \geq m(\mathbf{r}_l, \mathbf{s}_l; \hat{\mathbf{g}}_l), \tag{5.53}$$

where

$$m(\mathbf{r}_l, \mathbf{s}_l; \hat{\mathbf{g}}_l) = \sum_{i=1}^{l} \ln p_N(r_i|s_i, \hat{g}_i) \tag{5.54}$$

is the ML metric. Note that the ML metric defined in (5.54) is the same as (5.7) except that the fading amplitude $a_i$ in (5.7) is replaced by the complex channel gain estimate $\hat{g}_i$ in (5.54).

Using (5.47)-(5.52) and after some mathematical manipulation[12] the metric given in (5.54) can be simplified to

$$m(\mathbf{r}_l, \mathbf{s}_l; \hat{\mathbf{g}}_l) = -\sum_{i=1}^{l} \left\{ \ln(|s_i|^2\sigma_e^2 + \frac{N_0}{2}) + \frac{|r_i - s_i\hat{g}_i\beta|^2}{2|s_i|^2\sigma_e^2 + N_0} \right\}, \tag{5.55}$$

where

$$\beta \equiv \mu\frac{\sigma_g}{\sigma_{\hat{g}}}, \tag{5.56}$$

and

$$\sigma_e^2 \equiv \sigma_g^2(1 - |\mu|^2) = \frac{1}{2}(1 - |\mu|^2). \tag{5.57}$$

Note that $\sigma_e^2$, defined in (5.57), shows the variance of the channel gain estimation error $e_i \equiv g_i - \beta\hat{g}_i$. The metric given in (5.55) can be further simplified to $\sum_{i=1}^{l} |r_i - s_i\hat{g}_i\beta|^2$ for the constant envelope signals like MPSK where $|s_i| = |\hat{s}_i|$ or when the channel gain estimation error is negligible compared to the channel noise, i.e., $|s_i|^2\sigma_e^2 \ll N_0/2$.

Using (5.55) the condition for erroneous detection by an ML decoder, given in (5.53), can be simplified to

$$D^* = \sum_{i=1}^{l} D_i \leq \delta, \tag{5.58}$$

[12] In simplifying the metric we drop those terms which are independent of the signal $s_i$ and hence do not affect the decision.

where the decision variable $D_i$ is a quadratic form, in complex-valued Gaussian random variables $r_i$ and $\hat{g}_i$, given by

$$D_i = A_i|r_i|^2 + B_i|\hat{g}_i|^2 + C_i r_i \hat{g}_i^* + C_i^* r_i^* \hat{g}_i,$$

with

$$\begin{aligned}
A_i &= \frac{1}{2}\left(\frac{1}{2|\hat{s}_i|^2\sigma_e^2 + N_0} - \frac{1}{2|s_i|^2\sigma_e^2 + N_0}\right), \\
B_i &= |\beta|^2\left(\frac{|\hat{s}_i|^2}{2|\hat{s}_i|^2\sigma_e^2 + N_0} - \frac{|s_i|^2}{2|s_i|^2\sigma_e^2 + N_0}\right), \\
C_i &= \left(\frac{s_i^*\beta^*}{2|s_i|^2\sigma_e^2 + N_0} - \frac{\hat{s}_i^*\beta^*}{2|\hat{s}_i|^2\sigma_e^2 + N_0}\right), \\
\delta &= \ln\left(\prod_{i=1}^{l} \frac{2|s_i|^2\sigma_e^2 + N_0}{2|\hat{s}_i|^2\sigma_e^2 + N_0}\right).
\end{aligned}$$

Note that for constant envelope signals, such as MPSK, the coefficients $A_i$, $B_i$, and $\delta$ all become zero.

The probability that (5.58) satisfies is simply the pairwise error probability and can be expressed as

$$P_2(\mathbf{s}_l, \hat{\mathbf{s}}_l) = P(D^* \leq \delta) = \int_{-\infty}^{\delta} p(D^*)dD^*, \tag{5.59}$$

where $p(D^*)$ is the probability density function (pdf) of the variable $D^*$. To evaluate the pairwise error probability given in (5.59), we first find the two-sided Laplace transform of the pdf of $D^*$. Since the $D_i$'s are independent when conditioned on the codewords, the two-sided Laplace transform of the pdf of $D^*$, that is, $\phi_{D^*}(s)$, can be expressed as the product of the two-sided Laplace transform of the pdf of $D_i$'s (see Problem 5.23)

$$\phi_{D^*}(s) = \prod_{i \in \eta} \phi_{D_i}(s),$$

where $\eta$ is the set of $i$ for which $s_i \neq \hat{s}_i$.

The characteristic function of a quadratic form in complex Gaussian variates is evaluated in [59, Appendix B] and [60, Appendix 4B]. Using these results, with slight modification, the two-sided Laplace transform $\phi_{D_i}(s)$ is computed as [49] (see Problem 5.24)

$$\phi_{D_i}(s) = \frac{p_{1i}p_{2i}}{(s - p_{1i})(s - p_{2i})},$$

where

$$\begin{aligned}
p_{1i} &= w_i - \sqrt{w_i^2 - p_{1i}p_{2i}} \\
p_{2i} &= w_i + \sqrt{w_i^2 - p_{1i}p_{2i}}
\end{aligned}$$

are the left plane (LP) and the right plane (RP) poles, respectively,

$$\begin{aligned} p_{1i}p_{2i} &= -\frac{1+|s_i|^2(1-|\mu|^2)/N_0}{|\mu|^2 d_i^2/N_0}, \\ w_i &= -\frac{(1-|\mu|^2)(|s_i|^2-|\hat{s}_i|^2)-|\mu|^2 d_i^2}{2|\mu|^2 d_i^2}, \end{aligned}$$

and

$$d_i^2 = |s_i - \hat{s}_i|^2$$

is the squared Euclidean distance between the signals $s_i$ and $\hat{s}_i$.

To calculate the pairwise error probability, $p(D^*)$ can be obtained by computing $\phi_{D^*}(s)$ and taking its inverse Laplace transform by residues or partial fraction expansion. The integration is then performed to yield the pairwise error probability. An alternate, much simpler approach, however, is to calculate

$$P_2(\mathbf{s}_l, \hat{\mathbf{s}}_l) = \begin{cases} -\sum \text{Residue } [e^{s\delta}\phi_{D^*}(s)/s]_{\text{RPoles}}, & \delta \le 0 \\ \sum \text{Residue } [e^{s\delta}\phi_{D^*}(s)/s]_{\text{LPoles}}, & \delta > 0 \end{cases} \tag{5.60}$$

where $\phi_{D^*}(s)/s$ is the two-sided Laplace transform of the cumulative distribution function (cdf) of the random variable $D^*$, and the summation is taken over all distinct RP or LP poles.

**Example 5.16** Consider the case of coded PSK and perfect channel state information, where $|s_i| = |\hat{s}_i|$, and $|\mu|^2 = 1$. In this case

$$\begin{aligned} \delta &= 0 \\ p_{1i}p_{2i} &= -\frac{1}{d_i^2/N_0} \\ w_i &= \frac{1}{2}, \end{aligned}$$

and

$$\begin{aligned} p_{1i} &= \frac{1}{2} - \sqrt{\frac{1}{4} + \frac{1}{d_i^2/N_0}} \\ p_{2i} &= \frac{1}{2} + \sqrt{\frac{1}{4} + \frac{1}{d_i^2/N_0}}, \end{aligned}$$

where $d_i^2 = |s_i - \hat{s}_i|^2$. Using the above equations, the pairwise error probability, given in (5.60), for the case of coded PSK and perfect CSI is simplified to

$$\begin{aligned} P_2(\mathbf{s}_l,\ \hat{\mathbf{s}}_l) &= \left(\prod_{i\in\eta} \frac{1}{\frac{1}{4N_0}|s_i - \hat{s}_i|^2}\right) \cdot \\ &\quad \left(-\sum \text{Residue}\left[\frac{1}{s}\prod_{i\in\eta}\frac{-1}{4(s-p_{1i})(s-p_{2i})}\right]_{\text{RPoles}}\right). \end{aligned} \tag{5.61}$$

Note that the exact expression of the pairwise error probability of coded PSK schemes with perfect CSI, given in (5.61), has the form of the upper bound on the pairwise error probability, given in (5.13) for $K = 0$ (Rayleigh fading channel), multiplied by a correction factor. The correction factor depends on the locations of the poles, which in turn depend on the distance structure of the coded scheme.

### Bit Error Probability

Our ultimate goal in performance evaluation of the coded schemes is to calculate the bit error probability. Using the union bound, an upper bound on the bit error probability can be obtained by taking into consideration the effect of all possible error events of all lengths. Unfortunately, in enumerating such error events the generating function approach cannot be employed due to the form of the exact expression of $P_2(\mathbf{s}_l,\ \hat{\mathbf{s}}_l)$. Thus, the bit error probability can be estimated by considering a small number of dominant error events rather than by accounting for the error event paths of all lengths. In this case the bit error probability can be approximated as

$$P_b \approx \frac{1}{m} \sum_{l=1}^{\mathcal{L}} \sum_{\mathbf{s}_l \neq \hat{\mathbf{s}}_l} \overline{m_l} P_2(\mathbf{s}_l,\ \hat{\mathbf{s}}_l), \tag{5.62}$$

where $\overline{m_l}$ is the average number of bit errors associated with the error event $(\mathbf{s}_l,\ \hat{\mathbf{s}}_l)$ and $\mathcal{L}$ is a limit imposed on the actual length of the error events. Note that (5.62) is a truncation of the infinite series used in calculating the union upper bound on $P_b$. The value of $\mathcal{L}$ determines how accurately we can compute $P_b$. A larger value chosen for $\mathcal{L}$ yields a better accuracy at the expense of more computational complexity. In the following we consider two examples which reveal the details of the computation of $P_b$ given in (5.62).

**Example 5.17** Consider the 4-state 8PSK TCM scheme (4N) of Figure 5.6. To evaluate the bit error probability of this coded scheme we first compute the pairwise error probability for different error events. For this purpose we limit our calculation to the error events with $\mathcal{L} = 4$. To enumerate all error events with actual lengths less than or equal to 4 we use the error-state diagram of the 4N scheme, shown in Figure 5.13 with the labels given in Example 5.12.

Taking different powers of the error-state diagram transition matrix[13] $\mathbf{A}_e$, i.e., $\mathbf{A}_e^2$, $\mathbf{A}_e^3$, and $\mathbf{A}_e^4$, we can find the distance structure of all error events of lengths $l = 2$, 3, and 4, respectively, by considering the $a_{00}$ element of each matrix. This evaluation is done by hand for the 4N scheme and the results are shown in Table 5.6. In this table the effective length and the average number of bit errors associated with error events with effective lengths 2 and 3 and actual

[13] A $2^\nu \times 2^\nu$ matrix with the $ij^{th}$ element denoting the error weight profile of the branch from the $i$th state to the $j$th state of the error-state diagram , where $2^\nu$ is the number of states.

Table 5.6: Error Events for the 4N Scheme with Effective Lengths 2 and 3 and Actual Lengths Less Than or Equal to 4

| Error Event # | Actual Length $l$ | Effective Length $l_\eta$ | $\overline{m_l}$ | $\delta_0^2$ | $\delta_1^2$ | $\delta_2^2$ | $\delta_3^2$ |
|---|---|---|---|---|---|---|---|
| 1 | 2 | 2 | 0.5 | 1 | 0 | 0 | 1 |
| 2 | | 2 | 2.0 | 0 | 1 | 1 | 0 |
| 3 | | 2 | 0.5 | 0 | 0 | 1 | 1 |
| 4 | | 2 | 1.0 | 0 | 2 | 0 | 0 |
| 5 | 3 | 2 | 3.0 | 0 | 1 | 1 | 0 |
| 6 | | 3 | 3.5 | 1 | 1 | 1 | 0 |
| 7 | | 3 | 2.0 | 1 | 1 | 0 | 1 |
| 8 | | 3 | 1.0 | 1 | 0 | 1 | 1 |
| 9 | | 3 | 1.5 | 2 | 1 | 0 | 0 |
| 10 | | 3 | 0.5 | 2 | 0 | 0 | 1 |
| 11 | | 3 | 2.5 | 1 | 2 | 0 | 0 |
| 12 | | 3 | 2.5 | 0 | 2 | 1 | 0 |
| 13 | | 3 | 2.0 | 0 | 2 | 0 | 1 |
| 14 | | 3 | 2.0 | 0 | 1 | 2 | 0 |
| 15 | | 3 | 3.0 | 0 | 0 | 2 | 1 |
| 16 | 4 | 3 | 7.0 | 1 | 1 | 1 | 0 |
| 17 | | 3 | 4.0 | 1 | 0 | 1 | 1 |
| 18 | | 3 | 4.0 | 0 | 1 | 1 | 1 |
| 19 | | 3 | 2.0 | 2 | 1 | 0 | 0 |
| 20 | | 3 | 2.0 | 1 | 2 | 0 | 0 |
| 21 | | 3 | 2.0 | 0 | 2 | 1 | 0 |
| 22 | | 3 | 5.0 | 0 | 1 | 2 | 0 |

lengths 2, 3, and 4 are listed. The distance structure of each error event is denoted by the multiplicity of each distance.

Using the distance structure, the pairwise error probability of the error events in Table 5.6 can be computed. As sample calculations, we evaluate the pairwise error probability for error events #4 and # 5.

**Pairwise Error Probability of Error Event #4**

Using (5.61) the pairwise error probability of error event # 4 can be evaluated as

$$P_2(\# 4) = \frac{-1}{(\delta_1^2/4)^2}\left[\sum \text{Residue}\left(\frac{1}{16s(s-p_{11})^2(s-p_{21})^2}\right)\right]_{\text{RPoles}},$$

where

$$p_{11} = \frac{1}{2} - \sqrt{\frac{1}{4} + \frac{1}{\delta_1^2/N_0}},$$

$$p_{21} = \frac{1}{2} + \sqrt{\frac{1}{4} + \frac{1}{\delta_1^2/N_0}},$$

and $\delta_1^2 = 2E_s$. Note that error event # 4 has multiple poles of multiplicity of two. Using differentiation the residues in the above equation can be computed. The result of such a computation is

$$P_2(\# 4) = \frac{3p_{12} - p_{11}}{4(\frac{E_s}{N_0})^2 p_{12}^2 (p_{12} - p_{11})^3}.$$

**Pairwise Error Probability of Error Event #5**

In contrast to error event # 4 this error event has distinct poles (of multiplicity of one). Using (5.61) the pairwise error probability of error event # 5 is obtained as

$$P_2(\# 5) = \frac{-1}{(\delta_1^2/4)(\delta_2^2/4)} \left[ \sum \text{Residue} \left( \frac{1}{16s(s-p_{11})(s-p_{21})(s-p_{12})(s-p_{22})} \right) \right]_{\text{RPoles}},$$

where

$$p_{12} = \frac{1}{2} - \sqrt{\frac{1}{4} + \frac{1}{\delta_2^2/N_0}},$$

$$p_{22} = \frac{1}{2} + \sqrt{\frac{1}{4} + \frac{1}{\delta_2^2/N_0}},$$

and $\delta_2^2 = 3.414E_s$. The calculation of the residues in the above equation is straightforward and yields

$$P_2(\# 5) = \frac{1}{6.828(\frac{E_s}{N_0})^2} \left[ \frac{-1}{p_{12}(p_{12}-p_{11})(p_{12}-p_{21})(p_{12}-p_{22})} + \frac{-1}{p_{22}(p_{22}-p_{11})(p_{22}-p_{21})(p_{22}-p_{12})} \right].$$

The pairwise error probability of the other error events is calculated similarly.

Using (5.62) and the calculated pairwise error probabilities for different error events as given in Table 5.6, the bit error probability for the 4-state TCM (4N) scheme is plotted in Figure 5.19. The upper bound on $P_b$ and the simulation results are also included for comparison.

As shown in Figure 5.19 the bit error probability, given in (5.62), is evaluated for two cases; the first case considers the error events with an effective length

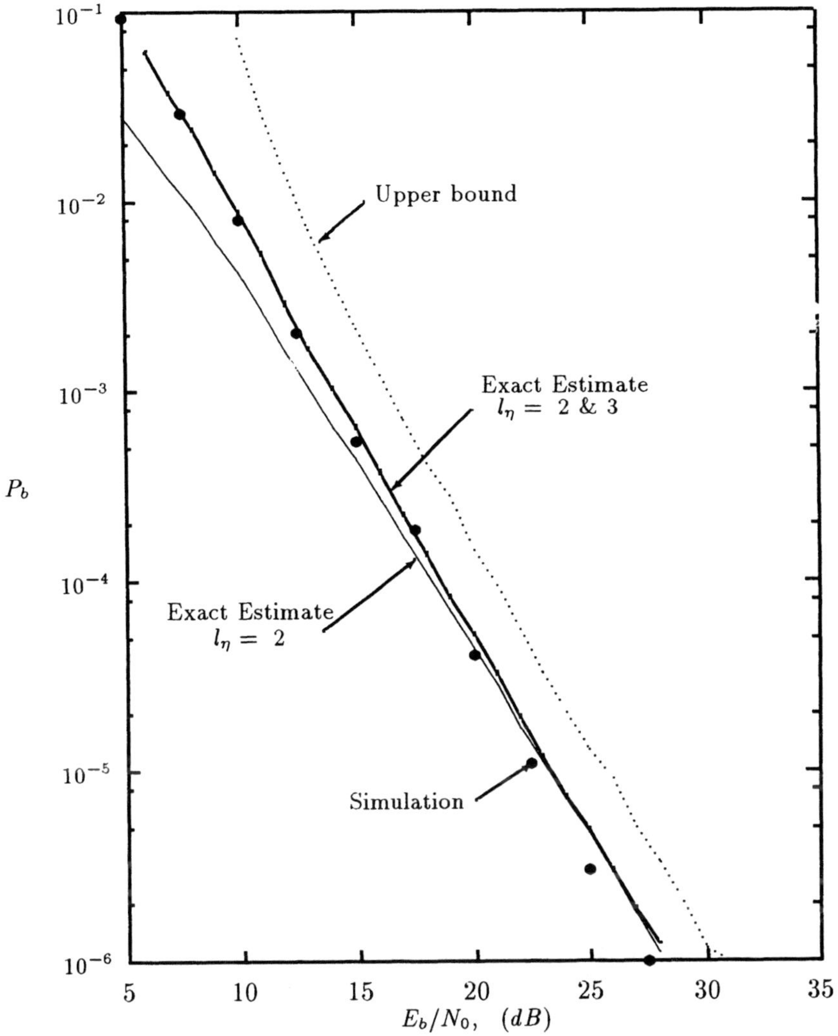


Figure 5.19: Bit error probability of the 4-state 8PSK (4N) TCM scheme in a Rayleigh fading channel with CSI.

Table 5.7: Error Events for the 8-state 8PSK TCM Scheme (8U) with Effective Lengths 2 and 3 and Actual Lengths Less Than or Equal to 4

| Error Event # | Actual Length $l$ | Effective Length $l_\eta$ | $\overline{m_l}$ | $\delta_0^2$ | $\delta_1^2$ | $\delta_2^2$ | $\delta_3^2$ |
|---|---|---|---|---|---|---|---|
| 1 | 2 | 2 | 1.0 | 0 | 1 | 0 | 1 |
| 2 | 3 | 3 | 2.0 | 1 | 1 | 0 | 1 |
| 3 | | 3 | 1.0 | 0 | 1 | 1 | 1 |
| 4 | | 3 | 1.5 | 1 | 2 | 0 | 0 |
| 5 | | 3 | 3.5 | 0 | 2 | 1 | 0 |
| 6 | | 3 | 2.0 | 0 | 2 | 0 | 1 |
| 7 | 4 | 3 | 3.0 | 0 | 1 | 1 | 1 |
| 8 | | 3 | 3.5 | 1 | 2 | 0 | 0 |
| 9 | | 3 | 2.0 | 1 | 0 | 0 | 2 |
| 10 | | 3 | 1.5 | 0 | 2 | 1 | 0 |

of $l_\eta = 2$ and the second case considers the error events with effective lengths $l_\eta = 2$, and 3. Comparison of these curves shows that at high signal-to-noise ratios only a small number of dominant error events is required to evaluate the bit error probability.

**Example 5.18** Consider the 8-state 8PSK TCM scheme of Figure 4.8. Using the error-state diagram of this scheme, shown in Figure 5.14 with branch labels given in Example 5.13, we find the distance structure for the error events with actual lengths less than or equal to 4. The distance structure for error events with effective lengths 2 and 3 and actual lengths less than or equal to 4 is listed in Table 5.7.

Using a similar approach to Example 5.17, the pairwise error probability for different error events, shown in Table 5.7, is found. The bit error probabilities of the 8-state TCM scheme considering the error event with $l_\eta = 2$ and the error events with $l_\eta = 2$, and 3 are plotted in Figure 5.20. Observe that at high signal-to-noise ratios the two curves approach each other. The upper bound on $P_b$ and the simulation results for this TCM scheme are also included in Figure 5.20 for comparison.

The results of the two foregoing examples show that when considering a small number of dominant error events (with short effective lengths) an estimate of the bit error probability is obtained by using (5.62) and the exact expression for the pairwise error probability. The simulation results show that such an estimate is accurate at high-signal-to-noise ratios. Comparing the analytical estimate and the upper bound of $P_b$ shows that the upper bound differs by about 2.2 dB from the analytical estimate at high signal-to-noise ratios.

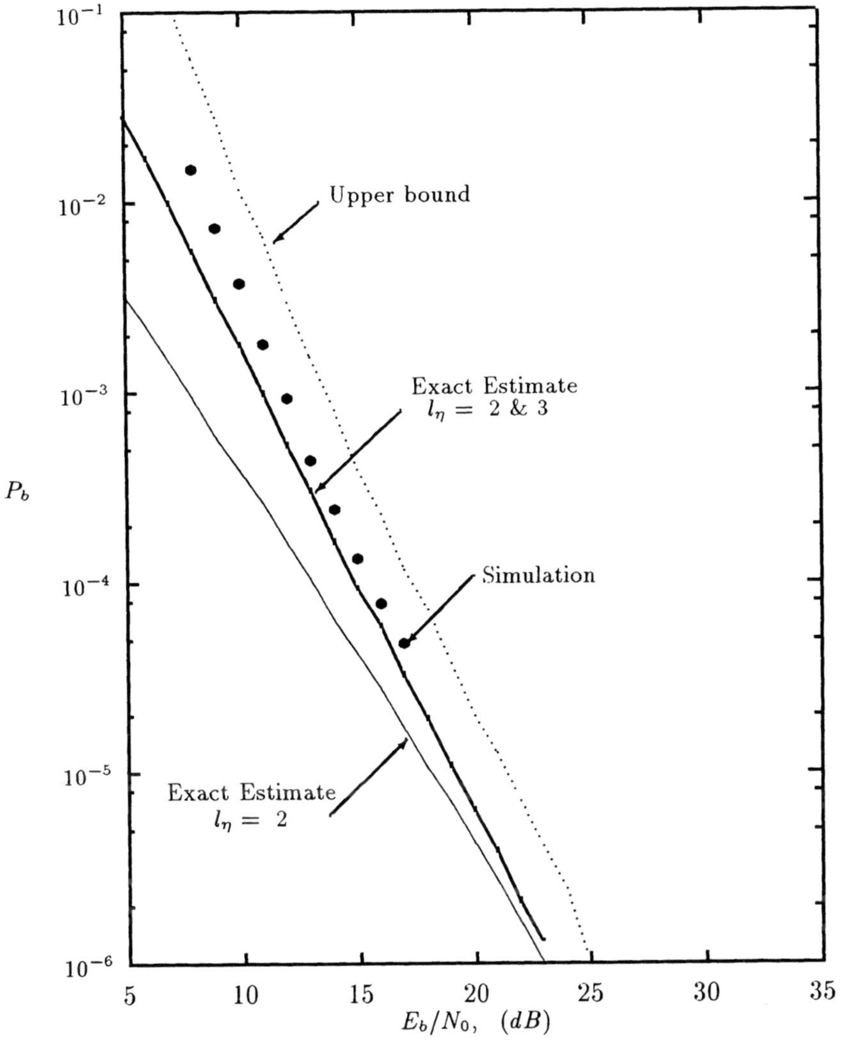


Figure 5.20: Bit error probability of the 8-state 8PSK (8U) TCM scheme in a Rayleigh fading channel with CSI.

### 5.3.3 Tighter Bounds on the Pairwise and Bit Error Probabilities in a Rayleigh Fading Channel with CSI

The Chernoff upper bound on the pairwise error probability in a Rayleigh fading channel with CSI can be derived by substituting (5.3) and (5.32) into (5.33) and performing the integration

$$P_2(\mathbf{s}_l, \hat{\mathbf{s}}_l) \leq \prod_{i \in \eta} \frac{1}{1 + d_i^2/4N_0}. \tag{5.63}$$

Recall that $d_i^2 = |s_i - \hat{s}_i|^2$ is the squared Euclidean distance between the signals $s_i$ and $\hat{s}_i$, and $\eta$ is the set of $i$ for which $s_i \neq \hat{s}_i$ as previously mentioned in Section 5.3.2.

The Chernoff bound is widely used for its simplicity. However as shown in the preceding sections, it could be loose for a wide range of signal-to-noise ratios of interest, particularly in Rayleigh fading channels. In the following we present the derivation and applications of tighter bounds on the pairwise and bit error probabilities in a Rayleigh fading channel with perfect CSI [123]. These bounds have a similar form to (5.63).

We assume the use of ideal interleaving so that the fading amplitudes $a_i$'s are considered to be statistically independent in the following derivation. We begin by using the exact expression for the conditional pairwise error probability with perfect CSI

$$P_2(\mathbf{s}_l, \hat{\mathbf{s}}_l | \mathbf{a}_l) = \frac{1}{2} \text{erfc}\left[ \mathrm{d_E}(\mathrm{s_l}, \hat{\mathrm{s}}_\mathrm{l} | \mathbf{a}_\mathrm{l}) \right], \tag{5.64}$$

where

$$d_E(\mathbf{s}_l, \hat{\mathbf{s}}_l | \mathbf{a}_l) = \left[ \frac{1}{4N_0} \sum_{i \in \eta} a_i^2 d_i^2 \right]^{1/2} \tag{5.65}$$

is the weighted Euclidean distance between sequences $\mathbf{s}_l$ and $\hat{\mathbf{s}}_l$.

To find the pairwise error probability in a Rayleigh fading channel with perfect CSI we average (5.64) over the pdf of $\mathbf{a}$, i.e.,

$$P_2(\mathbf{s}_l, \hat{\mathbf{s}}_l) = \frac{1}{2} \int_0^{+\infty} \cdots \int_0^{+\infty} \text{erfc}\left[ \mathrm{d_E}(\mathrm{s_l}, \hat{\mathrm{s}}_\mathrm{l} | \mathbf{a}_\mathrm{l}) \right] \mathrm{p_A}(\mathrm{a_1}) \cdots \mathrm{p_A}(\mathrm{a_l}) d\mathrm{a_1} \cdots d\mathrm{a_l}. \tag{5.66}$$

Define a new vector

$$\alpha_l = (\alpha_1, \alpha_2, ..., \alpha_l)$$

with

$$\alpha_i = a_i \sqrt{1 + d_i^2/4N_0} \tag{5.67}$$

It follows that

$$p_A(a_i) = p_A\left(\frac{\alpha_i}{\sqrt{1+d_i^2/4N_0}}\right)\frac{d\alpha_i}{\sqrt{1+d_i^2/4N_0}}$$
$$= \frac{1}{1+d_i^2/4N_0}e^{\alpha_i^2 x_i^2}p_A(\alpha_i)d\alpha_i, \tag{5.68}$$

where

$$x_i = \sqrt{\frac{d_i^2/4N_0}{1+d_i^2/4N_0}}. \tag{5.69}$$

Substituting (5.68) into (5.66) yields

$$P_2(\mathbf{s}_l, \hat{\mathbf{s}}_l) = \frac{1}{2}\prod_{i\in\eta}\frac{1}{1+d_i^2/4N_0}\int_0^{+\infty}\cdots\int_0^{+\infty} \operatorname{erfc}[\mathrm{d_E}(\mathbf{s}_l, \hat{\mathbf{s}}_l|\alpha_l)]$$
$$e^{d_E^2(\mathbf{s}_l,\hat{\mathbf{s}}_l|\alpha_l)}p_A(\alpha_1)\cdots p_A(\alpha_l)d\alpha_1\cdots d\alpha_l, \tag{5.70}$$

where $d_E(\mathbf{s}_l, \hat{\mathbf{s}}_l|\alpha_l)$ denotes the weighted Euclidean distance between two sequences $\mathbf{s}_l$ and $\hat{\mathbf{s}}_l$ in terms of $\alpha_i$'s, i.e., substituting (5.67) and (5.69) into (5.65) yields

$$d_E(\mathbf{s}_l, \hat{\mathbf{s}}_l|\alpha_l) = \sqrt{\sum_{i\in\eta}\alpha_i^2 x_i^2}. \tag{5.71}$$

From (5.69), for $d_{min} \le d_i \le d_{max}$, $i \in \eta$, we obtain

$$x_{min} = \sqrt{\frac{d_{min}^2/4N_0}{1+d_{min}^2/4N_0}} \le x_i \le x_{max} = \sqrt{\frac{d_{max}^2/4N_0}{1+d_{max}^2/4N_0}}. \tag{5.72}$$

It follows that the weighted Euclidean distance given by (5.71) can be bounded as

$$x_{min}^2 y \le d_E(\mathbf{s}_l, \hat{\mathbf{s}}_l|\alpha_l) \le x_{max}^2 y, \tag{5.73}$$

where

$$y = \sum_{i\in\eta}\alpha_i^2. \tag{5.74}$$

The pdf of $y$ can be derived as (see Problem 5.29)

$$p_Y(y) = \frac{1}{(l_\eta - 1)!}y^{l_\eta - 1}e^{-y}, \quad y > 0. \tag{5.75}$$

Recall that $l_\eta$ represents the number of elements of the set $\eta$, or the effective length of the error event $(\mathbf{s}_l, \hat{\mathbf{s}}_l)$.

Since the function $\mathrm{erfc}(x)e^{x^2}$ is monotonically decreasing with $x \geq 0$, from (5.70) and (5.73) we obtain (see Problem 5.30)

$$P_2(\mathbf{s}_l, \hat{\mathbf{s}}_l) \geq \mathcal{K}(l_\eta, x_{max}) \prod_{i \in \eta} \frac{1}{1 + d_i^2/4N_0} \tag{5.76}$$

and

$$P_2(\mathbf{s}_l, \hat{\mathbf{s}}_l) \leq \mathcal{K}(l_\eta, x_{min}) \prod_{i \in \eta} \frac{1}{1 + d_i^2/4N_0}, \tag{5.77}$$

where

$$\mathcal{K}(l_\eta, x) = \frac{1}{2} \int_0^{+\infty} \mathrm{erfc}(x\sqrt{y}) e^{x^2 y} p_Y(y) dy. \tag{5.78}$$

Substituting (5.75) into (5.78) and performing the integration, we obtain (see Problem 5.31)

$$\mathcal{K}(l_\eta, x) = \frac{1}{2^{l_\eta}(l_\eta - 1)!} \sum_{j=1}^{l_\eta} \frac{a_i^{(l_\eta)}}{(1+x)^j}, \tag{5.79}$$

where $a_j^{(l_\eta)}$'s are derived from the following recursive relations:

For $l_\eta = 1$, $a_1^{(1)} = 1$

For $l_\eta \geq 2$, $a_1^{(l_\eta)} = a_2^{(l_\eta)} = (2l_\eta - 3)!! = (2l_\eta - 3)(2l_\eta - 5)...3.1$

$a_j^{(l_\eta)} = 2\left[a_{j-1}^{(l_\eta)} - (l_\eta - 1)a_{j-2}^{(l_\eta)-1}\right], j = 3, 4, ..., l_\eta.$

As shown in (5.76) and (5.77), the lower and upper bounds on the pairwise error probability have a form equivalent to the Chernoff bound of (5.63) multiplied by a factor. This factor depends on $x_{max}$ and $x_{min}$ for lower and upper bounds, respectively. From (5.72), both $x_{min}$ and $x_{max}$ approach unity as the signal-to-noise ratio increases. Therefore, for high signal-to-noise ratios, the lower and upper bounds converge and the pairwise error probability can be approximated by

$$P_2(\mathbf{s}_l, \hat{\mathbf{s}}_l) \approx \mathcal{K}(l_\eta, 1) \prod_{i \in \eta} \frac{1}{1 + d_i^2/4N_0}, \tag{5.80}$$

where

$$\mathcal{K}(l_\eta, 1) = \frac{(2l_\eta - 1)!!}{2^{l_\eta+1} l_\eta!} = \frac{(2l_\eta - 1)!}{2^{l_\eta} l_\eta!(l_\eta - 1)!} \tag{5.81}$$

by performing the integration in (5.78) with $x = 1$ (see Problem 5.32 and 5.46).

**Example 5.19** Consider the case of uncoded BPSK. The performance of uncoded BPSK with perfect channel state information in a Rayleigh fading channel is (see Example 2.2)

$$P_b = \frac{1}{2}\left[1 - \sqrt{\frac{E_b/N_0}{1 + E_b/N_0}}\right]. \quad (5.82)$$

The Chernoff bound on the probability of bit error is bounded as

$$P_b \leq \frac{1}{2(1 + E_b/N_0)}. \quad (5.83)$$

Since uncoded BPSK uses symbol-by-symbol detection, the length of one error event path is exactly 1. The branch Euclidean distance through this path is given by

$$d_1^2 = E_b/N_0. \quad (5.84)$$

This means that

$$x_{min} = x_{max} = x = \sqrt{\frac{E_b/N_0}{1 + E_b/N_0}}. \quad (5.85)$$

Furthermore, when there is only one error event, the pairwise error probability is also the bit error probability of BPSK. Substituting (5.84) and (5.85) into (5.76) or (5.77) yields

$$\begin{aligned} P_b &= \frac{1}{2(1+x)} \frac{1}{1 + E_b/N_0} \\ &= \frac{1}{2}\left[1 - \sqrt{\frac{E_b/N_0}{1 + E_b/N_0}}\right], \end{aligned}$$

which is the exact expression given by (5.82).

**Example 5.20 Trellis Coded PSK Schemes:** Consider the 8-state 8PSK scheme shown in Figure 4.8. From Table 5.7, the length of the shortest error event path is 2.

With the assumption of perfect channel state information, the exact expression of the pairwise error probability of error event # 1 can be derived using a similar approach to Examples 5.17 and 5.18 (see Problem 5.27)

$$P_2(\# 1) = \frac{1}{\rho^2} \frac{1 + \rho\left(\sqrt{1 + 2\rho^{-1}} - \sqrt{1 + \rho^{-1}}\right)}{\left(1 + \sqrt{1 + 2\rho^{-1}}\right)\left(1 + \sqrt{1 + \rho^{-1}}\right)\left(\sqrt{1 + 2\rho^{-1}}\right)\left(\sqrt{1 + \rho^{-1}}\right)}, \quad (5.86)$$

where $\rho = E_s/N_0$. As we can see, even for a very short error event, the derivation of the exact expression of the pairwise error probability is quite time-consuming.

From Table 5.7 and Figure 4.1b, for error event # 1 we obtain

$$d_{min}^2 = \delta_1^2/4N_0 = 0.5\rho \text{ and } d_{max}^2 = \delta_3^2/4N_0 = \rho.$$

The Chernoff bound for error event # 1 is obtained by substituting the above values into (5.63)

$$P_2(\#1) \leq \frac{1}{(1+\rho)(1+0.5\rho)}. \tag{5.87}$$

We also obtain

$$x_{min} = \sqrt{\frac{0.5\rho}{1+0.5\rho}}, \text{ and } x_{max} = \sqrt{\frac{\rho}{1+\rho}}. \tag{5.88}$$

We get new lower and upper bounds as

$$P_2(\#1) \geq \frac{1}{4}\left[\frac{1}{1+x_{min}} + \frac{1}{(1+x_{min})^2}\right]\frac{1}{(1+\rho)(1+0.5\rho)}, \tag{5.89}$$

and

$$P_2(\#1) \leq \frac{1}{4}\left[\frac{1}{1+x_{max}} + \frac{1}{(1+x_{max})^2}\right]\frac{1}{(1+\rho)(1+0.5\rho)}. \tag{5.90}$$

The exact pairwise error probability, the Chernoff bound and the new bounds for error event # 1, given by (5.86), (5.87) and (5.89), (5.90), respectively, are shown in Figure 5.21. The bit energy $E_b$ is equal to $E_s/2$. It is observed from Figure 5.21 that even at a very low signal-to-noise ratio, both newly introduced upper and lower bounds are almost identical to the exact value of the pairwise error probability, while the Chernoff bound differs by about 2 dB.

Since factor $\mathcal{K}(l_\eta, x)$, given in (5.79), monotonically decreases with $l_\eta$, we obtain

$$\mathcal{K}(l_\eta, x_{min}) \leq \mathcal{K}(L, x_{min}),$$

where $L = min(l_\eta)$ is the effective length of the code as previously stated. It follows that

$$P_2(\mathbf{s}_l, \hat{\mathbf{s}}_l) \leq \mathcal{K}(L, x_{min}) \prod_{i=1}^{l} E_{a_i}\left[D^{d^2(s_i, \hat{s}_i)}\right], \tag{5.91}$$

where $\mathbf{s}_l = \mathcal{M}(\mathbf{C}_l)$ and $\hat{\mathbf{s}}_l = \mathcal{M}(\mathbf{c}_l \oplus \mathbf{E}_l)$. Recall that $E_{a_i}\left[D^{d^2(s_i,\hat{s}_i)}\right] = \left[1 + d_i^2/4N_0\right]^{-1}$ for the case of Rayleigh fading channels with perfect CSI.

Substituting (5.91) into (4.10) yields the upper bound on error event probability

$$P_e \leq \mathcal{K}(L, x_{min}) \sum_{l=1}^{+\infty} \sum_{C_l} \sum_{E_l \neq 0} P(\mathbf{C}_l) \prod_{i=1}^{l} E_{a_i}\left[D^{d^2(s_i,\hat{s}_i)}\right]. \tag{5.92}$$

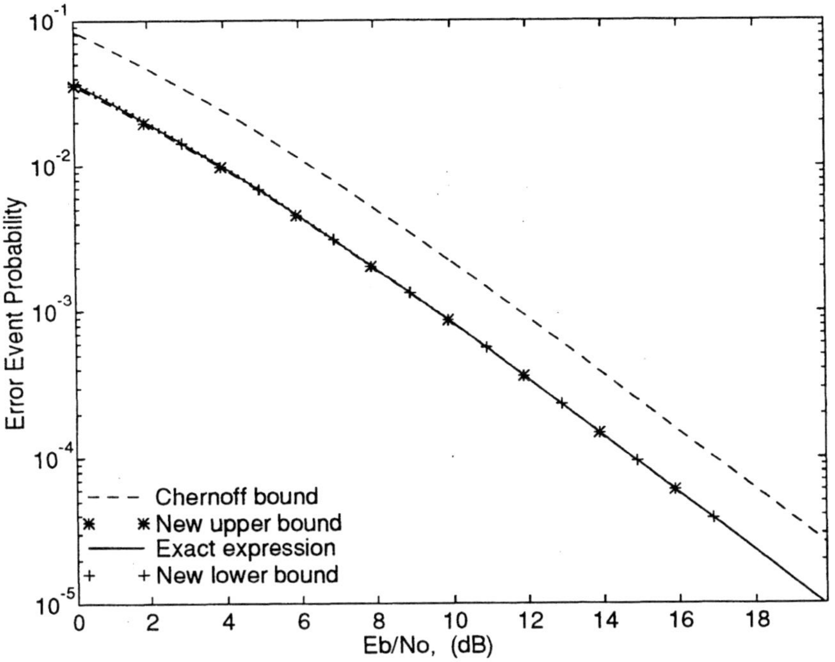


Figure 5.21: Bounds on the probability of error event # 1 of 8-state TCM 8PSK (8U) in a Rayleigh fading channel with perfect CSI.

Using the same argument as in Section 5.3.1 for the uniform TCM schemes, (5.92) can be rewritten as

$$P_e \leq \mathcal{K}(L, x_{min}) \sum_{l=1}^{+\infty} \sum_{E_l \neq 0} \prod_{i=1}^{l} \overline{\mathcal{F}}(\mathbf{e}_i, D) \quad (5.93)$$

$$= \mathcal{K}(L, x_{min}) T(D) \,|_{D=e^{-1/4N_0}} , \quad (5.94)$$

where $\overline{\mathcal{F}}(\mathbf{e}_i, D)$ is given by (5.36).

The upper bound on the bit error probability is subsequently obtained as

$$P_b \leq \frac{1}{m} \mathcal{K}(L, x_{min}) \frac{\partial T(D, I)}{\partial I} \,|_{I=1, D=e^{-1/4N_0}} , \quad (5.95)$$

where, for uniform TCM schemes, $T(D, I)$ is the generating function of the modified error-state diagram whose branches are modified as $I^r \overline{\mathcal{F}}(\mathbf{e}_i, D)$ and $r$ shows the number of incorrect input bits associated with $\mathbf{e}_i$.

It is observed that (5.95) is equivalent to the Chernoff bound of (5.39) multiplied by the factor $\mathcal{K}(L, x_{min})$. We will investigate the application and tightness of the new upper bound in the following examples.

**Example 5.21** Consider the 4-state 8PSK TCM schemes 4U (Fig. 4.5), 4WL (Fig. 5.3), and 4N (Fig. 5.6). Their effective length $L$ and corresponding value $x_{min}$ are

$$\begin{array}{llll} \text{4U:} & L=1, & d_{min}=\delta_3^2=4E_s & x_{min}=\sqrt{\frac{E_s/N_0}{1+E_s/N_0}} \\ \text{4WL:} & L=2, & d_{min}=\delta_0^2=0.568E_s & x_{min}=\sqrt{\frac{0.1465E_s/N_0}{1+0.1465E_s/N_0}} \\ \text{4N:} & L=2, & d_{min}=\delta_0^2=0.568E_s & x_{min}=\sqrt{\frac{0.1465E_s/N_0}{1+0.1465E_s/N_0}}. \end{array}$$

The generating function of the schemes 4U, 4WL and 4N are derived in Examples 5.10, 5.11 and 5.12, respectively. The new upper bound on the bit error probability of these schemes can be obtained by substituting the corresponding values of $L$, $x_{min}$ and the generating function into (5.95). Figure 5.22 shows the new upper bound on $P_b$ versus $E_b/N_0$ of the 4-state 8PSK TCM (4U) scheme in a Rayleigh fading channel with CSI. The new bound is very close to the simulation results, especially for $E_b/N_0 \geq 15$ dB, while the Chernoff bound is more than 3 dB away.

Figure 5.23 shows $P_b$ versus $E_b/N_0$ of the 4-state 8PSK TCM (4WL) scheme in a Rayleigh fading channel with CSI obtained by various techniques: simulation, Chernoff bound using (5.39), new bound using (5.95) and bound introduced in [65]. It is observed that the new bound using (5.95) is the closest to the simulation results. This improvement is obtained without increasing the complexity compared to the Chernoff bound. The bound introduced in [65] is derived by combining the exact pairwise error probabilities for a number of the most dominant error events (in terms of asymptotic contribution to $P_b$) with a weaker (but more readily determined) Chernoff bound for all remaining error events. The tightness of this bound and its complexity increase with the selected number of the most dominant error events. Seven dominant error events are used for the plot in Figure 5.23.

Figure 5.24 shows the new upper bound, Chernoff bound and simulation results on $P_b$ versus $E_b/N_0$ of the 4-state 8PSK TCM (4N) in a Rayleigh fading channel with CSI. It also indicates that the new upper bound is extremely close to the simulation results compared to the Chernoff bound.

**Example 5.22** Consider the 8-state 8PSK TCM (8U) scheme (Fig. 4.8). From Table 5.7, its effective length is $L = 2$ and $d_{min}^2 = \delta_1^2 = 2E_s$. Substituting the values of $L$ and $d_{min}^2$ into (5.95) yields the upper bound on its bit error probability as

$$P_b \leq \frac{1}{8}\left[\frac{1}{1+x_{min}} + \frac{1}{(1+x_{min})^2}\right]\frac{\partial T(D,I)}{\partial I},$$

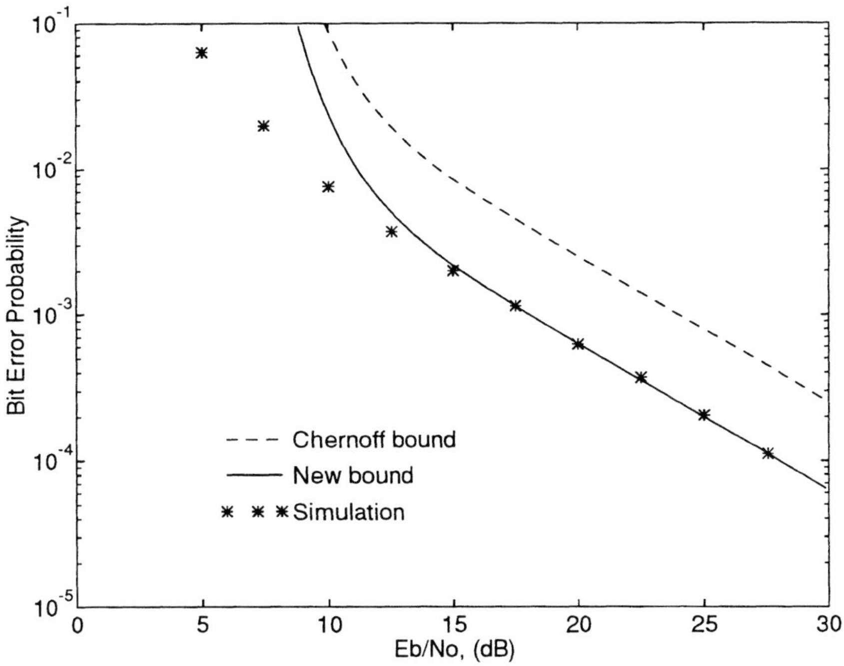


Figure 5.22: Upper bound on the bit error probability of the 4-state 8PSK TCM (4U) in a Rayleigh fading channel with perfect CSI.

where

$$x_{min} = \sqrt{\frac{E_s/2N_0}{1 + E_s/2N_0}}$$

and $T(D, I)$ is obtained as in Example 5.13.

Figure 5.25 shows the new upper bound, Chernoff bound, estimated and simulation results on $P_b$ versus $E_b/N_0$ of the 8-state 8PSK (8U) in a Rayleigh fading channel with CSI. The estimated results are obtained from Example 5.18 with $l_\eta = 3$. It is observed that for $E_b/N_0 \geq 15$ dB the new upper bound is extremely close to the simulation and estimated results.

The results of the above examples show that (5.95) provides a very tight upper bound on the bit error probability of TCM schemes in a Rayleigh fading channel with CSI. It is also observed that (5.95) has the same complexity as the Chernoff bound of (5.39).

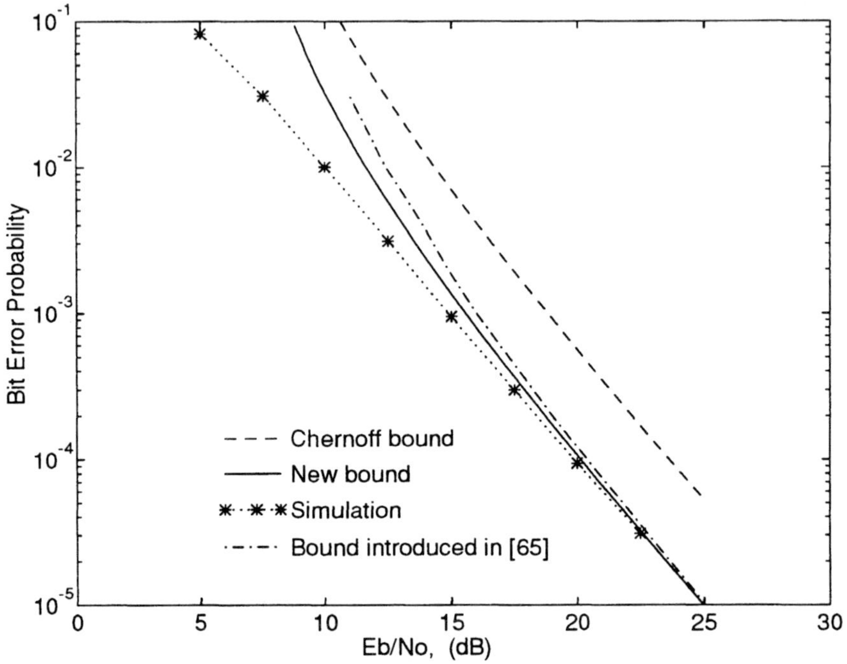


Figure 5.23: Upper bound on the bit error probability of the 4 state 8PSK TCM (4WL) in a Rayleigh fading channel with perfect CSI.

## 5.4 BIBLIOGRAPHICAL NOTES

Although TCM was originally developed for the AWGN channel, there has been considerable interest in recent years in applying TCM to fading channels. Wilson and Leung [48] and Divsalar and Simon [47] suggested that for a fading channel one has to use trellis codes which have higher diversity. The performance of the TCM schemes on a Rician fading channel was analyzed in [51] using the Chernoff bound technique.

The code design criteria for fading channels were formalized in [47] and the design of suitable multiple trellis-coded modulation (MTCM) schemes for fading channels was presented in [57] by Divsalar and Simon. The ability of conventional TCM schemes in providing time diversity in a Rician fading channel was exploited by Lodge and Moher [61]. In a later paper [62] they proposed a modulation and coding strategy for Rician fading channels, namely TCMP, which multiplexes a time pilot sequence with trellis-coded data to permit coherent detection. The performance of the TCM schemes with coherent detection

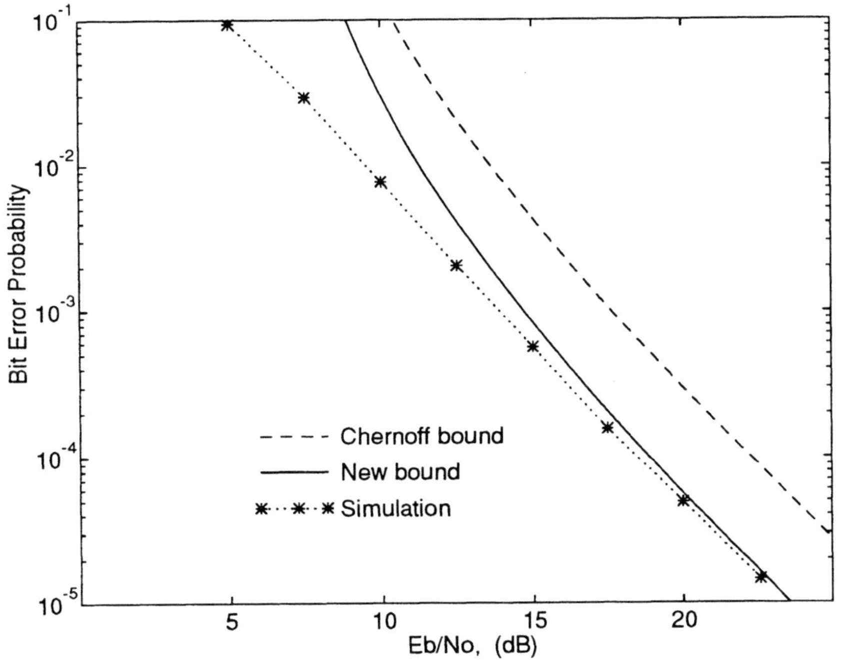


Figure 5.24: Upper bound on the bit error probability of the 4 state 8PSK TCM (4N) in a Rayleigh fading channel with perfect CSI.

on a Rician fading channel was also evaluated by Schlegel and Costello [55], [52] similar to the work of Divsalar and Simon [51].

Conventional TCM schemes suitable for fading channels were constructed by Wilson and Leung [48], Schlegel and Costello [55], [52], Jamali and Le-Ngoc [53], and Du and Vucetic [56]. Recently Zehavi [63] introduced a pragmatic approach in increasing the time diversity of the coded 8PSK scheme on a Rayleigh fading channel. His approach is based on using a convolutional code followed by three bit interleavers whose outputs are mapped to an 8PSK signal set. Since the interleaving is performed in bit level rather than symbol level, this approach provides more time diversity.

The performance analysis of TCM schemes on a Rayleigh fading channel based on finding an exact expression for the pairwise error probability was presented by Cavers and Ho [49]. McKay *et. al.* [64], [65] proposed improved upper bounds as well as lower bounds of TCM schemes on fading channels. Their technique involves computing error probability for a small number of pairwise error events. Tight upper and lower bounds on the pairwise error

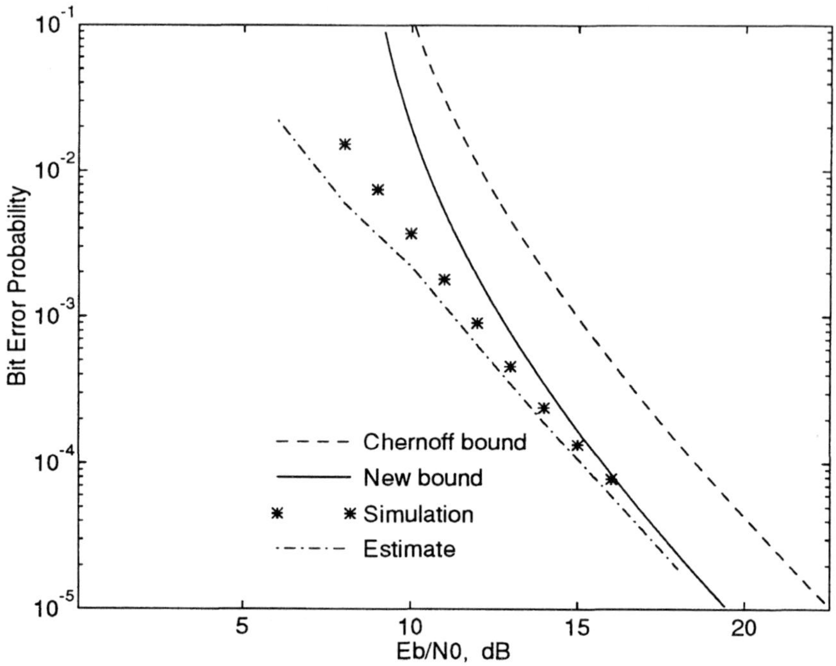


Figure 5.25: Performance of the 8-state 8PSK TCM (8U) in a Rayleigh fading channel with perfect CSI.

probability and bit error probability of TCM schemes in a Rayleigh fading channel with CSI were derived by Slimane and Le-Ngoc [123]. Their technique has a similar complexity as the Chernoff bound. The performance of trellis-coded MPSK schemes for different types of phase detection systems, such as pilot tone aided detection or pilot symbol aided detection, was evaluated in [66].

There are a number of other publications that deal with the performance of differential detection of TCM schemes on fading channels. For such references and also for those papers that deal with the other types of fading channels, e.g., shadowed Rician channels, the reader is referred to the bibliographical notes of the related chapters in the book.

# Problems

**5.1** Consider a BPSK scheme in an AWGN channel as shown in Figure P5.1-1. The transmitted BPSK signal can be represented by

$$s_{BPSK}(t) = \sum_{n=-\infty}^{+\infty} a_n p(t - nT) \cos w_c t,$$

where $T$ is the symbol (or bit) interval, $a_n = \pm 1$ with equal probability and $p(t)$ is the inverse Fourier transform of $P(f) = |P(f)|e^{-j\beta f}$, where $\beta$ is a constant ($\beta$ can be set to 0 without loss of generality),

$$P(f) = \begin{cases} \sqrt{T}, & 0 \leq |f| \leq (1-\alpha)/2T \\ \sqrt{\frac{T}{2}\left[1 - \frac{1}{\alpha}\sin \pi T(f - 1/2T)\right]}, & \frac{1-\alpha}{2T} \leq |f| \leq \frac{1+\alpha}{2T} \end{cases}$$

$0 \leq \alpha \leq 1.$

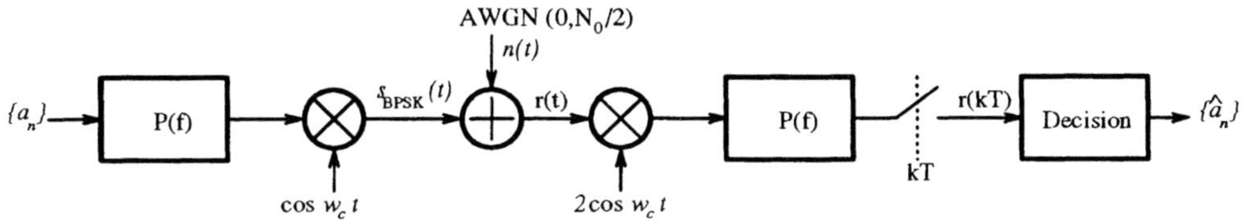


Figure P5.1-1: Block diagram of a BPSK scheme in an AWGN channel.

**a)** Show that the transmitted power spectrum has the same shape of $P(f)$ and that the symbol energy is

$$E_s = E_b = \int_{-\infty}^{+\infty} P(f)df = 1.$$

**b)** Show that the received sample $r(kT)$ in Figure P5.1-1 can be represented as

$$r(kT) = s(kT) + n(kT),$$

where $s(kT) = a_k$ (i.e., $s(kT)$ has no intersymbol interference and $|P(f)|^2$ meets the Nyquist criteria [60]) and $n(kT)$ is a Gaussian random variable with zero mean and variance

$$\sigma^2 = \frac{N_0}{2}\int_{-\infty}^{+\infty} P(f)df = \frac{N_0}{2}.$$

**c)** Show that the error probability of this scheme with a coherent receiver is

$$P_e = P_b = \frac{1}{2}\mathrm{erfc}\left(\frac{d}{\sqrt{8\sigma^2}}\right),$$

where $d = 2$.

**d)** Now, consider an MPSK scheme in an AWGN channel using the pulse-shape $p(t)$ where $T$ is the symbol interval, similar to the block diagram of Figure P5.1-1. The transmitted MPSK signal can be presented as

$$s_{MPSK}(t) = Re\left\{\sum_{n=-\infty}^{+\infty} \underline{a}_n p(t - nT)e^{-jw_c t}\right\},$$

where $Re\{\cdot\}$ is the real part of $\{\cdot\}$, $\underline{a}_n$ is the two-dimensional vector at time $nT$ chosen from an MPSK signal set $\{\underline{s}_0, \underline{s}_1, \cdots, \underline{s}_{M-1}\}$ (see Chapter 2) with $|\underline{s}_i| = \sqrt{E_s}$ for $i = 0, 1, \cdots, (M-1)$.

**i)** Find the transmitted power spectrum and average energy per symbol.

**ii)** Sketch the block diagram of the optimum receiver. Show that the received sample vector $\underline{r}(kT)$ can be represented as

$$\underline{r}(kT) = \underline{a}_k + \underline{n}(kT),$$

where $\underline{n}(kT)$ is a Gaussian random vector with zero mean and variance

$$\sigma^2 = \frac{N_0}{2}\int_{-\infty}^{+\infty} P(f)df = \frac{N_0}{2}.$$

**iii)** Show that the error probability of this MPSK scheme is given by

$$\frac{1}{2}\mathrm{erfc}\left(\sin(\pi/\mathrm{M})\sqrt{\frac{\mathrm{E_s}}{\mathrm{N_0}}}\right) \leq \mathrm{P_e} \leq \mathrm{erfc}\left(\sin(\pi/\mathrm{M})\sqrt{\frac{\mathrm{E_s}}{\mathrm{N_0}}}\right).$$

**5.2 a)** Show that at high signal-to-noise ratios, (5.17) can be written as

$$\begin{aligned}
P_2(\mathbf{s}_l, \hat{\mathbf{s}}_l) &\leq \min_{\lambda} \prod_{i\in\eta} e^{-K+\lambda^2|s_i-\hat{s}_i|^2/N_0}(1 - \sqrt{\pi}\mathrm{erfc}(\nu)e^{\nu^2}\nu) \\
&\leq \min_{\lambda}\left(\prod_{i\in\eta}[1 - \sqrt{\pi}\nu e^{\nu^2}\mathrm{erfc}(\nu)]\right) e^{\frac{\lambda^2}{N_0}\sum_{i\in\eta}|s_i-\hat{s}_i|^2 - l_\eta K} \\
&\leq \min_{\lambda} \frac{(1+K)^{l_\eta}}{2^{l_\eta}\left(\frac{\lambda}{2N_0}\right)^{2l_\eta} d_p^4(l_\eta)} e^{\frac{\lambda^2}{N_0}\sum_{i\in\eta}|s_i-\hat{s}_i|^2 - l_\eta K},
\end{aligned}$$

where

$$\nu = \frac{\lambda}{2N_0} \frac{|s_i - \hat{s}_i|^2}{\sqrt{1+K}}$$

**b)** From the above inequality, prove that the optimum value of $\lambda$ satisfies

$$\lambda_{opt}^2 = \frac{l_\eta}{\frac{1}{N_0}\sum_{i\in\eta} |s_i - \hat{s}_i|^2}$$

and

$$P_2(\mathbf{s}_l, \hat{\mathbf{s}}_l) \le d_p^{-4}(l_\eta) \left[\frac{2N_0(1+K)e^{1-K}(\sum_{i\in\eta} |s_i - \hat{s}_i|^2)}{l_\eta}\right]^{l_\eta}.$$

**5.3** **a)** For a large value of $K$ (i.e., the channel disturbance is dominated by AWGN) show that (5.12) can be written as

$$P_2(\mathbf{s}_l, \hat{\mathbf{s}}_l) \le \prod_{i=1}^{l} \exp\left(-\frac{|s_i - \hat{s}_i|^2}{4N_0}\right),$$

i.e., the dominant factor in the performance of TCM schemes is the free Euclidean distance.

**b)** For a low signal-to-noise ratio, show that (5.12) can also be written as

$$P_2(\mathbf{s}_l, \hat{\mathbf{s}}_l) \le \exp\left(-\frac{d_E^2(\mathbf{s}_l, \hat{\mathbf{s}}_l)}{4N_0}\right).$$

**5.4** A TCM scheme designed for Rician fading channels has an effective length $L$ and a minimum product distance $d_p^2(L)$. Suppose that this code is redesigned to provide an improvement of 3 dB in signal-to-noise ratio.

**a)** Compute the desired product distance $d_p^2(L)$ if we keep the same effective length $L$.

**b)** Compute the desired effective length $L$ if we keep the same $d_p^2(L)$.

**5.5** Using Figure 5.3, compute the free Euclidean distance of the Wilson-Leung code. Show that this code provides a gain of 1.1 dB compared to uncoded QPSK in an AWGN channel. Verify that this code is inferior to the Ungerboeck code by 1.9 dB in an AWGN channel.

**5.6** Follow the code design procedure of Example 5.1 and choose the signal point $s_1$ instead of $s_3$ in step 2. Show that the resulting code has the trellis structure of Figure 5.7. Compute the free Euclidean distance and the smallest squared product distance of this code. Show that this code has the same performance as the code of Example 5.1.

**5.7** Find the analytical description of the 4-state 8PSK TCM scheme of Example 5.2. Sketch the block diagram of the encoder of this scheme.

**5.8** Can the encoder of the 4-state 8PSK TCM scheme of Example 5.2 be realized by using the Ungerboeck representation?

*Hint*: Follow the procedure given in Example 5.4.

**5.9** Compute the approximated increase in coding gain of the rate 2/3 8PSK TCM schemes constructed by Schlegel and Costello (Table 5.3) compared to Ungerboeck codes of the same number of states in a Rician fading channel at a bit error rate of $10^{-4}$.

**5.10** Repeat Problem 5.9 for the rate 2/3 8PSK TCM schemes constructed by Du and Vucetic (Table 5.4).

**5.11** Draw the trellis diagram of the 4-state, rate 2/3 8PSK TCM code constructed by Du and Vicetic (Table 5.4). Derive the free Euclidean and minimum squared product distances.

**5.12** Compare the performance of the 4-state rate 2/3 8PSK TCM scheme of Problem 5.11 with that of Example 5.1.

**5.13** Show that if $\mu^*$ is a solution of (5.29) then $(M - \mu^*)$ is also a valid solution. Find the maximum solution(s), $\mu^*$, of (5.29) for $M = 4, 8, 16$. Verify your results with Table 5.5.

**5.14 a)** Following the procedure given in Example 5.6, derive the partition $\mathbf{A} \times \mathbf{B}_k, k = 0, 1, \cdots, 7$, with $\mu^* = 3$. Compute the intradistance and minimum squared product distance of each partition and the interdistance between the subsets.

**b)** Find the subsets of the second partitioning level. Compute the minimum squared product distance of each subset.

**c)** Draw the trellis diagram of the resulting 4-state, rate 4/6 8PSK MTCM scheme. Compute the free Euclidean minimum squared product distance of this MTCM scheme. Derive the coding gain of this MTCM scheme in an AWGN channel compared to uncoded QPSK.

**5.15** Compute the interdistance between the subsets $\mathbf{A} \times \mathbf{B}_k$ given in (5.30).

**5.16** Consider a QPSK signal set $\mathbf{A} = \{s_i, i = 0, 1, 2, 3\}$ as defined in Figure P5.16-1.

**a)** Find the subsets of $\mathbf{A} \times \mathbf{A}$ for the first two partitioning levels. At each partitioning level, calculate the intradistance of each subset and the interdistance between the subsets.

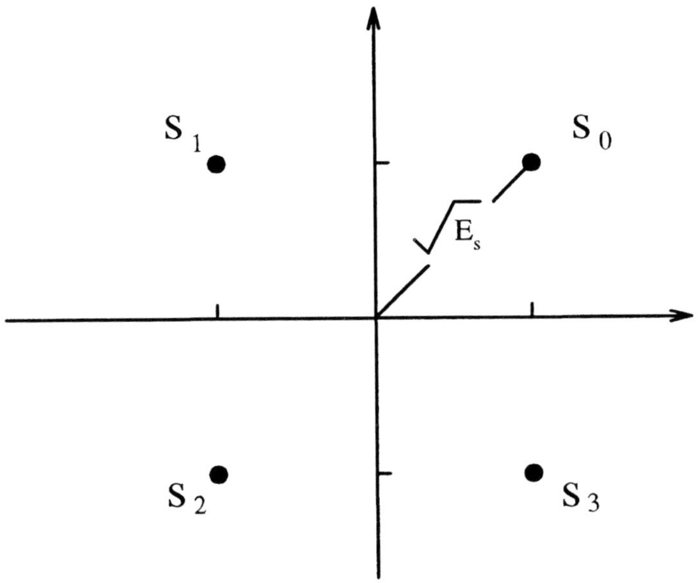


Figure P5.16-1: QPSK signal set.

b) Derive the trellis diagram of a 2-state, rate 2/4 QPSK MTCM scheme using the subsets obtained in a). Compute the free Euclidean distance and minimum squared product distance of this MTCM scheme. Derive the coding gain of this MTCM scheme in an AWGN channel compared to uncoded BPSK.

c) Repeat b) for a 4-state rate 2/4 QPSK MTCM scheme.

**5.17** Consider the design of a 2-state, rate 4/12 8PSK MTCM scheme with a multiplicity $k^* = 4$ and an 8PSK signal set $\mathbf{A} = \{s_i, i = 0, 1, \cdots, 7\}$.

a) Find the subsets of $\mathbf{A} \times \mathbf{A} \times \mathbf{A} \times \mathbf{A}$ for the first partitioning level. Calculate the intradistance of each subset and the interdistance between the subsets.

b) Construct the trellis diagram of the designed 2-state, rate 4/12 8PSK MTCM scheme. Compute the free Euclidean distance and minimum squared product distance of this code. Derive the coding gain of this MTCM scheme in an AWGN channel compared to uncoded BPSK.

**5.18** Draw the trellis diagram for the 4-state, rate 2/3 8PSK TCM scheme of Problem 5.6. Is this TCM scheme uniform?

**5.19** Are the rate 2/3 8PSK TCM schemes constructed by Du and Vucetic (Table 5.4) uniform? why?

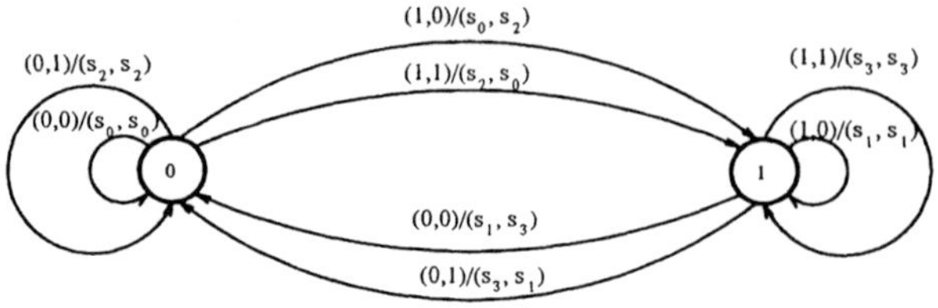


Figure P5.21-1: State-diagram of a 2-state, rate 2/4 QPSK MTCM scheme.

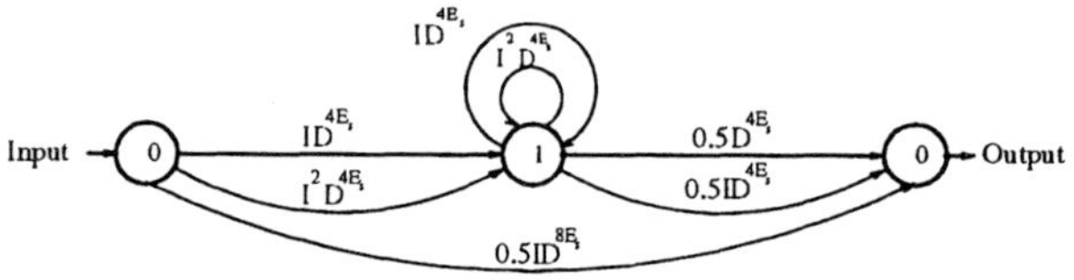


Figure P5.21-2: Error-state diagram (for an AWGN channel).

**5.20** Derive and plot the upper bound on the bit error probability of the 16-state, rate 2/3 8PSK TCM scheme constructed by Schlegel and Costello (Table 5.3).

**5.21** Consider the 2-state, rate 2/4 QPSK MTCM scheme obtained in Problem 5.16. Is it uniform? why?

**5.22** Repeat Problem 5.21 for the 4-state, rate 2/4 QPSK MTCM scheme obtained in Problem 5.16.

**5.23** The characteristic function of a random variable $X$ with pdf $p_X(x)$ is defined as

$$\Omega_X(jw) = \Phi_X(w) = E\left\{e^{jwX}\right\} = \int_{-\infty}^{+\infty} p_X(x)e^{jwx}dx$$

Show that

**a)** $\Phi_X(-w)$ is the Fourier transform of $p_X(x)$,

**b)** the moment function of $X$, $\Omega_X(s)$, is the Laplace transform of $p_X(x)$,

**c)** $|\Phi_X(w)| \leq |\Phi_X(0)| = 1$,

**d)** if $X$ is the sum of $n$ independent random variables $X_i$'s then

$$\Phi_X(w) = \prod_{i=1}^{n} \Phi_{X_i}(w)$$

and

$$\Omega_X(s) = \prod_{i=1}^{n} \Omega_{X_i}(s).$$

**5.24** Consider $Z = \sum_{i=1}^{n} Z_i$ where $Z_i$'s are independent random variables. Each individual random variable $Z_i$ is related to a pair of two correlated complex-valued Gaussian random variables $X_i$, $Y_i$ as

$$Z_i = A_i|X_i|^2 + B_i|Y_i|^2 + C_i X_i Y_i^* + C_i^* X_i^* Y_i$$

where $|C_i|^2 - A_i B_i > 0$.

**a)** Show that the characteristic function of $Z_i$ is

$$\Omega_{Z_i}(jw) = \frac{w_{1i}w_{2i}}{(w + jw_{1i})(w - jw_{2i})} \exp\left[\frac{w_{1i}w_{2i}(-a_{1i}w^2 + ja_{2i}w)}{(w + jw_{1i})(w - jw_{2i})}\right]$$

where

$$w_{1i} = \sqrt{w_{0i}^2 + w_{1i}w_{2i}} - w_{0i}$$

$$w_{2i} = \sqrt{w_{0i}^2 + w_{1i}w_{2i}} + w_{0i}$$

$$w_{1i}w_{2i} = \left(|C_i|^2 - A_iB_i\right)^{-1}\left(\mu_{xx_i}\mu_{yy_i} - |\mu_{xy_i}|^2\right)$$

$$w_{0i} = \frac{1}{2}w_{1i}w_{2i}(A_i\mu_{xx_i} + B_i\mu_{yy_i} + C_i\mu_{xy_i}^* + C_i^*\mu_{xy_i})$$

$$a_{1i} = (|C_i|^2 - A_iB_i)(|\mu_{x_i}|^2\mu_{yy_i} + |\mu_{y_i}|^2\mu_{xx_i} - \mu_{x_i}^*\mu_{y_i}\mu_{xy_i} - \mu_{x_i}\mu_{y_i}^*\mu_{xy_i}^*)$$

$$a_{2i} = A_i|\mu_{x_i}|^2 + B_i|\mu_{y_i}|^2 + C_i\mu_{x_i}^*\mu_{y_i} + C_i^*\mu_{x_i}\mu_{y_i}^*,$$

and $\mu_{x_i}$, $\mu_{y_i}$ are the means of $X_i$, $Y_i$, respectively. $\mu_{xx_i}$, $\mu_{yy_i}$, $\mu_{xy_i}$ are the central moments of $X_i$, $Y_i$, i.e., $\mu_{x_i} = E\{X_i\}$, $\mu_{y_i} = E\{Y_i\}$, $\mu_{xx_i} = E\{(X_i - \mu_{x_i})^2\}$, $\mu_{yy_i} = E\{(Y_i - \mu_{y_i})^2\}$, $\mu_{xy_i} = E\{(X_i - \mu_{x_i})(Y_i - \mu_{y_i})\}$.

**b)** Show that, for $\mu_{x_i} = \mu_{y_i} = 0$, we obtain

**i)**

$$\Omega_Z(jw) = \prod_{i=1}^{n} \frac{w_{1i}w_{2i}}{(w + jw_{1i})(w - jw_{2i})}$$

**ii)**

$$\Omega_Z(s) = \prod_{i=1}^{n} \frac{-w_{1i}w_{2i}}{(s + w_{1i})(s - w_{2i})}$$

**iii)**

$$
\begin{aligned}
Pr\{Z < 0\} &= \int_{-\infty}^{0} p_Z(z)dz \\
&= -\frac{1}{2\pi}\int_{-\infty+j\epsilon}^{+\infty+j\epsilon} \frac{\Omega_Z(jw)}{jw}dw,
\end{aligned}
$$

where $\epsilon$ is a small positive number inserted in order to move the path of integration away from the singularity at $w$

**iv)**

$$
Pr\{z < 0\} = -\sum_{RPoles} \text{residues}\left(\frac{\Omega_Z(s)}{s}\right)_{RPoles}.
$$

**5.25** Calculate the pairwise error probability of error events # 1, 2, 3 listed in Table 5.6 for the 4-state, 8PSK TCM scheme (4N) of Figure 5.6. (*Hint*: see Example 5.17.)

**5.26** Repeat Problem 5.25 for error events # 6, 7, 8.

**5.27** Compute the pairwise error probability of error event # 1 given in Table 5.7 and the estimated bit error probability with $l_\eta = 2$ for the 8-state 8PSK TCM scheme (8U). Compare your result with that given in Figure 5.20 and Example 5.20.

**5.28** Compare the pairwise error probability of error events # 2, 3, 4 given in Table 5.7 for the 8-state 8PSK TCM scheme (8U).

**5.29** Consider a random variable $Y$ defined as

$$
Y = \sum_{i=0}^{N-1} Y_i^2,
$$

where $Y_i$'s are independent and identically distributed random variables with pdf

$$
p_{Y_i}(y) = 2ye^{-y^2}, \quad y \geq 0.
$$

**a)** Prove that the characteristic function of $Y$ is given by

$$
\Phi_Y(w) = \frac{1}{(1-jw)^N}.
$$

**b)** Show that $\Phi_Y(w)$ can be rewritten as

$$
\Phi_Y(w) = \frac{1}{(N-1)!}\frac{d^{N-1}}{d(jw)^{N-1}}\left(\frac{1}{1-jw}\right),
$$

where

$$\frac{1}{1-jw} = E_{Y_i}\left\{e^{jwY_i^2}\right\}.$$

**c)** Show that the pdf of $Y$ is

$$p_Y(y) = \frac{1}{(N-1)!}y^{N-1}e^{-y}, \qquad y \geq 0.$$

**5.30** Show that $\mathrm{erfc(x)e^{x^2}}$ is monotonically decreasing with $x \geq 0$.

*Hint*: Take the derivative of $\mathrm{erfc(x)e^{x^2}}$ and show that this derivative is less than 0 for $x \geq 0$.

**5.31** Consider the following integration:

$$I_N = \int_0^{+\infty} \mathrm{erfc(x\sqrt{y})y^{N-1}e^{-y(1-x^2)}dy}, \qquad \mathrm{x < 1}.$$

**a)** By introducing $u = \sqrt{y}$ and performing the integration by part, show that

$$I_N = \frac{1}{1-x^2}\left[(N-1)I_{N-1} - \frac{2x}{\sqrt{\pi}}\int_0^{+\infty} u^{2N-2}e^{-u^2}du\right].$$

**b)** Prove that

$$\int_0^{+\infty} u^{2n}e^{-u^2}du = \frac{(2n-1)!!\sqrt{\pi}}{2^{n+1}}, \qquad n : \text{integer}$$

where $(2n-1)!! = (2n-1)(2n-3)\cdots 5.3.1$.

**c)** Using the result obtained in b), show that

$$I_N = \frac{1}{1-x^2}\left[(N-1)I_{N-1} - \frac{(2N-3)!!x}{2^{N-1}}\right], \qquad N \geq 2$$

with

$$I_1 = \frac{1}{2(1+x)}.$$

**d)** Prove that $I_N$ can be rewritten as

$$I_N = \frac{1}{2^N}\sum_{i=1}^{N}\frac{a_j^{(N)}}{(1+x)^j},$$

where $a_j^{(N)}$ is computed from the following recursive relation

$$\text{for } N = 1, \qquad a_1^{(1)} = 1,$$

$$\text{for } N \geq 2 \qquad a_1^{(N)} = a_2^{(N)} = (2N-3)!!$$

$$a_j^{(N)} = 2\left[a_{j-1}^{(N)} - (N-1)a_{j-2}^{(N-1)}\right], \qquad j = 3, 4, \cdots, N.$$

**5.32** Consider the following integration

$$J_N = \int_0^{+\infty} \mathrm{erfc}(\sqrt{\mathrm{y}})\mathrm{y}^{\mathrm{N}-1}\mathrm{dy}.$$

Show that

$$J_N = \frac{(2N-1)!}{2^{2N-1} \cdot N!},$$

**5.33** Using (5.76) and (5.77), derive the tight bounds on the pairwise error probability of error event # 4 for the 4-state 8PSK TCM scheme (4N) given in Table 5.6. Derive the approximated pairwise error probability of this error event for high signal-to-noise ratios using (5.80). Compare the results with that given in Example 5.17.

**5.34** Repeat Problem 5.33 for error event # 5.

**5.35** Using (5.80), derive the approximated pairwise error probability of error events # 1, 2, 3 listed in Table 5.6 for the 4-state, 8PSK TCM scheme (4N). Compare your results with those obtained in Problem 5.25.

**5.36** Compute the approximated pairwise error probability of error events # 6, 7, 8 listed in Table 5.6 for the 4-state, 8PSK TCM scheme (4N) using (5.80). Compare your results with those obtained in Problem 5.26.

**5.37** Derive the approximated pairwise error probability for error events # 2, 3, 4 given in Table 5.7 for the 8-state 8PSK TCM scheme (8U) using (5.81). Compare your results with those obtained in Problem 5.28.

**5.38** Derive the tight upper bound on the bit error probability of the 16-state rate 2/3 8PSK TCM scheme constructed by Schlegel and Costello (Table 5.3). Compare the result with that obtained in Problem 5.20.

**5.39** **Coded $\frac{\pi}{4}$-shift QPSK over fading channels**:
Derive the tight upper bound on the bit error probability of the 4-state, rate 2/3 trellis-coded $\frac{\pi}{4}$-shift QPSK scheme of Problem 4.24 in a Rayleigh fading channel.

**5.40** Repeat Problem 5.39 for the 8-state, rate 2/3 trellis-coded $\frac{\pi}{4}$-shift QPSK scheme of Problem 4.25.

**5.41** Consider the $\frac{\pi}{4}$-shift QPSK scheme discussed in Problem 2.22b). Observe that the signal set $\mathcal{S}_{16}$ is the Cartesian product of $\mathcal{S}_{8even}$ and $\mathcal{S}_{8odd}$. Using the set partitioning procedure for MTCM, construct a 2-state, rate 2/4 $\frac{\pi}{4}$-shift QPSK MTCM scheme. Derive the trellis diagram. Is this MTCM scheme uniform? why?

**5.42** Repeat Problem 5.41 for a 4-state, rate 2/4 $\frac{\pi}{4}$-shift QPSK MTCM scheme.

**5.43** **a)** Design a rate 1/2 2-state QPSK MTCM scheme for the following multiplicities:

$$k = 1, \qquad k = 2, \qquad k = 3.$$

**b)** Determine the effective code length, $L$, the squared minimum Euclidean distance, $d^2_{min}$, and the product of the squared Euclidean distances, $d^2(L)$, for each of the schemes designed in a).

**c)** Find the asymptotic coding gain of the schemes designed in a). Compare the performance of these schemes in AWGN and fading channels with interleaving/de-interleaving.

**d)** Derive the upper bound on $P_b$ for the TCM scheme with $k = 1$ on a Rayleigh fading channel with CSI.

**5.44** In the Nakagami-$m$ fading model the pdf of the fading amplitude is expressed as

$$p_A(a) = \frac{2m^m a^{2m-1}}{\Gamma(m)\Omega^m} e^{-ma^2/\Omega},$$

where $\Gamma(m)$ is the Gamma function with $\Gamma(m) = (m-1)!$ for integer $m$. Also $\Omega = E\{a^2\}$ is the average energy of the fading amplitude. For a normalized Nakagami-$m$ fading channel $\Omega = 1$ and

$$p_A(a) = \frac{2m^m a^{2m-1}}{\Gamma(m)} e^{-ma^2}.$$

The probability of bit error of a binary FSK scheme using envelope detection in an AWGN channel is given as

$$P_b = \frac{1}{2} e^{-E_b/2N_0}.$$

**a)** Derive the probability of bit error ($P_b$) of this scheme in a normalized Nakagami-$m$ fading channel.

**b)** Repeat a) for a normalized Rayleigh fading channel.

**c)** Using the results of a) and b) derive $P_b$ at a high signal-to-noise ratio for normalized Rayleigh and normalized Nakagami-$m$ fading channels. Compare and discuss the results.

**5.45** A rate 2/3 Ungerboeck 8AM TCM scheme is used in a Nakagami-$m$ fading channel with pdf

$$p_A(a) = \frac{2m^m a^{2m-1}}{\Gamma(m)} e^{-ma^2}$$

where $\Gamma(m) = (m-1)!$. For perfect CSI and ideal interleaving/de-interleaving, find the error weight profile for $\mathbf{e} = 101$ in terms of $E_b/N_0$.

**5.46 a)** Show that, for high signal-to-noise ratios, (5.61) can be approximated by

$$P_2(\mathbf{s}_l,\ \hat{\mathbf{s}}_l) \approx A(l_\eta) \prod_{i \in \eta} \frac{1}{d_i^2/4N_0}$$

where

$$A(l_\eta) = -\sum \text{Residue} \left[ \frac{1}{s} \prod_{i \in \eta} \frac{-1}{4s(s-1)} \right]_{\text{Rpoles}} .$$

*Hint*: For high signal-to-noise ratios, $p_{1i} \approx 0$ and $p_{2i} \approx 1$.

**b)** Show that $A(l_\eta) = \mathcal{K}(l_\eta,\ 1)$ of (5.81), and compare the expression for $P_2(\mathbf{s}_l,\ \hat{\mathbf{s}}_l)$ in a) with that of (5.80).

*Hint*: The residue of $f(z)$ at the pole $z_0$ of order $m$ is given by

$$b_0 = \frac{g^{(m-1)}(z_0)}{(m-1)!},$$

where $g^{(m-1)}(z)$ is the $(m-1)$th derivative of $g(z) = (z - z_0)^m f(z)$.

# Chapter 6

# REED-SOLOMON CODES

The purpose of this chapter is to give an overview of Reed-Solomon (RS) codes [67] as an important subclass of nonbinary Bose-Chaudhuri-Hocquenghem (BCH) codes. Our aim is in this chapter to cover the background which is required for the forthcoming sections related to RS codes. We start with definitions and operations in the finite field GF($2^m$). Then, after a brief introduction of cyclic and BCH codes we provide the definitions and encoding of the RS codes in GF($2^m$). Finally, we discuss various decoding algorithms of RS codes including the soft-decision decoding techniques.

## 6.1 BASIC DEFINITIONS AND OPERATIONS IN GF($2^m$)

To study the structure of Reed-Solomon codes, as a subclass of linear block codes, an algebraic background is required. We assume that the reader has such a background and is familiar with the notion of a *field* as a set of objects (numbers) which can be added, subtracted, multiplied, and divided using associative, distributive and commutative rules. Among the fields the finite fields, namely *Galois fields*, are of great interest in coding theory. We begin our discussion with the finite field GF(2).

The finite field GF(2) contains two elements 0 and 1. The two operations, namely, addition (+) and multiplication ($\cdot$) are defined in the following tables.

| Addition | | |
|---|---|---|
| + | 0 | 1 |
| 0 | 0 | 1 |
| 1 | 1 | 0 |

| Multiplication | | |
|---|---|---|
| $\cdot$ | 0 | 1 |
| 0 | 0 | 0 |
| 1 | 0 | 1 |

The elements 0 and 1 are the *additive identity element* and the *multiplicative identity element* of GF(2), respectively. The *characteristic* of GF(2) is 2 which is defined as the smallest number of times for which the multiplicative identity can be summed to produce the additive identity.

A polynomial of degree $m$ defined over GF(2) is represented as

$$p(x) = \sum_{i=0}^{m} p_i x^i,$$

where $p_i$'s are elements in GF(2), i.e., $p_i = 0$ or 1. A polynomial $p(x)$ of degree $m$ with coefficients in GF(2) is said to be *irreducible* if it is not divisible by any polynomial with coefficients in GF(2) of degree less than $m$ and greater than zero.

**Example 6.1** The polynomial $p(x) = x^3+x+1$ in GF(2) is an irreducible polynomial of degree 3.

Evaluating $p(x)$ at $x = 0$ or $x = 1$ we note that $p(0) = p(1) = 1$. Thus, $x^3+x+1$ is not divisible by either $x$ or $x+1$. We can easily argue that this polynomial is not divisible by any polynomial of degree 2. If it were divisible by a polynomial of degree 2, it would have a quotient of degree one, contradicting the fact that $x^3+x+1$ is not divisible by any polynomial of degree one. Therefore, $x^3+x+1$ is irreducible in GF(2).

The finite field GF($2^m$) is an *extension* of GF(2) generated from an irreducible polynomial of degree $m$ over GF(2). Equivalently, GF(2) is a *subfield* of GF($2^m$), i.e., the elements of GF(2), 0 and 1, are also the elements of GF($2^m$). The finite field GF($2^m$) also has its *primitive* element $\alpha$ which is the $(2^m-1)$th root of unity, that is,

$$\alpha^{(2^m-1)} = 1.$$

The finite field GF($2^m$) contains $2^m$ elements: $0, 1, \alpha, \alpha^2, \cdots, \alpha^{(2^m-2)}$. The irreducible polynomial of degree $m$ over GF(2), $p(x)$, has no roots in GF(2) but its roots lie in some extension field of GF(2). If $p(x)$ has the primitive element $\alpha$ of GF($2^m$) as a root, then it is called a *primitive* (irreducible) polynomial.[1] In this case, $\alpha^i$ with $i \leq m$ can be represented in terms of $1, \alpha, \alpha^2, \cdots, \alpha^{m-1}$, by using the fact that $\alpha^i$ is also a root of $p(x)$, that is,

$$p(\alpha^i) = 0.$$

[1] The list of irreducible and primitive polynomials of various degrees over GF(2) can be found in [68].

An element of GF($2^m$), $\beta$, can be represented as a binary $m$-tuple

$$\beta = (\beta_{m-1},\ \beta_{m-2},\ \cdots,\ \beta_1,\ \beta_0).$$

In this case, the zero element is represented as 0=(0, 0,$\cdots$,0). For $0 \le i \le (m-1)$, the binary components of the $m$-tuple representing $\beta = \alpha^i$ satisfy the following relation:

$$\beta_j = \begin{cases} 1, & \text{if } j = i, \\ 0, & \text{elsewhere,} \end{cases}$$

for $j = 0, 1, 2, \cdots, (m-1)$.

Let $\beta$ and $\gamma$ be two elements of GF($2^m$) represented in terms of the primitive element $\alpha$ and binary $m$-tuples as

$$\beta = \alpha^i = (\beta_{m-1},\ \beta_{m-2},\ \cdots,\ \beta_1,\ \beta_0),$$

and

$$\gamma = \alpha^j = (\gamma_{m-1},\ \gamma_{m-2},\ \cdots,\ \gamma_1,\ \gamma_0).$$

The two operations, additions (+) and multiplication ($\cdot$) in GF($2^m$), are defined as follows:

**Addition:**

$$\beta + \gamma = \theta = (\theta_{m-1},\ \theta_{m-2},\ \cdots,\ \theta_1,\ \theta_0), \tag{6.1}$$

where

$$\theta_i = \beta_i + \gamma_i, \qquad i = 0,\ 1,\ \cdots,\ (m-1). \tag{6.2}$$

**Multiplication**

$$\begin{aligned} \beta \cdot 0 &= 0, \\ \beta \cdot 1 &= \beta, \\ \beta \cdot \gamma &= \alpha^i \cdot \alpha^j = \alpha^{(i+j) \bmod (2^m-1)}. \end{aligned} \tag{6.3}$$

Note that 0 is the additive identity element of GF($2^m$) and every element $\beta$ is the *additive inverse* of itself, i.e., $\beta + \beta = 0$. Thus, subtraction is identical to addition. Also, 1 is the multiplicative identity of GF($2^m$) and $\alpha^{-i}$ is the *multiplicative inverse* of $\alpha^i$. Division $\div$ is defined as

$$\alpha^i \div \alpha^j = \alpha^i \cdot \alpha^{-j} = \alpha^{(i-j) \bmod (2^m-1)},$$

where

$$\alpha^j \neq 0.$$

It can be shown that for $n = 2^m - 1$ (see Problem 6.5)

$$\sum_{i=0}^{n-1} \alpha^{ij} = \begin{cases} 0, & \text{if } j \neq 0, \\ 1, & \text{if } j = 0. \end{cases}$$

**Example 6.2** Consider the finite field GF($2^3$) generated by the primitive polynomial $p(x) = x^3 + x + 1$. The elements of this finite field are

| Elements | | | Binary 3-tuple Representation | | |
|---|---|---|---|---|---|
| $0$ | $=$ | $0$ | 0 | 0 | 0 |
| $\alpha^0$ | $=$ | $1$ | 0 | 0 | 1 |
| $\alpha^1$ | $=$ | $\alpha$ | 0 | 1 | 0 |
| $\alpha^2$ | $=$ | $\alpha^2$ | 1 | 0 | 0 |
| $\alpha^3$ | $=$ | $\alpha + 1$ | 0 | 1 | 1 |
| $\alpha^4$ | $=$ | $\alpha^2 + \alpha$ | 1 | 1 | 0 |
| $\alpha^5$ | $=$ | $\alpha^3 + \alpha^2 = \alpha^2 + \alpha + 1$ | 1 | 1 | 1 |
| $\alpha^6$ | $=$ | $\alpha^4 + \alpha^3 = \alpha^2 + 1$ | 1 | 0 | 1. |

Using the definitions of addition and multiplication, given in (6.1)-(6.3), the addition and multiplication tables for GF($2^m$) are obtained as follows:

**Addition**

| $+$ | $0$ | $1$ | $\alpha$ | $\alpha^2$ | $\alpha^3$ | $\alpha^4$ | $\alpha^5$ | $\alpha^6$ |
|---|---|---|---|---|---|---|---|---|
| $0$ | $0$ | $1$ | $\alpha$ | $\alpha^2$ | $\alpha^3$ | $\alpha^4$ | $\alpha^5$ | $\alpha^6$ |
| $1$ | $1$ | $0$ | $\alpha^3$ | $\alpha^6$ | $\alpha$ | $\alpha^5$ | $\alpha^4$ | $\alpha^2$ |
| $\alpha$ | $\alpha$ | $\alpha^3$ | $0$ | $\alpha^4$ | $1$ | $\alpha^2$ | $\alpha^6$ | $\alpha^5$ |
| $\alpha^2$ | $\alpha^2$ | $\alpha^6$ | $\alpha^4$ | $0$ | $\alpha^5$ | $\alpha$ | $\alpha^3$ | $1$ |
| $\alpha^3$ | $\alpha^3$ | $\alpha$ | $1$ | $\alpha^5$ | $0$ | $\alpha^6$ | $\alpha^2$ | $\alpha^4$ |
| $\alpha^4$ | $\alpha^4$ | $\alpha^5$ | $\alpha^2$ | $\alpha$ | $\alpha^6$ | $0$ | $1$ | $\alpha^3$ |
| $\alpha^5$ | $\alpha^5$ | $\alpha^4$ | $\alpha^6$ | $\alpha^3$ | $\alpha^2$ | $1$ | $0$ | $\alpha$ |
| $\alpha^6$ | $\alpha^6$ | $\alpha^2$ | $\alpha^5$ | $1$ | $\alpha^4$ | $\alpha^3$ | $\alpha$ | $0$ |

**Multiplication**

| $\cdot$ | $0$ | $1$ | $\alpha$ | $\alpha^2$ | $\alpha^3$ | $\alpha^4$ | $\alpha^5$ | $\alpha^6$ |
|---|---|---|---|---|---|---|---|---|
| $0$ | $0$ | $0$ | $0$ | $0$ | $0$ | $0$ | $0$ | $0$ |
| $1$ | $0$ | $1$ | $\alpha$ | $\alpha^2$ | $\alpha^3$ | $\alpha^4$ | $\alpha^5$ | $\alpha^6$ |
| $\alpha$ | $0$ | $\alpha$ | $\alpha^2$ | $\alpha^3$ | $\alpha^4$ | $\alpha^5$ | $\alpha^6$ | $1$ |
| $\alpha^2$ | $0$ | $\alpha^2$ | $\alpha^3$ | $\alpha^4$ | $\alpha^5$ | $\alpha^6$ | $1$ | $\alpha$ |
| $\alpha^3$ | $0$ | $\alpha^3$ | $\alpha^4$ | $\alpha^5$ | $\alpha^6$ | $1$ | $\alpha$ | $\alpha^2$ |
| $\alpha^4$ | $0$ | $\alpha^4$ | $\alpha^5$ | $\alpha^6$ | $1$ | $\alpha$ | $\alpha^2$ | $\alpha^3$ |
| $\alpha^5$ | $0$ | $\alpha^5$ | $\alpha^6$ | $1$ | $\alpha$ | $\alpha^2$ | $\alpha^3$ | $\alpha^4$ |
| $\alpha^6$ | $0$ | $\alpha^6$ | $1$ | $\alpha$ | $\alpha^2$ | $\alpha^3$ | $\alpha^4$ | $\alpha^5$ |

For $n = 2^m - 1$, an $n$-element vector $\mathbf{a}$ in GF($2^m$) is represented as $\mathbf{a} = (a_{n-1}, a_{n-2}, \cdots, a_0)$ where $a_j$'s are elements in GF($2^m$). The vector $\mathbf{a}$ can be equivalently represented in polynomial form as

$$a(x) = \sum_{j=0}^{n-1} a_j x^j .$$

The $n$-element vector $\mathbf{a}'$ is called the $i$th *shift version* of $\mathbf{a}$ if

$$\mathbf{a}' = (a_{n-i-1},\ a_{n-i-2},\ \cdots,\ a_1,\ a_0,\ 0, 0, \cdots,\ 0).$$

The $n$-element vector $\mathbf{a}''$ is called the $i$th *cyclic shift version* of $\mathbf{a}$ if

$$\mathbf{a}'' = (a_{n-i-1},\ a_{n-i-2},\ \cdots,\ a_1,\ a_0,\ a_{n-1},\ a_{n-2},\ \cdots,\ a_{n-i}).$$

It can be shown that $a'(x)$ is the remainder of the division of $x^i a(x)$ by $x^n$ and $a''(x)$ is the remainder of the division of $x^i a(x)$ by $(x^n + 1)$ (see Problem 6.8), i.e.,

$$a'(x) = x^i a(x) \bmod x^n,$$

and

$$a''(x) = x^i a(x) \bmod (x^n + 1).$$

It follows that $a'(x) = a''(x)$ if $a_j = 0$ for $(n - i) \le j \le (n - 1)$.

## 6.2 BCH CODES

Bose-Chaudhuri-Hocquenghem (BCH) codes are an important subclass of *cyclic* codes. This class of codes was discovered by Hocquenghem in 1959 [69] and independently by Bose and Chaudhuri in 1960 [70].

Cyclic codes are linear codes having the property that for a given code any cyclic shift of a codeword is also a codeword. As a consequence of this property, they have a considerable amount of algebraic structure which can be employed in devising a number of efficient encoding and decoding algorithms. Cyclic codes are characterized by using a polynomial representation for codewords and manipulations of polynomials for encoding and decoding procedures.

Consider an $(n,\ k)$ binary cyclic code $C$. Let $a(x) = a_{k-1}x^{k-1} + \cdots + a_1 x + a_0$ be a polynomial of degree $k - 1$ or less corresponding to the $k$ information bits $a_{k-1},\ \cdots,\ a_1,\ a_0$, to be encoded. We call this polynomial the *information polynomial*. The *codeword polynomial* $c(x) = c_{n-1}x^{n-1} + \cdots + c_1 x + c_0$, corresponding to the information polynomial $a(x)$ can be represented as

$$c(x) = a(x) \cdot g(x), \tag{6.4}$$

where $g(x) = g_r x^r + \cdots + g_1 x + g_0$, called the *generator polynomial*, is a polynomial of degree $r = n - k$. The generator polynomial $g(x)$ has two properties: first, the coefficients $g_0$ and $g_r$ are always nonzero, i.e., $g_0 = g_1 = 1$; and second, it divides $x^n + 1$. Using a polynomial of degree $n - k$ which is a factor of $x^n + 1$ we can generate an $(n,\ k)$ cyclic code.

**Example 6.3** Consider a degree-3 polynomial,

$$g(x) = x^3 + x + 1,$$

which divides $x^7 + 1$. This polynomial can be used to generate a (7, 4) binary cyclic code. Using (6.4) the 16 codewords of this binary code can be obtained as follows:

| Information $a_3 a_2 a_1 a_0$ | Codeword Polynomial $c(x) = a(x) \cdot g(x)$ | Codeword $c_6 c_5 c_4 c_3 c_2 c_1\ c_0$ |
|---|---|---|
| 0 0 0 0 | $0$ | 0 0 0 0 0 0 0 |
| 0 0 0 1 | $x^3 + x + 1$ | 0 0 0 1 0 1 1 |
| 0 0 1 0 | $x^4 + x^2 + x$ | 0 0 1 0 1 1 0 |
| 0 0 1 1 | $x^4 + x^3 + x^2 + 1$ | 0 0 1 1 1 0 1 |
| 0 1 0 0 | $x^5 + x^3 + x^2$ | 0 1 0 1 1 0 0 |
| 0 1 0 1 | $x^5 + x^2 + x + 1$ | 0 1 0 0 1 1 1 |
| 0 1 1 0 | $x^5 + x^4 + x^3 + x$ | 0 1 1 1 0 1 0 |
| 0 1 1 1 | $x^5 + x^4 + 1$ | 0 1 1 0 0 0 1 |
| 1 0 0 0 | $x^6 + x^4 + x^3$ | 1 0 1 1 0 0 0 |
| 1 0 0 1 | $x^6 + x^4 + x + 1$ | 1 0 1 0 0 1 1 |
| 1 0 1 0 | $x^6 + x^3 + x^2 + x$ | 1 0 0 1 1 1 0 |
| 1 0 1 1 | $x^6 + x^2 + 1$ | 1 0 0 0 1 0 1 |
| 1 1 0 0 | $x^6 + x^5 + x^4 + x^2$ | 1 1 1 0 1 0 0 |
| 1 1 0 1 | $x^6 + x^5 + x^4 + x^3 + x^2 + x + 1$ | 1 1 1 1 1 1 1 |
| 1 1 1 0 | $x^6 + x^5 + x$ | 1 1 0 0 0 1 0 |
| 1 1 1 1 | $x^6 + x^5 + x^3 + 1$ | 1 1 0 1 0 0 1 |

The code generated in this example is the (7, 4) Hamming code.

To verify that the code generated by (6.4), with a generator polynomial being a factor of $x^n + 1$, is cyclic we argue as follows. Assume that $c(x)$ is a valid codeword, generated by (6.4), and $c''(x) = x^i c(x) \bmod (x^n + 1)$ is a cyclic shift version of $c(x)$. In order for $c''(x)$ to be a valid codeword it must be divisible by $g(x)$, that is, $(x^i c(x) \bmod (x^n + 1)\ ) \bmod g(x) = 0$. Since $g(x)$ divides $(x^n + 1)$ it suffices to show that $x^i c(x) \bmod g(x) = 0$ which follows immediately by noting that $c(x)$ divides $g(x)$. Therefore, $c''(x)$ is a valid codeword.

The foregoing example shows that the cyclic code, generated by using (6.4), is not in a systematic form, that is, with information and parity-check bits in separate positions. However, with a simple encoding operation we are able to put the cyclic code in a systematic form. For this purpose, the information polynomial $a(x)$ is multiplied by $x^{n-k}$. If we add the remainder of the division of $x^{n-k}a(x)$ by $g(x)$, that is $b(x) = x^{n-k}a(x) \bmod g(x)$ to $x^{n-k}a(x)$, then the resulting code is a systematic cyclic code.

**Example 6.4** The (7, 4) code of Example 6.3 can be generated by the same generator polynomial $g(x) = x^3 + x + 1$ in a systematic form as

$$c(x) = x^3 a(x) + (x^3 a(x) \bmod (x^3 + x + 1)\ ).$$

The resulting codewords are shown in the following table:

| Information $a_3a_2a_1a_0$ | Codeword Polynomial $c(x) = x^3a(x) + (x^3a(x) \bmod g(x))$ | Codeword $c_6c_5c_4c_3c_2c_1c_0$ |
|---|---|---|
| 0 0 0 0 | $0$ | 0 0 0 0 0 0 0 |
| 0 0 0 1 | $x^3 + x + 1$ | 0 0 0 1 0 1 1 |
| 0 0 1 0 | $x^4 + x^2 + x$ | 0 0 1 0 1 1 0 |
| 0 0 1 1 | $x^4 + x^3 + x^2 + 1$ | 0 0 1 1 1 0 1 |
| 0 1 0 0 | $x^5 + x^2 + x + 1$ | 0 1 0 0 1 1 1 |
| 0 1 0 1 | $x^5 + x^3 + x^2$ | 0 1 0 1 1 0 0 |
| 0 1 1 0 | $x^5 + x^4 + 1$ | 0 1 1 0 0 0 1 |
| 0 1 1 1 | $x^5 + x^4 + x^3 + x$ | 0 1 1 1 0 1 0 |
| 1 0 0 0 | $x^6 + x^2 + 1$ | 1 0 0 0 1 0 1 |
| 1 0 0 1 | $x^6 + x^3 + x^2 + x$ | 1 0 0 1 1 1 0 |
| 1 0 1 0 | $x^6 + x^4 + x + 1$ | 1 0 1 0 0 1 1 |
| 1 0 1 1 | $x^6 + x^4 + x^3$ | 1 0 1 1 0 0 0 |
| 1 1 0 0 | $x^6 + x^5 + x$ | 1 1 0 0 0 1 0 |
| 1 1 0 1 | $x^6 + x^5 + x^3 + 1$ | 1 1 0 1 0 0 1 |
| 1 1 1 0 | $x^6 + x^5 + x^4 + x^2$ | 1 1 1 0 1 0 0 |
| 1 1 1 1 | $x^6 + x^5 + x^4 + x^3 + x^2 + x + 1$ | 1 1 1 1 1 1 1 |

Note that the left-most 4 bits of each codeword, shown in the last column, are the unaltered corresponding information bits shown in the first column.

### 6.2.1 Binary BCH Codes

Binary BCH codes are a class of binary cyclic codes with block length[2] $n = 2^m - 1$, number of parity-check bits $n - k \leq mt$, and minimum Hamming distance $d \geq 2t + 1$, where $m \geq 3$ and $t < n/2$ are arbitrary positive integers. The code with foregoing specifications can correct any combination of $t$ or fewer errors per codeword and, hence, is a *t-error-correcting* code.

A $t$-error-correcting binary BCH code is constructed by using $2t$ consecutive powers of a primitive element of GF$(2^m)$. The generator polynomial of this code is the *lowest-degree* polynomial over GF(2) having the $2t$ consecutive powers $\alpha, \alpha^2, \alpha^3, \cdots, \alpha^{2t}$, as its roots where $\alpha$ is a primitive element in GF$(2^m)$. Mathematically speaking the generator polynomial $g(x)$ can be written as[3]

$$g(x) = \text{LCM}\{m_{\alpha}(x), m_{\alpha^2}(x), m_{\alpha^3}(x), \cdots, m_{\alpha^{2t}}(x)\}, \tag{6.5}$$

where LCM is the *least common multiple* and $m_{\alpha^i}(x)$ is the *minimal polynomial* of the element $\alpha^i$. The minimal polynomial of an element of GF$(2^m)$ is a polynomial of the *smallest degree* over GF(2) having that element as a root.

[2] This block length is restricted to a subclass of binary BCH codes, called *primitive* BCH codes. For nonprimitive BCH codes the block length $n$ is less than $2^m - 1$. In this section we consider only the primitive BCH codes.

[3] The most general definition of the generator polynomial of BCH permits the $2t$ consecutive powers of the primitive element to range over an arbitrary interval of consecutive values, say, $h, h+1, \cdots, h+2t-1$. In this section we use $h = 1$.

The minimal polynomials, originally published by Peterson [71], can be found in several references such as [68], [1], and [72].

The generator polynomial, given in (6.5), can be simplified by using the concept of *conjugate roots*. Let $f(x)$ be a polynomial with coefficients from GF(2). Let $\beta$ be an element of the extension field GF($2^m$). It is easy to verify that if $\beta$ is a root of $f(x)$, then for any $l \geq 0$, $\beta^{2^l}$ is also a root of $f(x)$ [1]. The element $\beta^{2^l}$ is called a conjugate of $\beta$. It follows that the conjugate roots have the same minimal polynomials. Consequently, the generator polynomial of (6.5) can be simplified to

$$g(x) = \text{LCM}\{m_\alpha(x), m_{\alpha^3}(x), \cdots, m_{\alpha^{2t-1}}(x)\}. \tag{6.6}$$

**Example 6.5** We design a 2-error-correcting primitive binary BCH code with $m = 4$. Let $\alpha$ be a primitive element of GF($2^4$). The minimal polynomials of $\alpha$ and $\alpha^3$ are obtained as [1]

$$\begin{aligned} m_\alpha(x) &= x^4 + x + 1, \\ m_{\alpha^3}(x) &= x^4 + x^2 + x + 1. \end{aligned}$$

It can be shown that $m_\alpha(x)$ and $m_{\alpha^3}(x)$ are two distinct irreducible polynomials. Using (6.6), the generator polynomial of this code is obtained as

$$\begin{aligned} g(x) &= \text{LCM}\ \{m_\alpha(x),\ m_{\alpha^3}(x)\} \\ &= m_\alpha(x) \cdot m_{\alpha^3}(x) \\ &= (x^4 + x + 1) \cdot (x^4 + x^2 + x + 1) \\ &= x^8 + x^7 + x^6 + x^4 + 1. \end{aligned}$$

The resulting BCH code is a (15, 7) cyclic code which can correct up to 2 errors in each codeword.

The parameter $d_D = 2t + 1$, which is used in determining the generator polynomial of a $t$-error-correcting BCH code, is called *design distance* of the code. The minimum Hamming distance of a BCH code may or may not be equal to its design distance. However, the minimum Hamming distance of BCH codes, $d$, is always lower-bounded by their design distance, i.e., $d \geq 2t + 1$.

### 6.2.2 Nonbinary BCH Codes

The concept and properties of binary BCH codes can be extended to nonbinary BCH codes where the symbols are chosen from GF($q$) rather than GF(2). The Galois field GF($q$) can be a prime field or some extension of a prime field.

Let $q$ be any power of a prime number. An $(n,\ k)$ $t$-error-correcting primitive nonbinary BCH code with $q$-ary symbols is generated on GF($q$) by constructing a generator polynomial with degree $r = n - k$. The generator polynomial of such a code is the lowest degree polynomial with coefficients from GF($q$)

and has $2t$ consecutive powers of a primitive element of $GF(q^{m^*})$ as roots. Assuming that $\alpha$ is a primitive element of $GF(q^{m^*})$, the generator polynomial $g(x)$ can be constructed by using (6.5). The block length of the primitive nonbinary BCH code is $n = q^{m^*} - 1$.

The most important subclass of the nonbinary BCH codes is Reed-Solomon codes. The properties, encoding and decoding procedures of these codes are considered in detail in the following sections.

## 6.3 DEFINITIONS OF REED-SOLOMON CODES

Reed-Solomon codes are an important subclass of non-binary Bose-Chaudhuri-Hocquenghem (BCH) codes [73]. They were introduced by Reed and Solomon in 1960 [67] independent of the work by Hocquenghem [69], and Bose and Chaudhuri [70].

As stated earlier, in nonbinary BCH codes the code symbols are chosen from $GF(q)$ with a generator polynomial having roots in GF $(q^{m^*})$, an extension field of $GF(q)$. To distinguish between these two fields, $GF(q)$ and $GF(q^{m^*})$ (extended field of $GF(q)$) are called *the symbol field* and *locator field*, respectively. Choosing the locator field to be the same as the symbol field (i.e., $m^* = 1$) the Reed-Solomon (RS) codes are obtained. Without loss of generality, in the following we restrict our discussion to the RS codes with code symbols from $GF(2^m)$, i.e., $q = 2^m$. Recall that in such a case the subtraction and addition operations are identical.

An $(n,\ k)$ primitive RS code defined over $GF(2^m)$ has a code length of $n = 2^m - 1$ and a minimum Hamming distance of $(n - k + 1)$. As its minimum distance equals $(n-k+1)$, an $(n,\ k)$ primitive RS code is a *maximum distance-separable* (MDS) code [72]. The number of information symbols is $k$ and the number of parity symbols is $(n - k)$. Each symbol is an element of $GF(2^m)$.

The generator polynomial of the RS code can be constructed as the least common multiple of the minimal polynomials of $n-k$ consecutive powers of the primitive elements of $GF(2^m)$, i.e., $\alpha^h$, $\alpha^{h+1}$, $\cdots$, $\alpha^{h+n-k+1}$, where $h$ is an integer. As a consequence of identical symbol and locator fields, the minimal polynomial of the $i$th power of $\alpha$ is simply $(x + \alpha^i)$. This means that the generator polynomial of RS code can be written as

$$g(x) = \prod_{i=0}^{n-k-1} (x + \alpha^{h+i}) = \sum_{j=0}^{n-k} g_j x^j .$$

With different values of $h$, $h$ different generator polynomials can be formed. The constant $h$ is usually, but not always, taken to be 1.

For $h = (k+1)/2$ with odd $k$, $g(x)$ is a *self-reciprocal* polynomial, i.e.,

$$g(x) = x^{n-k} g(\frac{1}{x}).$$

A self-reciprocal generator polynomial has its coefficients satisfying the following relation:

$$g_j = g_{n-k-j}.$$

**Example 6.6** The generator polynomial for RS(7, 5) codes defined over GF($2^3$) is

$$g(x) = (x + \alpha^h)(x + \alpha^{h+1}).$$

Using the tables of Example 6.2 the generator polynomial for two cases, $h = 1$ and $h = 3$, is obtained

$$\begin{array}{llcll} \text{for } h = 1 & g(x) & = & x^2 + \alpha^4 x + \alpha^3, & \\ \text{for } h = 3 & g(x) & = & x^2 + \alpha^6 x + 1. & \text{(self-reciprocal)} \end{array}$$

The generator polynomial for RS(7, 4) code with $h = 1$ is

$$\begin{aligned} g(x) &= (x + \alpha)(x + \alpha^2)(x + \alpha^3) \\ &= x^3 + \alpha^6 x^2 + \alpha x + \alpha^6. \end{aligned}$$

An $(n,\ k)$ RS code can correct $\epsilon$ symbol errors and $s$ symbol erasures as long as $(2\epsilon + s) \leq (n - k)$. Equivalently it can correct $l$ bursts of length $b_l = m(t - (2l - 1)) + (2l - 1)$ *bits*, where $t \leq (n-k)/2$.

An RS code can be shortened by omitting some of the information symbols. Such a shortening does not affect the minimum Hamming distance of the RS code, and hence, we can say that any shortened RS code is also an MDS code. However, in general, a shortened code is not necessarily a cyclic code.

## 6.4 ENCODING OF RS CODES

The encoding procedures stated earlier for cyclic codes are applicable to RS codes. Let $c(x)$ be the codeword polynomial corresponding to the information polynomial $a(x)$. One way to generate $c(x)$ is to multiply $a(x)$ by $g(x)$, i.e., $c(x) = a(x) \cdot g(x)$. The resulting code is not in a systematic form.

As part of the decoding process is to recover the information polynomial $a(x)$ from the decoded code polynomial, $\hat{c}(x)$, the aim of encoding is to facilitate the inverse operation. For this reason, the systematic codes are preferred.

To encode RS codes in a systematic manner a codeword polynomial is generated

$$c(x) = x^{n-k} a(x) + b(x),$$

where $b(x)$ is the *remainder* polynomial of degree less than $(n-k)$ of the division of $x^{n-k} a(x)$ by $g(x)$, i.e.,

$$b(x) = x^{n-k} a(x) \bmod g(x).$$

The polynomial $b(x)$ is generated by a feedback shift register (see Problem 6.10).

It can be seen that $c(x)$ contains $a(x)$ and $a(x)$ can be recovered by simply dropping $b(x)$. Systematic encoding is preferred as it provides a very simple way to recover $a(x)$ from $c(x)$.

**Example 6.7** For the RS(7, 5) code in Example 6.6 with $h = 1$, the code polynomial $c(x)$ corresponding to the information polynomial

$$a(x) = \alpha^2 x^4 + \alpha^5$$

is

$$\begin{aligned} c(x) &= a(x) \cdot g(x) \\ &= \alpha^2 x^6 + \alpha^6 x^5 + \alpha^5 x^4 + \alpha^5 x^2 + \alpha^2 x + \alpha, \end{aligned}$$

when it is encoded with a nonsystematic encoder, and

$$\begin{aligned} c(x) &= x^2 a(x) + (\ x^2 a(x) \bmod g(x)\ ) \\ &= (\alpha^2 x^6 + \alpha^5 x^2) + (x+1) \\ &= \alpha^2 x^6 + \alpha^5 x^2 + x + 1, \end{aligned}$$

when it is encoded systematically. Note that in the latter case the information symbols appear unchanged in the encoded codeword.

For another encoding technique of RS codes, namely, *transform encoding,* the reader is referred to Appendix C.

## 6.5 SINGLY-EXTENDED RS$(n+1,k)$ CODES

A singly-extended RS$(n+1,k)$ coder over GF$(2^m)$ can be formed from the original RS$(n,k)$ code by adding one additional parity symbol to the original codeword. The singly-extended RS$(n+1,k)$ code has a minimum distance of $(n-k+2)$ and can correct any combination of $\epsilon$ symbol errors and $s$ symbol erasures provided $2\epsilon + s \leq (n-k+1)$. The singly-extended RS$(n+1,k)$ code is suitable for applications requiring a block length of $2^m$.

For a given code polynomial $c(x)$ of the original RS$(n,k)$ code the additional parity-check symbol (or *extended symbol*) in the singly-extended RS$(n+1,k)$ code is defined as [74], [75], [76]

$$c_- = c(\alpha^{h-1}), \tag{6.7}$$

or

$$c_+ = c(\alpha^{h+n-k}). \tag{6.8}$$

The *augmented codeword* of the singly-extended RS$(n+1,\ k)$ code is

$$\mathbf{c}_- = (c_-,\ c_0,\ c_1,\ \cdots,\ c_{n-1}),$$

or

$$\mathbf{c}_+ = (c_0,\ c_1,\ \cdots,\ c_{n-1},\ c_+).$$

Note that the singly-extended code is not a cyclic code. The original $n$-component codeword $\mathbf{c}$ is called the *interior* [74].

For $h = 1$, the additional parity-check symbol $c_-$, given in (6.7) becomes

$$c_- = c(\alpha^0) = c(1) = \sum_{i=0}^{n-1} c_i.$$

In other words, $c_-$ represents the overall parity-check symbol on other code symbols for $h = 1$. Similarly, for $h = k$, the additional parity-check symbol $c_+$, given in (6.8), becomes

$$c_+ = c(\alpha^n) = c(1) = \sum_{i=0}^{n-1} c_i,$$

that is, $c_+$ is also the overall parity-check sum on other symbols for $h = k$.

## 6.6 ERRORS-ONLY DECODING OF RS(n, k) CODES BASED ON SYNDROMES

In this section, we consider errors- only decoding schemes for RS$(n, k)$ codes based on syndromes.

Consider a valid RS$(n, k)$ code polynomial $c(x)$ transmitted over a noisy channel. The received polynomial $r(x)$ can be represented as

$$r(x) = c(x) + e(x),$$

where $e(x) = \sum_{i=0}^{n-1} e_i x^i$ represents the error polynomial. The coefficient $e_i$ is an element of GF$(2^m)$ and denotes the *error value* at the $i$th *position* if $e_i \neq 0$. Suppose that there are $\epsilon$ errors where $0 < 2\epsilon \leq (n-k)$ at positions $i_l, l = 1, 2, , \cdots, \epsilon$. The error polynomial $e(x)$ has only $\epsilon$ non-zero coefficients and can be rewritten as

$$e(x) = \sum_{l=1}^{\epsilon} e_{i_l} x^{i_l}. \tag{6.9}$$

The objective of the RS decoder is to find the number of errors $\epsilon$, their positions $i_l$'s, and their values $e_{i_l}$'s.

Let us evaluate $r(x)$ at the roots of the generator polynomial,

$$\begin{aligned} r(\alpha^{h+i}) &= c(\alpha^{h+i}) + e(\alpha^{h+i}) \\ &= e(\alpha^{h+i}), \quad i = 0, 1, \cdots, (n-k-1). \end{aligned} \tag{6.10}$$

These $(n-k)$ values are known as the *syndromes* and are denoted by $S_{i+1}$, $i = 0, 1, \cdots, (n-k-1)$. Using (6.9) and (6.10) the syndromes can be represented as

$$\begin{aligned} S_{i+1} &= \sum_{l=1}^{\epsilon} e_{i_l} (\alpha^{h+i})^{i_l}, \\ &= \sum_{l=1}^{\epsilon} e_{i_l} (\alpha^{i_l})^{h+i}, \quad i = 0,\ 1,\ \cdots,\ (n-k-1). \end{aligned}$$

For convenience, let us define the *error value* as $Y_l = e_{i_l}$, and the *error locator* as $X_l = \alpha^{i_l}$ for $l = 1, 2, \cdots, \epsilon$. The syndromes can then be written in terms of error values and error locators

$$S_{i+1} = \sum_{l=1}^{\epsilon} Y_l X_l^{h+i}, \quad i = 0, 1, \cdots, (n-k-1). \tag{6.11}$$

**Example 6.8** Consider the case of $n-k=2$. This code is capable of correcting one symbol error. The two syndromes are $S_1$ and $S_2$. If both $S_1$ and $S_2$ are zero it is declared that no error occurs and the received vector $\mathbf{r}$ is a valid codeword.

If both $S_1$ and $S_2$ are nonzero, one error is assumed and the syndrome equations are $S_1 = Y_1 X_1^h$ and $S_2 = Y_1 X_1^{h+1}$. Solving the foregoing equations we obtain the error locator $X_1 = S_2/S_1$ and the error value $Y_1 = S_1/X_1^h$.

The syndrome equations represent a nonlinear relation between $S_{i+1}$, $Y_l$ and $X_l$. As $(n-k) > 2$, solving this set of simultaneous nonlinear equations is difficult and best avoided. It is possible to convert the syndrome equations into linear equations by introducing the *error-locator* polynomial $\Lambda(x)$ as follows [77], [73].

The error-locator polynomial $\Lambda(x)$ is defined as a polynomial of degree $\epsilon$ which has as roots the multiplicative inverses of the error-locators $X_l, l = 1, 2, \cdots, \epsilon$, i.e.,

$$\Lambda(x) = \prod_{l=1}^{\epsilon} (1 + X_l x) = 1 + \sum_{j=1}^{\epsilon} \Lambda_j x^j. \tag{6.12}$$

Multiplying both sides of (6.12) by a factor $Y_l X_l^{h+\epsilon+i}$ for $1 \le l \le \epsilon$ and $0 \le i \le (n-k-\epsilon-1)$, we obtain

$$Y_l X_l^{h+\epsilon+i} \Lambda(x) = Y_l X_l^{h+\epsilon+i} + \sum_{j=1}^{\epsilon} \Lambda_j Y_l X_l^{h+\epsilon+i} x^j. \tag{6.13}$$

Now, setting $x = X_l^{-1}$, root of $\Lambda(x)$, in (6.13), we get

$$0 = Y_l X_l^{h+\epsilon+i} + \sum_{j=1}^{\epsilon} \Lambda_j Y_l X_l^{h+\epsilon+i-j}, \quad 1 \le l \le \epsilon \text{ and } 0 \le i \le (n-k-\epsilon-1). \tag{6.14}$$

For a given value of $i$, summing (6.14) for all values of $l$ yields

$$0 = \sum_{l=1}^{\epsilon} Y_l X_l^{h+\epsilon+i} + \sum_{j=1}^{\epsilon} \Lambda_j \left( \sum_{l=1}^{\epsilon} Y_l X_l^{h+\epsilon+i-j} \right). \tag{6.15}$$

Combining (6.11) and (6.15), we obtain

$$0 = S_{\epsilon+i+1} + \sum_{j=1}^{\epsilon} \Lambda_j S_{\epsilon+i+1-j}, \quad 0 \le i \le (n-k-\epsilon-1),$$

or

$$S_i = \sum_{j=1}^{\epsilon} \Lambda_j S_{i-j}, \quad i = (\epsilon+1), (\epsilon+2), \cdots, (n-k). \tag{6.16}$$

The above relation represents a set of simultaneous linear equations with $\epsilon$ unknown coefficients, $\Lambda_j$'s, and $(n-k)$ known syndromes, $S_i$'s. If $\epsilon$ is known we only need a subset of $\epsilon$ linear equations rearranged in the following matrix form:

$$\begin{bmatrix} S_\epsilon & S_{\epsilon-1} & \cdots & S_1 \\ S_{\epsilon+1} & S_\epsilon & \cdots & S_2 \\ \vdots & \vdots & \vdots & \vdots \\ S_{2\epsilon-1} & S_{2\epsilon-2} & \cdots & S_\epsilon \end{bmatrix} \begin{bmatrix} \Lambda_1 \\ \Lambda_2 \\ \vdots \\ \Lambda_\epsilon \end{bmatrix} = \begin{bmatrix} S_{\epsilon+1} \\ S_{\epsilon+2} \\ \vdots \\ S_{2\epsilon} \end{bmatrix}.$$

Then, the coefficients $\Lambda_i$'s, $i = 1, 2, \cdots, \epsilon$ can be solved by using the matrix inversion. Unfortunately the value of $\epsilon$ is not known. As a matter of fact, finding $\epsilon$ is part of the decoding process.

The Peterson's direct solution method [77] is as follows. Starting with $\epsilon = t$ we verify the determinant of the **S** matrix. If it is zero, we reduce the value of $\epsilon$ by 1 and repeat the process until the first $\epsilon$ which yields a nonzero determinant. At this point, $\epsilon$ represents the assumed number of errors and the coefficients of the error-locator polynomial are derived by using matrix inversion.

**Example 6.9** Consider a 2-error-correcting (7, 3) RS code defined over GF($2^3$). We have the received polynomial $r(x) = \alpha x^5$. The syndrome values (with $h = 1$) are calculated as

$$\begin{aligned} S_1 &= r(\alpha) = \alpha^6, \\ S_2 &= r(\alpha^2) = \alpha^4, \\ S_3 &= r(\alpha^3) = \alpha^2, \\ S_4 &= r(\alpha^4) = 1. \end{aligned}$$

First, we assume that the number of errors is 2. In this case,

$$\begin{bmatrix} S_2 & S_1 \\ S_3 & S_2 \end{bmatrix} \begin{bmatrix} \Lambda_1 \\ \Lambda_2 \end{bmatrix} = \begin{bmatrix} S_3 \\ S_4 \end{bmatrix}.$$

Evaluating the determinant of the left-hand $\mathbf{S}$ matrix yields

$$\begin{vmatrix} S_2 & S_1 \\ S_3 & S_2 \end{vmatrix} = \begin{vmatrix} \alpha^4 & \alpha^6 \\ \alpha^2 & \alpha^4 \end{vmatrix} = \alpha - \alpha = 0,$$

where we have used $\alpha^7 = 1$. Thus, the number of errors cannot be two. Assuming that there is only one error, we obtain

$$S_2 = \Lambda_1 S_1,$$

or

$$\Lambda_1 = \frac{S_1}{S_2} = \alpha^2.$$

Thus, the error-locator polynomial $\Lambda(x)$ has a zero at $\alpha^2$, and the error location is at the reciprocal of the zero, that is $X_1 = \alpha^{i_1} = \alpha^5$. Using (6.11) the error value $Y_1$ (with $h = 1$) is obtained as

$$Y_1 = \frac{S_1}{X_1} = \frac{\alpha^6}{\alpha^5} = \alpha.$$

If we add the error polynomial $e(x) = \alpha x^5$ to the received signal $r(x)$ the all-zero transmitted codeword will be obtained.

The direct solution method requires a number of necessary computations, exponentially increased with $(n-k)$. For low error correction capability, it is not unreasonable. As $t$ increases beyond 3 this method becomes cumbersome and a computationally simpler technique is required. The Berlekamp-Massey [78], [79], Euclidean [80], [81] and Continued Fraction algorithms [82] are algorithms which are known to be computationally simpler than the direct solution. In the following we present the Berlekamp-Massey algorithm (BMA).

The Berlekamp-Massey algorithm can best be explained as a problem of designing a shortest linear feedback shift-register (LFSR) with tap weights as coefficients of the error-locator polynomial $\Lambda(x)$ which produces $(n-k)$ syndromes $S_1, S_2, \cdots, S_{n-k}$, according to (6.16).

The design approach is iterative and inductive. At each step $i$, $i = 1, 2, \cdots, (n-k)$, we search for a shortest LFSR which generates the first $i$ syndromes $S_1, S_2, \cdots, S_i$.

Let $\Lambda^{(i-1)}(x)$ be the polynomial representing the shortest LFSR, obtained in step $(i-1)$, and $L_{i-1}$ and $d_{i-1}$ be the length of this LFSR and the degree of $\Lambda^{(i-1)}(x)$, respectively. Note that $d_{i-1} \leq L_{i-1}$, where $L_{i-1}$ is the *expected* minimum length of the LFSR. It can happen that $d_{i-1} < L_{i-1}$ if $\Lambda_l^{(i-1)} =$

0 for $d_{i-1} < l \leq L_{i-1}$. This LFSR generates the first $(i-1)$ syndromes $S_1, S_2, \cdots, S_{i-1}$, i.e.,

$$S_j = \sum_{l=1}^{d_{i-1}} \Lambda_l^{(i-1)} S_{j-l}, \quad j-l \geq 1 \text{ and } j \leq (i-1).$$

At step $i$, we first use $\Lambda^{(i-1)}(x)$ with the first $i$ syndromes. We compute the discrepancy $\delta_i$ as

$$\delta_i = S_i + \sum_{l=1}^{d_{i-1}} \Lambda_l^{(i-1)} S_{i-l}. \tag{6.17}$$

If $\delta_i = 0$, $\Lambda^{(i-1)}(x)$ can generate the first $i$ syndromes. Therefore, we set $\Lambda^{(i)}(x) = \Lambda^{(i-1)}(x)$ and $L_i = L_{i-1}$. If $\delta_i \neq 0$, we have to find a new $\Lambda^{(i)}(x)$ with a minimum length $L_i$ to generate the first $i$ syndromes. This new $\Lambda^{(i)}(x)$ can have the same length as $\Lambda^{(i-1)}(x)$ or a different one. The first problem is therefore to establish the minimum possible length $L_i$. It can be shown that when $\delta_i \neq 0$, $L_i = \max(L_{i-1}, i - L_{i-1})$ [78, pp. 181-184], [74, pp. 180-181]. This implies that if $\delta_i \neq 0$ and $L_{i-1} \geq i - L_{i-1}$ then $L_i = L_{i-1}$, which indicates that we can still find another polynomial, of the same length $L_{i-1}$, to generate the first $i$ syndromes. Otherwise, if $\delta_i \neq 0$ and $L_{i-1} < i - L_{i-1}$ then $L_i > L_{i-1}$, which indicates that there is no polynomial of length $L_{i-1}$ that can produce the first $i$ syndromes.

The second problem is to find a new minimum length polynomial $\Lambda^{(i)}(x)$ of degree $d_i$ such that

$$S_j = \sum_{l=1}^{d_i} \Lambda_l^{(i)} S_{j-l}, \quad j-l \geq 1 \text{ and } j \leq i.$$

Let us consider $\Lambda^{(i)}(x)$ of the following form

$$\begin{aligned}\Lambda^{(i)}(x) &= \Lambda^{(i-1)}(x) + \delta_i A(x) \\ &= \sum_{l=1}^{d_i} \left(\Lambda_l^{(i-1)} + \delta_i A_l\right) x^l.\end{aligned}$$

From (6.17), the coefficients, $A_l$'s, of the modifying polynomial $A(x)$ have to satisfy the following relation:

$$\sum_{l=0}^{d_{i-1}} \Lambda_l^{(i-1)} S_{j-l} + \delta_i \sum_{l=0}^{d_i} A_l S_{j-l} = 0.$$

This implies that

$$\sum_{\substack{l=0 \\ l<j}}^{d_{i-1}} A_l S_{j-l} = \begin{cases} 1, & \text{for } j = i \\ 0, & \text{for } j < i \end{cases}. \tag{6.18}$$

Our objective is to find a modifying polynomial $A(x)$ which satisfies the above conditions and has a *minimum length* in a most computationally efficient way.

Let us first concentrate on the problem of finding $A(x)$ to satisfy (6.18) in a most computationally efficient way. Consider a previous step $p$ ($p < i$) at which $\delta_p \neq 0$ and $\Lambda^{(p-1)}(x)$ of degree $d_{p-1}$ can generate the first $(p-1)$ syndromes, i.e.,

$$\sum_{\substack{v=0 \\ v<q}}^{d_{p-1}} \Lambda_v^{(p-1)} S_{q-v} = \begin{cases} \delta_p, & \text{for } q = p, \\ 0, & \text{for } q < p. \end{cases}$$

We select $A(x)$ as

$$A(x) = \frac{1}{\delta_p} x^{(i-p)} \Lambda^{(p-1)}(x),$$

i.e.,

$$A_l = \begin{cases} \frac{1}{\delta_p} \Lambda_{l-i+p}^{(p-1)}, & \text{if l} = (i-p),\ (i-p+1),\ \cdots,\ (d_{p-1}+i-p), \\ 0, & \text{elsewhere.} \end{cases}$$

This yields

$$\begin{aligned} \sum_{\substack{l=0 \\ l<j}}^{d_i} A_l S_{j-l} &= \frac{1}{\delta_p} \sum_{\substack{l=(i-p) \\ l<j}}^{(d_{p-1}+i-p)} \Lambda_{l-i+p}^{(p-1)} S_{j-l} \\ &= \frac{1}{\delta_p} \sum_{v=0}^{d_{p-1}} \Lambda_v^{(p-1)} S_{j-v-i+p} \\ &= \begin{cases} 1, & \text{for } j = i, \\ 0, & \text{for } j < i, \end{cases} \end{aligned}$$

that is, $A(x)$ satisfies (6.18).

The next problem is to find the step $p$ such that $A(x)$ has a minimum length.

$\Lambda^{(p-1)}(x)$ is a minimum-length polynomial which generates the first $(p-1)$ syndromes $S_1$, $S_2$, $\cdots$, $S_{p-1}$. Its length is $L_{p-1}$. Therefore, the length of $A(x)$ is $(L_{p-1}+i-p)$. It is minimum if $(i-p)$ is the smallest. This implies that we should select the *most recent* step $p$ at which $\delta_p \neq 0$ *and* $L_p > L_{p-1}$.

A flowchart of the Berlekamp-Massey algorithm for errors-only decoding of RS codes is shown in Figure 6.1.

After obtaining the error-locator polynomial $\Lambda(x)$, the Chien search [83] can be used to find the roots of $\Lambda(x)$ and then the error locators $X_l, l = 1, 2, \cdots, \epsilon$. The Chien search simply consists of the computation of $\Lambda(\alpha^j)$ for

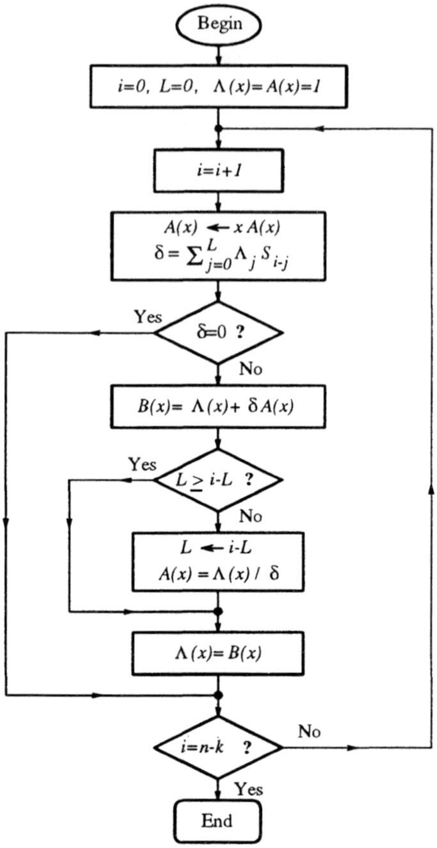


Figure 6.1: Berlekamp-Massey algorithm to find $\Lambda(x)$ for errors-only decoding.

$j = 0, 1, \cdots, (n-1)$, and checking for zero. This technique is feasible and efficient as the number of elements in $GF(2^m)$ is finite.

After the number of errors $\epsilon$ and $\epsilon$ error locators are found they can be substituted in (6.11). The resulting equation now represents a set of $\epsilon$ simultaneous *linear* equations with $\epsilon$ unknown error values $Y_l$'s. Matrix inversion can be evaluated to solve for the error values, $Y_l$'s.

Forney [84] introduced a much more computationally efficient way to find $Y_l$'s as follows.

Define the *syndrome polynomial* $S(x)$ of degree $(n-k-1)$ as

$$S(x) = \sum_{j=0}^{(n-k-1)} S_{j+1} x^j, \tag{6.19}$$

and the *error-evaluator polynomial* $V(x)$ of degree less than $(n-k)$ as

$$V(x) = S(x)\Lambda(x) \bmod x^{(n-k)}. \tag{6.20}$$

Substituting (6.11) and (6.19) into (6.20) and then simplifying the resulting equation we obtain (see Problem 6.17)

$$V(x) = \sum_{i=1}^{\epsilon} Y_i X_i \frac{\Lambda(x)}{1 + X_i x}.$$

The derivative of $\Lambda(x)$ is

$$\Lambda'(x) = \sum_{i=1}^{\epsilon} \frac{X_i \Lambda(x)}{1 + X_i x}.$$

Evaluating $V(x)$ and $\Lambda'(x)$ at $X_j^{-1}$, we obtain

$$V(X_j^{-1}) = Y_j \Lambda'(X_j^{-1}). \tag{6.21}$$

The above equation can be used to compute $Y_j$'s, $j = 1, 2, \cdots, \epsilon$. It can be seen that the Forney algorithm is much faster than matrix inversion (see Problem 6.18).

The syndrome-based decoding algorithm for RS$(n, k)$ codes is shown in Figure 6.2.

**Example 6.10** Consider the 2-error-correcting (7, 3) RS code given in Example 6.9. Assume that the received polynomial is $r(x) = \alpha x^6 + \alpha^4 x^2$. The syndrome values are calculated as

$$\begin{aligned} S_1 &= r(\alpha) = \alpha^2, \\ S_2 &= r(\alpha^2) = \alpha^5, \\ S_3 &= r(\alpha^3) = \alpha^2, \\ S_4 &= r(\alpha^4) = 1. \end{aligned}$$

Using the erlekamp-Massey algorithm, shown in Figure 6.1, we find the error-locator polynomial $\Lambda(x)$. The steps of this decoding procedure are shown in the following table:

| $i$ | $\delta$ | $A(x)$ | $\Lambda(x)$ | $L$ |
|---|---|---|---|---|
| 0 | | 1 | 1 | 0 |
| 1 | $\alpha^2$ | $\alpha^5$ | $1+\alpha^2 x$ | 1 |
| 2 | 1 | $1+\alpha^2 x$ | $1+\alpha^3 x$ | 1 |
| 3 | $\alpha^4$ | $\alpha^3+\alpha^6 x$ | $1+\alpha^6 x+\alpha^6 x^2$ | 2 |
| 4 | $\alpha^6$ | $\alpha+x+x^2$ | $1+x+\alpha x^2$ | 2 |
| 5 | STOP | | | |

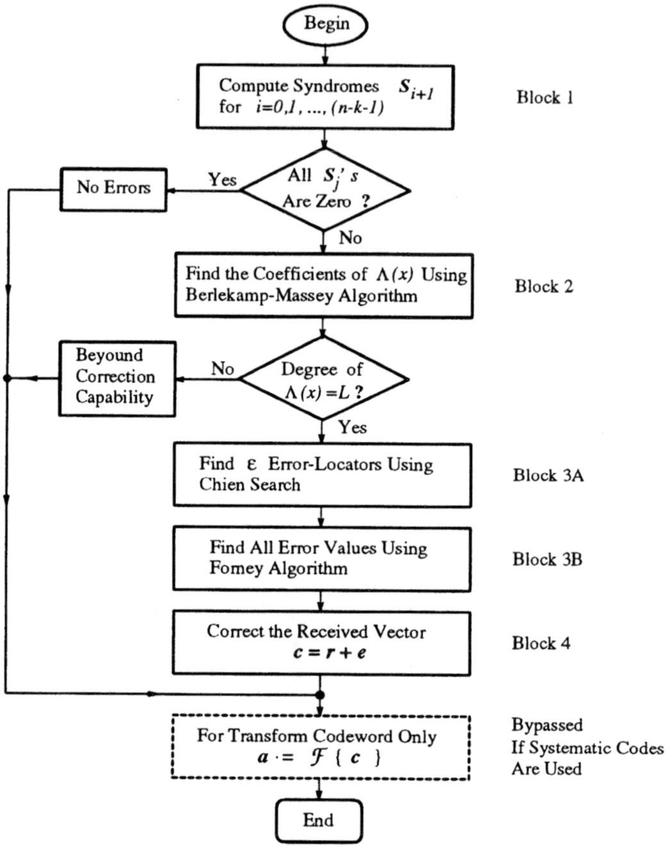


Figure 6.2: Syndrome-based decoding algorithm procedure for RS$(n, k)$ codes.

Thus, the error-locator polynomial is

$$\Lambda(x) = 1 + x + \alpha x = (1 + \alpha^2 x)(1 + \alpha^6 x),$$

which has $\frac{1}{\alpha^2} = \alpha^5$ and $\frac{1}{\alpha^6} = \alpha$ as roots. The reciprocals of these roots are $X_1 = \alpha^2$ and $X_2 = \alpha^6$ showing the error locations.

The error values can be found either by using (6.11) or the Forney algorithm. For example by using (6.11) we obtain the following equations:

$$\begin{aligned} \alpha^2 &= Y_1\alpha^2 + Y_2\alpha^6, \\ \alpha^5 &= Y_1\alpha^4 + Y_2\alpha^5, \end{aligned}$$

which yield

$$\begin{aligned} Y_1 &= \alpha^4, \\ Y_2 &= \alpha. \end{aligned}$$

Thus, the error polynomial is $e(x) = \alpha x^6 + \alpha^4 x^2$. Adding this polynomial to $r(x)$, results in the transmitted all-zero codeword.

## 6.7 ERRORS-AND-ERASURES DECODING OF RS(n, k) CODES BASED ON SYNDROMES

In errors-only decoding a *hard decision* is performed at the receiver prior to decoding. In other words, the receiver decides, on the basis of the received signal, which one of the symbols of the transmitter signal set was most likely to have been sent. Such a receiver discards part of the channel information about the reliability of the received symbols and this results in a performance degradation. One way of improving the performance is to allow the receiver to have the option of not deciding at all when the received signal does not clearly indicate one of the transmissions as the most probable. This extra possibility is called an *erasure*.

For receiver systems with the facility to provide erasure flags indicating unreliable (erased) symbols in a received polynomial $r(x)$, an RS$(n,k)$ code can correct $s$ erasure symbols and $\epsilon$ errors for $2\epsilon + s \leq (n-k)$.

The received polynomials $r(x)$ can be represented as

$$r(x) = c(x) + e^*(x) + e(x),$$

where $c(x)$ is a valid (transmitted) code polynomial and $e^*(x)$ and $e(x)$ are the *erasure* and *error* polynomials, respectively. Assume that there are $s$ erasures at positions $j_p$'s, $p = 1, 2, \cdots, s$, and $\epsilon$ errors at positions $i_l$'s, $l = 1, 2, \cdots, \epsilon$. Note that $j_p \neq i_l$ for all $p = 1, 2, \cdots, s$, and all $l = 1, 2, \cdots, \epsilon$, and all positions $j_p$'s are *known* and indicated by the erasure flags while $\epsilon$ and positions $i_l$'s, $l = 1, 2, \cdots, \epsilon$, are not known and are subjects to be found by the decoder. This allows us to generalize the case of both errors and erasures by using the term *errata* to indicate an error or an erasure [74]. We then redefine $e(x)$ as the *errata* polynomial

$$e(x) = \sum_{p=1}^{s} Y_p^* x^{j_p} + \sum_{l=1}^{\epsilon} Y_l x^{i_l},$$

where $Y_p^*$'s and $Y_l$'s denote erasure values at $j_p$'s and error values at $i_l$'s, respectively.

Evaluating $r(x)$ at roots $\alpha^{h+i}$'s, $i = 0, 1, \cdots, (n-k-1)$, of $c(x)$, we obtain syndromes

$$S_{i+1} = r(\alpha^{h+i}) = e(\alpha^{h+i})$$

$$
\begin{aligned}
&= \sum_{p=1}^{s} Y_p^*(\alpha^{j_p})^{h+i} + \sum_{l=1}^{\epsilon} Y_l(\alpha^{i_l})^{h+i} \\
&= \sum_{p=1}^{s} Y_p^*(X_p^*)^{h+i} + \sum_{l=1}^{\epsilon} Y_l(X_l)^{h+i}. \qquad (6.22)
\end{aligned}
$$

Let us define $\Lambda(x)$ as the *errata-locator polynomial* of degree $(\epsilon + s)$ as

$$
\begin{aligned}
\Lambda(x) &= \prod_{p=1}^{s}(1 + X_p^* x) \prod_{l=1}^{\epsilon}(1 + X_l x) \\
&= 1 + \sum_{i=1}^{\epsilon+s} \Lambda_i x^i, \qquad (6.23)
\end{aligned}
$$

where $X_p^* = \alpha^{j_p}, p = 1, 2, \cdots, s$, and $X_l = \alpha^{i_l}, l = 1, 2, \cdots, \epsilon$, are erasure locators and error locators, respectively.

It can be shown that the coefficients of the errata-locator polynomial $\Lambda(x)$ satisfy the following relation (see Problem 6.20):

$$
S_j + \sum_{i=1, i<j}^{\epsilon+s} \Lambda_i S_{j-i} = 0, \quad j = 1, 2, \cdots, (n-k). \qquad (6.24)
$$

As the erasure locations $j_p$'s are known, the $s$ erasure locators $X_p^*$'s are also known and in the first $s$ steps, $\Lambda(x)$ is computed as

$$
\begin{aligned}
\Lambda^{(0)}(x) &= 1, \\
\Lambda^{(i)}(x) &= \Lambda^{(i-1)}(x)(1 + X_i^* x), \quad i = 1,\ 2,\ \cdots,\ s.
\end{aligned}
$$

For the remaining $(n-k-s)$ steps, $\Lambda(x)$ is updated by using a modified version of the Berlekamp-Massey algorithm as shown by the flowchart in Figure 6.3. Note that for the step $i$, $s < i \leq (n-k)$, at which $\delta_i \neq 0$, and the minimum length $L_i$ of the LFSR is determined as

$$
L_i = \max\,(L_{i-1},\ i - (L_{i-1} - s)\ ).
$$

After the evaluation of the errata-locator polynomial $\Lambda(x)$, the Chien search can be performed to find the error locators, $X_l$'s, and then the Forney algorithm can be applied to solve for the $(\epsilon + s)$ errata values, $Y_p^*$'s and $Y_l$'s. The flowchart in Figure 6.2 can be used for errors-and-erasures decoding of an RS $(n,\ k)$ code except that the modified BMA of Figure 6.3 is used to find the errata-locator polynomial $\Lambda(x)$.

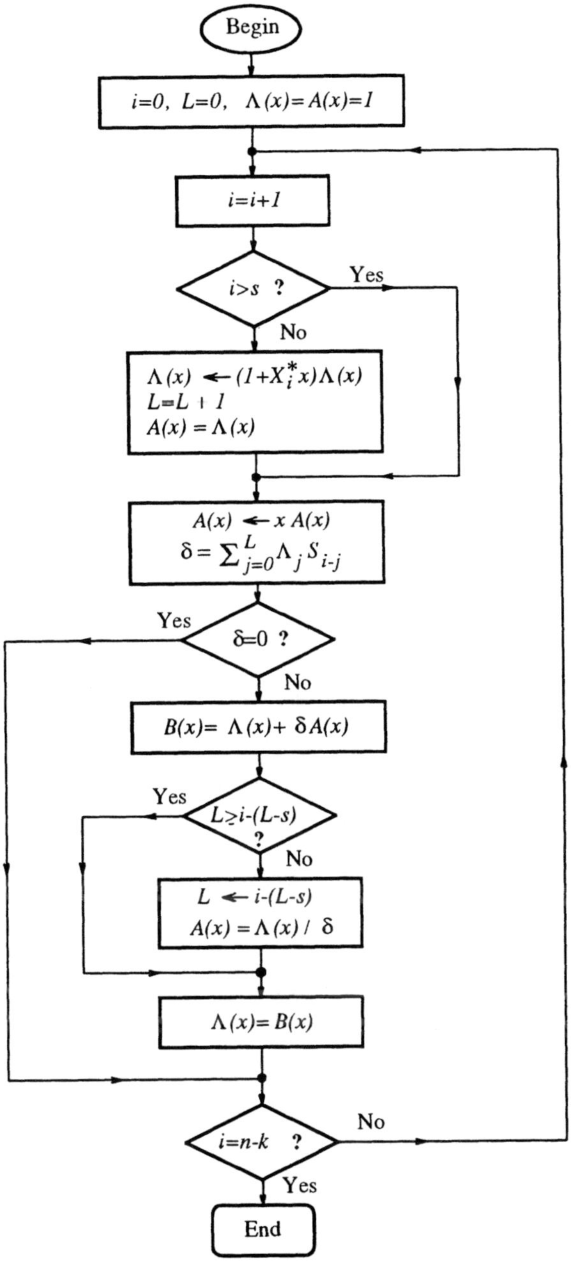


Figure 6.3: A modified Berlekamp-Massey algorithm to find $\Lambda(x)$ for errors-and-erasures decoding (note that $s < (n-k)$ ).

## 6.8 ERRORS-AND-ERASURES DECODING OF SINGLY EXTENDED RS(n+1, k) CODES BASED ON SYNDROMES

As mentioned earlier, by introducing an additional parity-check symbol $c_+$ (or $c_-$) to the original RS $(n,k)$ codeword (interior), a singly extended RS $(n+1,k)$ code is obtained which can correct $s$ erasures and $e$ errors provided $2\epsilon + s \leq (n-k+1)$.

Let $r_+$ (or $r_-$) be the received symbol corresponding to the additional parity-check symbol $c_+$ (or $c_-$). For the $(n+1)$-symbol received vector containing $s$ erasures and $\epsilon$ errors, provided $(2\epsilon + s) \leq (n-k+1)$, there are the following possibilities.

**(i)** If $(2\epsilon+s) \leq (n-k)$ then the $(n-k)$ syndromes calculated from the original $n$-symbol received vector (*received interior*) are sufficient to correct $s$ erasures and $\epsilon$ errors. The additional symbol $r_+$ (or $r_-$) does not need to be considered.

**(ii)** If $(2\epsilon + s) = n-k+1$ and the received symbol $r_+$ (or $r_-$) is declared as being erased, then there are only $s' = s-1$ erasures in the $n$-symbol received interior. As $2\epsilon + s' = n-k$, the original RS $(n,k)$ code is capable of correcting $s'$ erasures and $\epsilon$ errors in the original $n$-symbol received vector, i.e., we do not need to consider $r_+$ (or $r_-$).

**(iii)** If $(2\epsilon + s) = n-k+1$ and the received symbol $r_+$ (or $r_-$) is supposed to be in error, then there are only $\epsilon' = \epsilon - 1$ errors in the received interior. As $(2\epsilon'+s) = n-k-1 < (n-k)$ the original RS$(n,k)$ code is capable of correcting $s$ erasures and $\epsilon'$ errors in the received interior, i.e, we do not need to consider $r_+$ (or $r_-$)

**(iv)** If $2\epsilon + s = n-k+1$ and the received symbol $r_+$ (or $r_-$) is assumed to be correct , i.e., $r_+ = c_+$ or $r_- = c_-$, then there are $s$ erasures and $\epsilon$ errors in the received interior. As the RS $(n,k)$ code is not capable of correcting the received interior, we now need to consider the additional parity-check symbol $r_+$ (or $r_-$). From (6.7) and (6.8) we obtain

$$\begin{aligned} r_+ &= c_+ = c(\alpha^{h+n-k}), \\ r_- &= c_- = c(\alpha^{h-1}). \end{aligned}$$

The extra syndrome is calculated as

$$S_{n+1-k} = e(\alpha^{h+n-k}) = r(\alpha^{h+n-k}) + c(\alpha^{h+n-k}) = r(\alpha^{h+n-k}) + c_+,$$

or

$$S_0 = e(\alpha^{h-1}) = r(\alpha^{h-1}) + c(\alpha^{h-1}) = r(\alpha^{h-1}) + c_-.$$

The $(s+\epsilon)$ coefficients $\Lambda_1, \Lambda_2, \cdots, \Lambda_{s+\epsilon}$, of the errata polynomial $\Lambda(x)$ should satisfy the following relation:

$$S_j + \sum_{i=1}^{s+\epsilon} \Lambda_i S_{j-i} = 0,$$

for $i < j$ and $j \leq (n-k+1)$ if $c_+$ is used or for $i \leq j$ and $j \leq (n-k)$ if $c_-$ is used.

The BMA of Figure 6.1 needs to be modified [74] in order to perform $(n-k)$ iterative steps for possibilities (i) to (iii), and $(n-k+1)$ steps for possibility (iv). The decision on the number of steps is based on the discrepancy test at step $(n-k)$. For possibilities (i) to (iii), $\delta_{n-k}$ at step $(n-k)$ must be zero. If it is not zero, the BMA needs an extra step $(n-k+1)$. Furthermore, for possibility (ii) the erasure at $r_+$ (or $r_-$ ) must be disregarded in the initialization. The modified BMA for errata decoding of singly-extended RS $(n+1,\ k)$ codes is shown in Figure 6.4.

For other decoding techniques of RS codes, e.g., transform decoding, the reader should refer to Appendix C.

## 6.9 SOFT-DECISION DECODING OF RS CODES

The errors-and-erasure decoding technique, described in the previous sections, can be considered as a particular case of soft-decision decoding which utilizes the channel information in the decoding procedure. To implement maximum likelihood soft-decision decoding for block codes, in general, the received vector must be compared with each of the valid codewords and the closest codeword to the received one must be chosen as the transmitted one. However, when the number of codewords is large, implementing such a receiver is impractical. In such a case, soft-decision decoding of the block coded schemes can be implemented by employing some techniques which discard improbable codewords and hence avoid the comparison of the entire set of codewords with the received one. Several techniques for soft-decision decoding of block codes have been considered in the literature [85], [86], [87], [88], [89]. In the following we consider two soft-decision decoding techniques for RS codes; for the other types of soft-decision procedures we refer the reader to the aforementioned references.

### 6.9.1 Trellis-Decoding of RS Codes

In [89] Wolf shows that for any $(n,\ k)$ linear block code defined over GF$(q)$ there is a trellis structure having no more than $q^{n-k}$ states. For cyclic codes,

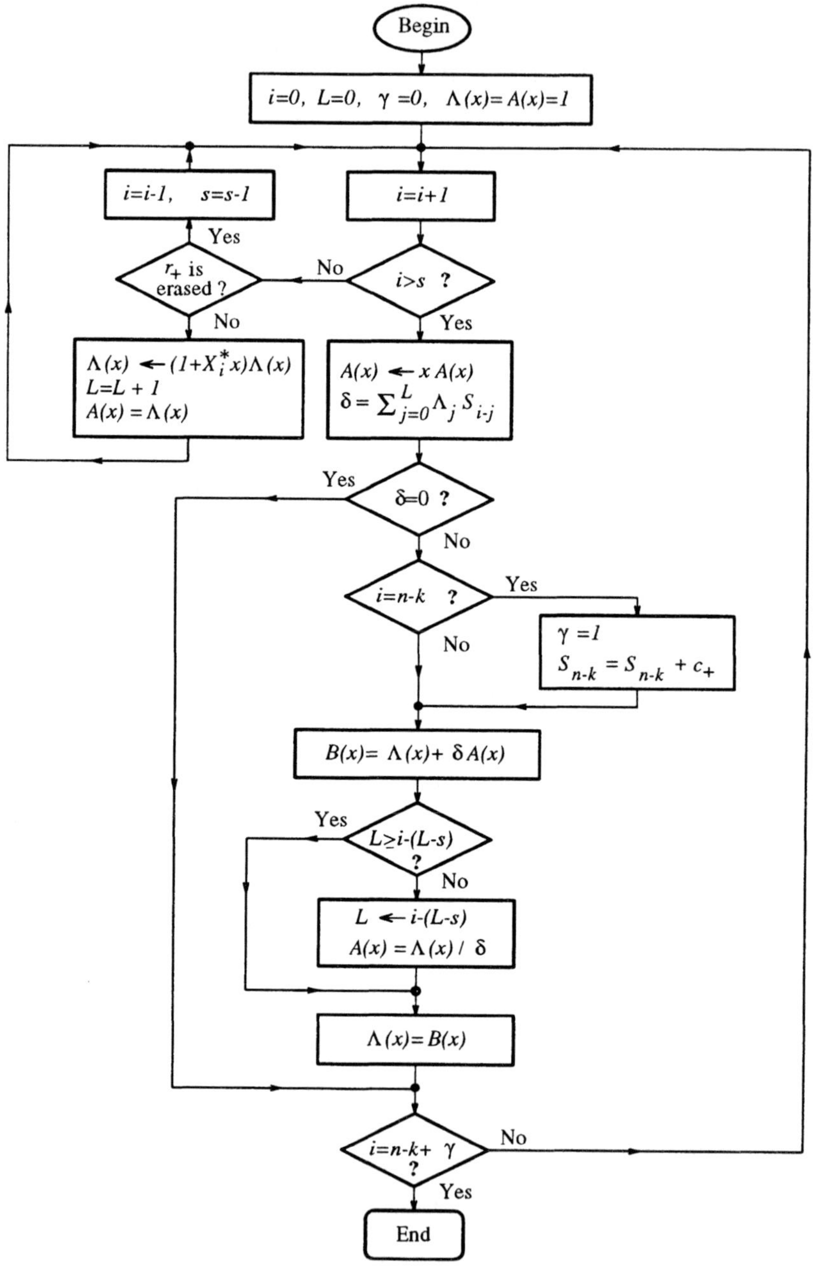


Figure 6.4: A modified Berlekamp-Massey algorithm to find $\Lambda(x)$ for errors-and-erasures decoding of singly-extended RS $(n+1,\ k)$ codes.

this trellis is periodic. Although this technique is, in general, applied to any linear block code, in this chapter we restrict our attention to RS codes.

Consider a linear $(n, k)$ cyclic code over GF $(q)$, e.g., an RS code. The encoder of this code has $n-k$ storage devices for elements from GF$(q)$. We construct a trellis for this code by associating the trellis states (nodes) with the $q^{n-k}$ possible states of the $(n-k)$-stage shift register used for encoding. Each state of the trellis is associated with an $(n-k)$-tuple number with digits chosen from GF$(q)$, namely, the state index. However, not all of these states are used at different depths of the trellis. The number of states at depth $j$ is given by [89]

$$N_s = \begin{cases} 1, & j = 0, \\ q^j, & j = 1,\ 2,\ \cdots,\ n-k-1, \\ q^{n-k}, & j = n-k,\ n-k+1,\ \cdots,\ k, \\ q^{n-j}, & j = k+1,\ k+2,\ \cdots,\ n. \end{cases} \tag{6.25}$$

Note that the trellis at depth $j = 0$ and depth $j = n$ contains only one node and it is repetitive for $j = n-k+1,\ \cdots,\ k$.

To interconnect states of the trellis a polynomial $p(x;\ j)$ in $x$ of degree $(n-k-1)$ with coefficients from GF$(q)$ is associated with each state at depth $j$. The polynomials at depth $(j+1)$ are then formed from the polynomial at depth $j$ in accordance with [89]

$$p_l(x;\ j+1) = (\ xp_i(x;\ j) + x^{n-k}\alpha_p) \bmod g(x), \tag{6.26}$$

where $g(x)$ is the generator polynomial of the code. In (6.26) $i$ shows the index of the starting state at depth $j$, $l$ shows the index of the terminating state at depth $j+1$, and $\alpha_p \in$ GF$(q)$ is the label of the connecting branch between these states.

**Example 6.11** As an example of constructing a trellis structure for an RS code we consider a (3, 2) RS code defined over GF $(2^2)$ which has the generator polynomial

$$g(x) = x + \alpha.$$

Using (6.25) the number of states of code trellis at different depths is

$$N_s = \begin{cases} 1, & j = 0, \\ 4, & j = 1,\ 2, \\ 1, & j = 3. \end{cases}$$

Note that this trellis has 4 states whose indices are 0, 1, $\alpha$, and $\alpha^2$. The branches of the code trellis are labeled using (6.26)with elements of GF$(2^2)$. This equation can be written as

$$p_l = (\ xp_i + x\alpha_p) \bmod (x + \alpha),$$

where, for convenience, we denote $p_l(x;\ j+1)$ and $p_i(x;\ j)$ by $p_l$ and $p_i$, respectively. Note that in this example both $p_l$ and $p_i$ are polynomials of zero degree.

Since $p_l$ is the remainder of the division of $(xp_i + x\alpha_p)$ by $(x+\alpha)$, we can write

$$(\,xp_i + x\alpha_p) = Q(x+\alpha) + p_l,$$

where $Q$ is the quotient of this division. Manipulating the foregoing equation yields

$$\begin{aligned} \alpha_p &= p_i + \alpha^{-1} p_l \\ &= p_i + \alpha^2 p_l, \end{aligned}$$

where the necessary arithmetic in GF($2^2$) is done by using $\alpha^2 = \alpha + 1$ over GF(2).

As a sample calculation the label of the interconnection (branch) between state 1 at depth 1 and state $\alpha^2$ at depth 2 is obtained as

$$\begin{aligned} \alpha_p &= 1 + \alpha^2 \alpha^2 \\ &= 1 + \alpha = \alpha^2. \end{aligned}$$

The code trellis of RS (3, 2) is shown in Figure 6.5. Note that each of the paths from state 0 at depth $j = 0$ to state 0 at depth $j = n$ corresponds to one valid codeword. As an illustrative example the path corresponding to the codeword $\alpha^2 0 \alpha$ is shown by bold lines in Figure 6.5. Considering all possible paths of the trellis diagram, originating from state 0 at depth 0 and arriving at state 0 at depth 3, the following 16 codewords of RS (3, 2) are obtained

| | | | | | | | | | | | |
|---|---|---|---|---|---|---|---|---|---|---|---|
| 0 | 0 | 0 | $\alpha^2$ | 1 | 0 | 1 | $\alpha$ | 0 | $\alpha$ | $\alpha^2$ | 0 |
| 0 | $\alpha^2$ | 1 | $\alpha^2$ | $\alpha$ | 1 | 1 | 1 | 1 | $\alpha$ | 0 | 1 |
| 0 | 1 | $\alpha$ | $\alpha^2$ | 0 | $\alpha$ | 1 | $\alpha^2$ | $\alpha$ | $\alpha$ | $\alpha$ | $\alpha$ |
| 0 | $\alpha$ | $\alpha^2$ | $\alpha^2$ | $\alpha^2$ | $\alpha^2$ | 1 | 0 | $\alpha^2$ | $\alpha$ | 1 | $\alpha^2$. |

Now the soft-decision decoding of RS code can be accomplished by using the Viterbi algorithm. The complexity of this decoding procedure grows with the number of code trellis states, that is $q^{n-k}$. Thus, this soft-decision decoding method is only applicable when $q$ and $n-k$ are small.

### 6.9.2 Successive-Erasure Decoding

Trellis decoding is applicable to short codes with a small amount of redundancy. For longer low rate codes this technique is impractical because of its high complexity. In this section we consider another soft-decision decoding approach which is applicable to all RS coded schemes. This approach, termed *successive-erasure decoding*, was introduced by G. D. Forney [90] in connection with *Generalized Minimum-Distance Decoding* (GMD). For binary signaling, this technique yields asymptotically the same performance as the maximum likelihood decoding (MLD) [85]. However, its performance for nonbinary codes

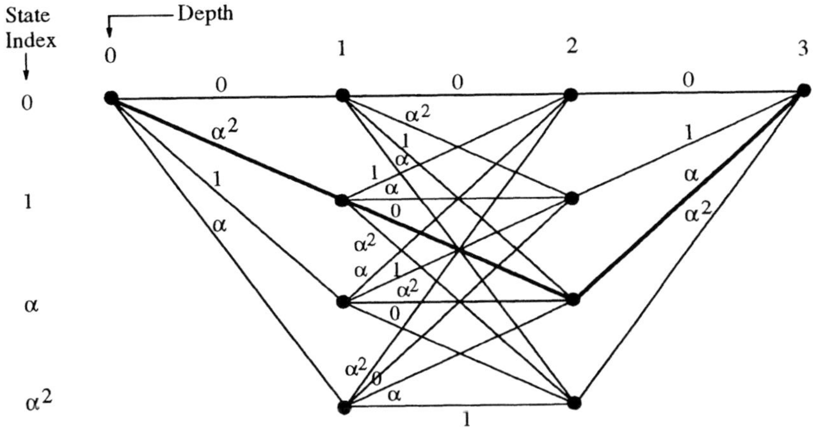

Figure 6.5: Trellis diagram for RS (3, 2) code of Example 6.11.

with multi level/phase signaling is inferior to that of MLD [91]. We begin our discussion with a brief review of GMD.

### Generalized Minimum-Distance Decoding

Consider an $(n, k)$ block code defined over GF$(q)$ with minimum Hamming distance $d$. Let $\mathbf{s} = (s_{n-1}, s_{n-2}, \cdots, s_0)$ be a transmitted codeword with symbols from GF$(q)$, and let $\mathbf{r} = (r_{n-1}, r_{n-2}, \cdots, r_0)$ be the corresponding output from an unquantized channel. Assume that the receiver consists of a detector and a decoder. The detector constructs two components corresponding to each received signal $r_i$, $i = 0, 1, , \cdots, n-1$, an estimate of $s_i$ denoted by $\hat{s}_i$, and a weight $w_i$, $0 \leq w_i \leq 1$, indicating the reliability of that estimate. The output of the detector will then be a sequence of $n$ estimates $\hat{s}_i$ with their $n$ corresponding weights $w_i$.

Assume a received vector $\mathbf{r}$ with a corresponding estimate vector $\hat{\mathbf{s}}$ and a reliability vector $\omega$. Then, it is shown that [90] there exists, at most, one codeword $\mathbf{s}_j$ for which

$$\omega \mathbf{s}_j \equiv \sum_{i=1}^{n} w_i f(\hat{s}_i, s_{ji}) > n - d, \tag{6.27}$$

where $0 \leq w_i \leq 1$ and

$$f(\hat{\mathbf{s}}, \mathbf{s}) = \begin{cases} +1, & \text{if} \quad \mathbf{s} = \hat{\mathbf{s}}, \\ -1, & \text{if} \quad \mathbf{s} \neq \hat{\mathbf{s}}. \end{cases}$$

It can be shown that [90] using an Errors-and-Erasures Decoding (EED) algorithm the unique codeword, satisfied in (6.27) can be decoded by no more than

$(d+1)/2$ trials if there is such a codeword. This decoding algorithm can be explained as follows.

First we start by applying the errors-and-erasures decoding to the hard-decision component of the received vector $\hat{\mathbf{s}}$. The decoded codeword is then examined by applying (6.27). If this codeword satisfies the condition then it is the estimate of the transmitted codeword. Otherwise, we erase the two symbols in $\hat{\mathbf{s}}$ with the lowest reliability weights and repeat the errors-and-erasures decoding process. Again, the decoded codeword is tested by the condition of (6.27). If it satisfies this condition the decoding ends, otherwise we erase the four symbols with the smallest reliability weights and continue the errors-and-erasures decoding.

This procedure continues and erases up to $d-1$ symbols, successively. The GMD procedure decodes successfully whenever any of the successive decoder outputs satisfies (6.27), otherwise a decoding failure is declared. When successful, a unique codeword is produced.

### Choice of Reliability Weight

The performance of GMD depends on the choice of the reliability weight, $w$. One way of obtaining an expression for the $w_i$'s is by optimizing the error probability bound for GMD at high SNR. This can be done by applying the Chernoff bound technique to the condition (6.27) as follows [90].

The probability that decoding is not done for GMD is obtained by considering condition (6.27)

$$P_{\text{ndc}} = \sum_{\mathbf{r},\mathbf{s}_j} P(\mathbf{r},\mathbf{s}_j) Pr\left\{ \sum_{i=0}^{n-1} w(r_i) f(\hat{s}(r_i), s_{ji}) \le n-d \right\},$$

where $w(r_i)$ and $\hat{s}(r_i)$ are the reliability function and the estimated symbol based on the received signal $r_i$, respectively. Using the Chernoff bound for an ensemble of codes, it is shown that [85], [90]

$$P_{\text{ndc}} \le e^{-n(\lambda d - \mu(\lambda))}, \tag{6.28}$$

where $\mu(\lambda) = \ln[g(\lambda)]$ and

$$g(\lambda) = E\left[ e^{\lambda[1-w(r)] f(\hat{s}(r),s)} \right].$$

Note that in the above equations the subscripts $i$ and $j$ have been deleted for simplicity. The upper bound (6.28) can be tightened by choosing an appropriate reliability weight $w$ and optimizing the Chernoff bound parameter $\lambda$. To choose $w(r)$, $g(\gamma)$ can be evaluated as

$$g(\gamma) = \frac{1}{q} \int \left[ P(r|\hat{s}(r)) e^{\lambda[1-w(r)]} + \sum_{s\neq\hat{s}} P(r|s) e^{\lambda[1+w(r)]} \right] dr. \tag{6.29}$$

To minimize the error bound, we minimize $g(\lambda)$ over $w(r)$. By differentiating (6.29) with respect to $w(r)$, we find that the $w(r)$ which minimizes $g(\lambda)$ is

$$w(r) = \frac{1}{2\lambda} \ln \frac{P(r|\hat{s}(r))}{\sum_{s \neq \hat{s}} P(r|s)} = \frac{1}{2\lambda} L.$$

Since $0 \leq w \leq 1$, we set $w = 1$ if $L \geq 2\lambda$ or $w = 0$ if $L \leq 0$. Thus

$$w(r) = \begin{cases} 1, & \text{if} \quad L \geq 2\lambda, \\ \frac{L}{2\lambda}, & \text{if} \quad 0 \leq L \leq 2\lambda, \\ 0, & \text{if} \quad L \leq 0, \end{cases}$$

where

$$L = \ln \frac{P(r|\hat{s}(r))}{\sum_{s \neq \hat{s}} P(r|s)}.$$

### SEMDD Algorithm

In GMD the decoder usually does not utilize the full error-correcting capability of the code since many correctable error patterns, in general, fall outside the error-correcting radius. In such a case a decoding failure is announced. To improve the performance of this decoding algorithm a decoding scheme closely related to GMD [90] is used as follows. Let the decoder be a full likelihood detector which stores a complete likelihood packet for each reception $r_i$, i.e., $P(r_i|s_j)$ for all $j$. Based on this information the detector also provides an estimate $\hat{s}_i$, for each received signal $r_i$, for which $P(r_i|s_j)$ is the greatest. Define the estimate log likelihood ratio as

$$L(\hat{s}_i) = \ln \frac{P(r_i|\hat{s}_i(r_i))}{\sum_{s_j \neq \hat{s}_i} P(r_i|s_j)}. \tag{6.30}$$

Note that this parameter is the same as the reliability weight, used in GMD, except that it is not bounded. Now, the same decoding procedure, as explained for GMD, can be applied by successive erasing pairs of unreliable symbols and utilizing errors-and-erasures decoding. In each step of this process an estimate of the transmitted codeword is obtained and stored. Up to $(d+1)/2$ estimated codewords may result. The decoder, then, chooses that one codeword for which $P(\mathbf{r}|\hat{\mathbf{s}}_j)$ is the greatest, i.e., the codeword closest to the received vector $\mathbf{r}$ in likelihood distance. For the AWGN channel this is equivalent to choosing an estimated codeword with minimum Euclidean distance from the received vector. This decoding algorithm is named, *Successive-Erasure Minimum Distance Decoding* (SEMDD) [91]. A block diagram of SEMDD is shown in Figure 6.6. Note that the set of estimated codewords generated in this algorithm is the same as those generated by GMD; the difference is that the choice between them is made by the likelihood measure rather than by testing the condition (6.27).

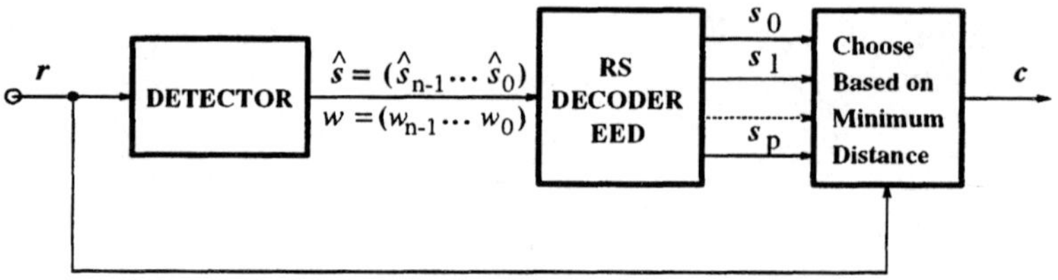


Figure 6.6: The block digram of SEMDD.

The SEMDD can be implemented by using an errors-and-erasures RS decoder. The block diagram of SEMDD, shown in Figure 6.6, shows that this can be done by adding some hardware to store the candidate codewords, resulting from successive use of EED block, and correlating these candidate codewords with received signal vector in order to select the codeword with minimum Euclidean distance from the received vector.

# Problems

**6.1** Show that the following polynomials are irreducible:

a) $p(x) = 1 + x + x^2 + x^3 + x^4$
b) $p(x) = 1 + x + x^2 + x^4 + x^6$
c) $p(x) = 1 + x^3 + x^6$

**6.2** Prove that the following polynomials are irreducible:

a) $p(x) = 1 + x + x^4$
b) $p(x) = 1 + x^2 + x^5$
c) $p(x) = 1 + x^2 + x^3 + x^4 + x^5$
d) $p(x) = 1 + x + x^6$
e) $p(x) = 1 + x + x^2 + x^5 + x^6$

**6.3** Show that the polynomials in Problem 6.1 are not primitive.

**6.4** Show that the polynomials in Problem 6.2 are primitive.

**6.5** Consider $2^m$ elements of GF($2^m$), $0, \alpha, \alpha^2, \cdots, \alpha^{n-1}$, where $n = 2^m - 1$. Show that

**a)** the sum of all $2^m$ elements is 0.

**b)**

$$\sum_{i=0}^{n-1} \alpha^{ij} = \begin{cases} 0, & \text{if } j \neq 0 \\ 1, & \text{if } j = 0 \end{cases} .$$

**6.6** Consider the finite field GF($2^4$) generated by the primitive polynomial $p(x) = 1 + x + x^4$. Find the $2^4$ elements of GF($2^4$) and establish the addition and multiplication tables.

**6.7** Repeat Problem 6.6 for the finite field GF($2^5$) generated by the primitive polynomial $p(x) = 1 + x^2 + x^5$.

**6.8** Consider an $n$-element vector $\mathbf{a}$ in GF($2^m$). $\mathbf{a}'$ and $\mathbf{a}''$ are the $i$th shift and $i$th cyclic shift versions of $\mathbf{a}$, respectively. Show that

**a)** $a'(x) + x^n q(x) = x^i a(x)$ where

$$q(x) = \sum_{k=0}^{i-1} a_{n-i+k} x^k$$

**b)** $a''(x) + (x^n + 1)q(x) = x^i a(x)$.

**6.9 a)** Show that

$$
\begin{aligned}
p_1(x) &= 1 + x^2 + x^5 \\
p_2(x) &= 1 + x^2 + x^3 + x^4 + x^5 \\
p_3(x) &= 1 + x + x^2 + x^4 + x^5
\end{aligned}
$$

are the minimal polynomials of $\alpha$, $\alpha^3$ and $\alpha^5$, respectively, where $\alpha$ is the primitive element in GF($2^5$).

**b)** Derive the generator polynomial of a primitive binary BCH(n, k) code based on the above minimal polynomials. Find the values of $n, k$ and the design distance of the code.

**6.10 a)** $g(x)$ and $a(x)$ are two polynomials of degree $(n-k)$ and $(n-1)$, respectively, where $k > 1$. Let $q(x)$ and $r(x)$ be the quotient and the remainder of $a(x)/g(x)$ respectively. Show that the coefficients of $g(x)$, $a(x)$, $q(x)$ and $r(x)$ satisfy the following relations:

$$r_l + \sum_{i=0}^{l} g_i q_{l-i} = a_l \quad \text{for } 0 \le l \le (n-k-1)$$

and

$$\sum_{i=0}^{l} g_i q_{l-i} = a_l \quad \text{for } (n-k) \le l \le (n-1).$$

**b)** If $a(x) = x^{n-k} I(x)$, where $I(x)$ is a polynomial of degree $(k-1)$, and $g_{n-k} = 1$, and the polynomials are defined in a finite-field GF($2^m$), prove that

**i)** the coefficients of $q(x)$ can be derived from the coefficients of $I(x)$ as

$$
\begin{aligned}
q_{k-1} &= I_{k-1} \\
q_{k-j} &= I_{k-j} + \sum_{i=1}^{j-1} g_{n-k-j+i} q_{k-i} \\
&= \sum_{i=1}^{j} g_{n-k-j+i} q_{k-i} \quad \text{for } j = 2, 3, \cdots, k.
\end{aligned}
$$

**ii)** the coefficients of $r(x)$ can be derived from the coefficients of $q(x)$ as

$$r_l = \sum_{i=0}^{l} g_i q_{l-i} \quad \text{for } 0 \le l \le (n-k-1).$$

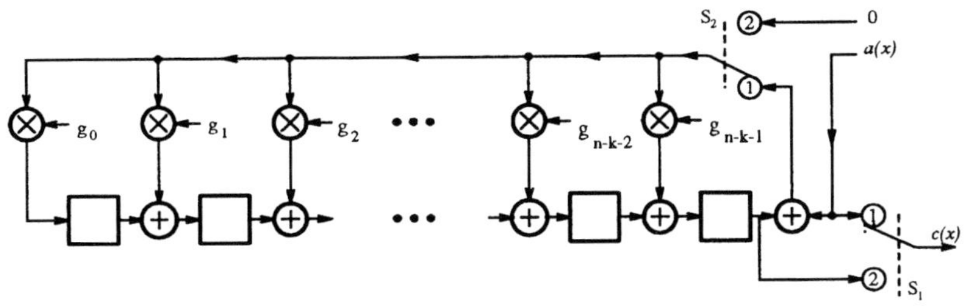


Figure P6.10-1: An RS$(n, k)$ systematic encoder.

c) Verify that the RS$(n, k)$ systematic encoder can be realized by the feedback shift-register circuit shown in Figure P6.10-1. The operation of the circuit is explained as follows:

**Step 1:** With switches $S_1$ and $S_2$ on position 1, clock $k$ information symbols.

**Step 2:** With switches $S_1$ and $S_2$ on position 2, clock $(n-k)$ parity symbols.

**6.11** Draw the block diagram of the systematic encoder of the RS$(7,5)$ code discussed in Example 6.7.

**6.12** Construct an RS$(7,4)$ code with $h = 1$ and draw the block diagram of the systematic encoder of this code.

**6.13** Find the codeword of the extended RS$(8,5)$ code with $h = 1$ for an information polynomial $a(x) = \alpha^2 x^4 + \alpha^5 x$.

*Hint:* Use the results of Example 6.7.

**6.15** Consider the RS$(7,5)$ systematic code discussed in Example 6.7. Decode the following received polynomials using the procedure given in Example 6.8:

a) $r(x) = \alpha^4 x^6 + \alpha^5 x^2 + x + 1$
b) $r(x) = \alpha^2 x^6 + \alpha^2 x^2 + x + 1$
c) $r(x) = \alpha^2 x^6 + \alpha^5 x^2 + 1$
d) $r(x) = \alpha^5 x^6 + \alpha^5 x^2$
e) $r(x) = \alpha^2 x^6 + \alpha^5 x^2 + x + 1$.

**6.16** Consider the RS$(7,3)$ code discussed in Example 6.9. Decode the following received polynomials:

a) $r(x) = \alpha^2 x^4$
b) $r(x) = \alpha^2 x^3 + \alpha^5 x$
c) $r(x) = \alpha x^5 + \alpha^3 x^4$
d) $r(x) = \alpha x^5 + \alpha^2 x^4 + \alpha$.

**6.17** From (6.11), (6.19) and (6.20) show that

$$V(x) = \sum_{i=1}^{\epsilon} Y_i X_i \frac{\Lambda(x)}{1 + xX_i}$$

and

$$V(X_j^{-1}) = Y_j \Lambda'(X_j^{-1}), \; j = 1, 2, \cdots, \epsilon,$$

where $\Lambda'(x)$ is the derivative of $\Lambda(x)$.

**6.18** **a)** Find the error values for the received polynomial of Example 6.10 by two techniques: matrix inversion using (6.11) and Forney algorithm using (6.21). Compare the complexity of the two techniques.

**b)** Show that for a large value of $\epsilon$ the Forney algorithm (6.21) is much faster than the matrix inversion (6.11). What is the crossing value of $\epsilon$ at which the two techniques have a similar complexity?

**6.19** Consider the 2-error-correcting RS(7,3) code given in Example 6.9. Decode the following received polynomials using the Berlekamp-Massey algorithm:

a) $r(x) = \alpha x^4 + \alpha^2 x^3$
b) $r(x) = \alpha x^3 + \alpha^2 x^2 + 1$
c) $r(x) = \alpha^5 x^2 + \alpha x$
d) $r(x) = \alpha x^2 + \alpha^3 x + \alpha^2$.

**6.20** From (6.22) and (6.23), complete the derivation of (6.24).

**6.21** Decode the received polynomial given in Problem 6.19b) if erasures are detected at symbols # 0 and # 2. Compare the result with that of Problem 6.19b) and discuss the effectiveness of errors-and-erasures decoding techniques.

**6.22** Decode the received polynomial given in Problem 6.19d) if the erasures are detected at symbols # 0 and # 1. Compare the result with that of Problem 6.19d) and discuss the effectiveness of errors-and-erasures decoding techniques.

**6.23** Consider the singly-extended RS(8,4) code with $h = 1$. A codeword is represented as $\mathbf{c}_- = (c_-, c_0, c_1, \cdots, c_6)$.

**a)** Derive the generator polynomial of the original RS(7,4) code.

**b)** Use the modified Berlekamp-Massey algorithm to decode the following received vectors:

i) $\mathbf{r} = (\alpha, 0, 0, 0, 0, \alpha^5, \alpha^3, \alpha^2)$
$\mathbf{f} = (1, 0, 0, 0, 0, 1, 1, 1)$

ii) $\mathbf{r} = (0, 0, 0, 0, \alpha, \alpha^2, 0, 0)$
$\mathbf{f} = (0, 0, 0, 0, 0, 0, 0, 0)$

iii) $\mathbf{r} = (\alpha, 0, 0, 0, 0, \alpha, \alpha^2, 0)$
$\mathbf{f} = (0, 0, 0, 0, 0, 1, 1, 0)$

iv) $\mathbf{r} = (0, 1, 0, \alpha, 0, \alpha^2, \alpha, 0)$
$\mathbf{f} = (0, 0, 0, 1, 0, 0, 0, 0)$

v) $\mathbf{r} = (\alpha^2, 0, 0, 0, \alpha, \alpha^3, 0, 0)$
$\mathbf{f} = (1, 0, 0, 0, 0, 0, 1, 0)$.

Note that $\mathbf{f}$ is a binary 8-element vector called the erasure flag vector. Its element is defined as

$$f_i = \begin{cases} 1, & \text{if symbol } r_i \text{ is erased} \\ 0, & \text{otherwise} \end{cases}$$

**6.24** Construct a trellis structure for an RS(3, 1) code with $h = 1$.

**6.25** The Berlekamp-Massey decoding algorithm discussed in Section 6.6 can be applied to binary BCH codes. In binary BCH decoding, the step to find the error values is not needed. Consider the binary BCH(15, 7) code discussed in Example 6.5. Decode the following received polynomials:

a) $r(x) = x^{13} + x^{12}$
b) $r(x) = x^{10} + x + 1$
c) $r(x) = x^2 + x + 1$
d) $r(x) = x^6 + x^5$.

**6.26** Derive the properties of the finite-field Fourier transform given in Section C.1 of Appendix C.

**6.27** Draw the block diagram of the RS$(n, k)$ transform encoder discussed in Section C.2 of Appendix C. Show that the structure of a transform encoder is regular and allows for the design of a programmable RS$(n, k)$ encoder defined in a given GF$(2^m)$ where $k$ is programmable.

**6.28** Repeat Problem 6.21 using the transform decoding algorithm discussed in Section C.3 of Appendix C.

**6.29** Repeat Problem 6.19 using the transform decoding algorithm discussed in Section C.3 of Appendix C.

**6.30** Using the Fourier transform properties in Section C.1 of Appendix C, find the inverse Fourier transform of $\Lambda^{(i)}(x)$.

**6.31** Repeat Problem 6.23 using the transform decoding algorithm discussed in Section C.3 of Appendix C.

**6.32** Repeat Problem 6.21 using the Time-Domain decoding algorithm discussed in Section C.4 of Appendix C.

**6.33** Repeat Problem 6.23 using the Time-Domain decoding algorithm discussed in Section C.5 of Appendix C.

**6.34** Consider two RS codes, RS$(n, k)$ and RS$(n, k')$, defined in the same GF$(2^m)$ with $h = 1$ and $k' = k - 1$. The generator polynomials of RS$(n, k)$ and RS$(n, k')$ are $g(x) = \sum_{j=0}^{n-k} g_j x^j$ and $g'(x) = \sum_{j=0}^{n-k'} g'_j x^j$, respectively.

**a)** Show that

$$\begin{aligned} g'_0 &= \alpha^{n-k} g_0, \\ g'_j &= \alpha^{n-k} g_j + g_{j-1}, \text{ for } j = 1, 2, \cdots, (n-k) \\ g'_{n-k'} &= g_{n-k}. \end{aligned}$$

**b)** Based on the equations obtained from a) draw the block diagram of the circuitry to generate the coefficients of $g(x)$ where $k$ is programmable.

**c)** Design a programmable RS$(n, k)$ based on Figure P6.10-1 so that $k$ can programmed from 1 to $(n-1)$.

**6.35** Consider an errors-and-erasures decoder for an RS$(n, k)$ code. Show that the decoder can be used as an RS$(n, k)$ systematic encoder by providing an input polynomial

$$r(x) = \sum_{j=0}^{n-1} r_j x^j,$$

where $r_j = I_{j-(n-k)}$, $j = (n-k), (n-k+1), \cdots, (n-1)$. $I_l$'s, $l = 0, 1, \cdots, (k-1)$, are the information symbols and $r_j$'s, $j = 0, 1, \cdots, (n-k-1)$, are declared erasures.

**6.36** Consider an RS$(n, k)$ codeword $\mathbf{c}$. A punctured codeword $\mathbf{c}_p$ of length $n'$, where $k < n' < n$, is generated from $\mathbf{c}$ by removing $(n - n')$ symbols at positions $l_1, l_2, \cdots, l_{n-n'}$. Assume that $\mathbf{c}_p$ is transmitted and positions $l_1, l_2, \cdots, l_{n-n'}$ are known by the errors-and-erasures decoder fo this RS$(n, k)$ code. Show that if the number of erasures $s$ and number of errors $\epsilon$ satisfy the relation $2\epsilon + s \leq (n' - k)$ then the received $n'$-element word can be correctly decoded by inserting additional erasures at positions $l_1, l_2, \cdots, l_{n-n'}$.

**6.37** Consider an RS$(n, k)$ systematic code. A shortened RS$(n', k')$ code is generated by using a shortened information polynomial of length $k' < k$. The corresponding codeword has a length $n' = n - (k - k')$. Show that a shortened RS$(n', k')$ can correct up to $\epsilon$ errors and $s$ erasures provided that $2\epsilon + s \leq (n - k)$.

# Chapter 7

# RS-CODED MPSK SCHEMES FOR FADING CHANNELS

In Chapter 5, the design and performance evaluation of the Trellis-Coded Modulation (TCM) schemes were presented for fading channels. In this chapter the issue of using Block-Coded Modulation (BCM) schemes, as a counterpart to trellis-coded modulation schemes, is addressed. Specifically, we combine Reed-Solomon (RS) codes with Multi-Phase Shift-Keying (MPSK) schemes, as possible joint power and bandwidth-efficient BCM schemes for fading channels. The performance of these schemes over a Rayleigh fading channel is evaluated for different algebraic decoding strategies; .i.e., *Errors-Only Decoding* (EOD) and *Errors-and-Erasures Decoding* (EED) techniques by means of lower-bounding the post decoding bit error probability. Next, we apply different soft-decision decoding techniques, introduced in Chapter 6, for decoding of RS coded MPSK schemes.

## 7.1 RS-CODED MPSK SCHEMES

Although, it is believed that: "Generally, anything that can be achieved with a block code can be achieved with somewhat greater simplicity with a convolutional code" [31], the BCM schemes may challenge their counterpart for the following reasons.

A feature of TCM schemes is that the transmitted symbol at any given time depends on the previous input symbol sequence. In the decoding process, hence, the decoder has to know the history of the coded sequences before being able to decode a particular symbol. Furthermore, the decoder has to look at the subsequence history of the sequence to examine the total influence of that symbol on the transmitted sequence. Therefore, if the decoder loses or makes a mistake in the history of the sequence, *errors will propagate*. This event can be considered as a drawback in some channels such as mobile radio channels with slow shadowing fading. In such a case, errors due to shadowing will affect the decoding process of the symbols in an unshadowed time and therefore may cause long-term error propagation. In such a scenario, block-coded modulation may be more advantageous because the decoding of a received code block is independent of any other blocks.

In Chapter 4 we briefly reviewed Block-Coded Modulation schemes which are designed for AWGN channels, based on maximizing the minimum Euclidean distance between codewords. Here, we address the issue of designing these schemes based on maximizing the time diversity (the effective length) of the code. For this purpose we introduce the idea of combining Reed-Solomon codes with MPSK signaling, as a *bandwidth-efficient* block-coded technique, over the *bandwidth-limited fading channels.*

There are many reasons for using RS codes. As stated in Chapter 6 these codes are *Maximum-Distance-Separable* (MDS) codes, and hence, they make highly efficient use of the redundancy. The block lengths and symbol sizes of RS codes can be readily adjusted to match the MPSK constellation. By choosing the same cardinality for the channel and code symbols, the minimum Hamming distance of the RS code will determine the effective length (time diversity) of the coded scheme. RS codes also provide a wide range of code rates that can be chosen such that the coded scheme has compatible bandwidth efficiency to the reference uncoded system. Finally, the available errors-and-erasures decoding techniques for these codes make them a suitable choice for channels with side information.

## 7.2 CONSTRUCTION OF RS-CODED MPSK SCHEMES FOR FADING CHANNELS

To develop an RS coded MPSK scheme we consider two approaches. First, an RS code, defined over GF($2^m$), is combined with an expanded $2^m$-PSK signal set. In this combination the code rate is chosen such that the rate of the coded scheme is the same as the uncoded one (usually $2^{m-1}$-PSK ). In this case the time diversity of the coded scheme is determined by the minimum Hamming distance of the RS code. In the second approach, the symbols of an RS code, defined over GF($2^{lm}$), are mapped to the signal points of a $2^m$-PSK signal set such that each symbol of the RS code consists of the concatenation of $l$ channel symbols. The following proposition indicates that the effective order of time diversity in such a mapping is at least $d$, the minimum Hamming distance of the RS code.

**Proposition 7.1** Assume that a code, defined over GF($2^m$), has a minimum Hamming distance $d$. Let the symbols of this code be transmitted using a signal set with a cardinality of $2^m$ in such a way that each code symbol consists of the concatenation of $l$ channel symbols. Then, the minimum Hamming distance between the coded signals is at least $d$.

**Proof** Let $\mathbf{c}_1 = (s_{11}, s_{12}, \cdots, s_{1n})$ and $\mathbf{c}_2 = (s_{21}, s_{22}, \cdots, s_{2n})$ be two code symbols with constituent channel symbol components $\{s_{ij}\}$. If $\mathbf{c}_1 \neq \mathbf{c}_2$, then *at least* one of the constituent channel symbols of $\mathbf{c}_1$ differs from that of $\mathbf{c}_2$. Thus, for two codewords which differ in $d$ code symbol positions, the corresponding channel coded sequences will differ in *at least* $d$ positions. The proposition, then, follows.

□

Note that the second approach allows us to enhance the order of time diversity by using more powerful low rate codes. Increasing $l$ provides more powerful coded schemes. However, for practical purpose, such as decoding delay, we restrict ourselves to $l = 2$. In the following we consider some examples of RS coded MPSK schemes.

**Example 7.1** As the first example we construct an RS coded scheme by using the 8PSK signal set. Applying the first approach, a $(7, 5)$ RS code, defined over GF($2^3$), is used with 8PSK signaling. The rate of RS code is chosen such that the overall coded 8PSK throughput is comparable to that of uncoded QPSK, i.e., 2 bits/s/Hz. The $(7, 5)$ RS code has a symbol rate $R_c = 5/7 > 2/3$, which translates into a throughput of slightly more than 2 bits/s/Hz when combined with 8PSK signal set. The minimum Hamming distance of the RS$(7, 5)$ coded 8PSK is $d = 3$.

**Example 7.2** The RS coded 8PSK scheme of Example 7.1 can be shortened by omitting one of the information symbols. The resulting scheme is the RS(6, 4) coded 8PSK.This coded scheme has a throughput of 2 bits/s/Hz which is comparable with the uncoded QPSK. Considering the fact that shortening does not affect the minimum Hamming distance of the RS code, the minimum Hamming distance of RS(6, 4) coded 8PSK is also $d = 3$.

To obtain a higher coding gain, minimum Hamming distance (time diversity) of the coded scheme should be increased. This can be done by using longer and lower rate RS codes. However, to compensate for the rate loss, the cardinality of the MPSK signal set should be increased.

**Example 7.3** As a third example we combine RS$(15, 8)$ code, defined over GF($2^4$), with 16PSK signal set. This yields a coded scheme with minimum Hamming distance 8 and a throughput of $4 \times 8/15 = 32/15 > 2$ bits/s/Hz with almost the same bandwidth efficiency as the uncoded QPSK.

Choosing a signal set with a large number of signal points has a drawback when the issue of the carrier recovery is concerned. For denser MPSK signal sets it is a difficult (if not impossible) task to maintain carrier synchronization during deep fades. An alternative is to use the second approach in constructing these schemes. Consider the following example.

**Example 7.4** As an example of constructing the RS coded MPSK schemes based on the second approach, we consider RS(63, 42) coded 8PSK scheme. The RS code is defined over GF($2^6$), and hence, each code symbol consists of two concatenated 8PSK symbols. The rate of RS code is 42/63 which translates into 2 bits/s/Hz throughput when combined with 8PSK signal set. The minimum Hamming distance of this coded MPSK scheme is lower bounded with that of the original RS code, i.e., 22.

In the following sections we evaluate the performance of RS coded MPSK schemes over a normalized Rayleigh fading channel for different decoding strategies. First, we consider the performance of algebraic decoding techniques, i.e., errors-only decoding and errors-and-erasures decoding. Then by means of analysis and simulation we evaluate the performance of soft-decision decoding techniques and compare them with algebraic decoding strategies.

## 7.3 PERFORMANCE EVALUATION OF ALGEBRAIC DECODING OF RS-CODED MPSK SCHEMES ON A RAYLEIGH FADING CHANNEL

The transmission model of the system under consideration is depicted in Figure 7.1. A single sample per symbol is assumed in representing this model. An RS encoder, defined over GF($2^m$), encodes the input bits into words of length $n$ with alphabet size $m$. The code symbols are interleaved and then used to modulate an M-ary PSK waveform. The channel corrupts the transmitted signal by introducing a fading gain and an additive Gaussian noise term. It is assumed that the receiver performs *coherent detection* and hence the channel phase shift is compensated for by the receiver. Also, we assume that ideal interleaving/de-interleaving is employed in order to destroy the memory of the fading channel. This allows us to consider the samples of the fading amplitude as being statistically independent.

The additive Gaussian noise is defined as having a two-sided spectral density, $N_0/2$. The multiplicative gain is a random variable having a normalized Rayleigh fading density function as (5.3).

In the case of errors-only decoding, the MPSK detector makes a hard decision of each of the received signals according to their phases. These decisions are de-interleaved and supplied to the RS decoder. If the decoder performs

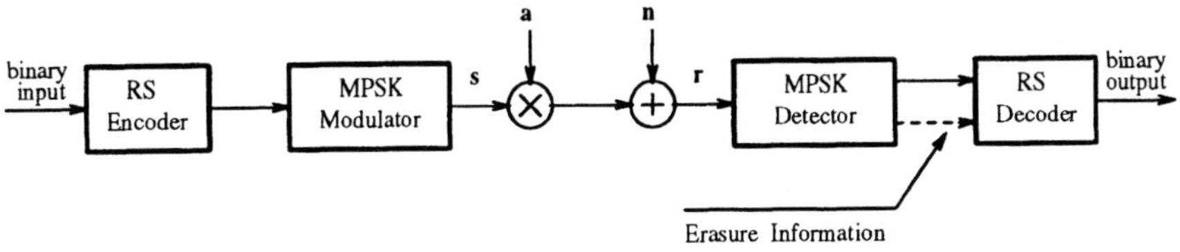


Figure 7.1: The baseband signal transmission model for RS-coded MPSK schemes on a fading channel.

errors-and-erasures decoding, the MPSK detector also provides the erasure location information to the RS decoder.

### 7.3.1 Uncoded MPSK Performance

The performance of MPSK signaling in the AWGN channel with an amplitude gain $a$ is derived in [60]. At high signal-to-noise ratios the *symbol-error probability* is well approximated [60, page 169] by (see Problem 2.10)

$$P_{sM}(a) \approx \text{erfc}\left(a\sqrt{\frac{\mathrm{E_s}}{\mathrm{N_0}}}\sin(\frac{\pi}{\mathrm{M}})\right), \tag{7.1}$$

where erfc($\cdot$) is the complementary error function.

As stated in Chapter 3, for the purpose of comparing coded systems with uncoded ones, we need to distinguish the signal energy per channel symbol, $E_s$, from the *energy per bit*, $E_b$, which is defined as the required energy per information bits. For MPSK signaling with $M = 2^m$, each M-ary symbol represents $m$ channel bits and therefore $E_s = mE_b$. In the case where a rate $R_c$ encoder is used, $m$ channel bits carry $mR_c$ information bits and the signal energy per channel symbol can be related to the energy per information bits as

$$E_s = mR_cE_b. \tag{7.2}$$

Substituting (7.2) into (7.1) yields

$$P_{sM}(a) \approx \text{erfc}\left(a\sqrt{\mathrm{mR_c\gamma_b}}\sin(\frac{\pi}{\mathrm{M}})\right), \tag{7.3}$$

where $\gamma_b \equiv E_b/N_0$. Equation (7.3) can be considered as the MPSK symbol-error probability conditioned on the fading amplitude, $a$.

The symbol error probability for MPSK signaling over a Rayleigh fading channel can be found by averaging the conditional symbol-error probability of (7.3) with respect to the fading amplitude $a$

$$P_{sM} = \int_0^{\infty} P_{sM}(a)\cdot p_A(a)\cdot da$$

$$= \int_0^\infty 2a \ \mathrm{erfc}\left(a\sqrt{\mathrm{mR_c\gamma_b}}\ \sin(\frac{\pi}{\mathrm{M}})\right) \mathrm{e}^{-a^2} \mathrm{d}a$$

$$= 1 - \sqrt{\frac{mR_c\gamma_b \sin^2(\frac{\pi}{M})}{1 + mR_c\gamma_b \sin^2(\frac{\pi}{M})}}. \tag{7.4}$$

The relationship between the symbol-error probability and *bit-error probability* of MPSK signals depends upon the mapping of the $m$-bit groups to the symbols. If one uses a *Gray code* for this purpose, in which the bit groups assigned to adjacent phases differ by only one bit, the bit-error probability is approximated by

$$P_{bM} \approx \frac{P_{sM}}{\log_2 M} = \frac{P_{sM}}{m}. \tag{7.5}$$

### 7.3.2 RS-Coded MPSK Performance

#### Errors-Only Decoding

Consider an $(n,\ k)$ RS code defined over GF$(2^m)$. The minimum Hamming distance $d$, and the error correction capability $t$ of this code are $d = n - k + 1$ and $t = \lfloor (d-1)/2 \rfloor = \lfloor (n-k)/2 \rfloor$, respectively (see Chapter 6), where $\lfloor x \rfloor$ is the largest integer less than or equal to $x$. If the number of errors in the received signal exceeds the error correction capability of the code, the decoder may produce an incorrect codeword. This is called a *decoding failure.* Summing up all possible decoding failure probabilities, the *post decoding block error probability* can be expressed as

$$P_{blc} = \sum_{i=t+1}^{n} \binom{n}{i} P_{sC}^i (1 - P_{sC})^{n-i}, \tag{7.6}$$

where $P_{sC}$ is the RS code symbol-error rate. Note that if the first approach in combining RS codes with MPSK signal set is used, then $P_{sC}$ is the same as $P_{sM}$, given by (7.4). However, in the second approach where each symbol of the RS code consists of two concatenated MPSK symbols, the probability of a correct reception of each code symbol is $(1 - P_{sM})^2$ and the error probability of the code symbol, $P_{sC}$, is derived as

$$P_{sC} = 1 - (1 - P_{sM})^2.$$

A lower bound on the *post decoding bit error probability* can be obtained by weighting the probabilities of all possible decoding failures, given in (7.6), by the corresponding bit error probabilities [95] (see Problem 7.5)

$$P_b \geq \frac{1}{n} \sum_{i=t+1}^{n} \binom{n}{i} P_{sC}^i (1 - P_{sC})^{n-i} (i\ \frac{P_{bC}}{P_{sC}}), \tag{7.7}$$

where $P_{bC}$ indicates the mean bit error probability of a code symbol.

### Errors-and-Erasures Decoding

Algorithms that decode both symbol errors and symbol erasures for RS codes, e.g., the Berlekamp-Massey decoding algorithm, are well known (see Chapter 6). As stated in Chapter 6, all patterns of $\epsilon$ symbol errors and $s$ symbol erasures can be corrected by using these decoding algorithms provided $2\epsilon + s \leq d - 1$.

To evaluate the performance of the RS coded schemes over fading channels with side information (errors-and-erasures decoding technique), a decision region for erasures should be defined. Depending on the existence or the lack of information about fading amplitude, CSI, two strategies to define the decision region for erasure are considered.

**Erasure Generation with CSI.** It is assumed that the value of the fading amplitude, $a$, is available at the receiver. In this case a simple erasure generation strategy is suggested by Hagenauer and Lutz [95], in which a threshold value, $a_T$, is defined for the fading amplitude. If the fading amplitude is less than this threshold value then the received symbol is declared to be an erasure. This definition is based on the value of the fading amplitude and it ignores the impact of the additive noise on the channel symbol.

To evaluate the performance of the RS coded schemes with erasure decoding, it is necessary to consider the probabilities of different events. For this erasure generation strategy the following probabilities are determined.

**i)** The probability of an unreliable (erased) MPSK symbol is

$$P_{EM} = \int_0^{a_T} p_A(a)\, da = 1 - e^{-a_T^2}.$$

**ii)** The probability of a reliable (not erased) MPSK symbol is

$$P_{\overline{E}M} = 1 - P_{EM} = e^{-a_T^2}.$$

**iii)** The probability of a reliable MPSK symbol being in error is (see Problem 7.8)

$$\begin{aligned} P_{sM|\overline{E}} &= \frac{1}{P_{\overline{E}M}} \int_{a_T}^{\infty} P_{sM}(a)\, p_A(a)\, da \\ &= \mathrm{erfc}(a_T \Gamma_b) - \frac{e^{a_T^2} \Gamma_b}{\sqrt{1+\Gamma_b^2}}\, \mathrm{erfc}\left(a_T \sqrt{1+\Gamma_b^2}\right), \end{aligned}$$

where

$$\Gamma_b \equiv \sqrt{mR_cE_b/N_0}\ \sin(\frac{\pi}{M}).$$

**iv)** The probability of an unreliable MPSK symbol being in error is

$$\begin{aligned} P_{sM|E} &= \frac{1}{P_{EM}}\int_0^{a_T} P_{sM}(a)\ p_A(a)\ da \\ &= \frac{1}{P_{EM}}\left[1 - e^{a_T^2}\mathrm{erfc}(a_T\Gamma_b) - \frac{\Gamma_b}{\sqrt{1+\Gamma_b^2}}\left(1 - \mathrm{erfc}(a_T\sqrt{1+\Gamma_b^2})\right)\right]. \end{aligned}$$

**Erasure Generation without CSI.** In the case of no channel state information, a circular region with radius $\rho_T$ in the MPSK signal space is considered for erasures [52]. This region, $\Lambda_e$, along with the angular decision regions, $\Lambda_l$, for an 8PSK signal set is shown in Figure 2.15. A channel symbol erasure is declared if the received signal falls into region $\Lambda_e$. Note that with this definition of the erasure region, the impact of the additive noise on the erasure declaration is taken into account in addition to that of fading. However, at high signal-to-noise ratios the effect of the additive noise is negligible, and, as will be shown, this erasure strategy is equivalent to the previous one.

Suppose the $i$th channel symbol is transmitted. The probability that the received signal $r$ falls into the decision region $\Lambda_l$ is expressed as

$$Q_i^{\overline{E}} = E_a\left[\int_{\Lambda_i} p(r|a,\ s_i)\ dr\right], \tag{7.8}$$

where $E_a$ represents the averaging operation over the fading amplitude $a$. Note that $Q_i^{\overline{E}}$ is the probability of a correct decision of a reliable MPSK symbol. Hence, the probability of a reliable MPSK symbol being in error is obtained as

$$P_{sM|\overline{E}} = 1 - \frac{1}{P_{\overline{E}M}}Q_i^{\overline{E}},$$

where

$$P_{\overline{E}M} = E_a\left[1 - \int_{\Lambda_l} p(r|a, s_i)dr\right]. \tag{7.9}$$

In general these expressions must be evaluated numerically. As we will see, in order to find the lower bounds of post-decoding bit error probabilities using errors-and-erasures decoding the threshold value for each signal-to-noise ratio has to be optimized. In order to avoid numerical evaluation of the above expressions in the optimization procedures, appropriate approximation that leads to a closed form expression is developed in Appendix D. Using the results of Appendix D, the probabilities of different events are given as follows.

**i)** The probability of a reliable (not erased) MPSK symbol is (see Problem 7.11)

$$P_{\overline{E}M} = e^{-\rho_{T_n}^2}[1 - \mathrm{erfc}(\sqrt{\mathrm{mR_c}\gamma_\mathrm{b}\rho_{\mathrm{T_n}}})] + \mathrm{erfc}(\sqrt{\mathrm{mR_c}\gamma_\mathrm{b}\rho_{\mathrm{T_n}}}).$$

**ii)** The probability of a reliable (not erased) MPSK symbol being in error is

$$\begin{aligned} P_{sM|\overline{E}} &= 1 - \frac{1}{P_{\overline{E}M}}\left[e^{-\rho_{T_n}^2} + \frac{\Gamma_b}{\sqrt{1+\Gamma_b^2}}\mathrm{erfc}(\sqrt{1+\Gamma_\mathrm{b}^2}\;\rho_{\mathrm{T_n}})\right] \\ &\quad - e^{-\rho_{T_n}^2}\;\mathrm{erfc}(\Gamma_\mathrm{b}\rho_{\mathrm{T_n}}). \end{aligned}$$

**iii)** The probability of an unreliable (erased) MPSK symbol being in error is

$$\begin{aligned} P_{sM|E} &= 1 - \frac{1}{1-P_{\overline{E}M}}\left[e^{-\rho_{T_n}^2}\mathrm{erfc}(\Gamma_\mathrm{b}\rho_{\mathrm{T_n}}) + \frac{\Gamma_\mathrm{b}}{\sqrt{1+\Gamma_\mathrm{b}^2}} - e^{-\rho_{\mathrm{T_n}}^2}\right. \\ &\quad \left. - \frac{\Gamma_b}{\sqrt{1+\Gamma_b^2}}\;\mathrm{erfc}(\sqrt{1+\Gamma_\mathrm{b}^2}\;\rho_{\mathrm{T_n}})\right]. \end{aligned}$$

In all of the foregoing equations $\rho_{T_n} = \rho_T/\sqrt{E_s}$ represents the normalized erasure threshold value.

Similar to the case of errors-only decoding the post-decoding block error probability of RS codes for the errors-and-erasures decoding technique can be computed as (see Problem 7.12)

$$\begin{aligned} P_{blc} &= \sum_{s=0}^{d-1} P_{EC}(s) \sum_{\epsilon=\lceil\frac{d-s}{2}\rceil}^{n-s} \binom{n-s}{\epsilon} P_{sC|\overline{E}}^{\epsilon}(1-P_{sC|\overline{E}})^{n-s-\epsilon} \\ &\quad + \sum_{s=d}^{n} P_{EC}(s), \end{aligned} \tag{7.10}$$

where $P_{EC}(s)$ is the probability of $s$ code symbols being in erasure, which can be expressed as

$$P_{EC}(s) = \binom{n}{s} P_{EC}^{s}(1-P_{EC})^{n-s}.$$

Again, for the first approach of RS coded schemes, $P_{EC}$ and $P_{sC|\overline{E}}$ can be replaced by $P_{EM}$ and $P_{sM|\overline{E}}$, respectively. For the second approach, we assume that the RS code symbol is erased if either one of the two MPSK signals belonging to that symbol lies in the erasure region. In this case $P_{EC}$ and $P_{sC|\overline{E}}$ are obtained as

$$\begin{aligned} P_{EC} &= 1-(P_{\overline{E}M})^2 \\ P_{sC|\overline{E}} &= 1-(1-P_{sM|\overline{E}})^2. \end{aligned} \tag{7.11}$$

To find a lower bound on $P_b$ for the errors-and-erasure decoding the probabilities of all possible decoding failures, given in (7.10), are weighted by the corresponding bit error probabilities [95] (see Problem 7.13)

$$\begin{aligned} P_b \geq & \frac{1}{n}\sum_{s=0}^{d-1} P_{EC}(s) \sum_{e=\lceil \frac{d-s}{2} \rceil}^{n-s} \binom{n-s}{e} P^e_{sC|\overline{E}} \\ & \cdot (1-P_{sC|\overline{E}})^{n-s-e}\left(sP_{bC|E} + e\frac{P_{bC|\overline{E}}}{P_{sC|\overline{E}}}\right) \\ & + \frac{1}{n}\sum_{s=d}^{n} P_{EC}(s)\left(sP_{bC|E} + (n-s)P_{bC|\overline{E}}\right). \end{aligned} \tag{7.12}$$

In the following we consider the performance of three RS coded MPSK schemes, constructed in Section 7.2, and compare them with a reference uncoded scheme over a normalized Rayleigh fading channel. The reference scheme for all examples is the uncoded QPSK.

**Example 7.5** As the first example of performance evaluation we consider the performance of the RS(7, 5) coded 8PSK scheme, constructed in Example 7.1, in a Rayleigh fading channel. The lower bounds of the post-decoding bit error probability of the EOD and EED techniques of this coded scheme are obtained by using (7.7) and (7.12). These bounds for RS(7, 5) coded 8PSK are shown in Figure 7.2. Simulation results and the bit-error rate of uncoded QPSK over a normalized Rayleigh fading channel are included for comparison.

In evaluating the lower bound for this coded scheme, the code symbol-error rates are replaced by the 8PSK symbol-error rates. Assuming that Gray mapping is used in the bit assignments of the code symbols, (7.5) is used in relating the mean bit error probability of a code symbol to its symbol-error rate.

The lower bounds for EED are evaluated for two erasure generation strategies: with and without CSI. The numerical results show that the performance of the coded scheme is slightly improved at low SNR's when CSI is utilized in the erasure generation process. However, this improvement is not significant enough to be shown in Figure 7.2. For high SNR's both erasure generation strategies yield the same performance.

Note that at bit-error rates around $10^{-5}$ RS(7,5) coded 8PSK, using errors-only decoding provides 17.6 dB coding gain compared to uncoded QPSK. By using errors-and-erasures decoding an additional 2.4 dB is obtained.

**Example 7.6** Consider the RS(15, 8) coded 16PSK scheme of Example 7.3. The lower bounds of the post-decoding bit error probability of this coded scheme for errors-only decoding and errors-and-erasures decoding are plotted in Figure 7.3 by using (7.7) and (7.12). In evaluating the error bounds of this coded scheme, the code-symbol error rates are replaced by the 16PSK symbol error rates. Figure 7.3 shows that at bit-error rates around $10^{-5}$ RS(15, 8) coded 16PSK provides 21 dB and 24 dB coding gain compared to uncoded QPSK by using errors-only decoding and errors-and-erasures decoding, respectively.

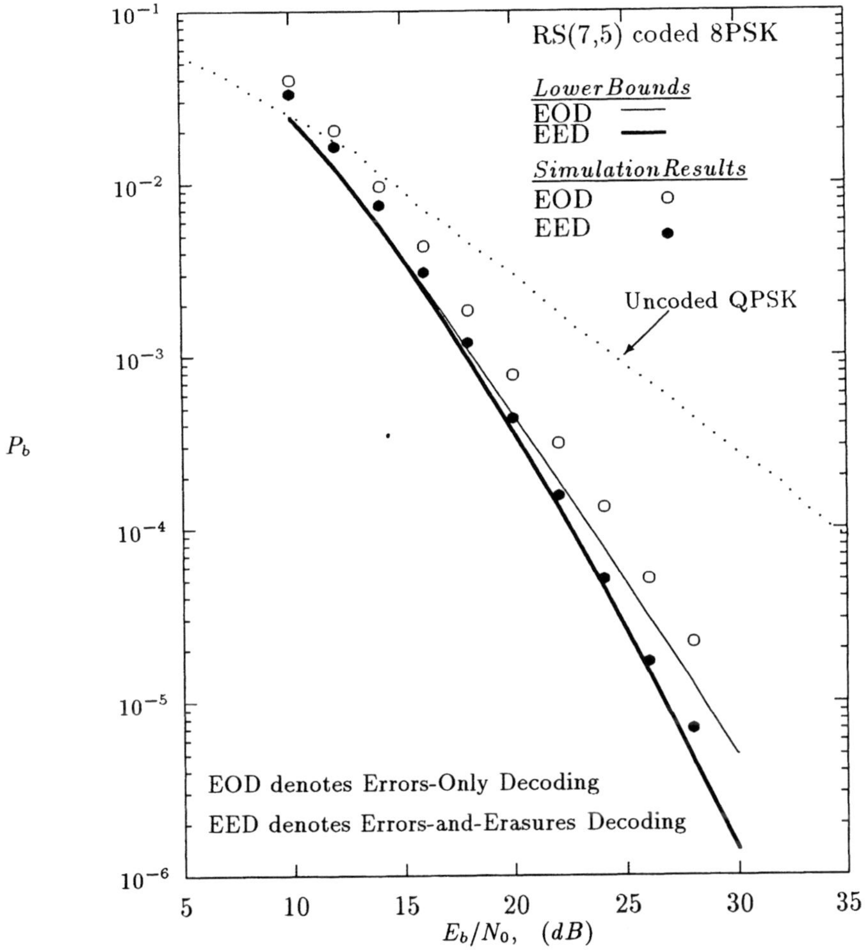


Figure 7.2: Performance of the RS(7, 5) coded 8PSK scheme in a normalized Rayleigh fading channel.

**Example 7.7** Now, we evaluate the performance of RS(63, 42) coded 8PSK, constructed in Example 7.4. As stated earlier, the second approach is used in constructing this coded scheme and each code symbol consists of two concatenated 8PSK symbols. Hence, in calculating the lower bounds for RS(63, 42) coded 8PSK, the code symbol-error rates are related to the 8PSK symbol-error rates using (7.11).

Again assume that a Gray mapping is used in assigning the bit groups to the 8PSK symbols. A simple argument shows that the mean bit error probability of a code symbol can be approximately related to the channel symbol-error rate by multiplying the latter by $2/m$, where $m$ is the code symbol size in bits, i.e., $m = 6$. This can be explained as follows. If a code symbol is in error then one of the following cases might occur. 1) Only one of the two concatenated 8PSK symbols is in error. 2) Both 8PSK symbols are in error. Weighting the probabilities of events (1) and (2) by $1/m$ and $2/m$, respectively, and summing them confirms the foregoing statement.

The lower bounds on $P_b$ along with simulation results for errors-only decoding and errors-and-erasures decoding of RS(63, 42) coded 8PSK are shown in Figure 7.4. The coding gains of this coded scheme compared to the uncoded QPSK are 28 dB and 29.6 dB for errors-only decoding and errors-and-erasures decoding, respectively.

In the foregoing examples the lower bounds on post-decoding bit error probability for the errors-and-erasures decoding technique have been found by optimizing the fading amplitude threshold, $a_T$, and the erasure region radius, $\rho_T$, for each value of $E_b/N_0$, based on the existence or the lack of CSI, respectively. Figure 7.5 represents the optimized fading amplitude threshold $a_T$ and the normalized erasure region radius $\rho_{T_n}$ as a function of $E_b/N_0$ for the different RS coded schemes considered in Figure 7.2, Figure 7.3 and Figure 7.4.

Note that at lower signal-to-noise ratios, the value of $\rho_{T_n}$ is slightly higher than that of $a_T$. This is due to the fact that in erasure decoding without CSI the impact of the additive noise is considered as well as the fading amplitude. As the signal-to-noise ratio increases these values get closer.

To compare the performance of the three RS coded schemes, i.e., RS(7, 5) coded 8PSK, RS(15, 8) coded 16PSK, and RS(63, 42) coded 8PSK, the corresponding error bounds are redrawn in Figure 7.6. It is observed that using longer and lower rate codes improves the performance of the coded schemes at the expense of the decoding complexity. Note that, for instance, the RS(15, 8) coded 16PSK scheme provides 3-4.5 dB more coding gain than the RS(7, 5) coded 8PSK at bit-error rates around $10^{-5}$. However, at low SNR's the latter scheme outperforms the former. This can be explained as follows.

By using 16PSK, the signal set becomes denser and the distance between channel symbols becomes shorter. Hence, at low signal-to-noise ratios for the second scheme, more errors occur during transmission and the decoder fails to correct more erasures/errors. In other words, at low error rates the high error correction capability of this scheme cannot compensate for the performance

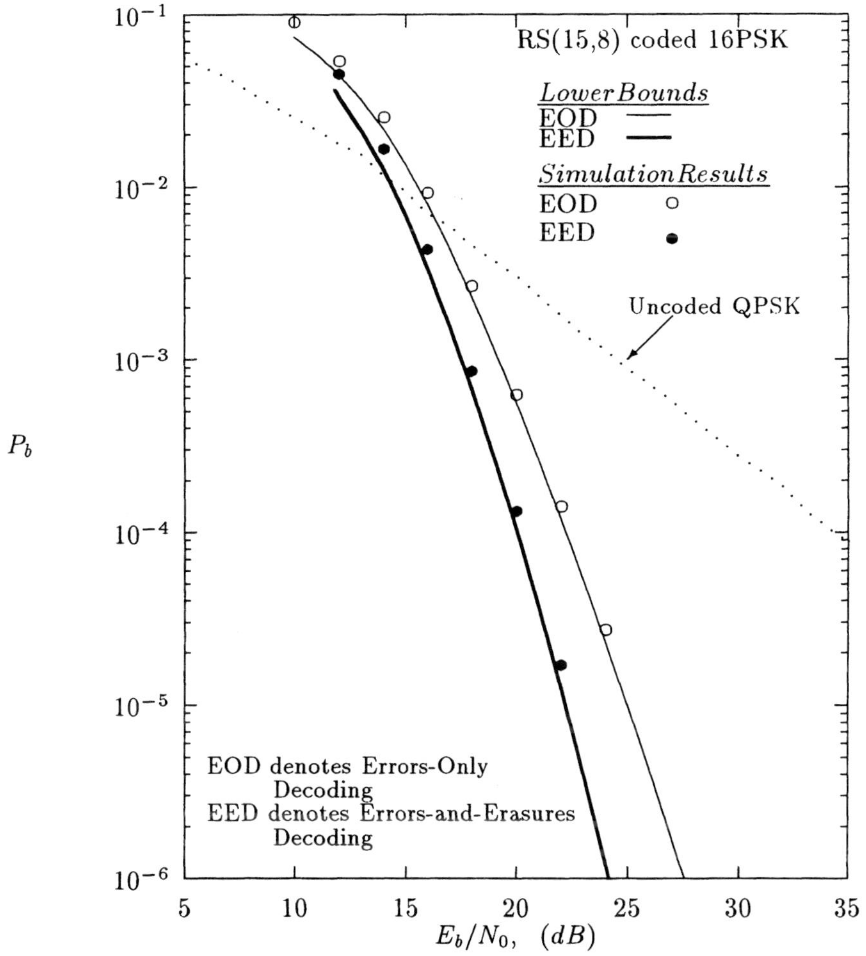


Figure 7.3: Performance of the RS(15, 8) coded 16PSK scheme in a normalized Rayleigh fading channel.

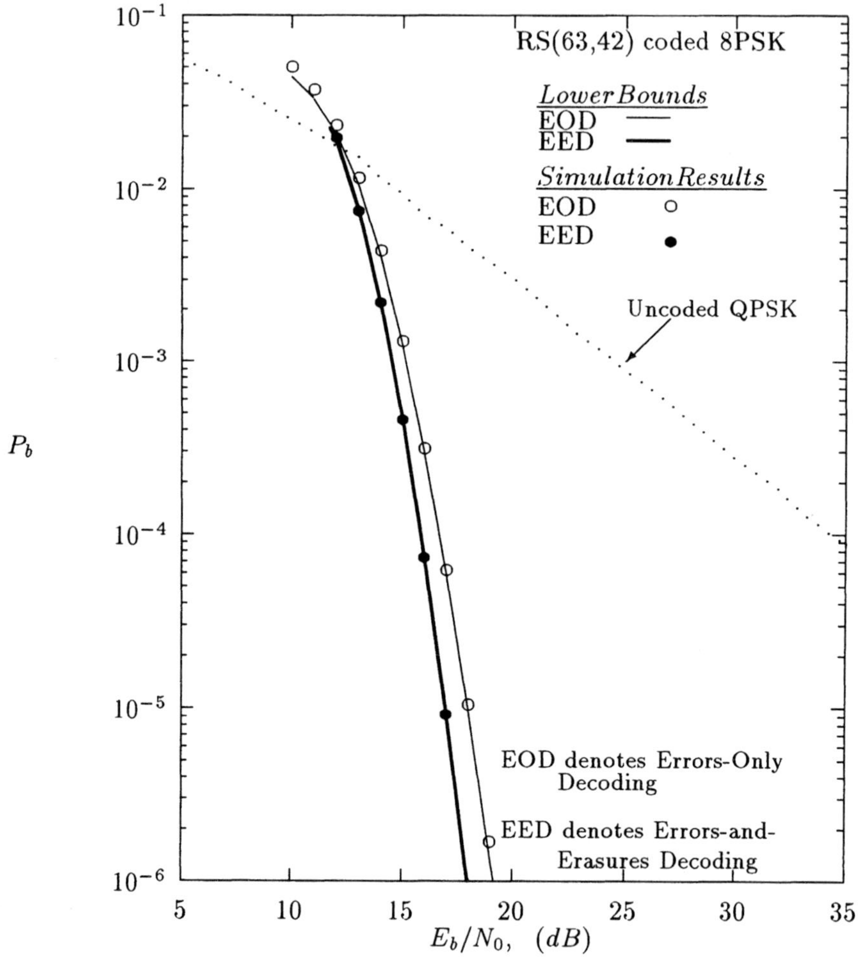


Figure 7.4: Performance of the RS(63, 42) coded 8PSK scheme in a normalized Rayleigh fading channel.

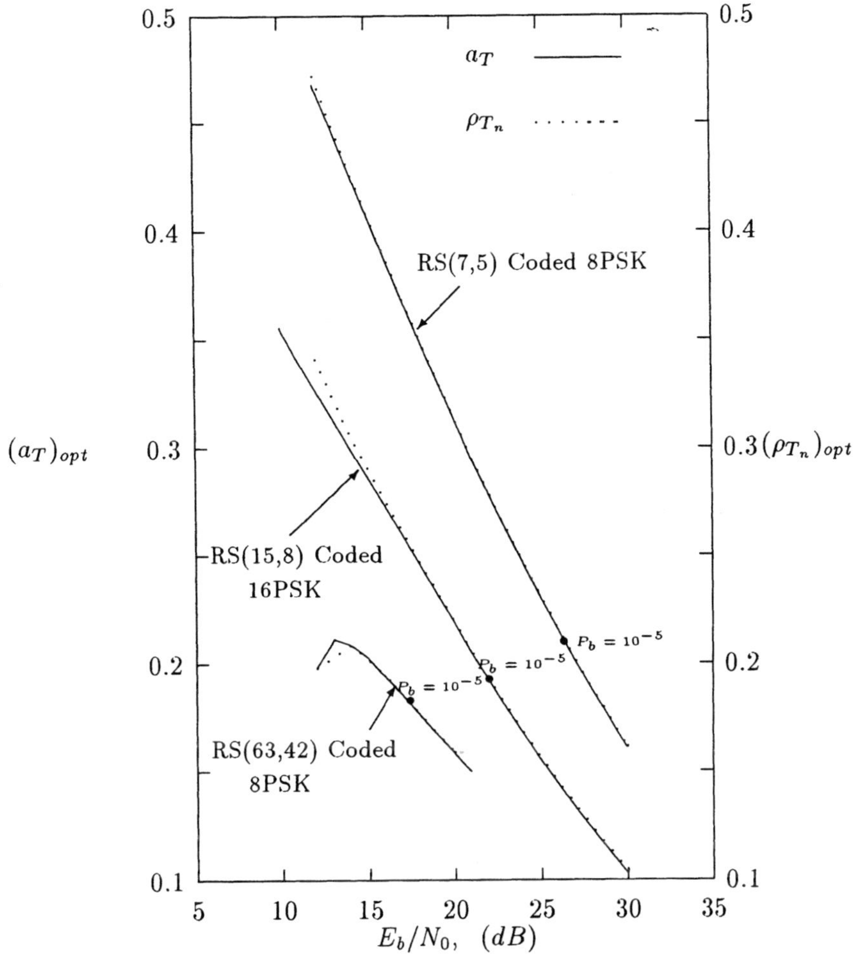


Figure 7.5: Optimum erasure threshold (with/without CSI) of the errors-and-erasures decoding of the three RS coded MPSK schemes as a function of the average bit signal-to-noise ratio.

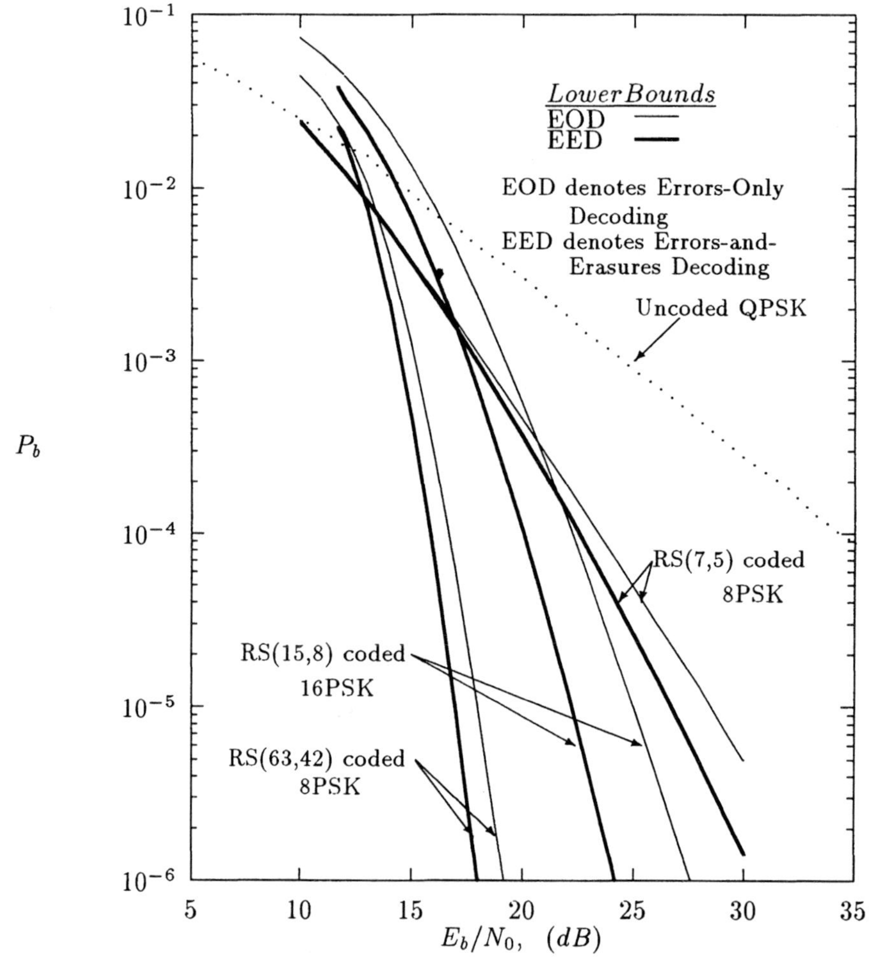


Figure 7.6: Comparison of the performance of the three different RS coded schemes in a normalized Rayleigh fading channel.

degradation due to the use of a denser signal set.

## 7.4 PERFORMANCE EVALUATION OF SOFT-DECISION DECODING OF RS CODED MPSK SCHEMES ON A RAYLEIGH FADING CHANNEL

In the preceding section we observed that the use of erasure information in the decoding process enhances the coding gain by an additional 2-3 dB at bit error rates around $10^{-5}$, depending on the RS coded scheme used. Indeed, this technique is a specific example of *soft-decision* decoding. Using more general unquantized soft-decision decoding, more coding gain is expected for these schemes on a fading channel. In general, the soft-decision decoding of RS codes is a complex task. However, for short RS coded schemes, it is practical to perform soft-decision decoding by introducing a trellis structure and using the Viterbi algorithm (see Chapter 6). Because of the higher complexity of the Viterbi algorithm, this decoding technique cannot be applied to long and/or low-rate RS coded schemes. In such cases, low-complexity soft-decision decoding techniques, such as *Successive Erasure Decoding* (SED) are used instead, at the expense of performance degradation (see Chapter 6). In the following we evaluate and compare the performance of soft-decision decoding techniques.

### 7.4.1 Upper Bound on $P_b$

In this section we derive the upper bound on the block-error rate as well as the bit-error rate for the RS coded MPSK schemes which employ the Viterbi algorithm for soft-decision decoding. In order to derive an upper bound, a union bound is first obtained on the *error event* probability. Recall that an error event starts where two paths in the trellis diagram diverge and ends where they remerge. A union bound for the error event probability can be obtained by summing the probabilities of the error events of all possible lengths given a particular transmitted sequence s and averaging this quantity over all possible correct sequences s (also see (4.8))

$$P_e \leq \sum_{\mathbf{s}} \sum_{\hat{\mathbf{s}} \in \mathbf{C}} P(\mathbf{s})\, P_2(\mathbf{s}, \hat{\mathbf{s}}), \tag{7.13}$$

where $\hat{\mathbf{s}}$ is the incorrect decoded codeword, and $\mathbf{C}$ is the set of all valid code sequences. In (7.13), $P_2(\mathbf{s}, \hat{\mathbf{s}})$ represents the pairwise error event probability for two codewords s and $\hat{\mathbf{s}}$.

By applying the Viterbi algorithm to the soft-decision decoding of block codes, the algorithm finds the closest codeword to the transmitted one for each received block. In this case the length of the error events is limited to the block length, $n$, and (7.13) is equivalent to an upper bound on the block error probability.

In order to evaluate the upper bound (7.13), the pairwise error probability $P_2(\mathbf{s},\ \hat{\mathbf{s}})$ should be found. The pairwise error probability of coded schemes in a Rayleigh fading channel is evaluated for cases with and without CSI in Chapter 5. Using the results of Section 5.1.1 we can write

With CSI

$$P_2(\mathbf{s},\ \hat{\mathbf{s}}) \leq \frac{1}{(\frac{1}{4N_0})^{l_\eta} \cdot d_p^2(l_\eta)}, \tag{7.14}$$

Without CSI

$$P_2(\mathbf{s},\ \hat{\mathbf{s}}) \leq \frac{(2e/l_\eta)^{l_\eta}}{(1/N_0)^{l_\eta}} \cdot \frac{(\sum_{i\in\eta} |s_i - \hat{s_i}|^2)^{l_\eta}}{d^4(l_\eta)}, \tag{7.15}$$

where $\eta$ is the set of all $i$ for which $s_i \neq \hat{s_i}$, $l_\eta$ shows the (effective) length of the error event $(\mathbf{s},\ \hat{\mathbf{s}})$, and $d_p^2(l_\eta)$ is the squared product of the distances defined as

$$d_p^2(l_\eta) = \prod_{i\in\eta} |s_i - \hat{s_i}|^2.$$

Substituting (7.14) and (7.15) into (7.13) gives the upper bound on the block error probability as (see Problem 7.18)

With CSI

$$\begin{aligned} P_{blc} &\leq \sum_{\mathbf{s}} \sum_{\hat{\mathbf{s}} \in \mathbf{C}} P(\mathbf{s}) \cdot \frac{1}{(\frac{1}{4N_0})^{l_\eta} \cdot d_p^2(l_\eta)} \\ &\leq \sum_{l_\eta = d} \sum_{d_p^2(l_\eta)} \frac{a(d_p^2(l_\eta))}{d_p^2(l_\eta)} \cdot \frac{1}{(\frac{1}{4N_0})^{l_\eta}}, \end{aligned} \tag{7.16}$$

where $d$ is the minimum Hamming distance of the code and $a(d_p^2(l_\eta))$ is the average number of codewords with a Hamming distance (effective length) of $l_\eta$ and a squared product of distance of $d_p^2(l_\eta)$ from the transmitted codeword.

If no CSI is used in the decoding process, we have

Without CSI

$$P_{blc} \leq \sum_{\mathbf{s}} \sum_{\hat{\mathbf{s}} \in \mathbf{C}} P(\mathbf{s}) \frac{(2e/l_\eta)^{l_\eta}}{(1/N_0)^{l_\eta}} \cdot \frac{(\sum_{i\in\eta} |s_i - \hat{s_i}|^2)^{l_\eta}}{d_p^4(l_\eta)}$$

$$\leq \sum_{l_\eta=d} \sum_{d_p^2(l_\eta)} a(d_p^2(l_\eta)) \cdot \frac{(2e/l_\eta)^{l_\eta}}{(1/N_0)^{l_\eta}} \cdot \frac{(d_E^2(d_p^2(l_\eta)))^{l_\eta}}{d_p^4(l_\eta)}, \tag{7.17}$$

where

$$d_E^2(d_p^2(l_\eta)) = \sum_{i \in \eta} |s_i - \hat{s}_i|^2$$

is the squared Euclidean distance between the two unmerged paths with a squared product distance of $d_p^2$ and a length of $l_\eta$. In deriving (7.17) it is assumed that all error events of length $l_\eta$ with squared product distance of $d_p^2(l_\eta)$ have the same Euclidean distance.

To derive the bit error probability we use the upper bounds of (7.17) and (7.16). Assuming that Gray mapping is used in assigning the bits to the $2^m$-ary PSK signal points, the bit error probability can be related to the probability of symbol error by dividing the latter by $m$ (see (7.5)). To find the symbol-error rate, we can weight the terms corresponding to the error events of effective length $l_\eta$ by $l_\eta/n$. Two paths with $l_\eta$ unmerged paths have $l_\eta$ different symbols. Thus, the probability of the bit error can be upper-bounded as

<u>With CSI</u>

$$P_b \leq \sum_{l_\eta=d} \sum_{d_p^2(l_\eta)} \frac{l_\eta}{mn} \cdot \frac{a(d_p^2(l_\eta))}{d_p^2(l_\eta)} \cdot \left(\frac{1}{\frac{1}{4N_0}}\right)^{l_\eta},$$

<u>Without CSI</u>

$$P_b \leq \sum_{l_\eta=d} \sum_{d_p^2(l_\eta)} \frac{a(d_p^2(l_\eta))}{mn} \cdot \frac{(2e/l_\eta)^{l_\eta}}{(1/N_0)^{l_\eta}} \cdot \frac{(d_E^2(d_p^2(l_\eta)))^{l_\eta}}{d_p^4(l_\eta)}.$$

For high signal-to-noise ratios these upper bounds can be simplified by considering only the terms corresponding to the shortest error event paths (the path with minimum Hamming distance $d$):

<u>With CSI</u>

$$P_b \leq \sum_{d_p^2(d)} \frac{d}{mn} \cdot \frac{a(d_p^2(d))}{d_p^2(d)} \cdot \left(\frac{1}{1/4N_0}\right)^d, \tag{7.18}$$

<u>Without CSI</u>

$$P_b \leq \sum_{d_p^2(d)} \frac{d}{mn} \cdot a(d_p^2(d)) \cdot \frac{(2e/d)^d}{(1/N_0)^d} \cdot \frac{(d_E^2(d_p^2(d)))^d}{d_p^4(d)}. \tag{7.19}$$

Table 7.1: Average number of codewords with Hamming distance 3 and squared product distance $d_p^2(3)$ for the RS(7, 5) coded 8PSK and RS(6, 4) coded 8PSK schemes.

| $d_p^2(3)/(E_s)^3$ | RS(7, 5) coded 8PSK $a(d_p^2(3))$ | RS(6, 4) coded 8PSK $a(d_p^2(3))$ |
|---|---|---|
| 0.201 | 10.5 | 6.0 |
| 0.686 | 10.5 | 6.0 |
| 1.172 | 17.5 | 10.0 |
| 1.373 | 7.0 | 4.0 |
| 2.343 | 14.0 | 8.0 |
| 4.000 | 35.0 | 20.0 |
| 4.686 | 28.0 | 16.0 |
| 6.828 | 17.5 | 10.0 |
| 8.000 | 28.0 | 16.0 |
| 13.657 | 14.0 | 8.0 |
| 23.314 | 10.5 | 6.0 |
| 27.314 | 28.0 | 16.0 |
| 39.799 | 10.5 | 6.0 |
| 46.627 | 7.0 | 4.0 |
| 64.000 | 7.0 | 4.0 |

**Example 7.8** As an illustrative example the upper bounds, given in (7.18) and (7.19), are evaluated for the RS(7, 5) coded 8PSK and the shortened RS(6, 4) coded 8PSK schemes, constructed in Example 7.1 and 7.2, respectively. To find the distance weight distribution of these coded schemes an exhaustive computer search is performed. As RS coded MPSK schemes are nonlinear the assumption of all-zero transmitted codeword is no longer correct and all possible codewords must be considered as the transmitted one. We tabulate the average number of codewords with Hamming distance 3 and squared product distance $d_p^2(3)$ for the RS(7, 5) coded 8PSK and the shortened RS(6, 4) coded 8PSK schemes in Table 7.1.

The upper bounds on $P_b$ for the RS(7, 5) coded 8PSK and the shortened RS(6, 4) coded 8PSK schemes are shown in Figure 7.7 and 7.8, respectively.

To examine the tightness of the upper bounds computer simulations are performed for these RS coded MPSK schemes. The soft-decision decoding is implemented by introducing a trellis structure and using the Viterbi algorithm (see Section 6.9.1). The trellis diagram of this coded scheme has $8^{(7-5)} = 8^2 = 64$ states (see (6.25)).

Note that by using the Viterbi decoding without CSI the RS(7, 5) coded 8PSK scheme provides a coding gain of about 24 dB compared to uncoded QPSK at bit-error rates around $10^{-5}$. The use of CSI in the Viterbi decoding improves this gain by an additional 1 dB. The coding gain of the shortened RS(6, 4) coded 8PSK is slightly more than that of RS(7, 5) coded 8PSK.

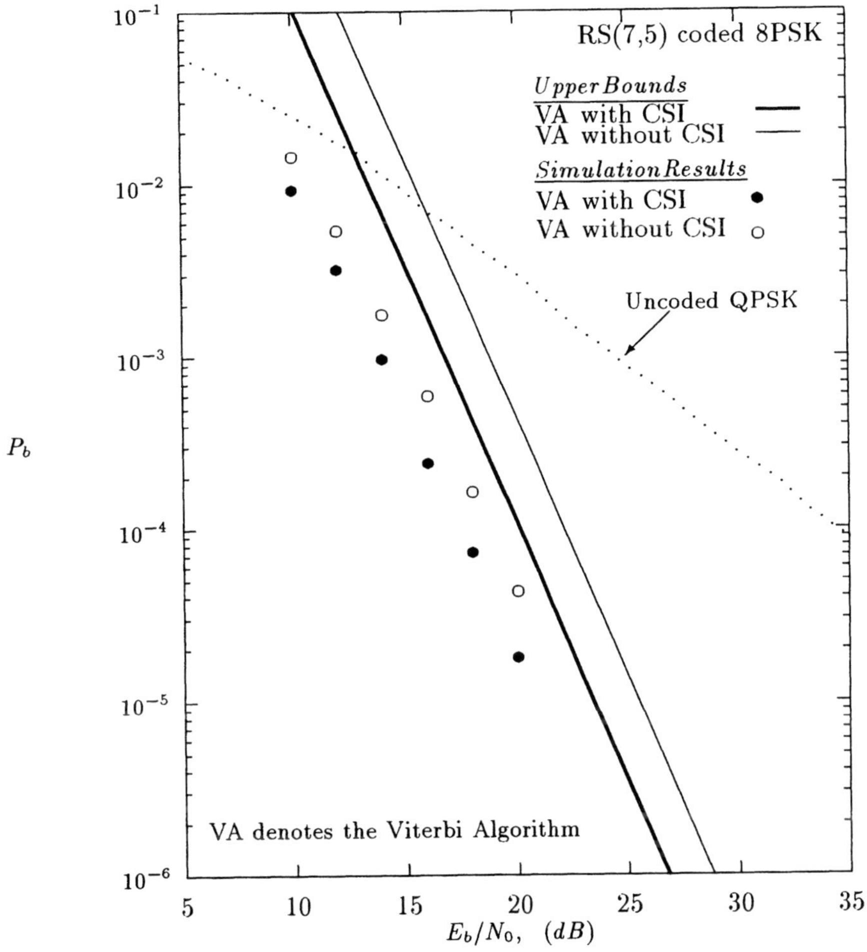


Figure 7.7: Performance of the RS(7,5) coded 8PSK scheme, using the VA, in a normalized Rayleigh fading channel.

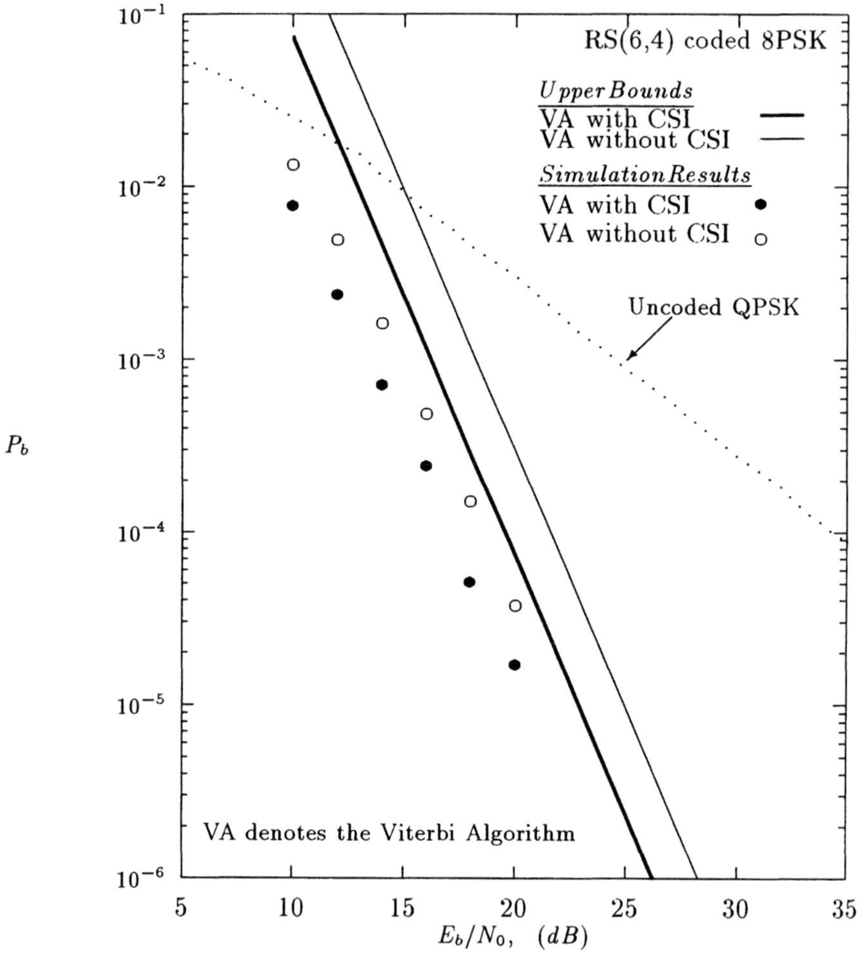


Figure 7.8: Performance of the RS(6, 4) coded 8PSK scheme, using the VA, in a normalized Rayleigh fading channel.

The results of Examples 7.5 and 7.8 show that Viterbi decoding provides coding gains of about 8.5 dB and 6.5 dB at a bit error probability of $10^{-5}$ compared to errors-only and errors-and-erasures decoding techniques respectively. This gain is improved by an additional 1 dB when CSI is used.

### 7.4.2 Successive-Erasure Decoding of RS-Coded MPSK Schemes

Th results of the preceding section reveal that the use of soft-decision information in the decoding process of RS coded MPSK schemes provides a remarkable coding gain over hard-decision decoding in Rayleigh fading channels. This gain, in general, is much larger than that achieved in the AWGN channel, due to the increase in time diversity when soft-decision decoding is employed.

In the foregoing examples, the implementation of maximum likelihood decoding is realized by introducing a trellis structure for RS codes and utilizing the VA. A measure of the complexity of this decoding technique is the number of states in the trellis diagram. This number for RS codes with symbol size $q$ is $q^{n-k}$, which grows exponentially with the number of redundant symbols. This decoding technique, hence, is only applicable to codes with small symbol size, $q$, and a small amount of redundancy.

Another implementation of the soft-decision decoding for RS codes is the Successive-Erasure Minimum Distance Decoding (SEMDD) algorithm (see Section 6.9.2). Although SEMDD is asymptotically optimum for binary codes with antipodal signaling [94], its performance for nonbinary codes, e.g., RS codes with multiphase signaling is inferior to that of MLD [91].

The performance of SEMDD is hard to analyze, particularly when the fading channel is considered. Hence, the performance of this decoding technique for different RS-coded MPSK schemes is evaluated by using computer simulations.

**Example 7.9** In this example we evaluate the performance of the SEMDD algorithm for three different RS coded MPSK schemes, i.e., RS(7, 5) coded 8PSK, RS(15, 8) coded 16PSK, and RS(63, 42) coded 8PSK. In our simulation we consider a normalized Rayleigh fading channel for cases where CSI is available at the receiver and where it is not. Depending on the existence or the lack of channel state information, the reliability function of (6.30) is (see Problem 7.22)

<u>With CSI</u>

$$L(\hat{s}_i) = \ln \frac{(M-1)\exp(-|r_i - a\hat{s}_i|^2/N_0)}{\sum_{s_j \neq \hat{s}_i} \exp(-|r_i - as_j|^2/N_0)},$$

<u>Without CSI</u>

$$L(\hat{s}_i) = \ln \frac{(M-1)\exp(-|r_i - \hat{s}_i|^2/N_0)}{\sum_{s_j \neq \hat{s}_i} \exp(-|r_i - s_j|^2/N_0)},$$

where $a$ shows the fading amplitude. In the foregoing equations the factor $(M-1)$ in the numerator ensures that the reliability function is always positive.

In the successive-erasure decoding process (see Figure 6.6), we start deleting 1, 3, $\cdots$, $d-1$, unreliable symbols successively when $d$ is even, while for odd $d$'s, 2, 4, $\cdots$, $d-1$, symbols are deleted in each decoding step. This ensures that the full erasure-correcting capability of the code will be employed.

Figure 7.9 shows the results of the simulation for the RS(7, 5) coded 8PSK scheme with and without CSI. It is observed that by using SEMDD without CSI for a $P_b = 10^{-5}$, a coding gain of about 3.5-6 dB is obtained compared to errors-only decoding and errors-and-erasures decoding, respectively. This coding gain is improved by an additional 0.7 dB when channel state information is used. To compare the performance of SEMDD with that of Viterbi decoding, the simulation results of Figure 7.7 are included in Figure 7.9. Such a comparison shows that the performance of Viterbi decoding is superior to that of the SEMDD technique.

The simulation results for the RS(15, 8) coded 16PSK and the RS(63, 42) coded 8PSK schemes are shown in Figure 7.10 and Figure 7.11, respectively. For these schemes, the use of SEMDD improves the performance compared to hard-decision decoding. However, this improvement is less than that of the RS(7, 5) coded 8PSK scheme.

Indeed the successive-erasure minimum distance decoding procedure bridges part of the gap between hard-decision decoding and MLD, but not the whole gap. For RS(7, 5) coded 8PSK at a $P_b = 10^{-5}$ almost 2/3 of the MLD performance is obtained by SEMDD. Due to lack of the results for the other codes such comparison cannot be done properly. But it is observed that by increasing the symbol size, SEMDD becomes less effective. This can be explained as follows.

For the nonbinary case the scalar reliability function, defined in (6.30), does not contain information about reliability in exactly the same way as in the binary case. In the binary case, there are only two signal points and one decision border. However, in the nonbinary case there is more than one decision border and a small reliability function does not determine which of these decision borders the received signal is closer to. In other words, it reveals that the hard-decided signal point is unreliable but does not say which one of the other signal points are most likely to be transmitted one. One way to improve this decoding algorithm is to use a reliability vector, rather than a reliability scalar, for each received signal and to produce more decoded codeword candidates than that of the SEMDD procedure. This may improve the performance at the expense of higher complexity.

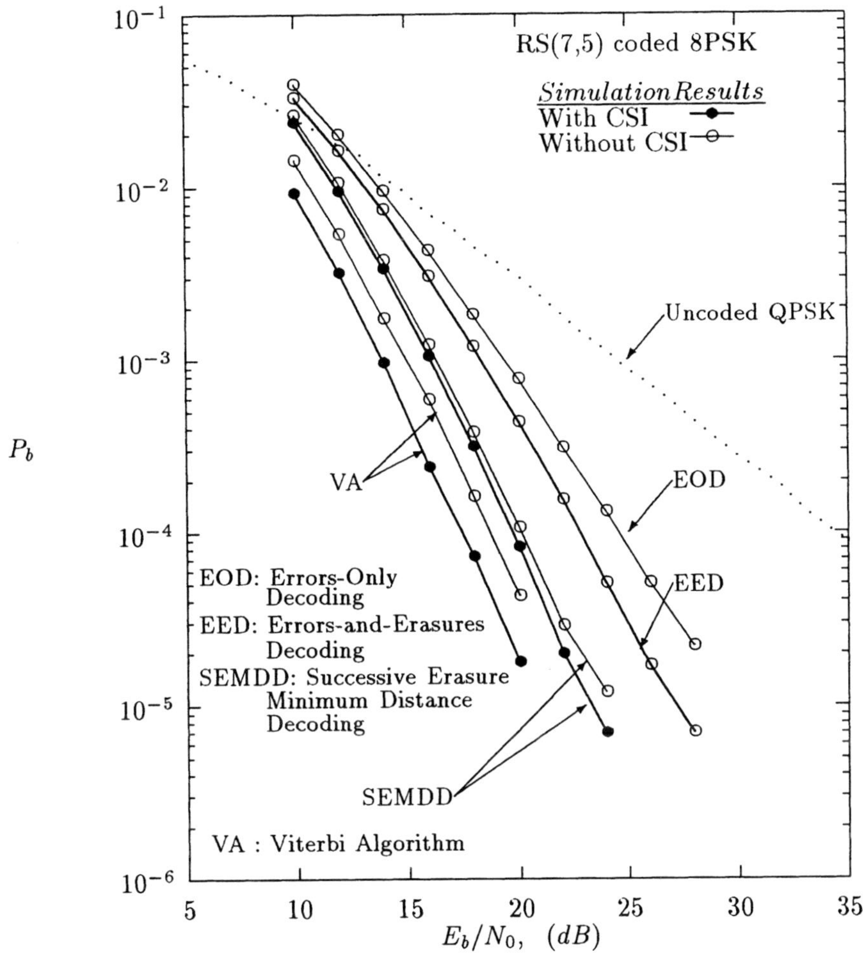


Figure 7.9: Comparison of the bit error probability of the RS(7, 5) coded 8PSK scheme for different decoding strategies in a normalized Rayleigh fading channel.

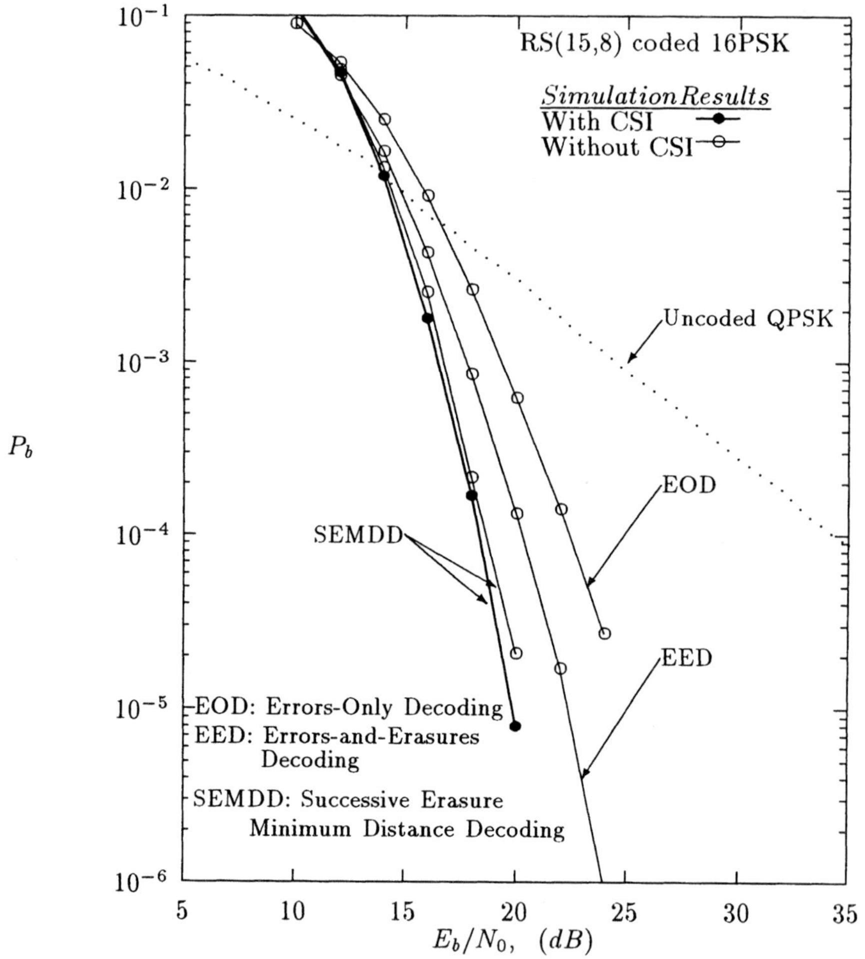


Figure 7.10: Comparison of the bit error probability of the RS(15, 8) coded 16PSK scheme for different decoding strategies in a normalized Rayleigh fading channel.

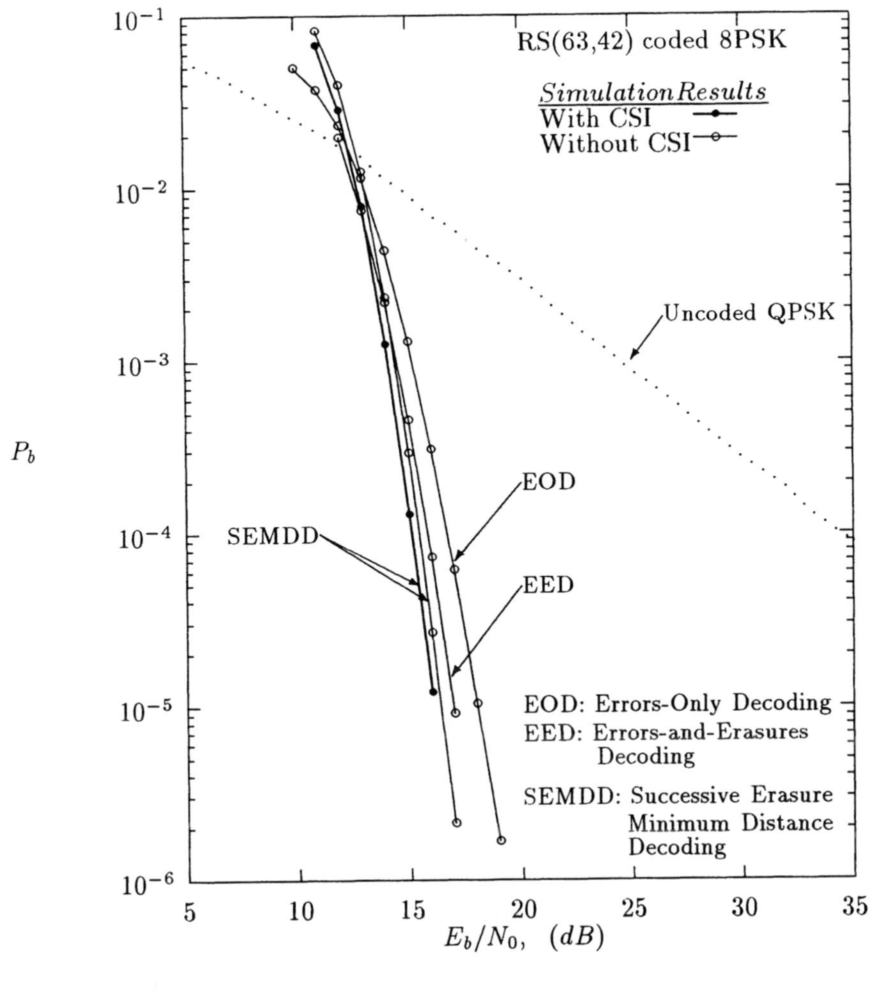


Figure 7.11: Comparison of the bit error probability of the RS(63, 42) coded 8PSK for different decoding strategies in a normalized Rayleigh fading channel.

## Problems

**7.1** Consider the symbol error probability of an MPSK scheme conditioned on the fading amplitude $a$ given by (7.3). Derive and plot the unconditional symbol error probability of an 8PSK scheme in a fading channel with a pdf as given in Problem 1.6, i.e.,

$$p_A(a) = \frac{a}{(1-\rho^2)\sigma^2} \exp\left(\frac{-a^2}{2(1-\rho^2)\sigma^2}\right) I_0\left(\frac{\rho a^2}{2(1-\rho^2)\sigma^2}\right),$$

for $\rho = 0, 0.25$ and $0.5$ and $\sigma^2 = 1/2$.

**7.2** Repeat Problem 7.1 for a 16PSK scheme.

**7.3** Repeat Problem 7.1 for a QPSK scheme.

**7.4** Repeat Problem 7.1 for a Nakagami fading channel with a pdf as given in Problem 1.7 and $m = 1, 2, \sigma^2 = m/2$.

**7.5** Derive (7.7).

**7.6** Assume a perfect knowledge of the fading amplitude $a$. If $a$ is smaller than a threshold value $a_T$, the received symbol is considered to be erased. Derive and plot the probability of an erased symbol for the fading channel discussed in Problem 1.6 with $\sigma^2 = 1/2$ and $\rho = 1/2$.

**7.7** Repeat Problem 7.6 for the Nakagami fading channel discussed in Problem 1.7 with $m = 1, 2, \sigma^2 = m/2$.

**7.8** Assume erasure generated with perfect CSI as discussed in Section 7.3.2. Derive and plot the error probabilities of a reliable MPSK symbol and of an unreliable MPSK symbol in a normalized Rayleigh fading channel.

**7.9** Repeat Problem 7.8 for the Nakagami fading channel discussed in Problem 1.7 with $m = 1, 2, \sigma^2 = m/2$.

**7.10** Repeat Problem 7.8 for the fading channel discussed in Problem 1.6 with $\sigma^2 = 1/2$ and $\rho = 1/2$.

**7.11** Using the results of Appendix D, derive the probability of a reliable MPSK symbol, and the error probability of reliable and unreliable MPSK symbols in a Rayleigh fading channel.

**7.12** Prove that the post-decoding block error probability of RS codes using an errors-and-erasures decoding technique is given as

$$P_{blc} = \sum_{s=0}^{d-1} P_{EC}(s) \sum_{\epsilon=\lceil \frac{d-s}{2} \rceil}^{n-s} \binom{n-s}{\epsilon} P^{\epsilon}_{sC|\overline{E}}(1 - P_{sC|\overline{E}})^{n-s-\epsilon}$$
$$+ \sum_{s=d}^{n} P_{EC}(s),$$

where

$$P_{EC}(s) = \binom{n}{s} P^{s}_{EC}(1 - P_{EC})^{n-s}.$$

**7.13** Using (7.10), derive the lower bound on the bit error probability of RS codes with an errors-and-erasures decoding technique.

**7.14** Using the results of Problems 7.7 and 7.9, derive and plot the lower bounds of the post-decoding bit error probability of the RS(7,5) coded 8PSK using EOD and EED techniques in a Nakagami fading channel with $m = 2$, and $\sigma^2 = 1$.

**7.15** Construct the extented RS(8,4) coded 8PSK scheme. Using (7.7) and (7.12), derive and plot the lower bounds of the post-decoding bit error probability of this coded scheme using EOD and EED techniques in a normalized fading channel.

**7.16** Repeat Problem 7.15 for an extended RS(16,4) coded 16PSK scheme.

**7.17** Repeat Problem 7.15 for an extended RS(16,8) coded 16PSK scheme.

**7.18** **a)** From (5.14), derive (7.14).

**b)** From the results of Problem 5.2, derive (7.15).

**c)** Complete the derivations for (7.16) and (7.17).

**d)** Complete the derivations for (7.18) and (7.19).

**7.19** Following the procedure given in Example 7.8, find the distance weight distribution of the extended RS(8,4) coded 8PSK scheme. Tabulate the average number of codewords with its minimum Hamming distance and squared product distance for this coded 8PSK scheme. Using (7.18), (7.19) and the obtained table, derive and plot the upper bounds on $P_b$ for this RS(8,4) coded 8PSK scheme in a Rayleigh fading channel. Compare the performance of this scheme with that of the coded schemes in Example 7.8.

**7.20** Repeat Problem 7.19 for the extended RS(8,6) coded 8PSK scheme.

**7.21** Using the results of Problems 7.15 and 7.19, compare the performance of EOD, EED and soft-decision decoding techniques for the extended RS(8, 4) coded 8PSK scheme in a normalized Rayleigh fading channel.

**7.22** Using (6.30), derive the reliability function used in the SEMDD algorithm for cases with and without CSI.

# Chapter 8

# DIFFERENTIAL DETECTION OF CODED-MODULATION SCHEMES ON A FADING CHANNEL

In the preceding chapters the performance of coded-modulation schemes was evaluated on fading channels. This evaluation considered the performance of coded schemes under the assumption of *ideal coherent* detection. Such an assumption neglects the effect of fading on the phase of the received signal and only reflects the performance degradation due to fading on the amplitude of the received signal. In this chapter we evaluate the performance of coded-modulation schemes on fading channel for *differential detection* technique. Although this technique degrades the performance of the system, it has a clear advantage over coherent detection in that it avoids the need for generating a carrier demodulation reference at the receiver.

In this chapter the cut-off rate of coded MDPSK schemes is derived for slow and fast Rayleigh fading channels. Next we evaluate the performance of differentially coherent detection of coded MPSK schemes by means of upper bound analysis. An exact solution to the pairwise error probability and an estimate of bit error probability, based on exact solution, are finally given.

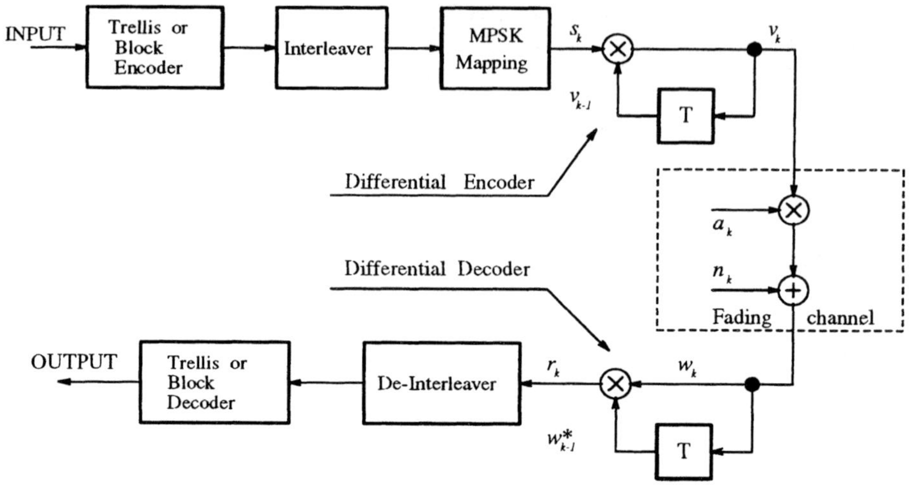


Figure 8.1: Block diagram of the baseband model of coded MDPSK schemes on a fading channel.

# 8.1 SYSTEM MODEL

The equivalent baseband transmission of the system under consideration is shown in Figure 8.1. A sequence of input bits is encoded, either by a trellis or block encoder, interleaved and mapped onto an MPSK signal set. The coded MPSK symbols are differentially encoded and transmitted over the fading channel. Representing the transmitted symbol at time interval $k$ by $v_k$, the differential encoding operation can be performed as

$$v_k = v_{k-1} \cdot s_k,$$

where $s_k$ is the coded MPSK signal at time interval $k$ and $v_{k-1}$ is the transmitted signal at time interval $k-1$. Using phasor notation, $v_k$ and $v_{k-1}$ are represented as

$$\begin{aligned} v_k &= e^{j\theta_{k-1}}, \\ v_{k+1} &= e^{j\theta_k} = e^{j(\theta_{k-1}+\Delta\theta_k)}, \end{aligned}$$

where we assume that $s_k = e^{\Delta\theta_k}$ is the phasor notation of the corresponding MPSK symbol at time interval $k$. For convenience we use the normalized energy $E_s = 1$ for MDPSK signals throughout this chapter.

The received signal is the corrupted and distorted version of the transmitted signal, which can be expressed as

$$w_k = g_k \cdot v_k + n_k,$$

where $n_k$ represents a complex additive white Gaussian noise component with zero mean and normalized variance of $\sigma_n^2 = \frac{N_0}{2E_s}$, and $g_k$ shows the complex fading process. For a Rician fading channel $g_k$ is a sample of a complex Gaussian process with

$$\begin{aligned} E[g_k] &= \eta, \\ E[|g_k|^2] - |\eta|^2 &= 2\sigma_g^2. \end{aligned} \tag{8.1}$$

The Rician model is often normalized by $|\eta|^2 + 2\sigma_g^2 = 1$ and characterized by the Rician parameter $K = \frac{|\eta|^2}{2\sigma_g^2}$. For a normalized Rayleigh fading channel, $\eta = 0$ and $\sigma_g^2 = 1/2$.

At the receiver the received signal passes through the differential decoder. The operation of this decoder can be written as

$$\begin{aligned} r_k &= w_{k-1}^\star \, w_k \\ &= (g_{k-1}^\star \, v_{k-1}^\star + n_{k-1}^\star)(g_k v_k + n_k) \\ &= g_{k-1}^\star \, g_k \, s_k + \text{ noise terms}, \end{aligned} \tag{8.2}$$

where $|v_{k-1}|^2 = 1$ is used in deriving the last line of the above equation. The differentially decoded signal is de-interleaved and finally decoded by a trellis or block decoder. In the following sections we will assume that the interleaving and de-interleaving operations are ideal, and hence, the coding channel is memoryless.

In general, the samples of fading process $g_{k-1}$ and $g_k$ are correlated with the coefficient $\xi$ defined as

$$\xi = \frac{\frac{1}{2}\, E[(g_{k-1} - \eta)(g_k - \eta)^\star]}{\sigma_g^2} = \frac{R_g(T_s) - 0.5|\eta|^2}{\sigma_g^2}, \tag{8.3}$$

where

$$R_g(T_s) = \frac{1}{2} E[g_{k-1} g_k^\star] = \frac{1}{2} E[g(t) g^\star(t + T_s)] \tag{8.4}$$

is the autocorrelation of the complex fading process $g(t)$ and $T_s$ is the symbol interval. Note that $R_g(T_s)$ is equivalent to the spaced-tone autocorrelation function $R_H(\Delta f; \Delta t)$ previously defined in Section 1.2.2 with $\Delta f = 0$ and $\Delta t = T_s$.

In a fading environment where the fading process varies very slowly compared to the transmitted signal rate the correlation coefficient is very close to unity. This is called *slow-fading*. An example is mobile fading channel when the mobile unit is moving very slowly, yet not stopping. For a slow-fading channel we have $\xi \approx 1$ which implies that $g_{k-1} \approx g_k$. On the other hand if the variations of the fading sample relative to the transmitted signal rate are rapid then the fading is called *fast-fading*. Considering again the mobile fading channel, fast-fading occurs if the speed of the vehicle increases. In this case the correlation coefficient $\xi$ is less than unity and approaches zero when the speed of the vehicle goes to infinity.

# 8.2 DIFFERENTIAL DETECTION OF CODED-MPSK SCHEMES

In order to evaluate the performance of the coded schemes we need to find the maximum likelihood metric. In [96] and [27, page 373], Divsalar and Simon derive a maximum likelihood metric for trellis-coded MDPSK schemes which is quite complicated to implement. Furthermore, performance analysis of the coded schemes, based on this optimum metric, is a prohibitive task. Therefore, following the approach of Divsalar and Simon [96], we use the squared Euclidean distance metric as the soft decoding metric in our analysis. Although this metric is suboptimum for MDPSK it is considerably easier to implement than the optimum metric and hence of significant practical interest.

Before evaluating the performance of any specific coded scheme we derive the cut-off rate of a Rayleigh fading channel for MDPSK modulation with soft-decision and hard-decision decoding.

## 8.2.1 Cut-Off Rate

As stated in Chapter 2 the cut-off rate is one of the best criteria for characterizing a digital communication channel. One of the important features of the cut-off rate is that it characterizes the coding channel independent of the specific code employed.

The cut-off rate for an unquantized channel is derived in Chapter 2

$$R_0 = 2\log_2 q - \log_2 \min_{\lambda} \sum_{n=0}^{q-1} \sum_{m=0}^{q-1} C(s_n, s_m, \lambda), \tag{8.5}$$

where $C(s_n, s_m, \lambda)$ shows the Chernoff factor of signals $s_n$ and $s_m$ and $\lambda$ is the Chernoff parameter. Assuming $s_k$ is transmitted and $\hat{s}_k \neq s_k$ is decoded, the Chernoff factor of signals $s_k$ and $\hat{s}_k$ is defined as

$$C(s_k, \hat{s}_k, \lambda) \equiv E_g[E_n[\exp(\lambda(m(r_k, \hat{s}_k) - m(r_k, s_k)))]], \tag{8.6}$$

where $r_k$ is the received signal at time interval $k$ and $m(\cdot)$ shows the decoding metric. Also, $E_g$ and $E_m$ represent the expectation operations over the fading and additive noise processes, respectively.

In the following we evaluate the Chernoff factor for a Rayleigh fading channel without channel state information (CSI). For the case of fading channel with CSI a closed form expression cannot be obtained for the Chernoff factor [27, page 377]. Hence we restrict our analysis to the case of fading channel without CSI.

### Chernoff Factor without Use of CSI

To derive the Chernoff factor, defined in (8.6), we use the Euclidean distance as a decoding metric, that is

$$m(r_k, s_k) = -|r_k - s_k|^2.$$

Using $r_k = w_{k-1}^\star w_k$ and noting that for MDPSK signals $|s_k| = |\hat{s}_k|$, the difference of metrics in (8.6) becomes

$$m(r_k, \hat{s}_k) - m(r_k, s_k) = w_{k-1}^\star w_k(\hat{s}_k^\star - s_k^\star) + w_{k-1} w_k^\star(\hat{s}_k - s_k).$$

The above equation can be written in matrix form as

$$m(r_k, \hat{s}_k) - m(r_k, s_k) = \mathbf{W}_k^{\star\ T}(\hat{\mathbf{S}}_k - \mathbf{S}_k)\ \mathbf{W}_k, \tag{8.7}$$

where

$$\mathbf{W}_k = \begin{bmatrix} w_{k-1} \\ w_k \end{bmatrix} = \begin{bmatrix} g_{k-1}v_{k-1} + n_{k-1} \\ g_k v_k + n_k \end{bmatrix},$$

$$\mathbf{S}_k = \begin{bmatrix} 0 & s_k^\star \\ s_k & 0 \end{bmatrix}, \quad \hat{\mathbf{S}}_k = \begin{bmatrix} 0 & \hat{s}_k^\star \\ \hat{s}_k & 0 \end{bmatrix}.$$

Now consider a Rician fading channel. As stated earlier, $g_k$ and $n_k$ are complex Gaussian variables in this channel. Therefore, $\mathbf{W}_k$ is a Gaussian vector and the difference of metrics in (8.7) is a quadratic form in Gaussian random variables. The characteristic function of a quadratic form in Gaussian random variables is derived in [59, Appendix B]. Using the characteristic function of the quadratic form of (8.7) and following [59, Appendix B], [98], and [27] the Chernoff factor of (8.6) can be written as (see Problem 8.1)

$$C(s_k,\ \hat{s}_k,\ \lambda) = \frac{\exp\{\lambda\ \mu_k^{\star T}\ \mathbf{F}_k(\ \mathbf{I} - 2\lambda\ \mathbf{R}_k^\star\ \mathbf{F}_k)^{-1}\ \mu_k\}}{\det(\ \mathbf{I} - 2\lambda\ \mathbf{R}_k^\star\ \mathbf{F}_k)}, \tag{8.8}$$

where $\mathbf{I}$ is a $2 \times 2$ identity matrix, $\mathbf{F}_k = \hat{\mathbf{S}}_k - \mathbf{S}_k$, and

$$\mu_k = \begin{bmatrix} \mu_{k-1} \\ \mu_k \end{bmatrix} = \begin{bmatrix} E_g[E_n[w_{k-1}]] \\ E_g[E_n[w_k]] \end{bmatrix},$$

$$\mathbf{R}_n = \tfrac{1}{2} E_g[E_n[(\mathbf{W}_k - \mu_k)^\star (\mathbf{W}_k - \mu_k)^T]]$$

$$= \frac{1}{2}\begin{bmatrix} E_g[E_n[|w_{k-1} - w_k|^2]] & E_g[E_n[(w_{k-1} - \mu_{k-1})^\star (w_k - \mu_k)]] \\ E_g[E_n[(w_{k-1} - \mu_{k-1})(w_k - \mu_k)^\star]] & E_g[E_n[\ |w_k - \mu_k|^2]] \end{bmatrix}.$$

In general, we assume that the fading samples $g_k$ and $g_{k-1}$ are correlated with correlation coefficient $\xi$, as given in (8.3). In such a case, after some manipulation, $\mu_k$ and $\mathbf{R}_k$ are obtained as

$$\begin{aligned} \mu_k &= \eta \begin{bmatrix} v_{k-1} \\ v_k \end{bmatrix}, \\ \mathbf{R}_k &= \frac{1}{2} \begin{bmatrix} \sigma_g^2 + \sigma_n^2 & \xi \sigma_g^2 s_k \\ \xi \sigma_g^2 s_k^\star & \sigma_g^2 + \sigma_n^2 \end{bmatrix}, \end{aligned} \tag{8.9}$$

where $\eta$ and $\sigma_g^2$ are the mean and the variance of fading process $g_k$, given in (8.1), and $\sigma_n^2 = \frac{N_0}{2E_s}$ is the variance of the additive noise.

Substituting (8.9) into (8.8) and expressing $\sigma_g^2$ and $\eta$ in terms of the Rician parameter $K$ as

$$\begin{aligned} |\eta|^2 &= \frac{K}{1+K}, \\ 2\sigma_g^2 &= \frac{1}{1+K}, \end{aligned}$$

the Chernoff factor for a Rician fading channel is obtained as

$$C(s_k, \hat{s}_k, \lambda_0) = \frac{1}{\Delta} \exp\left\{ -\frac{\lambda_0 K}{\Delta(1+K)} \frac{E_s}{N_0} |\hat{s}_k - s_k|^2 \left( 1 - 2\lambda_0 \left[ \frac{1-\xi}{1+K} \frac{E_s}{N_0} + 1 \right] \right) \right\}, \tag{8.10}$$

where

$$\Delta = 1 + |\hat{s}_k - s_k|^2 \left\{ \lambda_0 \frac{E_s}{N_0} \left( \frac{\xi - 2\lambda_0}{1+K} \right) - \lambda_0^2 \left[ 1 + \frac{(1-\xi^2)}{(1+K)^2} \left( \frac{E_s}{N_0} \right)^2 \right] \right\},$$

and $\lambda_0 \equiv \lambda N_0 / 2E_s$ is the normalized Chernoff parameter.

For a Rayleigh fading channel with $K = 0$, the Chernoff factor, given in (8.8) simplifies to

$$C(s_k, \hat{s}_k, \lambda_0) = \frac{1}{1 + \lambda_0 |\hat{s}_k - s_k|^2 \left\{ \frac{E_s}{N_0} (\xi - 2\lambda_0) - \lambda_0 \left[ 1 + (1-\xi^2) (\frac{E_s}{N_0})^2 \right] \right\}}. \tag{8.11}$$

In this case, the Chernoff factor, given in (8.11) can be optimized over the Chernoff parameter $\lambda_0$ to tighten the Chernoff bound. The value of $\lambda_0$ which minimizes the Chernoff factor can be obtained as (see Problem 8.2)

$$\lambda_{opt} = \frac{\xi \frac{E_s}{N_0}}{4 \frac{E_s}{N_0} + 2 \left[ 1 + (\frac{E_s}{N_0})^2 (1 - \xi^2) \right]},$$

which, when substituted into (8.11), yields the Chernoff factor

$$C(s_k, \hat{s}_k, \lambda_{opt}) = \frac{1}{1 + |\hat{s}_k - s_k|^2 \left\{ \frac{\xi^2}{4} (\frac{E_s}{N_0})^2 \left[ 2\frac{E_s}{N_0} + 1 + (1-\xi^2)(\frac{E_s}{N_0})^2 \right]^{-1} \right\}}. \tag{8.12}$$

### Slow-Fading Channel

As stated earlier for a slow-fading channel the correlation coefficient is unity. Substituting $\xi = 1$ in (8.12) yields

$$C(s_k, \hat{s}_k, \lambda_{opt}) = \frac{1}{1 + \left|\frac{\hat{s}_k - s_k}{2}\right| \frac{(E_s/N_0)^2}{1+2E_s/N_0}}. \tag{8.13}$$

Introducing

$$\nu = \frac{E_s/N_0}{1 + E_s/N_0}, \tag{8.14}$$

the Chernoff factor, given in (8.13), can be expressed as

$$C(s_k,\ \hat{s}_k,\ \lambda_{opt}) = \frac{1-\nu^2}{1 - \nu^2\left(1 - \left|\frac{\hat{s}_k - s_k}{2}\right|^2\right)}. \tag{8.15}$$

Substituting (8.15) into (8.5) yields

$$R_0 = 2\log_2 q - \log_2\,(1-\nu^2) \sum_{n=0}^{q-1} \sum_{m=0}^{q-1} \frac{1}{1-\nu^2\left(1 - \left|\frac{s_n - s_m}{2}\right|^2\right)}.$$

For MPSK signals, using $s_n = \cos\frac{2\pi n}{q} + j\sin\frac{2\pi n}{q}$, we can write

$$\begin{aligned} R_0 &= 2\log_2 q - \log_2\,(1-\nu^2) \sum_{n=0}^{q-1}\sum_{m=0}^{q-1} \frac{1}{1-\nu^2\cos^2\frac{\pi}{q}(n-m)} \\ &= \log_2 q - \log_2(1-\nu^2)\sum_{i=0}^{q-1}\frac{1}{1-\nu^2\cos^2\frac{\pi i}{q}}, \end{aligned} \tag{8.16}$$

where $\nu$ is defined in (8.14). Similar expression for $R_0$ is found in [98].

**Example 8.1** As an illustrative example, using (8.16), we calculate the cut-off rates of QDPSK, 8DPSK, and 16DPSK schemes for a Rayleigh slow-fading channel. The results of these calculations are shown in Figure 8.2. Observe that the cut-off rate increases with $E_s/N_0$ and at high signal-to-noise ratios it achieves the maximum values of 2, 3, and 4 for QDPSK, 8DPSK, and 16DPSK, respectively. This also can be shown by examining (8.16) at high SNR's.

Rewrite (8.16) as

$$R_0 = \log_2 q - \log_2\left\{1 + \sum_{i=1}^{q-1}\frac{1-\nu^2}{1-\nu^2\cos\frac{\pi i}{q}}\right\}.$$

For high signal-to-noise ratios the parameter $\nu$ approaches unity, and hence, the summation term becomes zero. Thus,

$$\lim_{E_s/N_0\to\infty} R_0 = \log_2 q.$$

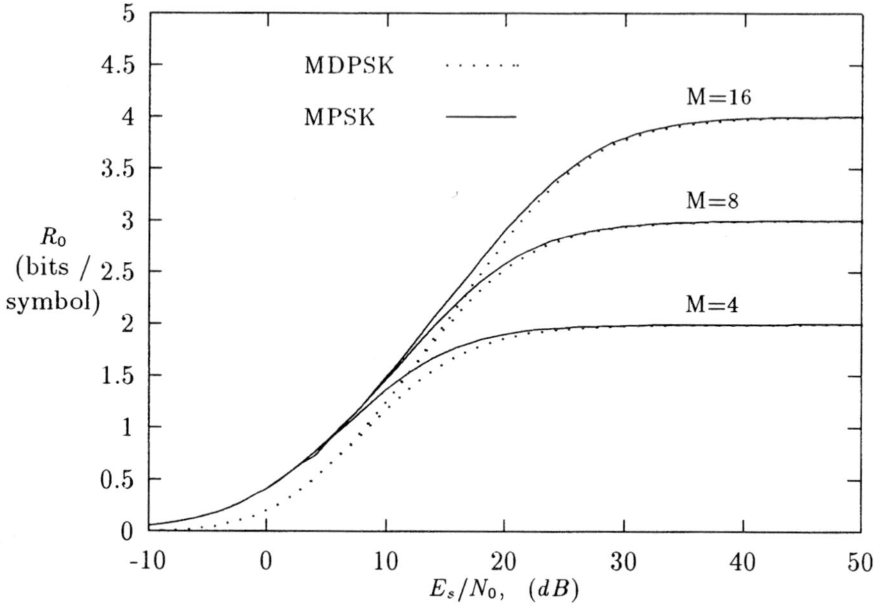


Figure 8.2: Cut-off rates of differential detection and coherent detection of QPSK, 8PSK, and 16PSK signaling on an unequalized Rayleigh slow-fading channel without CSI.

As we will show later, the asymptotic value of $R_0$ is less than $\log_2 q$ for the fast fading channel and there exists an irreducible loss in $R_0$ due to fast fading.

The cut-off rate of coherent detection of MPSK signals is also included in Figure 8.2 for comparison.

### Quantized Slow-Fading Channel

In this section we calculate the cut-off rate of MDPSK signals on a quantized (hard-decision) Rayleigh slow-fading channel. For this case the cut-off rate can be obtained as (see Chapter 2)

$$R_0 = \log_2 q - 2\log_2 \left[\sum_{l=0}^{q-1} \sqrt{q_l}\right], \tag{8.17}$$

where $q_l$ is the probability that $s_0$ is transmitted and the received signal falls in the $l$th decision region (probability of decoding $s_l$ given that $s_0$ is transmitted).

To find $q_l$ we use expressions derived by Proakis [99], [60, Appendix 7A], for probabilities of error of $q$-phase signaling over an L-diversity-branch Rayleigh slow-fading channel. We use $L = 1$ for first-order diversity and in the differential detection technique. In this case the probability density function of the received signal phase $\theta$, that is $p(\theta)$, can be expressed as (see Problem 8.3)

$$p(\theta) = \frac{1-\nu^2}{2\pi}\left[\frac{1}{1-\nu^2\cos^2\theta} + \frac{\nu\cos\theta}{(1-\nu^2\cos^2\theta)^{3/2}}\cos^{-1}(-\nu\cos\theta)\right],$$

where $\nu$ is defined as (8.14). To find $q_l$, we have

$$\begin{aligned} P[\theta_1 \le \theta < \theta_2] &= \int_{\theta_1}^{\theta_2} p(\theta)\, d\theta \\ &= \frac{1}{2\pi}\left\{\nu\sqrt{1-(\frac{1}{\nu^2}-1)x^2}\ \cot^{-1}x - \cot^{-1}\left(\frac{x/\nu}{\sqrt{1-(\frac{1}{\nu^2}-1)x^2}}\right)\right\}\Bigg|_{x_2}^{x_1}, \end{aligned}$$

where

$$x_i = \frac{-\nu\cos\theta_i}{\sqrt{1-\nu^2\cos^2\theta_i}}, \qquad i = 1,\ 2.$$

Replacing $\theta_2 = (2l+1)\pi/q$ and $\theta_1 = (2l-1)\pi/q$, $l = 0, 1, \cdots, q-1$, and after some manipulation, we obtain

$$\begin{aligned} q_l = \frac{1}{2\pi}\Bigg\{ & \frac{2\pi}{q} + \nu\frac{[\pi - \cos^{-1}(\nu\cos(\frac{2l+1}{q}\pi))]}{\sqrt{1-\nu^2\cos^2(\frac{2l+1}{q}\pi)}}\sin(\frac{2l+1}{q}\pi) \\ & -\nu\frac{[\pi - \cos^{-1}(\nu\cos(\frac{2l-1}{q}\pi))]}{\sqrt{1-\nu^2\cos^2(\frac{2l-1}{q}\pi)}}\sin(\frac{2l-1}{q}\pi)\Bigg\}, \\ & l = 0,\ 1,\ \cdots,\ q-1. \end{aligned} \tag{8.18}$$

Substituting (8.18) into (8.17) yields the cut-off rate of MDPSK schemes on a Rayleigh slow-fading channel for hard-decision decoding.

**Example 8.2** Using (8.18) and (8.17) the cut-off rates of QDPSK, 8DPSK, and 16DPSK schemes on a Rayleigh slow-fading channel are shown in Figure 8.3. For comparison the cut-off rates of these schemes for soft-decision decoding,

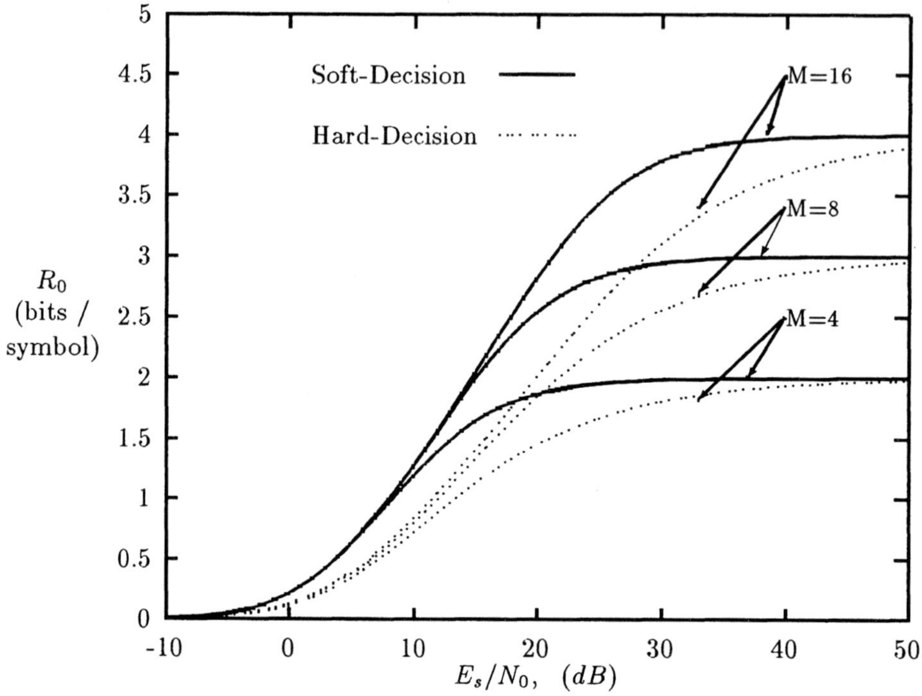


Figure 8.3: Cut-off rates of MDPSK schemes on a Rayleigh slow-fading channel for soft-decision and hard-decision decoding.

evaluated in Example 8.1, are also included in this figure. It is observed that the performance degradation of hard-decision decoding, compared to soft-decision decoding, is significant. For example, for the 8DPSK scheme at $R_0 = 2.25$ bits/symbol, hard-decision decoding requires about 7.5 dB more SNR than soft-decision decoding.

## Fast-Fading Channel

In this section we evaluate the cut-off rate for a Rayleigh fast-fading channel. In a fast-fading channel the Chernoff factor, given in (8.12), is a function of the correlation coefficient $\xi$ which, in turn, depends on the autocorrelation function $R_g(T_s)$, defined in (8.4).

To find the cut-off rate for a Rayleigh fast-fading channel the Chernoff factor, given in (8.12), can be rewritten as

$$C(s_k, \hat{s}_k, \lambda_{opt}) = \frac{1}{1 + \frac{1}{4}\nu_\xi |\hat{s}_k - s_k|^2}, \tag{8.19}$$

where, for convenience, we have introduced

$$\nu_\xi = \frac{\xi^2 (\frac{E_s}{N_0})^2}{1 + 2(\frac{E_s}{N_0}) + (1 - \xi^2)(\frac{E_s}{N_0})^2}. \tag{8.20}$$

Substituting (8.19) into (8.5) and using $s_n = \cos\frac{2\pi n}{q} + j\sin\frac{2\pi n}{q}$, and after some manipulation, we obtain (see Problem 8.4)

$$R_0 = \log_2 q - \log_2 \sum_{i=0}^{q-1} \frac{1}{1 + \nu_\xi \cos^2(\frac{\pi i}{q})}. \tag{8.21}$$

**Example 8.3** Using (8.21) we can calculate the cut-off rate for MDPSK over a Rayleigh fast-fading channel. From Section 1.3, the correlation coefficient $\xi$ for a land-mobile radio channel is

$$\xi = J_0(2\pi B T_s).$$

Figure 8.4 shows the cut-off rates of QDPSK and 8DPSK, as functions of $E_s/N_0$, for various values of normalized fading bandwidth $b = BT_s$, over a land-mobile Rayleigh fast-fading channel. The cut-off rate for $b = 0$ corresponds to a slow-fading channel. Observe that for nonzero $b$ the asymptote $R_0$ for both QDPSK and 8DPSK does not reach $\log_2 q$, $q = 4, 8$. Instead, for nonzero $b$, say $b = 0.05$, the limiting values of $R_0$ for QDPSK and 8DPSK tend to 1.70 and 2.10 bits/symbol, respectively. Indeed, using(8.20) and (8.21), we can show that the limiting value of $R_0$ is

$$\lim_{\frac{E_s}{N_0} \to \infty} R_0 = \log_2 q - \log_2 \sum_{i=0}^{q-1} \frac{1 - \xi^2}{1 - \xi^2 \sin^2(\frac{\pi i}{q})},$$

which shows an irreducible loss in $R_0$ for nonzero values of $b$ ($\xi < 1$). For $b = 0$ ($\xi = 1$), the asymptote value of $R_0$ reaches $\log_2 q$.

Figure 8.5 shows the limiting value of $R_0$ versus normalized fading bandwidth for QDPSK, 8DPSK, and 16DPSK schemes over a land-mobile Rayleigh fading channel. It is observed that the relative degradation of the asymptote $R_0$ (irreducible loss in $R_0$) increases with MDPSK alphabet size. For example, the irreducible losses for QDPSK, 8DPSK, and 16DPSK with $b = 0.05$ are 15.1%, 29.9%, and 45.5% of $\log_2 q$, respectively. The irreducible loss in cut-off rate of MDPSK due to fast-fading ($b = 0$) as well as to its sensivity to alphabet size is caused by a differentially coherent detection technique in which the preceding symbol is used as reference. As the fading gain rapidly varies the reference symbol becomes less reliable. Also, an increase in the alphabet size shrinks the distance between the symbols, and hence, enhances the effect of the gain variation.

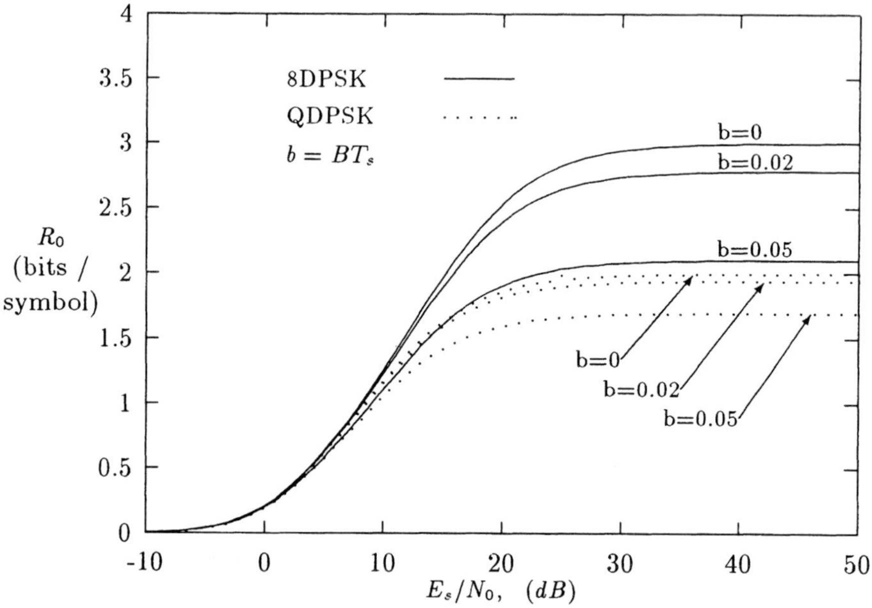


Figure 8.4: Cut-off rates of QDPSK and 8DPSK on a land-mobile Rayleigh fast-fading channel for various values of normalized fading bandwidth.

### 8.2.2 Error-Bounds of Differential Detection of TCM on a Fading Channel

In this section we evaluate the performance of differentially coherent detection of TCM on a fading channel. As in Section 8.2.1 we only consider the squared Euclidean distance metric without CSI. Also, we assume that ideal interleaving is employed and thus the coding channel is memoryless.

Let $\mathbf{s}_l = (s_1, s_2, \cdots, s_l)$ and $\mathbf{r}_l = (r_1, r_2, \cdots, r_l)$ be the transmitted sequence and the received sequence, respectively. Based on the Euclidean decoding metric, the receiver makes a decision on the transmitted sequence $\mathbf{s}_l$, namely $\hat{\mathbf{s}}_l$. The probability that $\hat{\mathbf{s}}_l \neq \mathbf{s}_l$ is decoded if $\mathbf{s}_l$ is sent, namely the pairwise error probability, is given by

$$P_2(\mathbf{s}_l, \hat{\mathbf{s}}_l) = E_{\mathbf{g}_l}[P_2(\mathbf{s}_l, \hat{\mathbf{s}}_l|\mathbf{g}_l)],$$

where

$$P_2(\mathbf{s}_l, \hat{\mathbf{s}}_l|\mathbf{g}_l) = Pr[\, m(\mathbf{r}_l, \hat{\mathbf{s}}_l; \mathbf{g}_l) \geq m(\mathbf{r}_l,\ \mathbf{s}_l;\ \mathbf{g}_l)|\mathbf{g}_l\,]$$

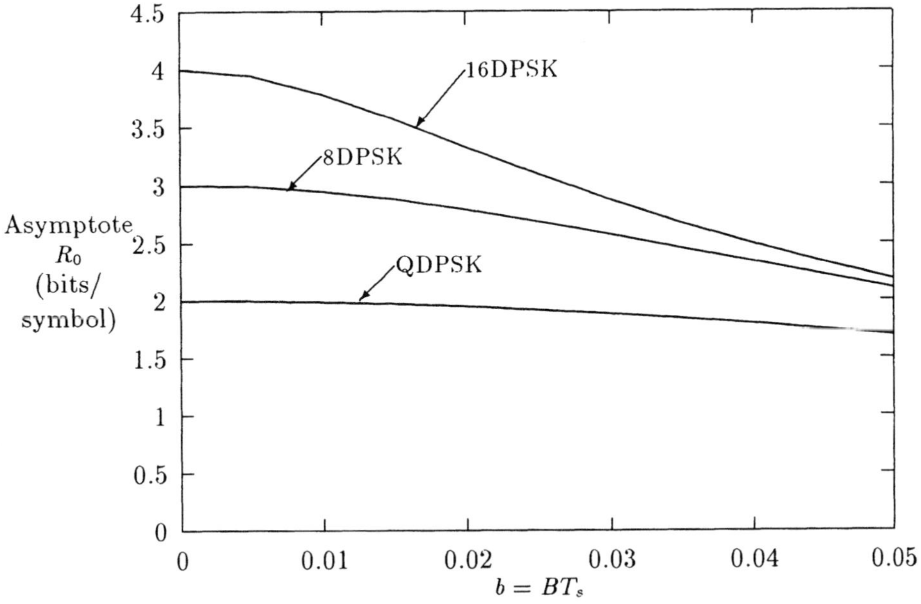


Figure 8.5: Asymptote cut-off rates of MDPSK schemes versus normalized fading bandwidth.

is the conditional pairwise error probability and $m(\cdot)$ shows the squared Euclidean distance metric. Also, $\mathbf{g}_l = (g_1, g_2, \cdots, g_l)$ shows the fading sequence. Using the Chernoff bound technique and considering the coding channel to be memoryless the pairwise error probability is upper-bounded as

$$P_2(\mathbf{s}_l,\ \hat{\mathbf{s}}_l) \ \leq\ \min_{\lambda} \prod_{k=1}^{l} C(s_k,\ \hat{s}_k,\ \lambda),$$

where $C(\hat{s}_k,\ s_k,\ \lambda)$ is the Chernoff factor evaluated in the previous section. For example, using (8.13) the pairwise error probability on a Rayleigh slow-fading channel can be upper-bounded as

$$P_2(\mathbf{s}_l,\ \hat{\mathbf{s}}_l)\ \leq\ \prod_{k=1}^{l} \frac{1}{1+\left|\frac{\hat{s}_k - s_k}{2}\right|^2 \frac{(E_s/N_0)^2}{1+2E_s/N_0}}. \qquad \text{Slow} - \text{Fading}$$

Similarly, using (8.12), the upper bound on pairwise error probability on a

Rayleigh fast-fading channel is expressed as

$$P_2(\mathbf{s}_l,\ \hat{\mathbf{s}}_l) \ \leq \ \prod_{k=1}^{l} \frac{1}{1+|\hat{s}_k - s_k|^2 \left\{ \frac{\xi^2}{4}(\frac{E_s}{N_0})^2 \left[ 2\frac{E_s}{N_0} + 1 + (1-\xi^2)(\frac{E_s}{N_0})^2 \right]^{-1} \right\}}.$$

$$\text{Fast} - \text{Fading}$$

Once again we observe that for slow-fading ($\xi = 1$) the pairwise error probability tends to zero as $E_s/N_0$ approaches infinity. However, for fast-fading ($\xi < 1$) the asymptote pairwise error probability is a nonzero value, that is

$$\lim_{(\frac{E_s}{N_0}) \to \infty} P_2(\mathbf{s}_l,\ \hat{\mathbf{s}}_l) \ \leq \ \prod_{k=1}^{l} \frac{1-\xi^2}{1-\xi^2+\frac{\xi^2}{4}\ |\hat{s}_k - s_k|^2}.$$

This nonzero error probability is referred to as an *error floor.*

## Upper Bound on $\mathbf{P_b}$

Like the case of coherent detection, an upper bound on the bit error probability is obtained by using the generating function (see Section 5.3.1)

$$P_b \ \leq \ \frac{1}{2m} \ \frac{\partial T(D,I)}{\partial I}\bigg|_{I=1,\ D=e^{-\frac{1}{4N_0}}}, \tag{8.22}$$

where, for uniform TCM schemes, $T(D,I)$ is the generating function of the modified error-state diagram whose branches are modified as $I^r\overline{\mathcal{F}}(\mathbf{e}_i, D)$ and $r$ shows the number of incorrect input bits associated with the binary error vector $\mathbf{e}_i$. For differentially coherent detection

$$\overline{\mathcal{F}}(\mathbf{e}_i, D) = \sum_{\mathbf{c}_i = \mathbf{c}_i^{\kappa}} C\left(\mathcal{M}(\mathbf{c}_i), \mathcal{M}(\mathbf{c}_i \oplus \mathbf{e}_i), \lambda_{opt}\right),$$

where $C\left(\mathcal{M}(\mathbf{c}_i), \mathcal{M}(\mathbf{c}_i \oplus \mathbf{e}_i), \lambda_{opt}\right)$ is the Chernoff factor of signals $\mathcal{M}(\mathbf{c}_i)$ and $\mathcal{M}(\mathbf{c}_i \oplus \mathbf{e}_i)$ and $\mathcal{M}$ is a one-to-one mapping as defined in (4.3).

**Example 8.4** Consider the 4-state TCM scheme (4N) of Figure 5.6. The performance of this scheme is evaluated for coherent detection in Example 5.12. Here, we obtain an upper bound on $P_b$ for differential detection of this scheme on a Rayleigh fading channel.

The modified error-state diagram of the 4-state TCM is shown in Figure 5.13 with the branch labels given in Example 5.12, where $f(\delta_i^2)$ is specified by (5.38) for coherent detection. For differentially coherent detection of this scheme on a Rayleigh fading channel without CSI, we simply replace $f(\delta_i^2)$; $i = 0, 1, 2, 3,$ in branch labels by the corresponding Chernoff factor

$$f(\delta_i^2) = \frac{1}{1+\delta_i^2 \left\{ \frac{\xi^2}{4}(\frac{E_s}{N_0})^2 \left[ 2(\frac{E_s}{N_0}) + 1 + (1-\xi^2)(\frac{E_s}{N_0})^2 \right]^{-1} \right\}}, \tag{8.23}$$

where $\delta_0^2 = 0.586$, $\delta_1^2 = 2$, $\delta_2^2 = 3.414$, and $\delta_3^2 = 4$.

The generating function $T(D, I)$ can be obtained as shown in Example 5.12. Using this generating function and (8.22) the upper bound on $P_b$ on a land-mobile Rayleigh fading channel is plotted in Figure 8.6 for various values of the normalized fading bandwidth $b$. Note that $b = 0$ corresponds to a slow-fading channel. The upper bound on $P_b$ for coherent detection is also included in this figure for comparison.

Note that as normalized fading bandwidth increases the error floor decreases. For $b = 0$ (slow-fading) the error floor approaches zero. In this case the performance of differentially coherent detection is about 3 dB inferior to that of coherent detection, asymptotically.

**Example 8.5** Next we consider the 8-state rate 2/3 Ungerboeck TCM scheme (8U) of Figure 4.8. The performance of this scheme for coherent detection is evaluated in Example 5.13. Using the generating function of this scheme, given in Example 5.13, and substituting $f(\delta_i^2)$, $i = 0, 1, 2, 3$, with (8.23) the upper bound on $P_b$ on a land-mobile Rayleigh fading channel is calculated and plotted in Figure 8.7 for various values of the normalized fading bandwidth $b$.

### 8.2.3 Performance Evaluation of Differential Detection of TCM on a Fading Channel Using Exact Expression for the Pairwise Error Probability

In Section 5.3.2 an exact analytical expression for pairwise error probability of TCM schemes on a Rayleigh fading channel was derived (see (5.60)). Using (5.60), for PSK signals ($\delta = 0$), the pairwise error probability can be expressed as

$$P_2(\mathbf{s}_l, \hat{\mathbf{s}}_l) = -\sum \text{Residue}\left[\frac{\Phi_D(s)}{s}\right]_{\text{RPoles}}, \tag{8.24}$$

where

$$\Phi_D(s) = \prod_{i \in \eta} \phi_{D_i}(s) = \prod_{i \in \eta} \frac{p_{1i}p_{2i}}{(s - p_{1i})(s - p_{2i})}$$

is the two-sided Laplace transform of the pdf of the decision variable D (see (5.58)) and

$$\begin{aligned} p_{1i} &= w_i - \sqrt{w_i^2 - p_{1i}p_{2i}}, \\ p_{2i} &= w_i + \sqrt{w_i^2 - p_{1i}p_{2i}} \end{aligned}$$

are the left plane (LP) and the right plane (RP) poles, respectively, with[1]

$$p_{1i}p_{2i} = -\frac{1 + (1 - |\mu|^2)E_s/N_0}{|\mu|^2 d_i^2 E_s/N_0}$$

[1] As stated earlier, we assume that $|s_i|^2 = 1$ for MPSK signals in this chapter. Based on this assumption the expression for $p_{1i}p_{2i}$ is modified slightly.

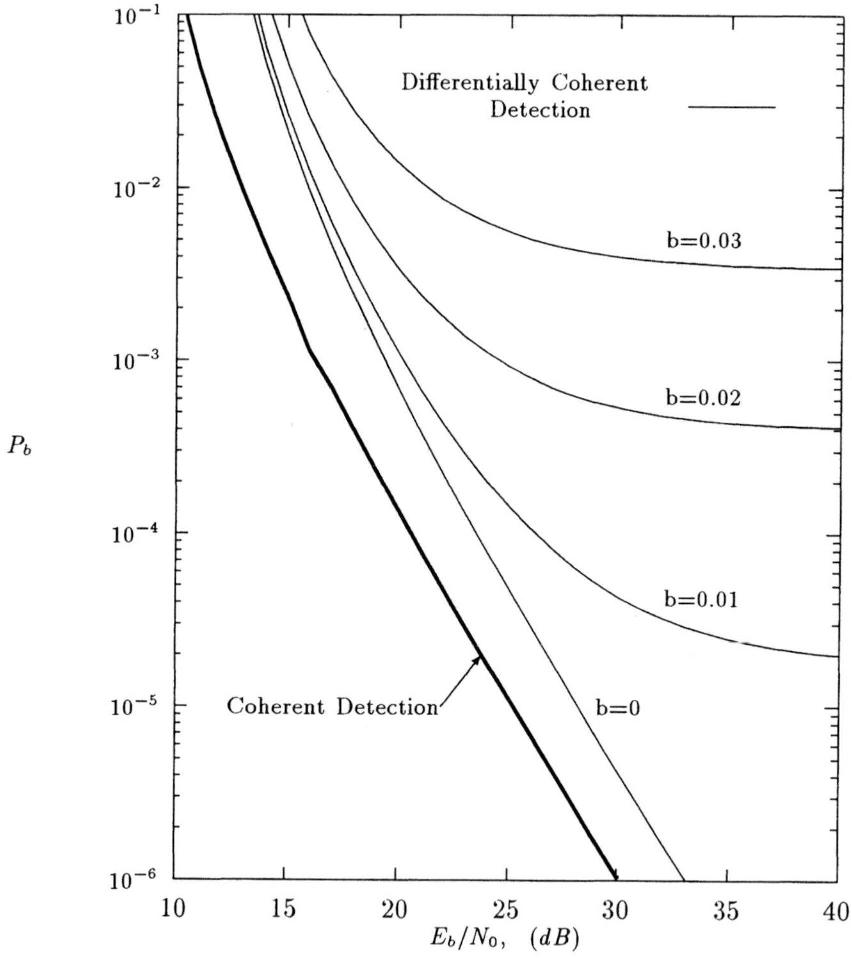


Figure 8.6: Performance of differentially coherent detection of a 4-state rate 2/3 8PSK TCM scheme (4N) on a Rayleigh fast-fading channel without CSI.

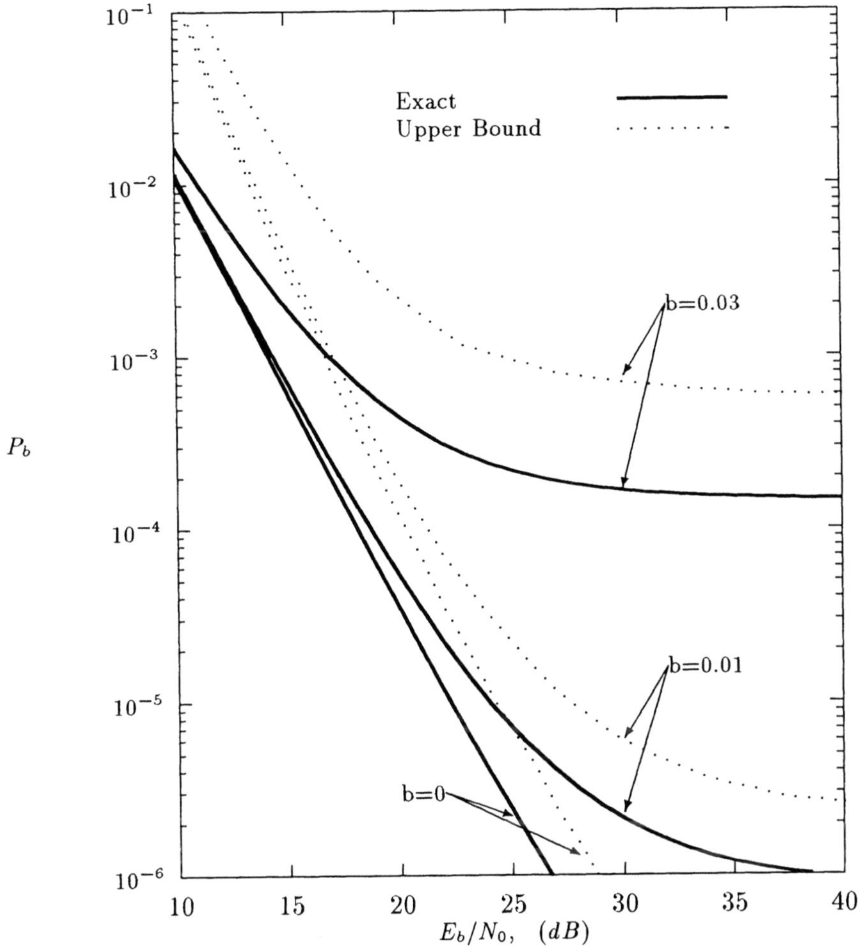


Figure 8.7: Performance of differentially coherent detection of an 8-state rate 2/3 8PSK TCM scheme (8U) on a Rayleigh fast-fading channel.

$$w_i = -\frac{1}{2}$$

and

$$d_i^2 = |s_i - \hat{s}_i|^2,$$

is the squared Euclidean distance between the signals $s_i$ and $\hat{s}_i$. In the foregoing expression, $\mu$ is the correlation coefficient between the complex channel gain $g$ and its estimate $\hat{g}$, given in (5.48).

A differentially coherent detector uses the signal received in the $(k-1)$th symbol interval as a phase reference for the received signal in the $k$th symbol interval. In this case the received signal in the $(k-1)$th symbol interval can be considered as an estimate of complex channel gain in the $k$th symbol interval, that is $\hat{g}_k = g_{k-1}v_{k-1} + n_{k-1}$. This implies that

$$\begin{aligned} \delta_{\hat{g}}^2 &= \delta_g^2 + \frac{N_0}{2E_s} \\ &= \frac{1}{2}(1 + \frac{N_0}{E_s}) \end{aligned} \tag{8.25}$$

and

$$\delta_{g\hat{g}}^2 = R_g(T_s) \cdot v_k, \tag{8.26}$$

where $R_g(T_s)$ is the autocorrelation function of the fading process, defined in (8.4). For a land-mobile radio system the autocorrelation function can be expressed as

$$R_g(T_s) = \delta_g^2 \cdot J_0(2\pi BT_s), \tag{8.27}$$

where $J_o(\cdot)$ is the zero-order Bessel function, $B$ is the fading bandwidth, and $T_s$ is the symbol interval.

Substituting (8.25) and (8.26) into (5.48) and using $|v_k|^2 = 1$, the squared magnitude of the correlation coefficient between $g$ and $\hat{g}$, for a land-mobile Rayleigh channel, becomes

$$|\mu|^2 = \frac{J_0^2(2\pi BT_s)}{1 + (E_s/N_0)^{-1}}. \tag{8.28}$$

The term $|\mu|^2$ can be used to obtain the poles of $\Phi_D(s)$, which subsequently will be used to obtain the pairwise error probability.

As in the case of coherent detection, by using the exact expression of pairwise error probability, an estimate of the bit error probability can be obtained as (5.62). In the following example, we evaluate the performance of a TCM scheme using this method and then compare it with the upper bound analysis.

**Example 8.6** Once again we use the 8-state rate 2/3 Ungerboeck TCM scheme (8U) of Example 8.5. The error events for this scheme, with effective lengths 2 and

3, and actual lengths less than or equal to 4 are already tabulated in Table 5.7 (see Example 5.18).

We evaluate the bit error probability for the differentially coherent detection of this scheme by calculating the exact expression of pairwise error probability for the error events as given in Table 5.7. As a sample calculation we find the pairwise error probability for error event # 1.

**Pairwise Error Probability of Error Event # 1**

Using (8.24) the pairwise error probability of error event # 1 can be evaluated as

$$P_2(\#1) = \frac{\Gamma_s^2}{\delta_1^2 \cdot \delta_3^2}\left[-\sum \text{Residue}\left(\frac{1}{s(s-p_{11})(s-p_{21})(s-p_{13})(s-p_{23})}\right)\right]_{\text{RPoles}},$$

where

$$\begin{aligned} \Gamma_s &= \frac{|\mu|^2 E_s/N_0}{1+(1-|\mu|^2)E_s/N_0}, \\ p_{1i} &= \frac{1}{2} - \sqrt{\frac{1}{4} + \frac{1}{\Gamma_s \delta_i^2}}, \\ p_{2i} &= \frac{1}{2} + \sqrt{\frac{1}{4} + \frac{1}{\Gamma_s \delta_i^2}}, \qquad i = 1,\ 3 \end{aligned}$$

and $\delta_1^2 = 2.000$ and $\delta_3^2 = 4.000$. The calculation of the residues in the foregoing equation yields

$$\begin{aligned} P_2(\#1) = \; & \frac{1}{8\Gamma_s^2}\left[\frac{-1}{p_{21}(p_{21}-p_{11})(p_{21}-p_{13})(p_{21}-p_{23})}\right. \\ & \left. + \frac{-1}{p_{23}(p_{23}-p_{11})(p_{23}-p_{21})(p_{23}-p_{13})}\right]. \end{aligned}$$

Similarly, we can calculate the pairwise error probability of the other error events. Note that as $E_s/N_0$ goes to infinity, $\Gamma_s$ approaches $\frac{|\mu|^2}{1+|\mu|^2}$, which is nonzero for $b \neq 0$. Subsequently for nonzero values of $b$ asymptote of pairwise error probability is nonzero (error floor).

Calculating the pairwise error probability for error events given in Table 5.7, and using (5.62), an estimate of $P_b$ for differentially detection of 8U-TCM scheme is plotted in Figure 8.7 for various values of the normalized fading bandwidth $b$.

# 8.3 BIBLIOGRAPHICAL NOTES

In [98] Johnston and Jones computed the cut-off rate of coded MDPSK schemes on a Rayleigh fading channel to show that these schemes can provide significant coding gain without expanding the bandwidth of transmitted signal.

The generalized cut-off rate of time and frequency-selective fading channel was later evaluated for MDPSK signals as well as for M-ary frequency shift keying (MFSK) by Rainish and Perl [101].

Simon and Divsalar evaluated the upper bound of trellis-coded MDPSK schemes over slow [96] and fast [102] Rician fading channels (see also [27]). The exact pairwise error probability of these schemes in Rayleigh fading channels was derived by Cavers and Ho [49]. Chan and Bateman [66] found a good approximation to the exact pairwise error probability expression, suggested by Cavers and Ho, which gives rise to a simpler expression under certain conditions, valid at medium to high signal-to-noise ratios. The performance of trellis-coded MDPSK schemes was also evaluated in correlated and shadowed Rician fading channels by Huang and Campbell [103]. The pairwise error probabilities were calculated directly and approximated asymptotically. The asymptotic expressions were used in a generating function approach to estimate the bit error probabilities. Edbauer [104] also investigated the performance of a trellis-coded 8DPSK with differential detection over Rayleigh and Rician fading channels for various values of fading bandwidths and interleaver sizes.

## Problems

**8.1** **a)** Derive the characteristic function of the quadratic form of (8.7).
*Hint:* Read Appendix B of [59].

**b)** Show that the Chernoff factor of signals $s_k$ and $\hat{s}_k$ is

$$C(s_k, \hat{s}_k, \lambda) \equiv E_g\left[E_n\left(e^{\lambda[m(r_k,\hat{s}_k)-m(r_k,s_k)]}\right)\right] = \frac{\exp\{\lambda\ \mu_k^{\star T}\ \mathbf{F}_k(\ \mathbf{I}-2\lambda\ \mathbf{R}_k^{\star}\ \mathbf{F}_k)^{-1}\ \mu_k\}}{\det(\ \mathbf{I}-2\lambda\ \mathbf{R}_k^{\star}\ \mathbf{F}_k)}.$$

**8.2** From (8.11),

$$\min_{\lambda_0} C(s_k, \hat{s}_k, \lambda_0) \text{ is equivalent to}$$

$$\max_{\lambda_0}\left\{1+\lambda_0|\hat{s}_k - s_k|^2\left\{\frac{E_s}{N_0}(\xi-2\lambda_0)-\lambda_0\left[1+(1-\xi^2)(\frac{E_s}{N_0})^2\right]\right\}\right\}.$$

**a)** Show that the optimum value of $\lambda_0$ is

$$\lambda_0 = \frac{\xi E_s/N_0}{2\left[1+(1-\xi^2)\left(\frac{E_s}{N_0}\right)^2\right]+4E_s/N_0}.$$

**b)** Derive (8.12).

**8.3** **a)** Show that, for $L = 1$ (first-order diversity) and differential detection, the probability density function of the received signal phase is given by

$$p(\theta) = \frac{1-\nu^2}{2\pi}\left[\frac{1}{1-\nu^2\cos^2\theta}+\frac{\nu\cos\theta}{(1-\nu^2\cos^2\theta)^{3/2}}\cos^{-1}(-\nu\cos\theta)\right],$$

where

$$\nu = \frac{E_s/N_0}{1+E_s/N_0}.$$

*Hint:* Read Appendix 7A of [60].

**b)** Derive (8.18).

**8.4** Derive the cut-off rate for a Rayleigh fast-fading channel (Equation (8.21)).

**8.5** An aeronautical channel can be characterized by its Doppler power spectrum with a Gaussian shape [129]

$$S_H(\lambda) = \frac{A}{\sqrt{2\pi}B} \exp\left(-\frac{\lambda^2}{2B^2}\right),$$

where $A$ and $B$ are constant. $B$ is the standard deviation of the Doppler shift, $\lambda$, and represents the Doppler spread.

a) Show that the correlation coefficient $\xi$ is

$$\xi = e^{-2\pi^2 B^2 T_s^2}.$$

b) Plot the cut-off rates of QDPSK and 8DPSK versus $E_s/N_0$ for $BT_s =$ 0, 0.02, 0.05, over an aeronautical Rayleigh fast-fading channel.

**8.6** A Rayleigh fading channel is characterized by its Doppler power spectrum with a rectangular shape

$$S_H(\lambda) = \begin{cases} \frac{A}{B}, & |\lambda| \leq B \\ 0, & \text{elsewhere} \end{cases},$$

where $A$ and $B$ are constant. $B$ represents the maximum value of the Doppler shift $\lambda$. Find the correlation coefficient and plot the cut-off rates of QDPSK and 8DPSK versus $E_s/N_0$ for $BT_s = 0$, 0.025, 0.50 over this channel.

**8.7** Consider the 4-state 8PSK TCM scheme (4U) of Figure 4.5. Following the procedure given in Example 8.4, derive and plot the upper bound on $P_b$ of this scheme using differential detection over a land-mobile Rayleigh fading channel for $BT_s = 0$, 0.01, 0.02, 0.03. Compare the results to the performance of coherent detection.

**8.8** Repeat Problem 8.7 with the 4-state 8PSK TCM scheme (4WL) shown in Figure 5.3.

**8.9** Repeat Problem 8.7 for an aeronautical Rayleigh fading channel with the Doppler power spectrum shown in Problem 8.5.

**8.10** Consider the 4-state 8PSK TCM scheme (4WL) shown in Figure 5.3. Derive and plot the upper bound on $P_b$ of this scheme using differential detection over an aeronautical Rayleigh fading channel for $BT_s = 0$, 0.01, 0.02, 0.05.

**8.11** Consider the 4-state 8PSK TCM scheme (4N) shown in Figure 5.6. Following the procedure given in Example 8.6, evaluate the bit error probability for differentially coherent detection of this TCM scheme over a land-mobile Rayleigh fading channel for $BT_s = 0.01$, 0.02. Compare the results with the upper bound shown in Figure 8.6.

**8.12** Repeat Problem 8.11 with the 4-state 8PSK TCM scheme (4WL) shown in Figure 5.3. Compare the results with the upper bound obtained from Problem 8.8.

**8.13** Show that the bit error probability of the binary DPSK in a Gaussian channel is

$$P_b = \frac{1}{2} e^{-E_b/N_0}.$$

*Hint*: Follow the derivation in [17, Chapter 7] or [60, Chapter 4].

**8.14** From Problem 8.13, the conditional bit error probability of the binary DPSK for a given attenuation $a$ is

$$P_{b|a} = \frac{1}{2} e^{-a^2 E_b/N_0}$$

and the average bit error probability of the binary DPSK scheme over a frequency-nonselective, slowly fading channel is

$$P_b = E_a\{P_{b|a}\} = \int_0^{+\infty} P_{b|a} p_A(a)\, da, \quad a \geq 0.$$

Derive the average bit error probability of the binary DPSK scheme in a Rayleigh fading channel.

**8.15** Repeat Problem 8.14 for a Rician fading channel.

# Chapter 9

# PERFORMANCE OF CODED MODULATION SCHEMES ON SHADOWED RICIAN FADING CHANNELS

In preceding chapters, two classes of bandwidth-efficient coded-modulation schemes, namely, TCM and BCM schemes were presented. The performance of these schemes was evaluated for a Rayleigh fading channel. The purpose of this chapter is to evaluate the performance of these schemes on a shadowed Rician fading channel. With this objective in mind, we first briefly review the statistical model of the shadowed Rician fading channel. Then, the performance of the coded schemes is evaluated based on the assumption of ideal interleaving. Finally, we study the effect of fading bandwidth and imperfect interleaving on performance degradation of the coded schemes.

## 9.1 DESCRIPTION OF THE CHANNEL MODEL

As described in Chapter 1, for mobile fading channels, the fading amplitude is modeled as shadowed Rician fading. This model is a special case of Rician fading where the line-of-sight (LOS) component is not constant but rather a lognormally distributed random variable. In this channel the fading amplitude may be expressed as

$$a = \gamma + \eta_0 + j\eta_1$$

with

$$\gamma = e^{\eta_2 + m_\gamma},$$

where $\eta_0, \eta_1$, and $\eta_2$ are three independent Gaussian random variables with zero mean and variances $\sigma_a^2, \sigma_a^2$, and $\sigma_\gamma^2$, respectively. Considering $\eta_2$ to be a zero-mean Gaussian random variable with variance $\sigma_\gamma^2$, the probability density function of the LOS component, $\gamma$, can be expressed as

$$p_\Gamma(\gamma) = \frac{1}{\sqrt{2\pi}\sigma_\gamma \gamma} e^{-\frac{(\ln \gamma - m_\gamma)^2}{2\sigma_\gamma^2}}, \quad \gamma \geq 0, \tag{9.1}$$

which is a lognormal density function.

Keeping the value of $\gamma$ constant, the conditional probability density function of the fading amplitude $a$ is expressed as

$$p_{A|\Gamma}(a|\gamma) = \frac{a}{\sigma_a^2} e^{-\frac{a^2+\gamma^2}{2\sigma_a^2}} I_0\left(\frac{a\gamma}{\sigma_a^2}\right), \quad \gamma \geq 0, \tag{9.2}$$

where $I_0(\cdot)$ is the zero-order modified Bessel function of the first kind. The unconditional probability density function of fading amplitude is obtained by averaging (9.2) over the pdf of $\gamma$, given in (9.1). Thus,

$$\begin{aligned} p_A(a) &= \frac{a}{\sqrt{2\pi\sigma_a^2}\sigma_\gamma} \int_0^\infty \frac{1}{\gamma} e^{-\left[\frac{(\ln \gamma - m_\gamma)^2}{2\sigma_\gamma^2} + \frac{a^2+\gamma^2}{2\sigma_a^2}\right]} I_0\left(\frac{a\gamma}{\sigma_a^2}\right) d\gamma \\ &= \frac{a}{\sqrt{2\pi\sigma_a^2}\ \sigma_\gamma} \int_{-\infty}^\infty e^{-\frac{a^2}{2\sigma_a^2} - \frac{e^{2m_\gamma+2x}}{2\sigma_a^2}} I_0\left(\frac{a}{\sigma_a^2}\ e^{x+m_\gamma}\right) dx, \end{aligned} \tag{9.3}$$

where the second line in (9.3) is obtained by assuming that $x = \ln \gamma - m_\gamma$.

The integrals in (9.3) cannot be evaluated in a closed form, and hence, numerical methods are required for their evaluation. Using the series expansions for both $e^x$ and $I_0(x)$

$$e^x = \sum_{k=0}^\infty \frac{x^k}{k!}, \tag{9.4}$$

and

$$I_0(x) = \sum_{n=0}^\infty \frac{(x/2)^{2n}}{(n!)^2}, \tag{9.5}$$

the probability density function of $a$ can also be expressed as [122]

$$p_A(a) = \sum_{k=0}^\infty \sum_{n=0}^\infty \left(-\frac{e^{2m_\gamma}}{2\sigma_a^2}\right)^k \left(\frac{e^{m_\gamma}}{2\sigma_a^2}\right)^{2n} \frac{e^{2\sigma_\gamma^2(k+n)^2}}{\sigma_a^2 k!(n!)^2} a^{2n+1} e^{-\frac{a^2}{2\sigma_a^2}}. \tag{9.6}$$

It is interesting to note that the pdf of a Rayleigh fading or of a Rician fading channel can be obtained as a special case from the pdf of shadowed Rician fading channel.

**Example 9.1** A Rayleigh fading channel is a special case of a shadowed Rician fading channel with parameters $\sigma_a^2 = 1/2$, $\sigma_\gamma^2 = 0$, and $m_\gamma = -\infty$. We rewrite (9.6) as

$$\begin{aligned} p_A(a) &= \frac{a}{\sigma_a^2} e^{-\frac{a^2}{2\sigma_a^2}} + \\ & \sum_{k=1}^{\infty}\sum_{n=1}^{\infty}\left(-\frac{e^{2m_\gamma}}{2\sigma_a^2}\right)^k \left(\frac{e^{m_\gamma}}{2\sigma_a^2}\right)^{2n} \frac{e^{2\sigma_\gamma^2(k+n)^2}}{\sigma_a^2\, k!\,(n!)^2} a^{2n+1} e^{-a^2/2\sigma_a^2}. \end{aligned}$$

Setting $\sigma_a^2 = 1/2$, $\sigma_\gamma^2 = 0$, and $m_\gamma = -\infty$ in the above equation, and noting that the second term vanishes, we obtain

$$p_A(a) = 2a\, e^{-a^2}, \quad a \geq 0.$$

**Example 9.2** A Rician fading channel with Rician parameter $K$ can be considered to be a special case of a shadowed Rician fading channel when we set $\sigma_a^2 = \frac{A^2}{2K}$, $\sigma_\gamma^2 = 0$, and $m_\gamma = \ln A$. For a normalized Rician channel $A^2 + 2\sigma_a^2 = 1$, and hence, the parameters $A$ and $\sigma_a^2$ can be expressed in terms of the Rician parameter $K$

$$\begin{aligned} A &= \sqrt{\frac{K}{1+K}}, \\ \sigma_a^2 &= \frac{1}{2(1+K)}. \end{aligned}$$

Substituting the values of $\sigma_a^2$, $\sigma_\gamma^2$, and $m_\gamma$ of the Rician fading channel into (9.6), yields

$$\begin{aligned} p_A(a) &= \sum_{k=0}^{\infty}\sum_{n=0}^{\infty}\left(-\frac{A^2}{2\sigma_a^2}\right)^k \left(\frac{A}{2\sigma_a^2}\right)^{2n} \frac{1}{2\sigma_a^2\, k!\,(n!)^2} a^{2n+1} e^{-\frac{a^2}{2\sigma_a^2}} \\ &= \frac{a\, e^{-\frac{a^2}{2\sigma_a^2}}}{\sigma_a^2} \sum_{i=0}^{\infty} \frac{(-A^2/2\sigma_a^2)^i}{i!} \sum_{j=0}^{\infty} \frac{(aA/2\sigma_a^2)^{2n}}{(n!)^2} \\ &= \frac{a\, e^{-\frac{a^2}{2\sigma_a^2}}}{\sigma_a^2} e^{-A^2/2\sigma_a^2} I_0\left(\frac{aA}{\sigma_a^2}\right). \end{aligned}$$

Using the values of $\sigma_a^2$ and $A$ in terms of the Rician parameter $K$ we obtain

$$p_A(a) = 2(1+K)\, e^{-[K+(1+K)a^2]}\, I_0(2a\sqrt{K(1+K)}).$$

In mobile radio channels the fading amplitudes are correlated due to the Doppler shift associated with relative motion of the transmitter and receiver antennas. The Doppler shift depends upon the mobile speed, the carrier frequency, and the angle between the propagation vector and the mobile vector (see Chapter 1, Section 1.3). For reasonable bit rates and mobile velocity, fading varies slowly compared to bit rate. This results in a channel with memory where the errors occur in bursts. The error bursts degrade the performance of the coded schemes. To transform the bursts into statistically independent errors, interleaving/de-interleaving is commonly applied. Sufficiently large interleaving destroys the memory of the channel. Considering the block diagram of Figure 5.1 for transmission of coded schemes over a shadowed Rician fading channel, we evaluate the performance of coded schemes over such a channel through upper bound analysis. In this evaluation, ideal interleaving and the use of channel state information (CSI) by the receiver are assumed.

## 9.2 PERFORMANCE OF TRELLIS-CODED MPSK SCHEMES ON A SHADOWED RICIAN FADING CHANNEL

In determining the upper bound on bit error probability of trellis-coded schemes, it is useful to recall the state diagram and the associated generating function approach described in Chapter 5. Using the Chernoff bound, the conditional pairwise error probability, given $\mathbf{s}_l = (s_1, s_2, \cdots, s_l)$, is sent and $\hat{\mathbf{s}}_l = (\hat{s}_1, \hat{s}_2, \cdots, \hat{s}_l)$ is decoded, conditioned on the fading amplitude $\mathbf{a}_l = (a_1, a_2, \cdots, a_l)$ and upper bounded as (see (5.11))

$$P_2(\mathbf{s}_l, \hat{\mathbf{s}}_l|\mathbf{a}_l) \leq \prod_{i=1}^{l} \exp(-\frac{1}{4N_0} a_i^2 |s_i - \hat{s}_i|^2). \tag{9.7}$$

The unconditional pairwise error probability upper bound can be determined by averaging (9.7) over the fading amplitude vector $\mathbf{a}_l$ with the result

$$P_2(\mathbf{s}_l, \hat{\mathbf{s}}_l) \leq E_{\mathbf{a}_l} \left[ \prod_{i=1}^{l} \exp(-\frac{1}{4N_0} a_i^2 |s_i - \hat{s}_i|^2) \right], \tag{9.8}$$

where $E_{\mathbf{a}_l}$ denotes the statistical averaging with respect to $\mathbf{a}_l$. Recall that a union bound on the bit error probability can be obtained by summing the pairwise error probabilities weighted by the number of nonzero information bits along the corresponding incorrect path. For a TCM scheme this information is implicit in the generating function $T(D,\ I)$ associated with the error state diagram (see Chapter 5).

To evaluate the pairwise error probability (9.8) we use the fact that the fading channel has zero memory. This is achieved by assuming suitably large interleaving. Because of ideal interleaving the fading amplitude, although constant within a signaling interval, is independent from interval to interval. In this case (9.8) can be written as

$$P_2(s_l,\ \hat{s}_l) \ \leq \ \prod_{i=1}^{l} E_{a_i} \ \left[ e^{-\frac{1}{4N_0} \ a_i^2 \ |s_i - \hat{s}_i|^2} \right]. \tag{9.9}$$

Using (9.3), the expectation term in (9.9) is evaluated as

$$\begin{aligned} E_{a_i} \ [ e^{-a_i^2 \ d_i^2}] \ &= \ \frac{1}{\sqrt{2\pi} \ \sigma_a^2 \ \sigma_\gamma} \int_0^\infty \frac{1}{\gamma} \ e^{-\left[\frac{(\ln \gamma - m_\gamma)^2}{2\sigma_\gamma^2} + \frac{\gamma^2}{2\sigma_a^2}\right]} \\ &\cdot \left( \int_0^\infty a_i \ e^{-\frac{a_i^2 (1+2\sigma_a^2 \ d_i^2)}{2\sigma_a^2}} \ I_0(\frac{a_i \gamma}{\sigma_a^2}) \ da_i \right) d\gamma, \end{aligned} \tag{9.10}$$

where, for simplicity, we redefine[1]

$$d_i^2 \equiv \frac{\delta_i^2}{4N_0} = \frac{1}{4N_0} |s_i - \hat{s}_i|^2.$$

The second integral in (9.10) can be simplified using [92, Page 717]

$$\int_0^\infty t \ e^{-\alpha t^2} \ J_0(\beta t) \ dt = \frac{1}{2\alpha} \ e^{-\beta^2/4\alpha},$$

where [92, Page 952]

$$J_0(jt) \ = \ I_0(t).$$

Using the above equation, and after some mathematical manipulations, (9.10) can be expressed as (see Problem 9.1)

$$f(\delta_i^2) \equiv E_{a_i} \ [ e^{-a_i^2 \ d_i^2}] = \frac{1}{(1 + 2\sigma_a^2 d_i^2)\sqrt{\pi}} \int_{-\infty}^\infty e^{-t^2 - \frac{d_i^2}{1+2\sigma_a^2 d_i^2} e^{2(\sqrt{2} \ \sigma_\gamma t + m_\gamma)}} dt, \tag{9.11}$$

where, for high signal-to-noise ratios, i.e., $\frac{|s_i - \hat{s}_i|^2}{4N_0} \equiv d_i^2 \gg \frac{1}{2\sigma_a^2}$, it can be simplified to

$$f(\delta_i^2) \equiv E_{a_i} \ [ e^{-a_i^2 \ d_i^2}] \approx \frac{1}{2\sqrt{\pi}\sigma_a^2 \ d_i^2} \int_{-\infty}^\infty e^{-t^2 - \frac{1}{2\sigma_a^2} e^{2(\sqrt{2} \ \sigma_\gamma t + m_\gamma)}} dt. \tag{9.12}$$

Alternatively, we may express $E_{a_i}[e^{-a_i^2 d_i^2}]$ in a series expansion form. Using (9.6), we can write

$$\begin{aligned} E_{a_i} \ [ e^{-a_i^2 \ d_i^2}] \ &= \ \sum_{k=0}^\infty \sum_{n=0}^\infty \left( -\frac{e^{2m_\gamma}}{2\sigma_a^2} \right)^k \left( \frac{e^{m_\gamma}}{2\sigma_a^2} \right)^{2n} \frac{e^{2\sigma_\gamma^2 (k+n)^2}}{\sigma_a^2 k! (n!)^2} \\ & \int_0^\infty a_i^{2n+1} e^{-\frac{a_i^2}{2\sigma_a^2}(1+2\sigma_a^2 d_i^2)} da_i. \end{aligned} \tag{9.13}$$

[1] In Chapter 5, we used $d_i^2 = |s_i - \hat{s}_i|^2$.

The integral in (9.13) is then evaluated as

$$\int_0^\infty a_i^{2n+1} e^{-\frac{a_i^2}{2\sigma_a^2}(1+2\sigma_a^2 d_i^2)} da_i = \frac{n!\sigma_a^2}{(1+2\sigma_a^2 d_i^2)^{n+1}}.$$

Substituting the above equation into (9.12) yields

$$f(\delta_i^2) \equiv E_{a_i}\left[e^{-a_i^2 d_i^2}\right] = \frac{1}{1+2\sigma_a^2 d_i^2} \sum_{k=0}^{\infty}\sum_{n=0}^{\infty}\left(-\frac{e^{2m_\gamma}}{2\sigma_a^2}\right)^k \left(\frac{e^{2m_\gamma}}{2\sigma_a^2(1+2\sigma_a^2 d_i^2)}\right)^n \frac{e^{2\sigma_\gamma^2(k+n)^2}}{k!n!}. \tag{9.14}$$

It is seen that the calculation of $f(\delta_i^2)$ involves the evaluation of an infinite series when (9.14) is used or the numerical integration when (9.11) is used. Both methods are time consuming. To avoid this problem we may approximate the function $f(\delta_i^2)$ as [122] (see Problem 9.2)

$$f(\delta_i^2) = \frac{1}{1+2\sigma_a^2 d_i^2} e^{-\frac{C_i}{1+0.2\sigma_\gamma C_i}}, \tag{9.15}$$

where

$$C_i = \frac{d_i^2}{1+2\sigma_a^2 d_i^2} e^{2m_\gamma}. \tag{9.16}$$

Using (9.15), the upper bound on the pairwise error probability is written as

$$P_2(\mathbf{s}_l, \hat{\mathbf{s}}_l) \le \prod_{i=1}^{l} \frac{1}{1+2\sigma_a^2 d_i^2} e^{-\frac{C_i}{1+0.2\sigma_\gamma C_i}}. \tag{9.17}$$

**Example 9.3** The parameters of a normalized Rician fading channel are given in Example 9.2 as

$$\sigma_a^2 = \frac{1}{2(1+K)}, \qquad \sigma_\gamma^2 = 0 \qquad e^{m_\gamma} = A = \sqrt{\frac{K}{1+K}}.$$

Substituting these values into (9.17) the pairwise error probability is upper bounded as

$$P_2(\mathbf{s}_l, \hat{\mathbf{s}}_l) \le \prod_{i=1}^{l} \frac{1+K}{1+K+\frac{1}{4N_0}|s_i - \hat{s}_i|^2} \exp\left[-\frac{K\frac{1}{4N_0}|s_i - \hat{s}_i|^2}{1+K+\frac{1}{4N_0}|s_i-\hat{s}_i|^2}\right],$$

which is the same obtained in Chapter 5 (see (5.12)).

For uniform TCM scheme the upper bound on the bit error probability $P_b$ is obtained as

$$P_b \le \frac{1}{2m} \frac{\partial T(D,I)}{\partial I}\bigg|_{I=1, D=e^{-\frac{1}{4N_0}}},$$

where $m$ is the number of input bits and $T(D, I)$ is the generating function of the modified error-state diagram (see Chapter 4). The branches of the modified error-state diagram are labeled as $I^r\overline{\mathcal{F}}(\mathbf{e}_i, D)$ where $r$ shows the number of incorrect input bits associated with $\mathbf{e}_i$ and $\overline{\mathcal{F}}(\mathbf{e}_i, D)$ is the error weight profile of $\mathbf{e}_i$ (see Chapter 5).

**Example 9.4** The upper bounds for two rate 2/3 trellis coded 8PSK schemes, namely, the Ungerboeck 4-state scheme (4U) and the 4-state scheme without parallel paths (4N), designed for fading channels in Chapter 5 (see Figure 5.6), are evaluated on shadowed Rician fading channels in this example. The transfer function $T(D, I)$ is obtained for 4U and 4N schemes using the results of Examples 5.10 and 5.12, respectively, with

$$f(\delta_i^2) = \frac{1}{1 + \sigma_a^2 \delta_i^2 / 2N_0} \; e^{-\frac{C_i}{1+0.2\,\sigma_\gamma\, C_i}},$$

and

$$C_i = \frac{\delta_i^2 / 4N_0}{1 + \sigma_a^2\; \delta_i^2 / 2N_0} \; e^{2m_\gamma},$$

where $\delta_0^2 = 0.586\ E_s$, $\delta_1^2 = 2.000\ E_s$, $\delta_2^2 = 3.414\ E_s$, and $\delta_3^2 = 4.000\ E_s$. The upper bounds on $P_b$ for the 4-state schemes 4U and 4N are shown in Figures 9.1, 9.2, and 9.3 for light, average, and heavy shadowed Rician fading channels, respectively. The parameters $\sigma_a^2$, $\sigma_\gamma$, and $m_\gamma$ for light, average, and heavy shadowed Rician channels are given in Table 1.1. The simulation results are included in these figures in order to examine the tightness of the upper bounds.

# 9.3 PERFORMANCE OF RS-CODED MPSK SCHEMES ON A SHADOWED RICIAN FADING CHANNEL

In this section we evaluate the performance of coherent detection of RS-coded MPSK schemes on a shadowed Rician fading channel. First we obtain an upper bound on bit error probability for soft-decision decoding and then we find lower bounds on $P_b$ for errors-only and errors-and-erasures decoding techniques.

The upper bound on bit error probability for RS-coded MPSK schemes which employ the Viterbi algorithm for soft-decision decoding is obtained using the analytical approach of Chapter 7 (see Section 7.4). For a time-varying fading channel, the upper bound on the bit error probability at high SNR's is determined to be

$$P_b \le \sum_{d_p^2(d)} \frac{d}{mn} \frac{a(d_p^2(d))}{d_p^2(d)} \prod_{i \in \eta} \delta_i^2 f(\delta_i^2), \tag{9.18}$$

where $f(\delta_i^2)$ is given by (9.12) or (9.15) and the other parameters are defined as in Section 7.4.

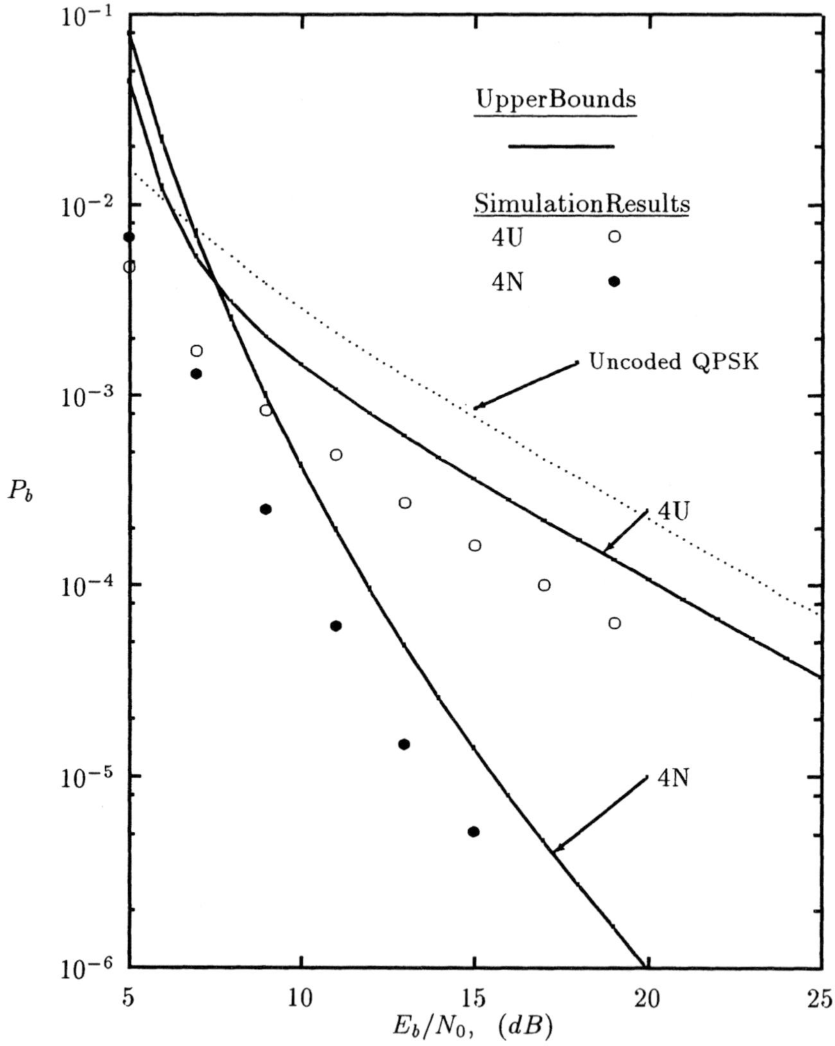


Figure 9.1: Performance of the 4-state 8PSK TCM schemes on the light shadowed Rician fading channel with CSI and ideal interleaving.

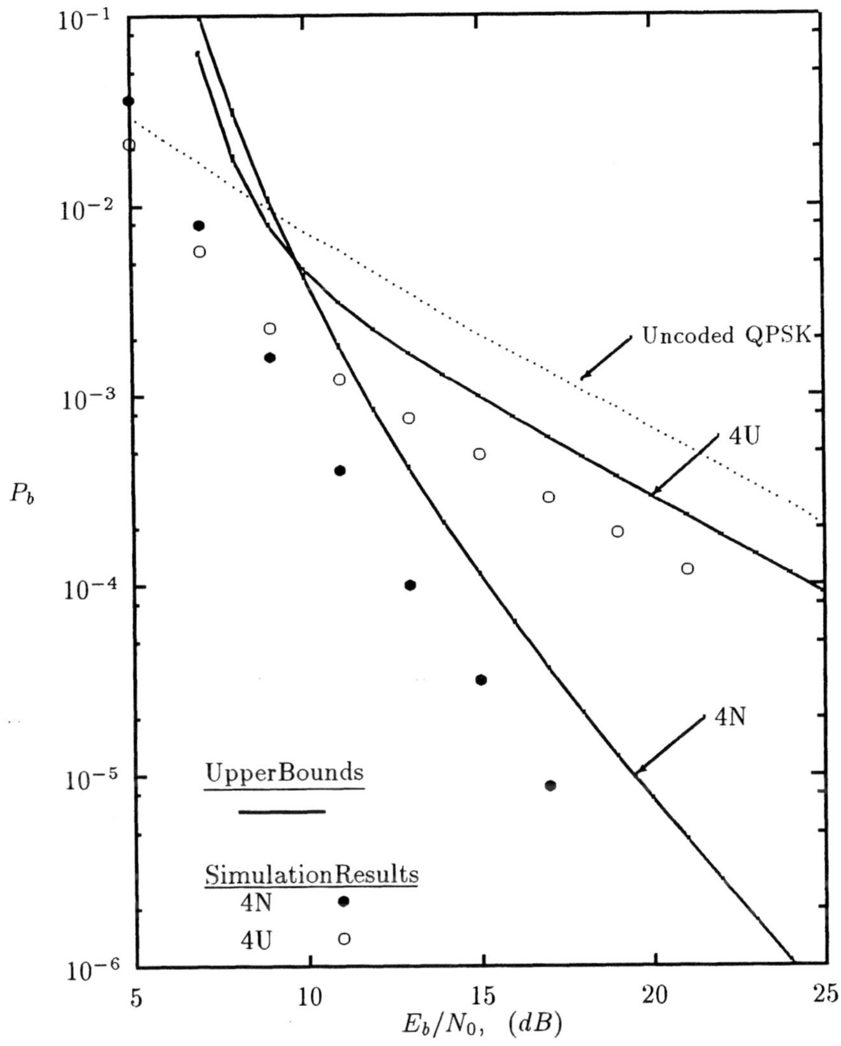


Figure 9.2: Performance of the 4-state 8PSK TCM schemes on the average shadowed Rician fading channel with CSI and ideal interleaving.

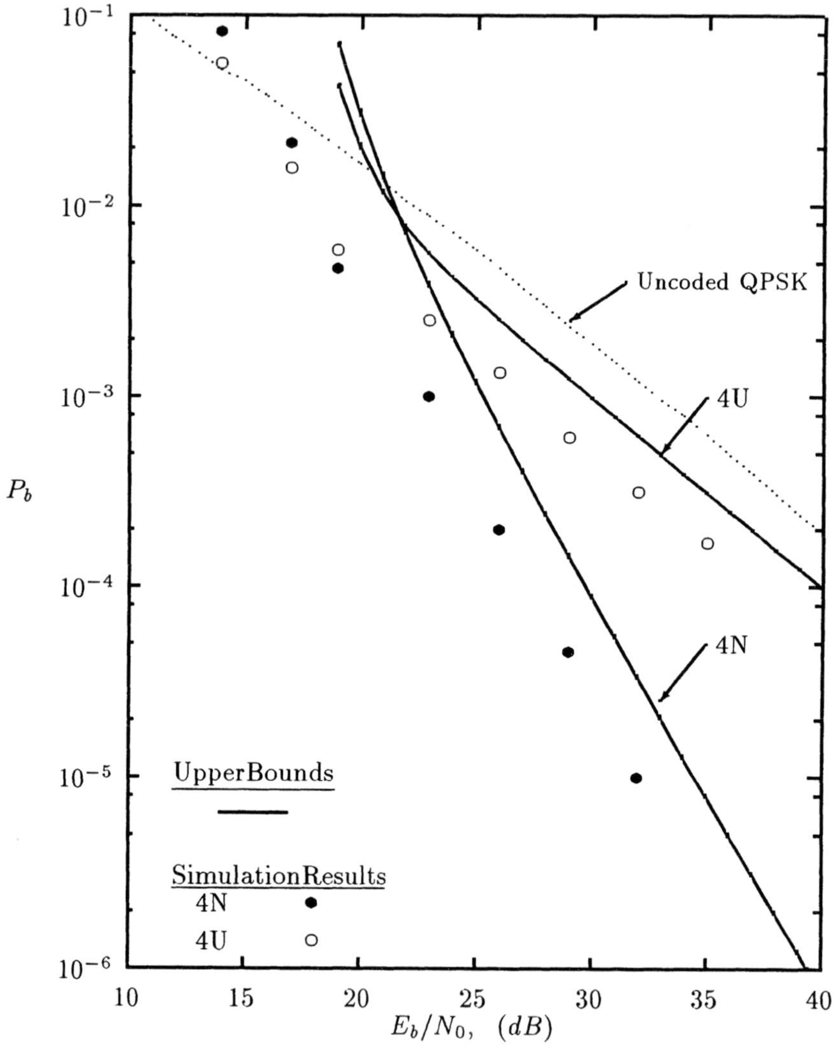


Figure 9.3: Performance of the 4-state 8PSK TCM schemes on the heavy shadowed Rician fading channel with CSI and ideal interleaving.

We can further simplify (9.18) for high SNR's when $\frac{1}{N_0} \gg 2/(\sigma_a^2 \delta_i^2)$. In such a case $f(\delta_i^2)$ can be expressed as (see Problem 9.8)

$$f(\delta_i^2) = \frac{1}{\sigma_a^2 \delta_i^2 / 2N_0} e^{-\frac{C_\infty}{1+0.2\sigma_\gamma C_\infty}}, \tag{9.19}$$

where

$$C_\infty = \frac{1}{2\sigma_a^2} e^{2m_\gamma}. \tag{9.20}$$

Substituting (9.19) into (9.18) yields

$$\begin{aligned} P_b &\leq \sum_{d_p^2(d)} \frac{d}{mn} \frac{a(d_p^2(d))}{d_p^2(d)} \left( \frac{1}{\sigma_a^2 / 2N_0} e^{-\frac{C_\infty}{1+0.2\sigma_\gamma C_\infty}} \right)^d \\ &= \frac{d}{mn} \left( \frac{1}{\sigma_a^2 / 2N_0} e^{-\frac{C_\infty}{1+0.2\,\sigma_\gamma\, C_\infty}} \right)^d \sum_{d_p^2(d)} \frac{a(d_p^2(d))}{d_p^2(d)}. \end{aligned} \tag{9.21}$$

The upper bound given in (9.21) gives an asymptotic estimate of bit error probability at high signal-to-noise ratios.

To evaluate the performance of the RS-coded schemes on shadowed Rician fading channels for errors-only and errors-and-erasure decoding the lower bounds on the bit error probability of (7.7) and (7.12) are utilized. The associated probabilities, however, can no longer be expressed as closed forms, as they were in the Rayleigh fading channel. In this case, these probabilities are obtained by numerical integration.

**Example 9.5** Consider the RS(7, 5) coded 8PSK scheme introduced in Chapter 7. The lower bounds of this coded scheme for different degrees of shadowing are represented in Figures 9.4, 9.5, and 9.6. The simulation results for different decoding strategies, i.e, errors-only, errors-and-erasures, SEMDD, and Viterbi decoding, are also included in these figures.

In the foregoing example the lower bounds of the post decoding bit error rate for errors-and-erasures decoding techniques have been found by optimizing the fading amplitude threshold, $a_T$, for each value of $E_b/N_0$. Figure 9.7 represents the optimized fading amplitude threshold $a_T$ versus $E_b/N_0$ for three different shadowed channels. The optimum threshold decreases with increasing signal-to-noise ratio, $E_b/N_0$. A comparison between optimum threshold values for different channels shows that this value for light shadowed channels is higher than for the other ones. However, the slope of the optimum threshold variation of the heavy shadowed channel is smaller than the light and average shadowed channels.

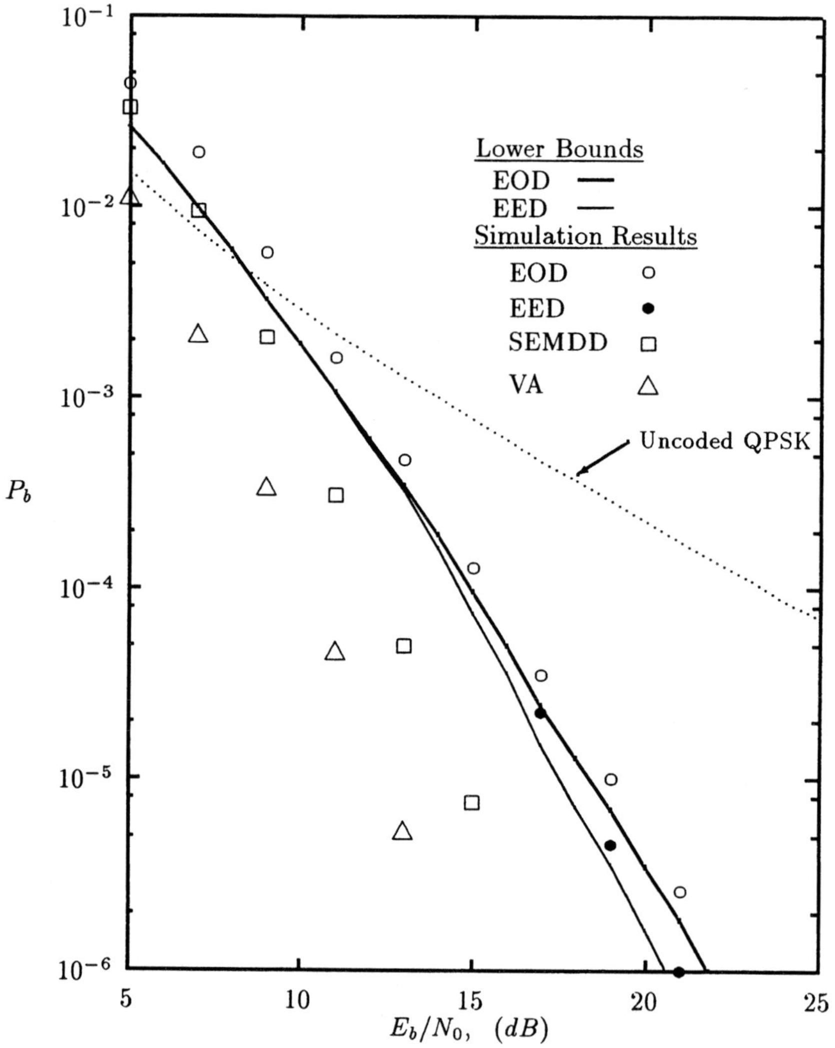


Figure 9.4: Comparison of the bit error probability of RS(7, 5) coded 8PSK for different decoding strategies on a light shadowed Rician fading channel.

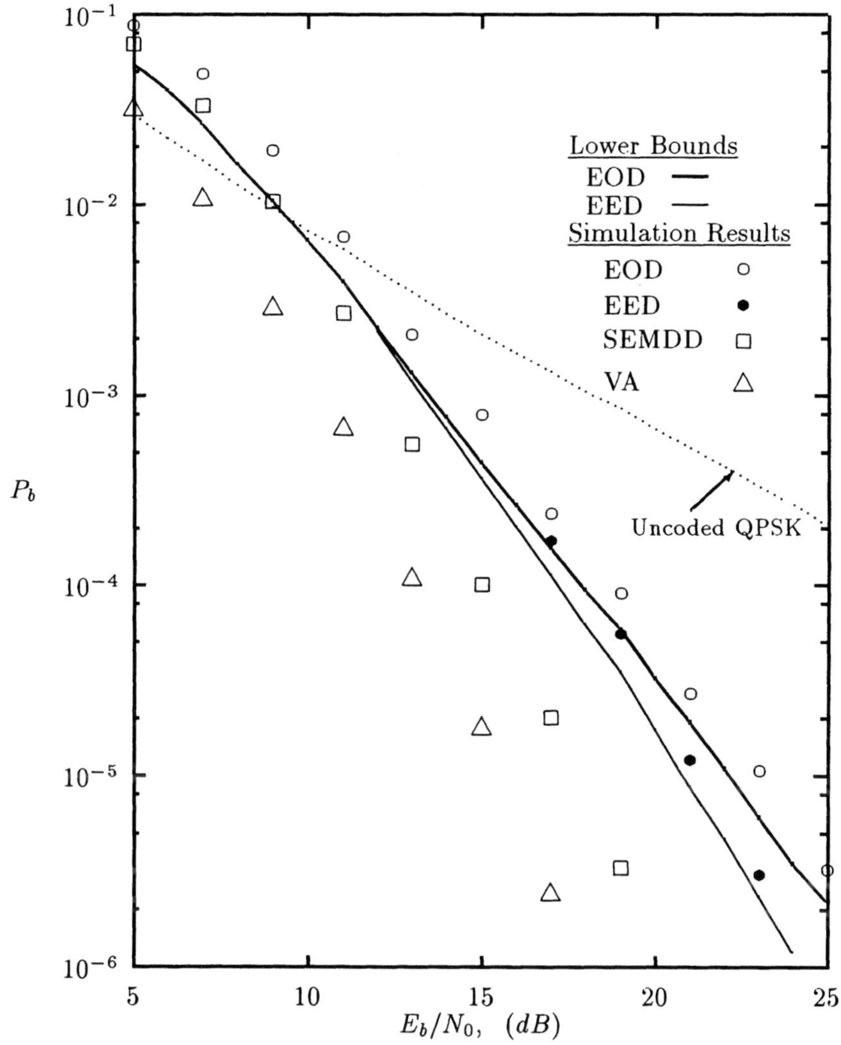


Figure 9.5: Comparison of the bit error probability of RS(7, 5) coded 8PSK for different decoding strategies on an average shadowed Rician fading channel.

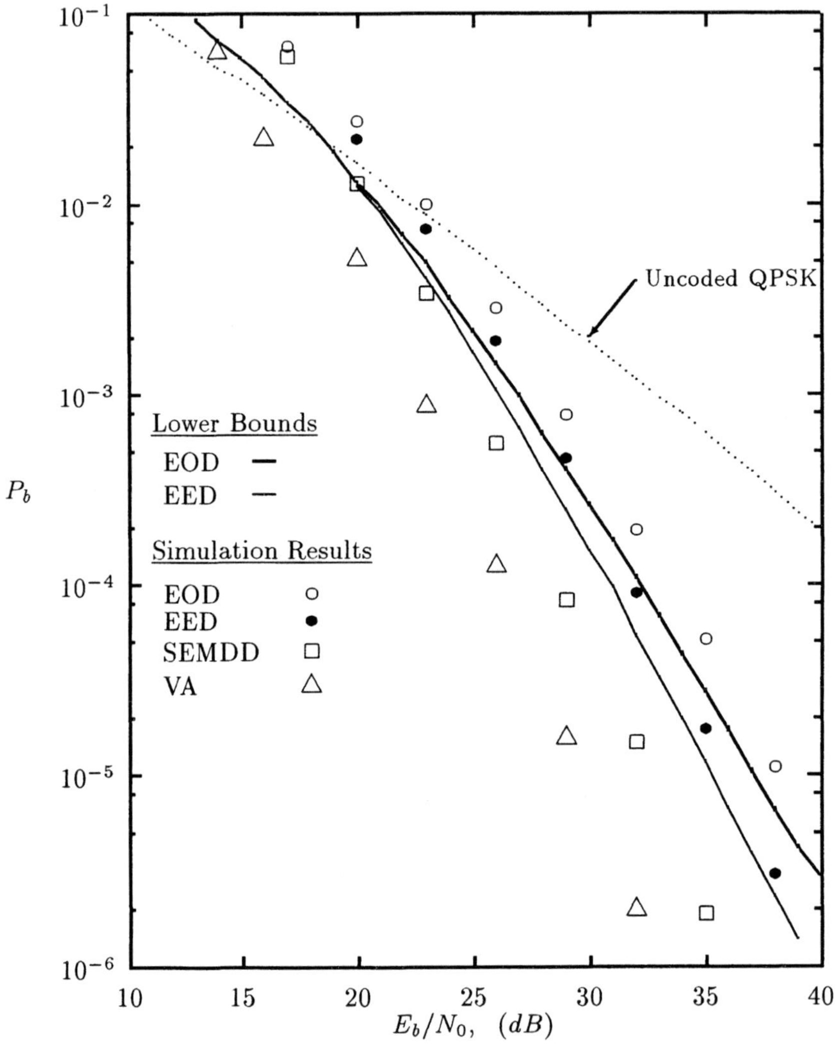


Figure 9.6: Comparison of the bit error probability of RS(7, 5) coded 8PSK for different decoding strategies on a heavy shadowed Rician fading channel.

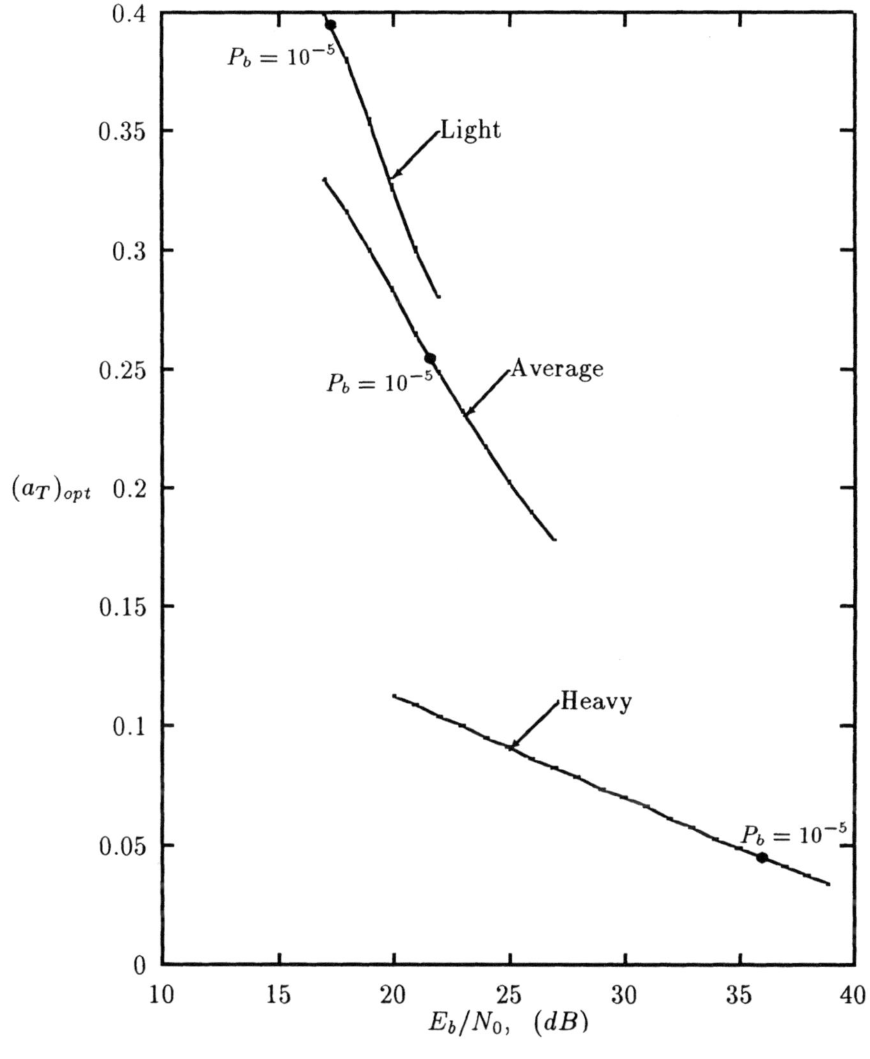


Figure 9.7: Optimum fading amplitude threshold for errors-and-erasure decoding of RS(7, 5) coded 8PSK for three different shadowed Rician fading channels.

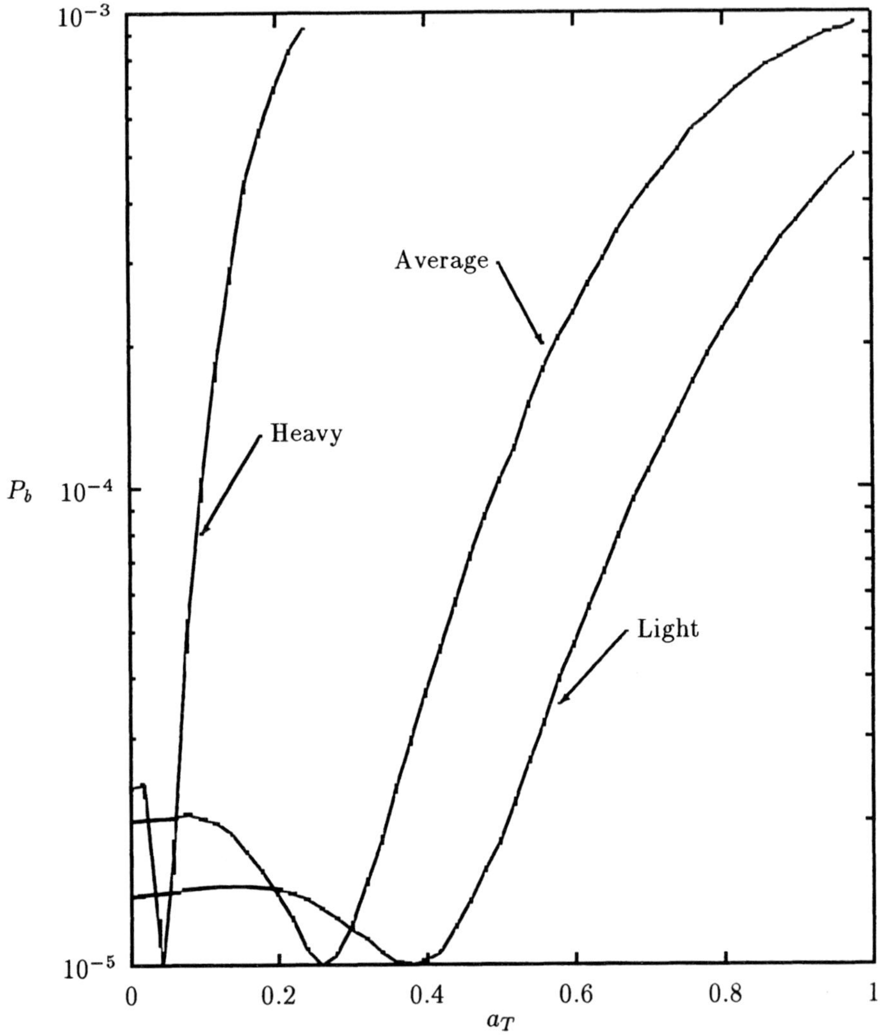


Figure 9.8: Bit error probability of RS(7, 5) coded 8PSK as a function of erasure threshold, $a_T$, for different shadowed Rician fading channels.

In order to study the effect of the erasure threshold $a_T$ on bit error probability, this probability is evaluated for different threshold values. The signal-to-noise ratio is chosen to yield $P_b = 10^{-5}$ for optimum threshold decoding. Figure 9.8 represents the bit error probability as a function of the threshold value $a_T$ for different degrees of shadowing. It is observed that for values below and above the optimum threshold the bit error probability increases. This can be explained as follows. For threshold values below the optimum one, the decoder ignores too many erroneous symbols as erasures and hence $P_b$ increases. On the other hand, for values above the optimum threshold the decoder erases too many correct symbols which again increases $P_b$.

It is interesting to note that the performance improvement due to errors-and-erasure decoding depends on the degree of shadowing. This can be seen from Figure 9.8. In this figure the intersection of the curves with $a_T = 0$ denotes the bit error probability corresponding to the errors-only decoding. It is observed that this value becomes larger when the degree of shadowing increases. Hence, the errors-and-erasures decoding in heavy shadowed channels outperforms the errors-only decoding.

## 9.4 THE EFFECT OF FADING BANDWIDTH ON THE PERFORMANCE OF THE CODED SCHEMES

The preceding sections have been concerned with the evaluation of the performance of coded schemes in time-varying shadowed Rician fading channels under the assumption of ideal interleaving/de-interleaving. The present section will be concerned with the effect of limited interleaving/de-interleaving. In order to study performance degradation of coded schemes due to imperfect interleaving, we first change the normalized bandwidth of the fading process. With this approach the effect of interleaving is taken into account by assuming that the results of higher fading bandwidth are equivalent to those when the actual fading bandwidth is low and interleaving is utilized.

In this case the performance is not amenable to analytical evaluation so that only computer simulations have been used in the study. Before presenting simulation results for some examples of coded schemes a brief description of the simulation model is required.

To simulate a shadowed Rician channel with limited fading bandwidth the components of the line-of-sight and multipath fading are filtered to provide the desired Doppler bandwidth.

The block diagram of the simulated channel is shown in Figure 1.4. The complex channel $\underline{a} = a\, e^{j\theta}$ is generated by three independent Gaussian Random Variables (GRV). These random variables are filtered by a third order Butterworth filter of 3 dB normalized bandwidth $BT_s$, where $B$ is the Doppler frequency of the channel and $T_s$ denotes the symbol interval. The parameters $\sigma_a^2$, $m_\gamma$, and $\sigma_\gamma^2$ determine the degree of shadowing. In this study we use light shadowed channels with parameters as given in Table 1.1. The parameter $\rho$ depends on the bandwidth of the fading process and its value for different normalized bandwidth $BT_s$ is given in Table 9.1 [53].

**Example 9.6** The effects of fading bandwidth on the performance of two rate 2/3 trellis coded 8PSK schemes, namely the 4-state scheme without parallel paths (4N) (see Figure 5.6), and the Ungerboeck 8-state (8U) (see Figure 4.8), on light shadowed Rician fading channels are shown in Figure 9.9 and Figure 9.10, respectively.

**Example 9.7** The effects of fading bandwidth on the performance of RS(7, 5) coded 8PSK and RS(63, 42) coded 8PSK schemes, on light shadowed Rician fading channels are shown in Figures 9.11, 9.12, and 9.13. For RS(7, 5) coded 8PSK results are given for both hard-decision decoding (Figure 9.11) and soft-decision decoding (Figure 9.12) while for RS(63, 42) coded 8PSK only errors-only decoding (Figure 9.13) is considered.

As one would expect the performance of coded schemes gets better as the fading bandwidth increases, that is, when channel errors become less bursty. Indeed, for small normalized fading bandwidths the fading is slow enough to be considered constant over a number of symbol intervals in a decoding span. Using the *slow-fading* model the pairwise error probability, given in (9.8), is written as

$$P_2(\mathbf{s}, \hat{\mathbf{s}}) \leq E_a \left[ \exp(-\frac{a^2}{4N_0} |\mathbf{s} - \hat{\mathbf{s}}|^2) \right],$$

which can be expressed as

$$P_2(\mathbf{s}, \hat{\mathbf{s}}) \leq \frac{1}{\sqrt{2\pi\sigma_a^2}\, d_E^2(\mathbf{s},\, \hat{\mathbf{s}})(\frac{1}{4N_0})} \int_{-\infty}^{\infty} e^{-t^2 - \frac{1}{\sigma_a^2} e^{2(\sqrt{2}\sigma_\gamma t + m_\gamma)}} dt,$$

where $d_E^2(\mathbf{s}, \hat{\mathbf{s}}) \equiv |\mathbf{s} - \hat{\mathbf{s}}|^2$ shows the squared Euclidean distance between $\mathbf{s}$ and $\hat{\mathbf{s}}$. This equation shows that the pairwise error probability varies inversely with $1/N_0$, and hence, the coded scheme only provides time diversity of order one. This contrasts with time-varying fading where the time diversity of the coded schemes depends upon the length of the shortest error event path. Considering that the uncoded schemes also provide time diversity of order one, we may find that the coded schemes on slow-fading channels perform worse than uncoded ones (see, for example, Figures 9.9, 9.10, 9.11). To get the benefit of coding in fading channels with memory, therefore, the use of sufficient interleaving is required.

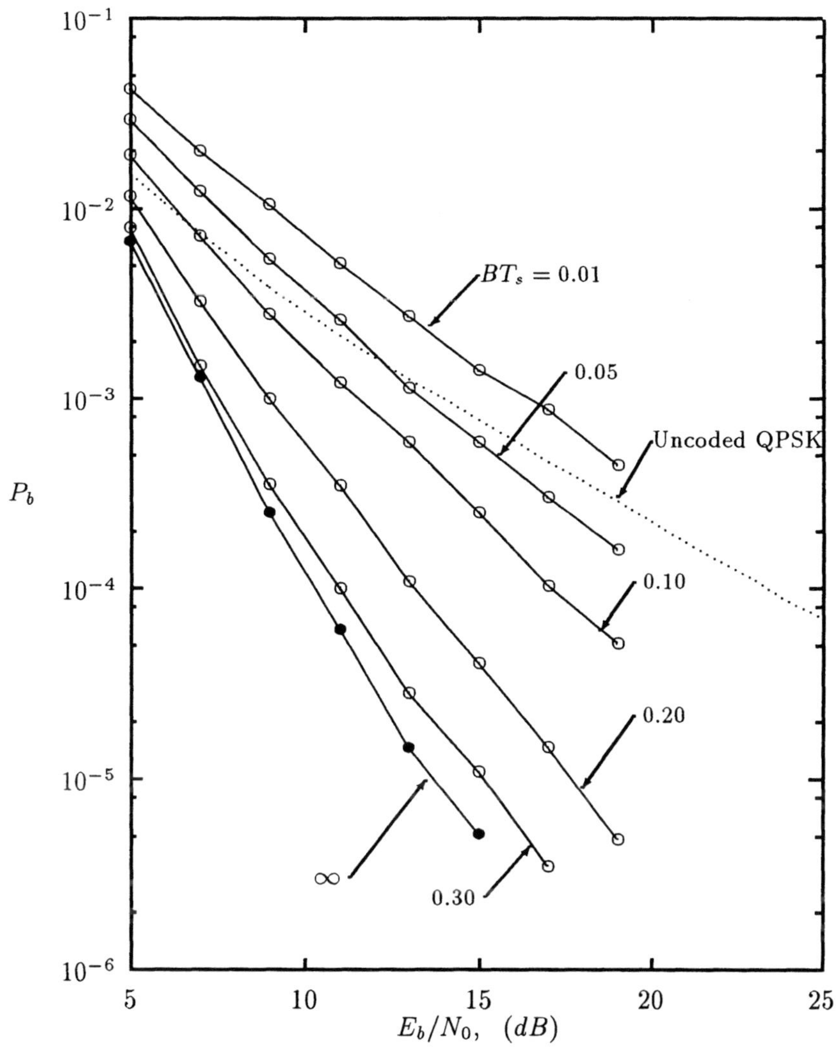


Figure 9.9: Effect of fading bandwidth on the performance of the 4-state TCM scheme without parallel paths (4N) on light shadowed Rician fading channels.

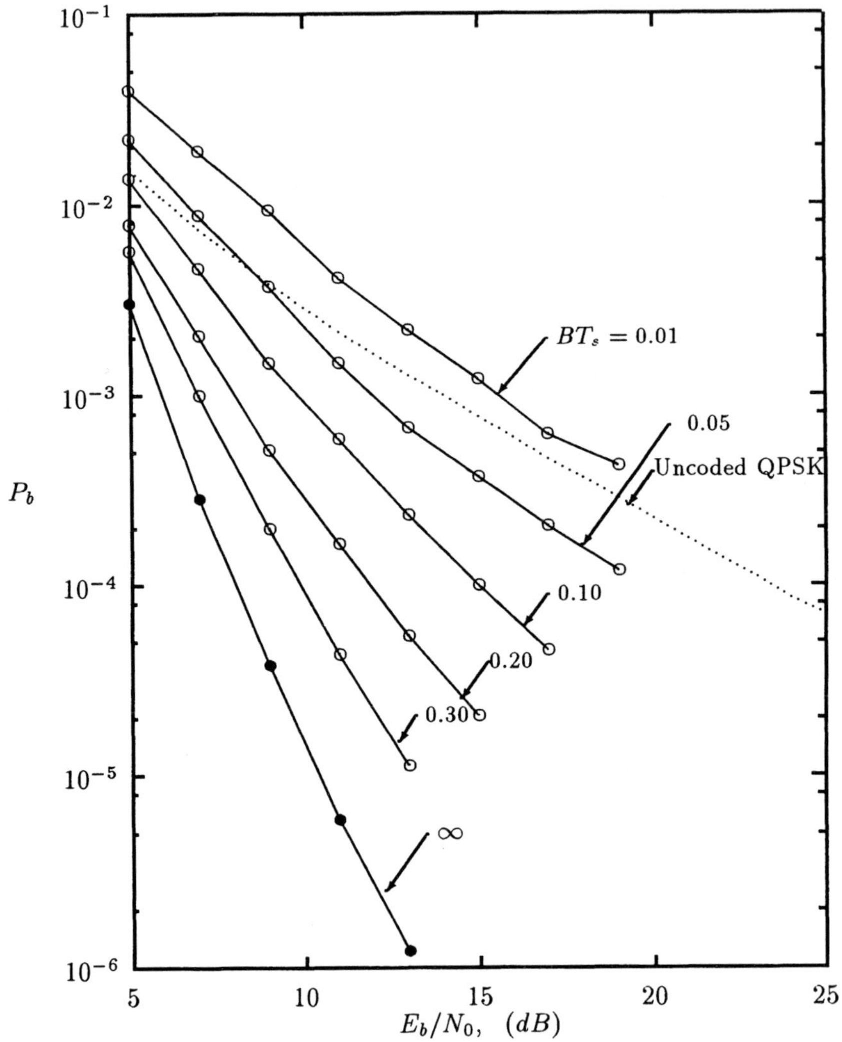


Figure 9.10: Effect of normalized fading bandwidth on the performance of the 8-state TCM scheme (8U) on light shadowed Rician fading channels.

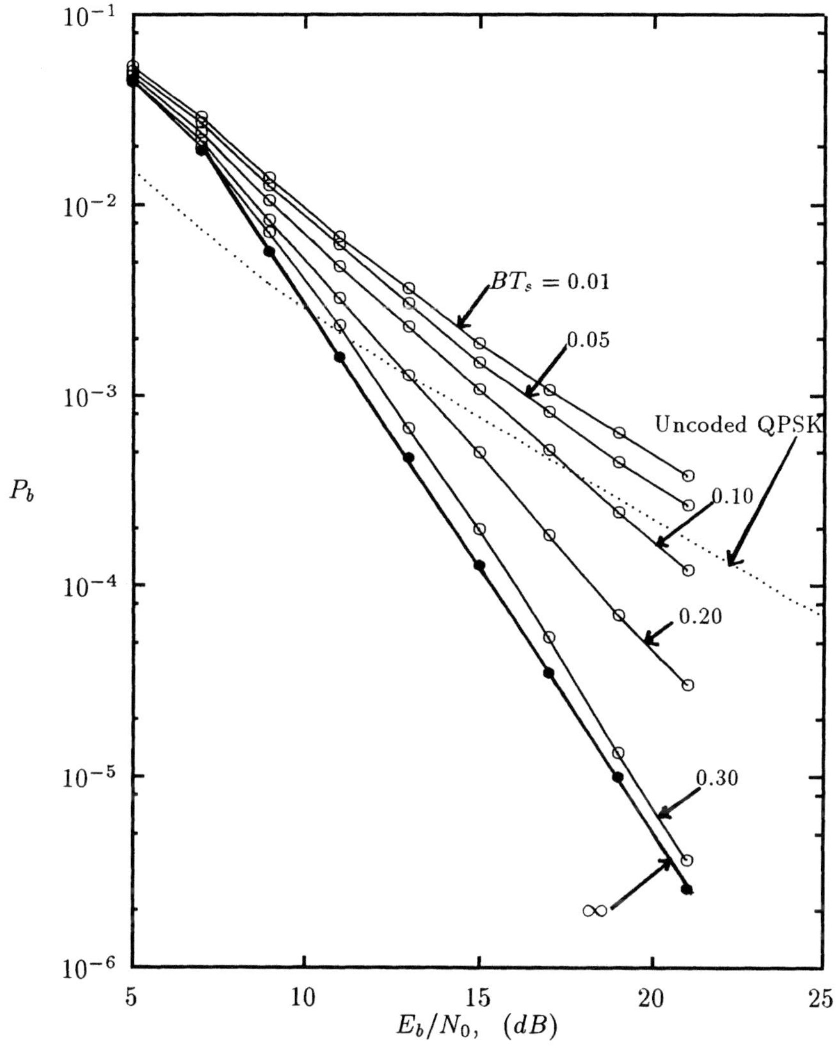


Figure 9.11: Effect of normalized fading bandwidth on the performance of RS(7,5)-coded 8PSK in light shadowed Rician fading channels using errors-only decoding.

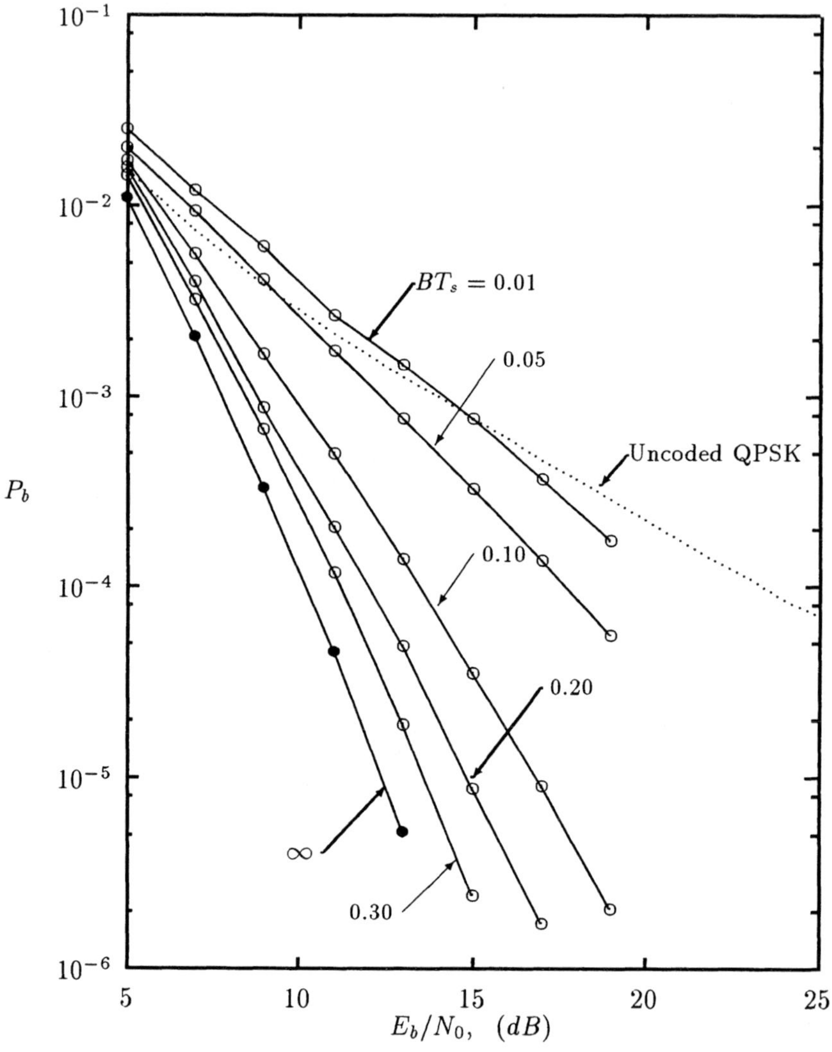


Figure 9.12: Effect of normalized fading bandwidth on the performance of RS(7, 5) coded 8PSK on light shadowed Rician fading channels using Viterbi decoding.

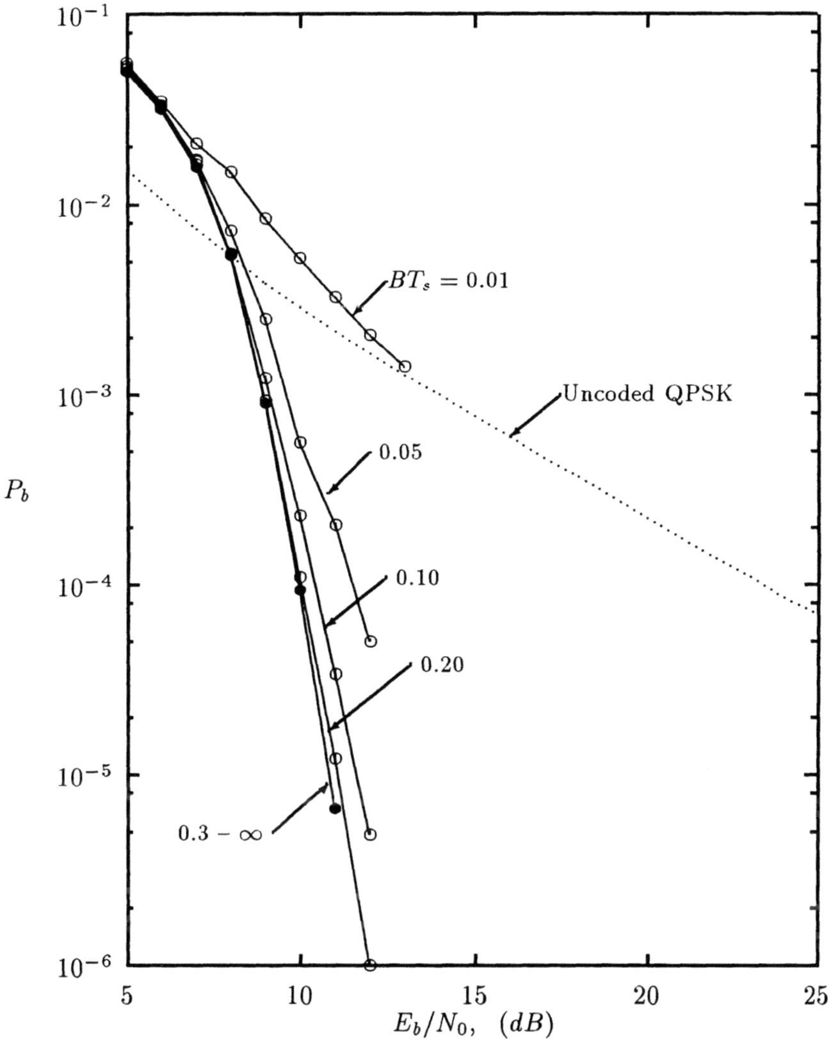


Figure 9.13: Effect of normalized fading bandwidth on the performance of RS(63, 42) coded 8PSK on light shadowed Rician fading channels using errors-only decoding.

Table 9.1: The value of $\rho$ for different normalized bandwidths.

| $BT_s$ | 0.30 | 0.20 | 0.10 | 0.05 | 0.01 |
|---|---|---|---|---|---|
| $\rho$ | 0.978 | 0.701 | 0.466 | 0.324 | 0.145 |

It is observed that the performance degradation of RS(63, 42) coded 8PSK is much less than that of RS(7, 5) coded 8PSK. For the former code, with $BT_s \geq 0.2$, there is little difference from the limiting random channel, i.e., $BT_s = \infty$. One might expect this by considering the high error correction capability of this code. Also, because of its structure this coded scheme treats burst errors of two channel symbols as one error. Also of interest is the difference in performance degradation of the RS(7, 5) coded 8PSK scheme for different decoding strategies. The sensitivity of the Viterbi decoding to variation of fading bandwidth (imperfect interleaving) is more than the errors-only decoding technique.

Here we compare the effect of the fading bandwidth on TCM and RS coded schemes. The average bit error probability, for different coded schemes, in terms of the normalized fading bandwidth is plotted in Figure 9.14. These results are given for $E_b/N_0 = 10$ dB. It is observed that, in general, RS coded MPSK schemes are less sensitive to variation of fading bandwidth than TCM schemes. For example, for RS(63, 42) coded 8PSK the bit error rate degradation for the fading bandwidth range of 0.1-0.3 is less than 2 times, while this degradation for an 8-state TCM is about 10 times.

The performance degradation of TCM schemes is due to the sensitivity of the Viterbi decoding to bursty errors. The performance degradation of this algorithm is due to two problems. First, error propagation occurs more frequently when the fading bandwidth is small and the channel creates burst errors. Second, the metric becomes non-optimum when samples of the fading process are correlated. Recall that in applying the Viterbi algorithm it is assumed that the channel is memoryless, therefore, the maximum likelihood rule results in an equivalent Euclidean distance as an additive metric. For a channel with memory the use of such a metric may not be the optimum choice. Note that in Viterbi decoding of RS codes, since each block is decoded independently, the error propagation does not exist, while the second problem can be considered to be a source of performance degradation.

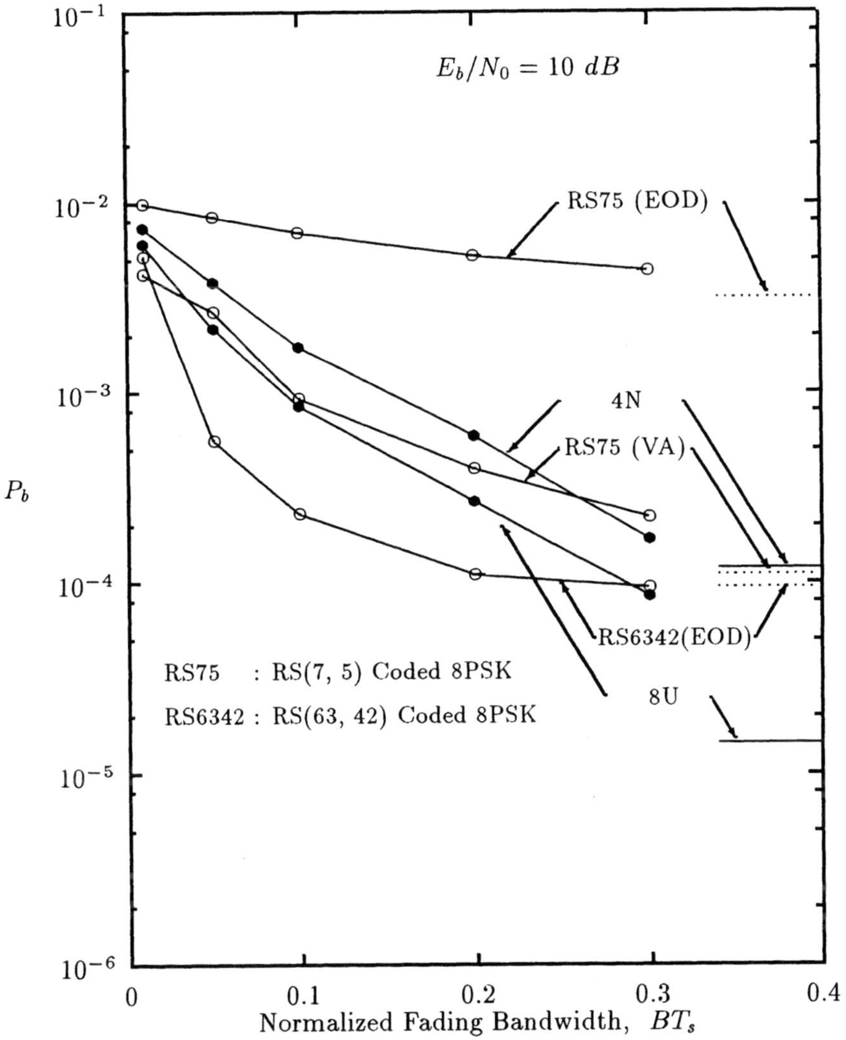


Figure 9.14: Performance of different coded schemes versus normalized fading bandwidth on a light shadowed Rician fading channel without interleaving.

## 9.5 THE PERFORMANCE OF THE INTERLEAVED CODED SCHEMES ON A SLOW SHADOWED RICIAN FADING CHANNEL

In the preceding section, in order to study the effect of imperfect interleaving, we followed the approach of taking the increased fading bandwidth as an effect caused by interleaving[124]. This approach does not explicitly reveal the role of interleaver size. Using computer simulations, the effect of interleaver size on the performance of coded schemes on a slow shadowed Rician fading channel is evaluated in the following example.

**Example 9.8** In this example we evaluate the performance of the 4-state 8PSK TCM scheme without parallel paths (4N) (see Figure 5.6) on a light shadowed Rician fading channel using a block interleaver. The block interleaver is modeled as a buffer with $d$ rows (depth of interleaving) and $s$ columns (span of interleaving). Thus, the size of interleaver (in symbols) is $d \times s$. Data are written into the buffer in successive rows and read out of the buffer in columns. At the receiver, the block de-interleaver performs the reverse operation, i.e., the received soft quantized symbols are written into the buffer in successive columns and read out in rows. Ideally, the interleaver buffer size should be chosen as infinite. However, in practice, constraint on allowable signal time delay limits the buffer size of the interleaver/de-interleaver. In this example the depth and the span of interleaving are chosen as 16 and 8 symbols, respectively.

The simulation results for the interleaved 4N TCM scheme on a light shadowed Rician fading channel with a normalized fading bandwidth $BT_s = 0.05$ are shown in Figure 9.15. Simulation results for ideal interleaving and no interleaving cases are also included for comparison.

Another interleaving technique which is well suited for trellis coded modulation schemes is convolutional interleaving. Such a technique may be implemented using the system shown in Figure 9.16. The convolutional interleaver consists of a bank of $I$ shift registers with an increasing length. The coded symbols out of the trellis encoder are inserted into the shift registers of the interleaver bank through a commutator. The output commutator transmits the contents of the interleaver shift registers through a discrete channel. The input and output commutators of the interleaver operate synchronously and return to the 0th shift register after $I$th time. The convolutional de-interleaver, which must invert the operation of the interleaver, is clearly just its converse. The de-interleaver consists of the same bank of I shift registers, with the reverse order, and two commutators which operate synchronously with those of the interleaver.

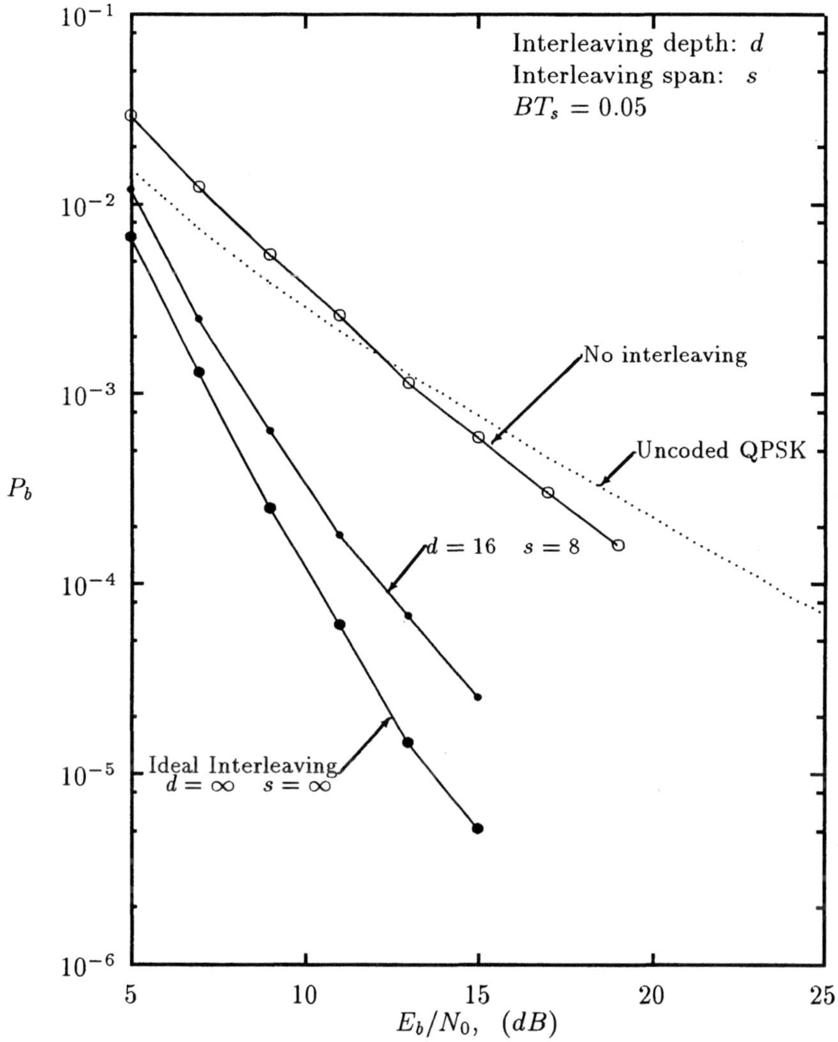


Figure 9.15: Performance of the 4-state 8PSK TCM (4N) scheme on a light shadowed Rician fading channel with block interleaver.

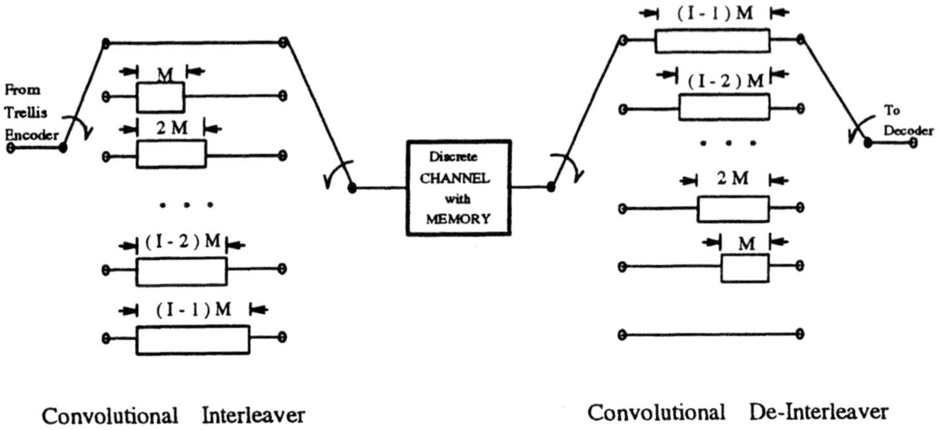


Figure 9.16: Structure of a convolutional interleaver and de-interleaver.

The interleaving degree for a convolutional interleaver is defined as $\mu \equiv M \times I$, where $M$ and $I$ are parameters of the interleaver depicted in Figure 9.16. Note that the delay introduced by this interleaving technique is equal to $MI(I-1)$ which increases proportionally with interleaving degree $\mu$. Although the performance improves with increasing interleaving degree, the time delay also increases. An appropriate value for $\mu$ is often chosen after considering the delay constraint of the communication system.

**Example 9.9** In this example the performance of the convolutionally interleaved 8-state 8PSK TCM (8U) (see Figure 4.8) is evaluated on the light shadowed Rician fading channel with a normalized fading bandwidth $BT_s = 0.05$ [125]. In this evaluation, the parameter M is set to 1 and the interleaving degree $\mu$ is varied. Figure 9.17 shows the probability of bit error for different values of interleaving degree. Bit error rates for cases with ideal interleaving and without interleaving are also included in this figure for comparison.

# 9.6 BIBLIOGRAPHICAL NOTES

The performance of coherent and differentially coherent 8PSK trellis coded schemes was presented by McLane et. al. [124] for a class of fast fading shadowed Rician fading channel through computer simulations. Later Lee and McLane [125] studied the effect of imperfect interleaving, using a computer simulation-based performance analysis, for convolutionally interleaved 8PSK TCM schemes. The effect of interleaver size on the performance of some 8PSK and 8DPSK schemes is reported in this reference. The upper bound on the probability of bit error of TCM schemes in a shadowed Rician fading channel for the case of coherent detection and ideal interleaving was presented by McKay et.

al. [65]. Huang and Campbell [103] evaluated the performance of trellis coded multilevel differential phase shift keying (MDPSK) signals in correlated and shadowed Rician fading channels by means of upper bound analysis. The error bounds of RS-coded schemes in shadowed Rician fading channels for coherent detection [126] and differentially coherent detection [127] were presented by Jamali and Le-Ngoc.

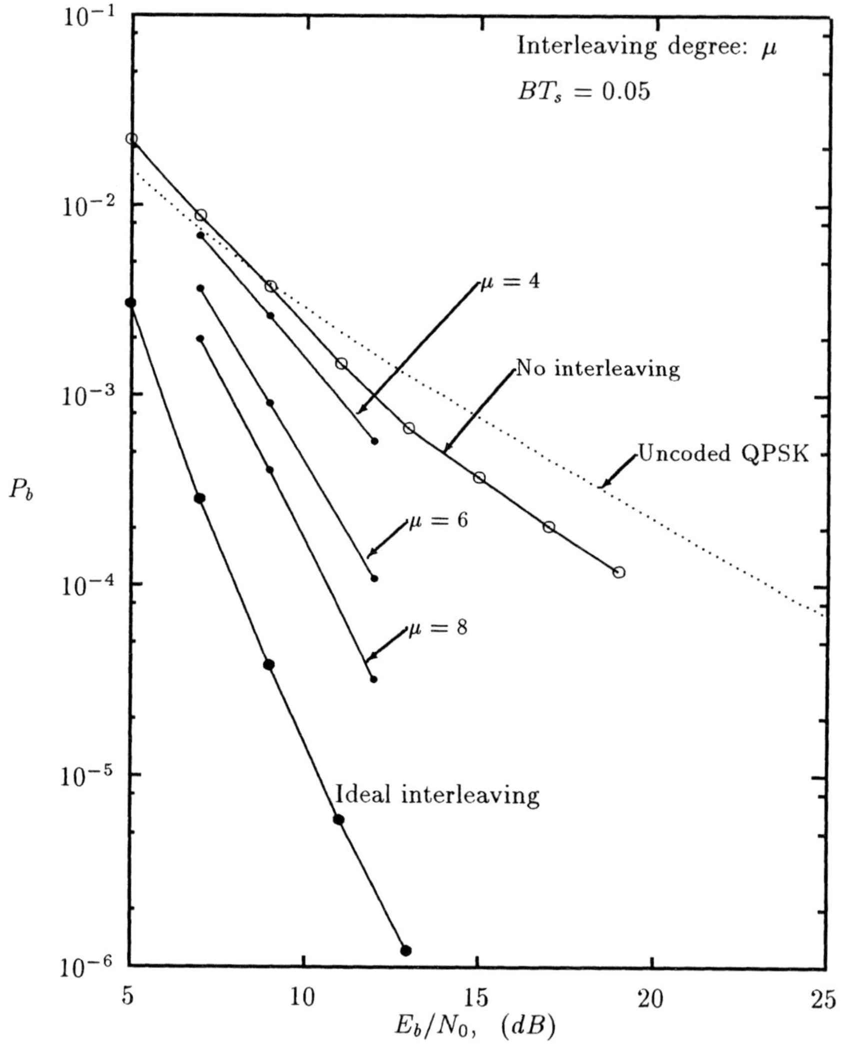


Figure 9.17: Performance of the 8-state 8PSK TCM scheme (8U) on a light shadowed Rician fading channel with convolutional interleaver.

# Problems

**9.1** Complete the derivations for (9.12), (9.13), and (9.14).

**9.2** a) Perform numerical calculations and plot $f(\delta_i^2)$ versus $E_b/N_0$ using (9.12) and (9.14).

b) Plot $f(\delta_i^2)$ versus $E_b/N_0$ using (9.15). Compare the results with those obtained in a).

c) Compare your results with those in Figure P9.2-1

**9.3** Following the procedure given in Example 9.4, derive the upper bound on the bit error probability of the 4-state, rate 2/3 8PSK TCM (WL) scheme (Figure 5.3) in a shadowed Rician fading channel. Plot the upper bound versus $E_b/N_0$ for light, average and heavy shadowing conditions (see Table 1.1).

**9.4** Repeat Problem 9.3 for $\sigma_a^2 = 0.1$, $m_\gamma = 0$ and $\sigma_\gamma = 0.1$.

**9.5** Repeat Problem 9.3 for the 16-state, rate 2/3 8PSK TCM scheme constructed by Schlegel and Costello (Table 5.3).

**9.6** Repeat Problem 9.3 for the 2-state, rate 2/4 QPSK MTCM scheme discussed in Problem 5.16b).

**9.7** Repeat Problem 9.3 for the 4-state, rate 2/4 QPSK MTCM scheme discussed in Problem 5.16c).

**9.8** From (9.16) and (9.15) derive (9.20) and (9.19) for $2d_i^2\sigma_a^2 \gg 1$.

**9.9** Compute the upper bound on the bit error probability of the extended RS(8,6) coded 8PSK scheme using EOD, EED and soft-decision decoding techniques in a shadowed Rician fading channel. Plot the upper bound versus $E_b/N_0$ for various degrees of shadowing (see Table 1.1).

**9.10** Repeat Problem 9.9 with extended RS(8,4) coded 8PSK scheme.

**9.11** Using computer simulations, evaluate the effect of fading bandwidth on the bit error probability of the 4-state, rate 2/3 8PSK TCM scheme constructed by Du and Vucetic (Table 5.4). Use the simulation channel model of Figure 1.4 with $BT_s = 0.01, 0.10, 0.30$ for a light shadowed Rician fading channel without interleaving. Plot the bit error probability versus $E_b/N_0$. Discuss the results.

**9.12** Repeat Problem 9.11 for the 16-state, rate 2/3 8PSK TCM scheme constructed by Schlegel and Costello (Table 5.3).

**9.13** Repeat Problem 9.11 with the extended RS(8,6) coded 8PSK scheme using EOD and SEMDD decoding techniques.

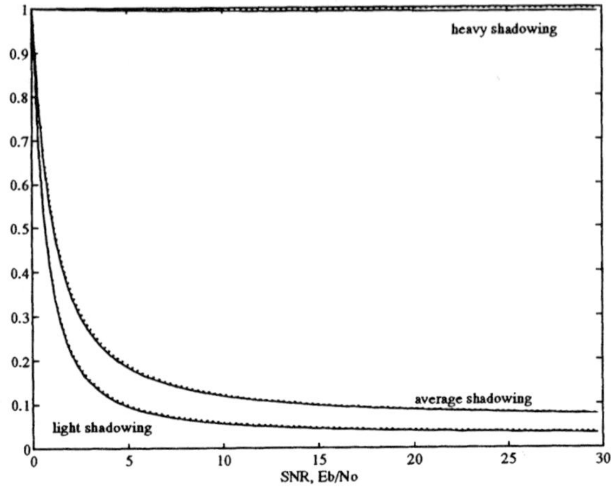


Figure P9.2-1: $f(\delta_i^2)$ versus $E_b/N_0$ for various shadowing conditions (Table 1.1) using (9.15) (dashed line) and (9.12) (solid line).

**9.14** Repeat Problem 9.13 with the extended RS(8, 4) coded 8PSK scheme.

**9.15** Using computer simulations, evaluate the effect of interleaver size on the bit error probability of the 4-state, rate 2/3 8PSK TCM scheme constructed by Du and Vucetic (Table 5.4). Use the simulation channel model of Figure 1.4 with $BT_s = 0.05$ for a light shadowed Rician fading channel. Plot the bit error probability versus $E_b/N_0$ for three cases: without interleaving, with ideal interleaving and with ($d = 16, s = 4$) block interleaver.

**9.16** Repeat Problem 9.15 with the shortened RS(6, 4) coded 8PSK scheme using Viterbi decoder.

**9.17** Repeat Problem 9.15 with the 16-state, rate 2/3 8PSK TCM scheme constructed by Schlegel and Costello (Table 5.3).

**9.18** Repeat Problem 9.15 with the extended RS(8, 6) coded 8PSK scheme using Viterbi decoder.

# Chapter 10

# QUADRATURE PULSE-OVERLAPPING MODULATION TECHNIQUES

With a special contribution by Dr. S. BEN SLIMANE

Concordia University, Canada

Pulse shaping has been widely used to provide modulated signals with compact spectra. The bandwidth of a modulated signal can be reduced by the use of multiple-interval shaping pulses. Due to pulse overlapping, intersymbol interference (ISI) is introduced in the baseband components of the modulated signal. This controlled ISI can be properly used in the design of the optimum receiver based on maximum likelihood sequence estimation (MLSE). In this chapter, we present the structure, properties and performance of Quadrature Pulse-Overlapping Modulation (QPOM) schemes over AWGN and fading channels. The QPOM is decomposed into a linear convolutional encoder followed by a memoryless one-to-one mapper (modulator). The trellis diagram representing its inherent, non-redundant coding structure is used to construct its MLSE receiver. Results obtained from the previous chapters are used to analyze the performance of QPOM schemes. Variants of QPOM schemes with good spectral properties suitable to applications using nonlinear or saturated power amplifiers are also introduced.

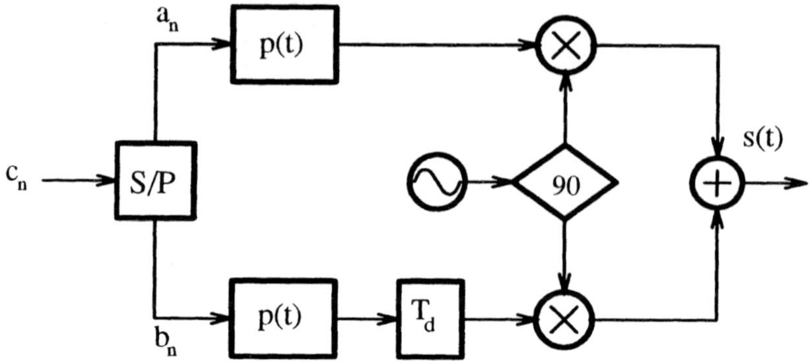


Figure 10.1: Block diagram of a quadrature amplitude modulation scheme.

# 10.1 QUADRATURE PULSE OVERLAPPING MODULATION

The Quadrature Pulse-Overlapping Modulation (QPOM) scheme can be represented by an encoder structure similar to that of TCM schemes. Using this representation, an optimum receiver using MLSE is constructed for QPOM signals. This representation also facilitates the performance analysis of the QPOM schemes.

## 10.1.1 Generation and Representation of QPOM Signals

Figure 10.1 shows the block diagram of the QPOM scheme. The input binary data stream with a bit rate of $1/T_b$ is demultiplexed into two data streams $\{a_n\}$ and $\{b_n\}$ by the serial-to-parallel (S/P) converter. Each data stream with a symbol rate of $1/2T_b$ is passed through a lowpass filter with an impulse response $p(t)$. $T_d$ represents the time offset between the in-phase and quadrature components. For $T_d = 0$, the QPOM signal $s(t)$ is represented by

$$s(t) = s_I(t) \cos w_c t + s_Q(t) \sin w_c t, \tag{10.1}$$

where $w_c$ is the angular frequency, $s_I(t)$ and $s_Q(t)$ are the in-phase and quadrature components of $s(t)$, respectively. They are represented as

$$s_I(t) = \sum_{n=-\infty}^{+\infty} A_n p(t - nT)$$

$$s_Q(t) = \sum_{n=-\infty}^{+\infty} B_n p(t - nT),$$

where

$$A_n = 2a_n - 1, \quad a_n = 0 \text{ or } 1 \text{ with probability of } 1/2,$$
$$B_n = 2b_n - 1, \quad b_n = 0 \text{ or } 1 \text{ with probability of } 1/2,$$

$T$ is the symbol interval, $T = 2T_b$.

It can be shown that the spectrum of the QPOM signal has the same shape of $|P(f)|^2$ where $P(f)$ is the Fourier transform of $p(t)$. In this chapter we consider a double-interval overlapping pulse $p(t)$ with the following characteristics:

$$\begin{array}{lll} p(t) = 0, & |t| \geq T & \text{(double-interval pulse)} \\ p(t) = p(-t), & |t| \leq T & \text{(symmetry).} \end{array} \tag{10.2}$$

**Example 10.1** Since the spectrum of the QPOM signal has the same shape as that of $|P(f)|^2$, QPOM signals with compact spectra can be generated by a proper selection of $p(t)$. Consider a class of double-interval pulse shapes $p(t)$ defined as [132]

$$p(t) = \begin{cases} [\cos(\pi t/2T)]^{M+1} & |t| \leq T \\ 0, & \text{elsewhere.} \end{cases} \tag{10.3}$$

As we can see this class of pulse shapes has $M$ continuous derivatives over the interval $]-\infty, +\infty[$. Hence its Fourier transform $P(f)$ decays asymptotically as $|f|^{M+2}$ and its power spectral density $|P(f)|^2$ as $|f|^{-2(M+2)}$ [144].

The power spectral density of the QPOM signal using this class of pulse shapes is derived in [132], and is given by

$$S_s(x) = \begin{cases} \left|\frac{\sin(2\pi x)}{2\pi x}\prod_{i=1}^{(M+1)/2}\frac{i^2}{i^2-4x^2}\right|^2, & \text{odd } M \\ \left|\cos(2\pi x)\prod_{i=1}^{1+M/2}\frac{(2i-1)^2}{(2i-1)^2-16x^2}\right|^2, & \text{even } M, \end{cases} \tag{10.4}$$

where $x = (f - f_0)T$ and $S_s(0)$ is normalized to one. Figure 10.2 shows the normalized power spectral density of a number of double-interval overlapping pulse shapes, $p(t)$. This figure indicates that as $M$ increases, the spectral roll-off of the corresponding QPOM signal becomes faster for high frequencies while the main lobe becomes wider. As a result, the choice of the QPOM signal depends on the trade-off between main lobe occupancy and side lobes roll-off. In many practical applications, a spectral density of more than 40 dB below the highest spectral density (near 0 Hz) is considered to be negligible. For this reason, pulse shapes corresponding to $M = 0$ and $M = 1$ are more attractive. It is also seen that the spectrum is improved at all frequencies compared to QPSK.

The double-interval pulse shapes $p_0(t)$, $p_1(t)$ corresponding to $M = 0$ and $M = 1$ are given by

$$p_0(t) = \begin{cases} \cos(\pi t/2T), & |t| \leq T \\ 0, & \text{elsewhere.} \end{cases} \tag{10.5}$$

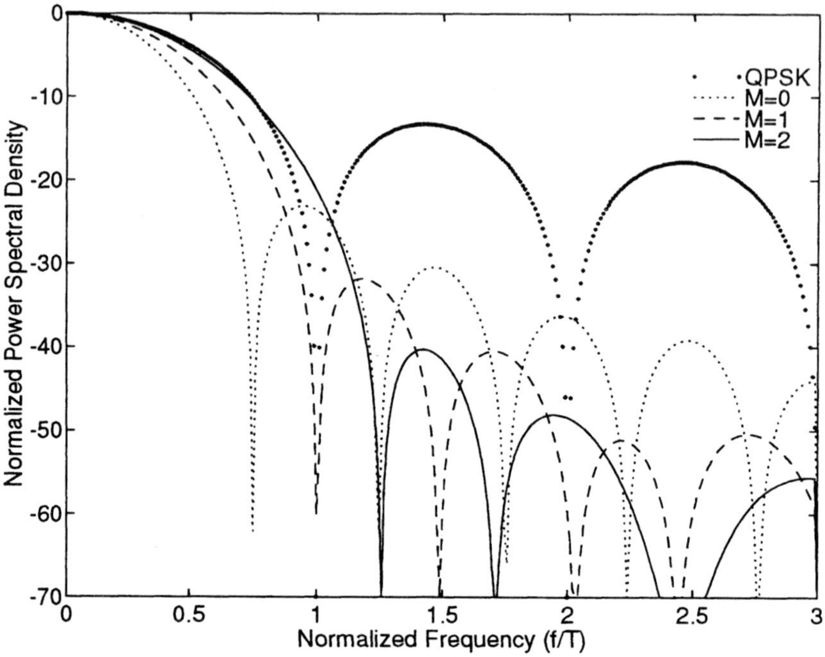


Figure 10.2: Normalized power spectral density of QPOM signals with double-interval pulse $p(t) = [\cos(\pi t/2T)]^{M+1}$.

$$p_1(t) = \begin{cases} (1+\cos(\pi t/T))^2, & |t| \leq T \\ 0, & \text{elsewhere.} \end{cases} \tag{10.6}$$

As shown in Figure 10.2, the first null frequency of $p_0(t)$ is at $0.75/T$, while that of $p_1(t)$ is at $1/T$. However, the asymptotic spectral roll-off of $p_0(t)$ is of the order $|f|^{-4}$ while that of $p_1(t)$ is of the order $|f|^{-6}$. Both show spectral advantages over conventional QPSK and MSK schemes. Note that $p_1(t)$ is also called overlapped raised-cosine pulse shape and it is used in the Quadrature Overlapped Raised-Cosine (QORC) modulation scheme [128].

**Example 10.2** Amoroso [130] presented a class of quasi-band-limited double-interval pulse shapes $p(t)$. This class of pulse shapes is defined as

$$p(t) = \begin{cases} \left(\frac{\sin(\pi t/T)}{\pi t/T}\right)^M, & |t| \leq T \\ 0, & \text{elsewhere.} \end{cases} \tag{10.7}$$

These pulse shapes are more and more band-limited by increasing $M$ because the truncated energy outside the interval $[-T,\ +T]$ is a rapidly decreasing

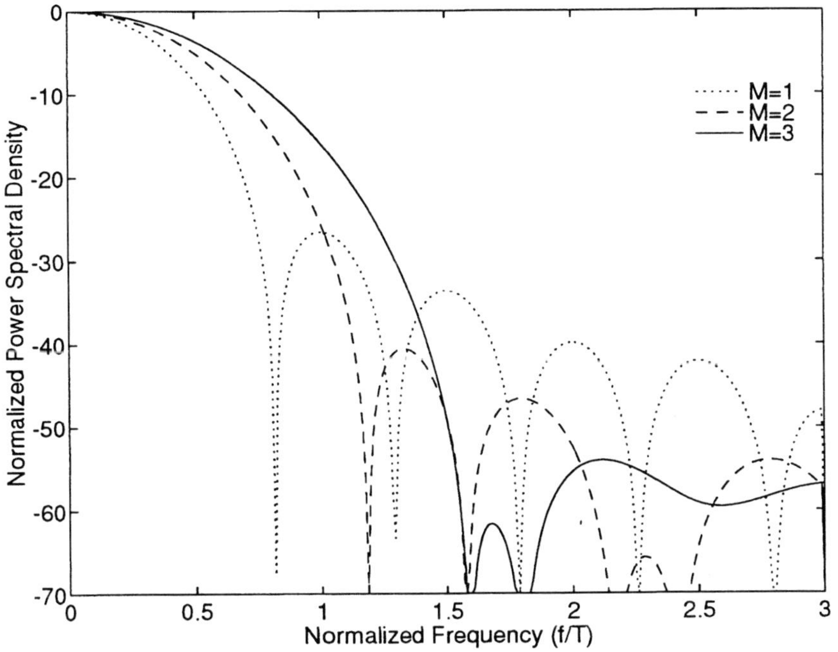


Figure 10.3: Normalized power spectral density of QPOM signals with double-interval pulse $p(t) = [\sin(\pi t/T)/(\pi t/T)]^M$.

function of $M$. This can be seen by noting that the unlimited pulse shape $(\sin \pi t/T)/(\pi t/T)$ is strictly band-limited.

The power spectral density of the QPOM signal using some of these pulse shapes is shown in Figure 10.3. As seen from the figure, good spectral properties are obtained with this kind of pulse shape. Also by increasing $M$, very low side lobes are obtained.

The double-interval pulse $p(t)$ introduces intersymbol interference to the in-phase and quadrature components $s_I(t)$ and $s_Q(t)$. Over one symbol interval, $nT \leq t \leq (n+1)T$, the two components can be rewritten as

$$s_I(t) = A_n p(t - nT) + A_{n-1} p(t - (n-1)T) \tag{10.8}$$

$$s_Q(t) = B_n p(t - nT) + B_{n-1} p(t - (n-1)T). \tag{10.9}$$

It is observed that in each symbol interval, the two components $s_I(t)$ and $s_Q(t)$

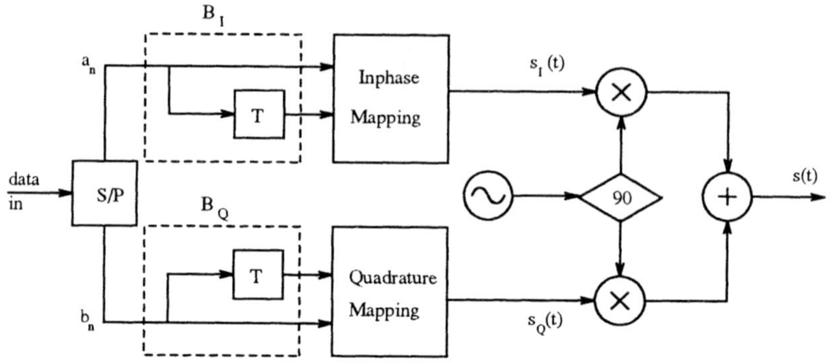


Figure 10.4: An alternative representation of linear QPOM schemes.

take one of four possible single-interval waveforms:

$$\begin{aligned} s_0(\tau) &= p(\tau) + p(\tau - T) \\ s_1(\tau) &= p(\tau) - p(\tau - T) \\ s_2(\tau) &= -p(\tau) + p(\tau - T) = -s_1(\tau) \\ s_3(\tau) &= -p(\tau) - p(\tau - T) = -s_0(\tau), \end{aligned} \qquad (10.10)$$

where $0 \leq \tau \leq T$ and $\tau = t - nT$.

The in-phase component $s_I(t)$ is a function of $(a_n,\ a_{n-1})$, and the quadrature component $s_Q(t)$ is a function of $(b_n,\ b_{n-1})$:

$$\begin{aligned} s_I(\tau) &= \mathcal{M}([a_n,\ a_{n-1}]) \\ s_Q(\tau) &= \mathcal{M}([b_n,\ b_{n-1}]), \end{aligned}$$

where

$$\begin{aligned} \mathcal{M}(00) &= s_3(\tau) = -s_0(\tau) \\ \mathcal{M}(01) &= s_2(\tau) = -s_1(\tau) \\ \mathcal{M}(10) &= +s_1(\tau) \\ \mathcal{M}(11) &= +s_0(\tau). \end{aligned}$$

As $\{a_n\}$ and $\{b_n\}$ are statistically independent, $s_I(t)$ and $s_Q(t)$ are also independent.

Therefore, the QPOM modulator can be modeled by an equivalent block diagram as shown in Figure 10.4. The two blocks $B_I$ and $B_Q$ are identical and resemble a convolutional encoder represented by the trellis diagram and state diagram shown in Figure 10.5. The states of this trellis diagram are represented by the contents of the shift registers of the encoder, and all possible outputs by the set $\mathbf{A} = \{00, 01, 10, 11\}$. This set is partitioned into two subsets $\mathbf{A}_0 = \{00, 10\}$ and $\mathbf{A}_1 = \{01, 11\}$. For the set of signals $s_i(\tau)$, we notice that

$$\int_0^T s_0(t)s_1(t)dt = \int_0^T [p^2(t) - p^2(t-T)]dt = 0.$$

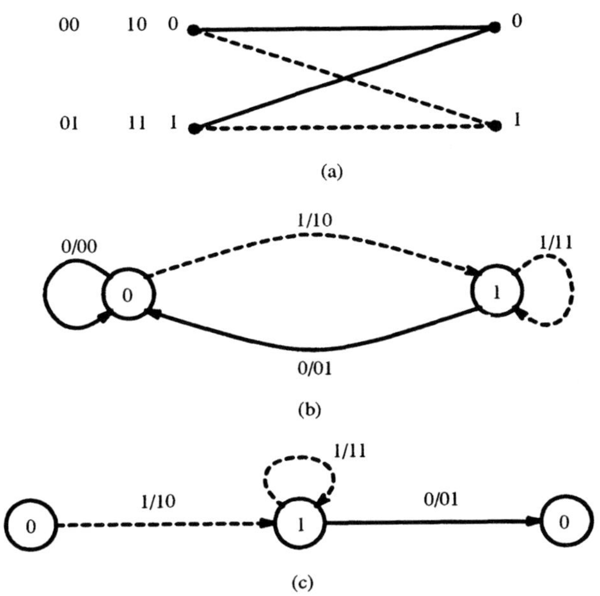


Figure 10.5: Diagrams representing the inphase (or quadrature) encoder of the linear QPOM scheme:
(a) trellis diagram, (b) state diagram, (c) modified state diagram.

This means that $s_0(\tau)$ and $s_1(\tau)$ are orthogonal. Note that $s_2(\tau)$ and $s_3(\tau)$ are antipodal signals to $s_0(\tau)$ and $s_1(\tau)$, respectively. Hence, a 2-dimensional orthonormal basis $\{\phi_j(\tau),\ j = 0,\ 1\}$ can be used to represent the *four* signals of (10.10). This orthonormal basis is easily obtained from $s_0(\tau)$ and $s_1(\tau)$ as

$$\phi_0(\tau) = \frac{s_0(\tau)}{\sqrt{E_0}}, \qquad E_0 = \int_0^T s_0^2(\tau)d\tau$$
$$\phi_1(\tau) = \frac{s_1(\tau)}{\sqrt{E_1}}, \qquad E_1 = \int_0^T s_1^2(\tau)d\tau,$$

where $E_i$ is the energy of the waveform $s_i(\tau)$. $E_i$ can be expressed in terms of the bit energy $E_b$ as (see Problem 10.1)

$$E_0 = (1+\gamma)E_b$$
$$E_1 = (1-\gamma)E_b,$$

where $\gamma$ is a coefficient depending on the double-interval overlapping pulse shape $p(t)$ and is given by

$$\gamma = \frac{2}{E_b}\int_0^T p(t)p(t-T)dt.$$

Once an orthonormal basis $\phi_j(\tau)$, $j = 0,\ 1$, has been adopted, each waveform $s_i(\tau)$, $i = 0,\ ...,\ 3$, is completely determined by its vector $\underline{s}_i$ in the signal space $(\phi_0(\tau),\ \phi_1(\tau))$ as shown in Figure 10.6, with

$$\underline{s}_i = (s_{i0},\ s_{i1}), \tag{10.11}$$

and

$$s_{ij} = \int_0^T s_i(\tau)\phi_j(\tau)d\tau, \qquad j = 0,\ 1.$$

Consequently, over one symbol interval, $nT \leq t \leq (n+1)T$, the transmitted QPOM signal can be represented by two vectors, $\underline{s}_I$ for its in-phase component and $\underline{s}_Q$ for its quadrature component.

### 10.1.2 MLSE Receiver for QPOM Signals and its Performance in an AGWN Environment

Figure 10.7 shows the MLSE receiver for QPOM signals. Consider the case when the QPOM signal $s(t)$ is transmitted over an Additive White Gaussian Noise (AWGN) channel. The received signal can be written as

$$r(t) = s(t) + n(t), \tag{10.12}$$

where $n(t)$ is the bandpass AWGN process represented by

$$n(t) = n_c(t)\cos 2\pi f_c t + n_s(t)\sin 2\pi f_c t, \tag{10.13}$$

and $n_c(t)$ and $n_s(t)$ are independent, identically distributed (i.i.d.) Gaussian random processes with zero mean and a variance of $N_0/2$. The received signal can be rewritten as

$$r(t) = [s_I(t) + n_c(t)]\cos 2\pi f_c t + [s_Q(t) + n_s(t)]\sin 2\pi f_c t.$$

The first task of an optimum receiver will be to extract the pair of waveforms, $(s_I(t), s_Q(t))$, by using a coherent quadrature demodulator. However, due to the noise introduced by the channel, the obtained pair of waveforms becomes $(r_I(t), r_Q(t))$, where

$$r_I(t) = s_I(t) + n_c(t)$$
$$r_Q(t) = s_Q(t) + n_s(t).$$

Since $s_I(t)$, $s_Q(t)$, $n_c(t)$ and $n_s(t)$ are statistically independent, $r_I(t)$ and $r_Q(t)$ are also statistically independent. Passed through the bank of two matched filters with impulse responses $\{\phi_j(T-t),\ j = 0, 1\}$, these components are converted into $\{\underline{r}_I\}$ and $\{\underline{r}_Q\}$, respectively, where

$$\underline{r}_I = \underline{s}_I + \underline{n}_I$$
$$\underline{r}_Q = \underline{s}_Q + \underline{n}_Q.$$

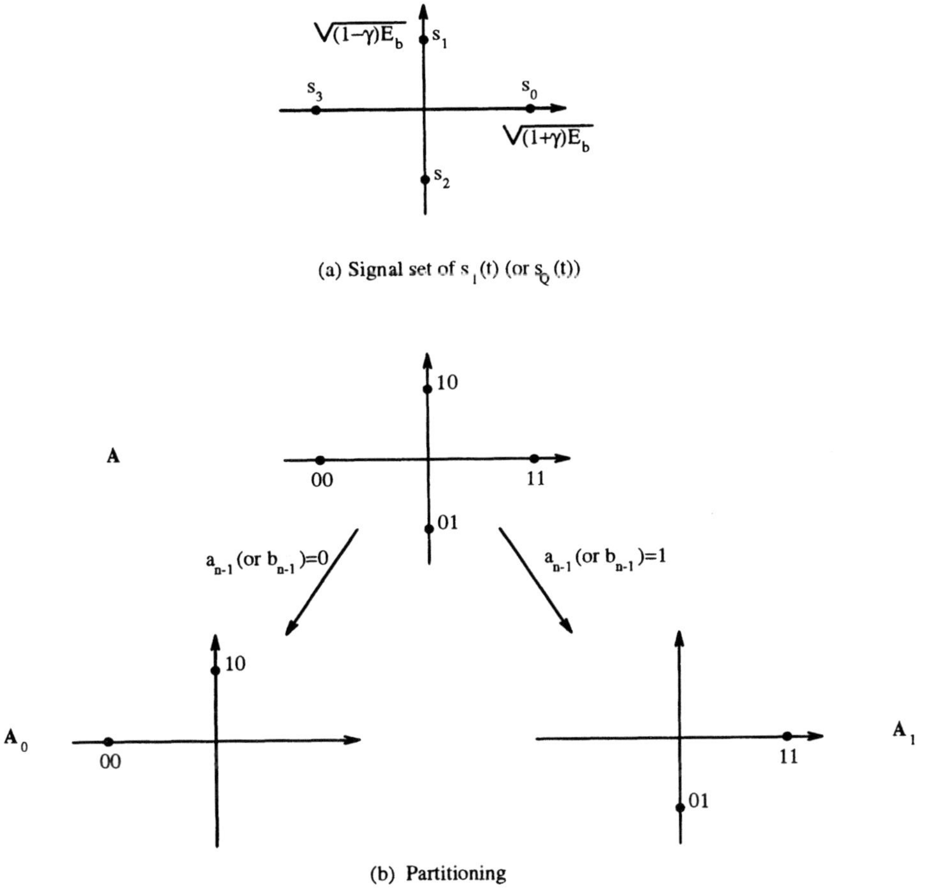


Figure 10.6: Representation of $s_I(t)$ (or $s_Q(t)$) in $\{\phi_0(t),\ \phi_1(t)\}$.

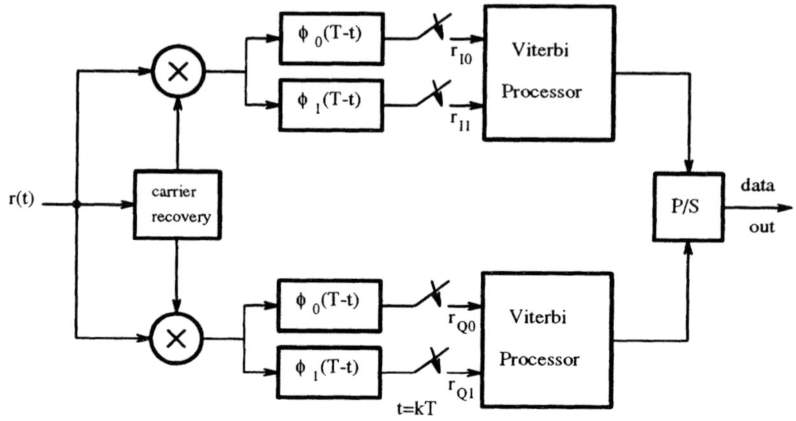


Figure 10.7: MLSE receiver of linear QPOM signals.

The elements of the vector $\underline{n}_I$ ($\underline{n}_Q$) are i.i.d. zero-mean Gaussian random variables with variance $N_0/2$. The sequences $\{\underline{r}_I\}$ and $\{\underline{r}_Q\}$ are passed through two independent and identical Viterbi processors which perform the maximum likelihood sequence estimation (MLSE) based on the trellis structure of the QPOM encoder.

Consider a segment of the transmitted QPOM signal. Its in-phase (or quadrature) component, $\{\underline{s}_I\}$ (or $\{\underline{s}_Q\}$), can be denoted as a sequence of $N$ vectors

$$\mathrm{s} = \{\underline{s}_0, \underline{s}_1, \cdots, \underline{s}_{N-1}\}.$$

The corresponding received inphase (or quadrature) component $\{\underline{r}_I\}$ (or $\{\underline{r}_Q\}$) is denoted by the received sequence of $N$ vectors

$$\mathbf{r} = \{\underline{r}_0, \underline{r}_1, \cdots, \underline{r}_{N-1}\}.$$

The MLSE decodes the received sequence into a transmitted sequence which maximizes the log-likelihood function $\ln p\{\underline{\mathbf{r}}|\underline{\mathrm{s}}\}$. As shown in Section 3.3, when the signals are transmitted with equal probability in an AWGN, the maximum likelihood decoder is equivalent to one which selects the transmit sequence s with the minimum Euclidean distance from the received sequence $\mathbf{r}$.

Using the same approach as that used for TCM in Chapter 4, we can derive the upper bound on the average bit error probability of the QPOM scheme in an AWGN environment with MLSE receiver.

From (4.20), the error probability of the inphase (or quadrature) component is upper bounded as

$$P_I(\text{or } P_Q) \leq \frac{1}{2}\mathrm{erfc}\left(\sqrt{\frac{\mathrm{d}^2_{\mathrm{free}}}{4\mathrm{N}_0}}\right) \mathrm{e}^{\mathrm{d}^2_{\mathrm{free}}/4\mathrm{N}_0} \frac{\partial \mathrm{T(D,I)}}{\partial \mathrm{I}}\Bigg|_{I=1,\, D=e^{-1/4N_0}} \quad (10.14)$$

where $d^2_{free}$ is the minimum squared Euclidean distance through the trellis diagram. Using Figures 10.5 and 10.6, we can prove that

$$d^2_{free} = |\underline{s}_1 - \underline{s}_3|^2 + |\underline{s}_2 - \underline{s}_3|^2 = 4E_b.$$

Furthermore, since the in-phase and quadrature components of the transmitted signal are statistically independent with equal probability, the bit error probability of the QPOM scheme is

$$P_b = \frac{1}{2}(P_I + P_Q) = P_I = P_Q.$$

Similar to coded modulation schemes, the QPOM is nonlinear and the assumption of transmitting the all-zero codeword cannot be used in the evaluation of the error bound on $P_b$. Instead, every possible code word should be considered as a transmitted code word. This requires an error evaluation algorithm which has a computational complexity increasing with $2^{2v}$, where $v = 2$ is the total memory of the QPOM encoder. However, it is possible to reduce the computational complexity of the error bound to $2^v$ as the QPOM encoder has a Uniform Error Property (UEP).

**Example 10.3** The generator of the in-phase (or quadrature) component $s_I(t)$ (or $s_Q(t)$) of the QPOM signal consists of a linear convolutional encoder followed by a memoryless one-to-one mapper. The set partitioning of the $s_I$ (or $s_Q$) signal set is shown in Figure 10.6. Using the results of Chapter 4, the error weight profile of these signal subsets $\mathbf{A}_0$ and $\mathbf{A}_1$ is derived as

$$\mathcal{F}(\mathbf{A}_i, \underline{e}_j, D) = \frac{1}{2} \sum_{\underline{a} \in \mathbf{A}_i} D^{d_E^2[\mathcal{M}(\underline{a}), \mathcal{M}(\underline{a} \oplus \underline{e}_j)]}. \tag{10.15}$$

Using Figures 10.5 and 10.6, the error weight profiles for all possible error vectors $\underline{e}_j$ are obtained as:

$$\begin{aligned}
\mathcal{F}(\mathbf{A}_0, 00, D) &= \mathcal{F}(\mathbf{A}_1, 00, D) = 1 \\
\mathcal{F}(\mathbf{A}_0, 01, D) &= \mathcal{F}(\mathbf{A}_1, 01, D) = D^{2E_b} \\
\mathcal{F}(\mathbf{A}_0, 10, D) &= \mathcal{F}(\mathbf{A}_1, 10, D) = D^{2E_b} \\
\mathcal{F}(\mathbf{A}_0, 11, D) &= \mathcal{F}(\mathbf{A}_1, 11, D) = \tfrac{1}{2}(D^{4(1+\gamma)E_b} + D^{4(1-\gamma)E_b}).
\end{aligned}$$

As the subsets $\mathbf{A}_0$ and $\mathbf{A}_1$ have the same error weight profiles, the QPOM encoder has a UEP.

The error state diagram of Figure 10.8 is derived from the modified state diagram of the QPOM encoder by labeling its branches with $I^r \mathcal{F}(\mathbf{A}_i, \underline{e}_j, D)$, where $r$ represents the number of *incorrect input* bits associated with each error vector $\underline{e}_j$. The generating function of the error state diagram is obtained as

$$T(D, I) = \frac{I D^{4E_b}}{1 - 0.5I(D^{4(1+\gamma)E_b} + D^{4(1-\gamma)E_b})} \tag{10.16}$$

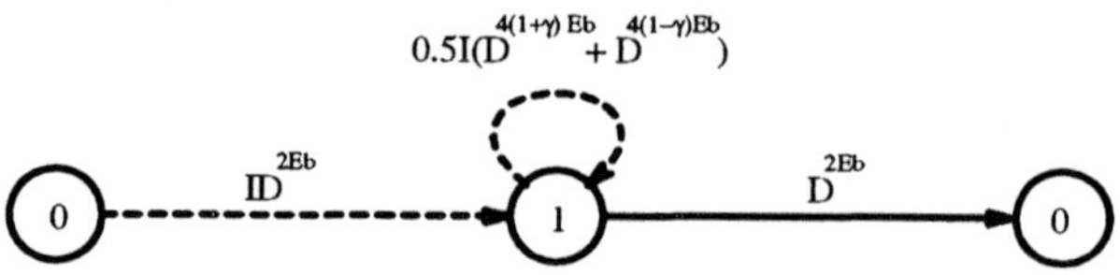


Figure 10.8: Error-state diagram for AWGN channels.

Note that $d_{free}^2$ can also be obtained from the transfer function as follows:

$$d_{free}^2 = \lim_{D\to 0} \frac{\delta T(D,I)/\delta D}{T(D,I)/D}\Bigg|_{I=1} = 4E_b. \tag{10.17}$$

Taking the derivative of (10.16) and then combining it with (10.14), the upper bound on the bit error probability of linear QPOM signals is given by

$$P_b \le \frac{1}{2}\mathrm{erfc}\left(\sqrt{\frac{E_b}{N_0}}\right)\frac{1}{\left[1-0.5e^{-(1+\gamma)E_b/N_0}-0.5e^{-(1-\gamma)E_b/N_0}\right]^2}. \tag{10.18}$$

**Example 10.4** At high signal-to-noise ratio $E_b/N_0$, the coefficient multiplying the complementary error function in (10.18) converges to one. In this situation the upper bound on the average bit error probability becomes

$$P_b \le \frac{1}{2}\mathrm{erfc}\left(\sqrt{E_b/N_0}\right),$$

which is equivalent to the performance of ideal coherent QPSK. Therefore, it is possible to improve the performance of quadrature pulse overlapping modulated signals simply by using the memory introduced by the overlapping pulse. Figure 10.9 shows a plot of this upper bound for $p(t) = 0.5(1+\cos(\pi t/T))$ along with the performance of the ideal coherent QPSK. Simulation results are also given to verify the tightness of the upper bound. The effect of the double-interval overlapping pulse on the performance is negligible especially at high $E_b/N_0$.

By spreading the pulse shape over more than *two symbol* intervals, QPOM signals with more compact spectra can be generated. Consider the case when the pulse shape is spread over the interval $[-KT, KT]$, i.e.,

$$\begin{aligned} p(t) &= p(-t), & |t| &\le KT \\ p(t) &= 0, & |t| &\ge KT. \end{aligned}$$

Over one symbol interval, the two components of the transmitted signal become

$$s_I(t) = \sum_{i=0}^{2K-1} A_{n-i}p(t-(n+i-K+1)T)$$

$$s_Q(t) = \sum_{i=0}^{2K-1} B_{n-i}p(t-(n+i-K+1)T).$$

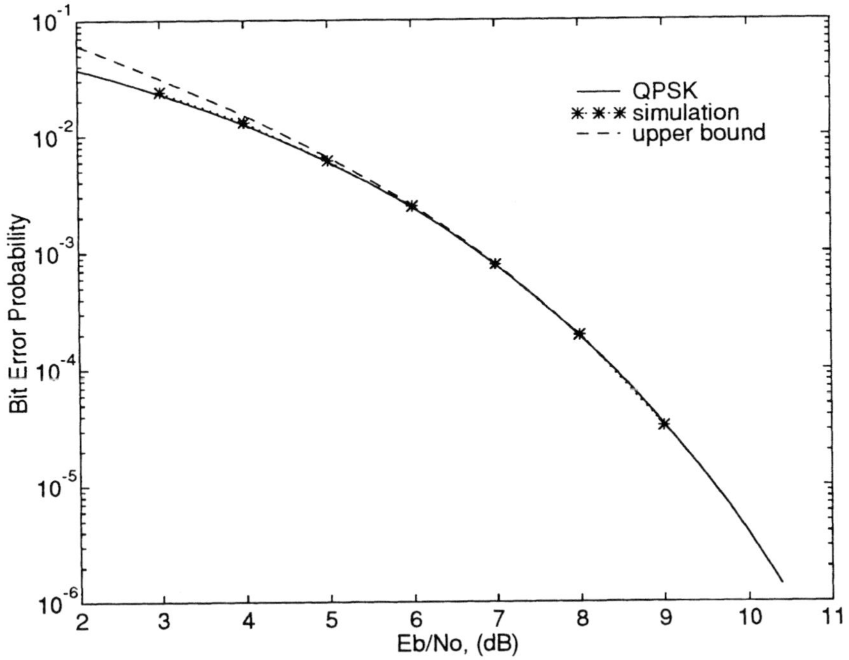


Figure 10.9: Performance of linear QPOM signals in AWGN channels.

It is observed that the in-phase component becomes a function of $\{a_n, a_{n-1}, \ldots, a_{n-2K+1}\}$, and the quadrature component becomes a function of $\{b_n, b_{n-1}, \ldots, b_{n-2K+1}\}$. In such a situation, the QPOM encoder of each component will have a total of $2^{2K-1}$ states.

Following the same procedure used for double interval pulse shapes, it can be shown that the minimum squared Euclidean distance for the in-phase (or quadrature) component is given by (see Problem 10.2)

$$d_{free}^2 = 8 \int_0^{KT} p^2(t - iT)dt = 4E_b,$$

which is equivalent to (10.17). Therefore, spreading the pulse shape over more than two symbol intervals does not affect the performance of QPOM signals. The only effect is the increase in the number of states of the overall encoder.

### 10.1.3 Performance of QPOM Signals in Fading Channels

Quadrature pulse overlapping modulation schemes have a structure similar to coded modulation schemes although no extra redundancy is used. The

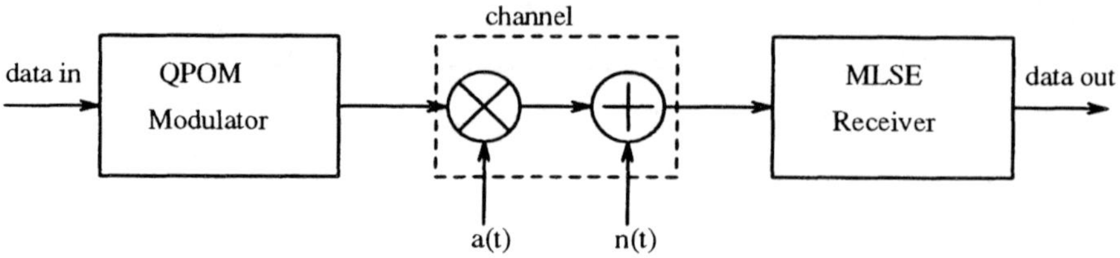


Figure 10.10: Block diagram of a QPOM system in a fading channel.

coding structure of QPOM signals is due to the pulse overlapping. The trellis representing the QPOM encoder (Figure 10.5) does not contain parallel paths between two adjacent states. In fact, the effective length of the QPOM code is 2. From the results of Chapter 5, it is expected that the QPOM schemes outperform the conventional QPSK under fading channel conditions.

Consider the block diagram of a QPOM system over a fading channel shown in Figure 10.10. We assume that the channel fading is sufficiently slow so that the phase shift can be estimated from the received signal without error, and the fading amplitude is constant over a symbol interval. In this case ideal coherent detection of the received signal can be achieved. Under this assumption in-phase and quadrature components of the received signal can be expressed as

$$\begin{aligned} \underline{r}_I &= a\underline{s}_I + \underline{n}_I \\ \underline{r}_Q &= a\underline{s}_Q + \underline{n}_Q, \end{aligned}$$

where $a$ is a random variable representing the fading amplitude.

Using the results of Chapter 5 (Example 5.9), Chapter 9 and Example 10.3, the error weight profiles for a Rayleigh fading channel are obtained as

$$\begin{aligned} &\mathcal{F}(00) = 1 \\ &\mathcal{F}(01) = f(2) \\ &\mathcal{F}(10) = f(2) \\ &\mathcal{F}(11) = \tfrac{1}{2}\left[f(4(1+\gamma)) + f(4(1-\gamma))\right], \end{aligned}$$

with

$$f(x) = \begin{cases} \frac{1}{1+xE_b/4N_0} & \text{with CSI} \\ e^{\lambda^2 xE_b/4N_0}\left[1 - \sqrt{\pi} erfc\left(\frac{\lambda x E_b}{2N_0}\right) e^{(\lambda x E_b/2N_0)^2}\right] & \text{without CSI} \end{cases} \tag{10.19}$$

for a Rayleigh fading channel, and

$$f(x) = \frac{1}{1 + x\sigma_a^2 E_b/2N_0} e^{-\frac{C(x)}{1+0.2\sigma_\gamma C(x)}}, \tag{10.20}$$

where

$$C(x) = \frac{xE_b/4N_0}{1 + x\sigma_\gamma^2 E_b/2N_0} e^{2m_\gamma} \tag{10.21}$$

for a shadowed Rician fading channel. The parameters $\sigma_a$, $\sigma_\gamma$ and $m_\gamma$ (defined in Sections 1.3.1 and 9.1) represent the degree of shadowing.

The branches of the error-state diagram of the QPOM encoder are modified as $I^r \mathcal{F}(\underline{e}_i)$ where $r$ shows the number of incorrect input bits associated with $\underline{e}_i$. The generating function of the modified error-state diagram becomes

$$T(I) = \frac{I f^2(2)}{1 - 0.5I[f(4[1+\gamma]) - f(4[1-\gamma])]}.$$

The Chernoff bound (5.39) on the bit error probability is obtained as

$$P_b \leq \frac{1}{2}\frac{\partial T(I)}{\partial I}\bigg|_{I=1} = \frac{f^2(2)}{2[1 - 0.5f(4[1+\gamma]) - 0.5f(4[1-\gamma])]^2}. \quad (10.22)$$

**Example 10.5** For high $E_b/N_0$, the denominator approaches 1 and the upper bound on $P_b$ reduces to

$$P_b \leq \frac{1}{2\left[1 + 2E_b/N_0\right]^2}$$

for a Rayleigh fading channel with CSI.

The performance of linear QPOM signals is inversely proportional to the square of the signal-to-noise ratio $E_b/N_0$. Furthermore, we notice that for high $E_b/N_0$, the effect of the overlapping pulse shape is negligible. Hence, based on the new representation of QPOM schemes, it is possible to select bandwidth efficient pulse shapes without affecting the performance of the overall system. Figure 10.11 shows the performance of this scheme over Rayleigh fading channels. It is observed that QPOM schemes offer better performance than the conventional QPSK scheme and TCM 8PSK (4U). Their performance is comparative to that of the TCM 8PSK (4N).

**Example 10.6** For a tighter upper bound on the bit error probability of the linear QPOM in a Rayleigh fading channel with CSI, we use (5.96) defined as

$$P_b \leq \frac{1}{2}\mathcal{K}(L, x_{min})\frac{\partial T(I)}{\partial I}\bigg|_{I=1}, \quad (10.23)$$

where $\frac{\partial T(I)}{\partial I}$ is represented by (10.22) with $f(x) = 1/(1 + xE_b/N_0)$.
$L = 2$ is the effective length of the QPOM code and $x_{min}$ corresponds to $d_{min} = \sqrt{2E_b}$, obtained from Figures 10.5 and 10.6.

It follows that

$$x_{min} = \sqrt{\frac{E_b/2N_0}{1 + E_b/2N_0}}$$

and

$$\mathcal{K}(2, x_{min}) = \frac{1}{2}\left(\frac{1}{1 + x_{min}} + \frac{1}{(1 + x_{min})^2}\right).$$

Figure 10.12 shows that the new bound is much closer to the simulation results than the Chernoff bound.

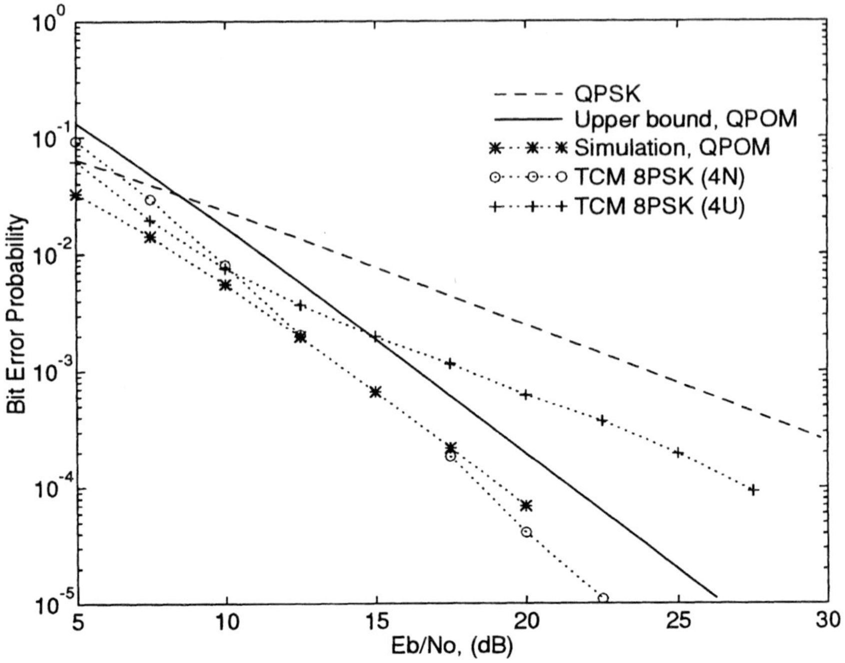


Figure 10.11: Performance of QPOM signals in a Rayleigh fading channel.

## 10.2 HARDLIMITED QPOM SCHEMES

Consider the simplified block diagram of a typical digital radio transmitter shown in Figure 10.13. To achieve its maximum power efficiency, the transmit power amplifier has to operate in a nonlinear (or saturated) mode. In this case, a modulated signal with *constant envelope* is preferred to avoid spectral spreading and distortion due to the AM/AM and AM/PM conversion effects of the nonlinear power amplifier. In this section, we present a technique to remove the amplitude fluctuation of the QPOM signals by using a hardlimiter. Hardlimiter introduces crosstalk between the in-phase and quadrature components of the QPOM signals. Using the approach introduced in Section 10.1, we represent both pulse overlapping and crosstalk effects by an encoder structure to develop the performance analysis of the hardlimited QPOM signals over AWGN and fading channels.

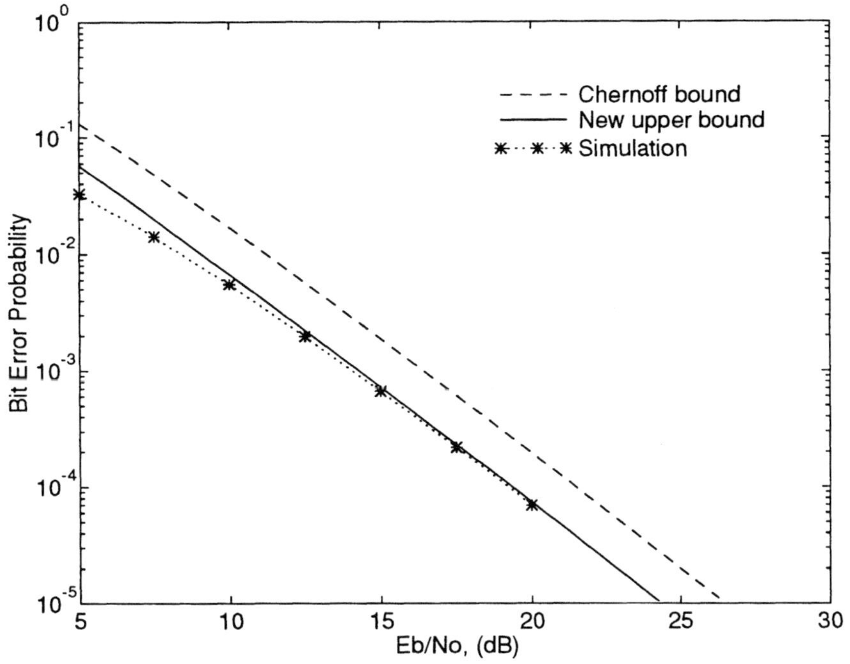


Figure 10.12: Performance of linear QPOM signals over a Rayleigh fading channel with perfect CSI.

### 10.2.1 Generation and Representation of Hardlimited QPOM Signals

Consider the QPOM scheme followed by an ideal hardlimiter as shown in Figure 10.14. The ideal hardlimiter produces a constant envelope with no change to the phase of the QPOM transmitted signal. The transfer function characteristic of the ideal hardlimiter is shown in Figure 10.15. The hardlimited QPOM signal $z(t)$ can be represented as

$$\begin{aligned} z(t) &= z_I(t)\cos(2\pi f_c t) + z_Q(t)\sin(2\pi f_c t) \\ &= \cos(2\pi f_c t + \phi_z(t)), \end{aligned} \tag{10.24}$$

where $z_I(t)$ and $z_Q(t)$ are in-phase and quadrature equivalent baseband components of the hardlimited QPOM signal with

$$z_I^2(t) + z_Q^2(t) = 1$$

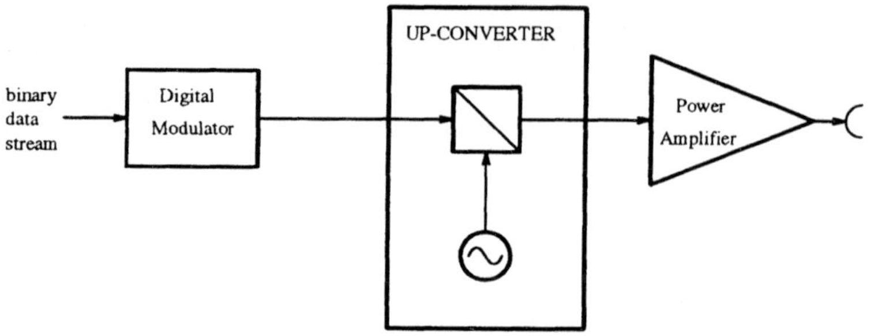


Figure 10.13: Simplified block diagram of a typical digital radio transmitter.

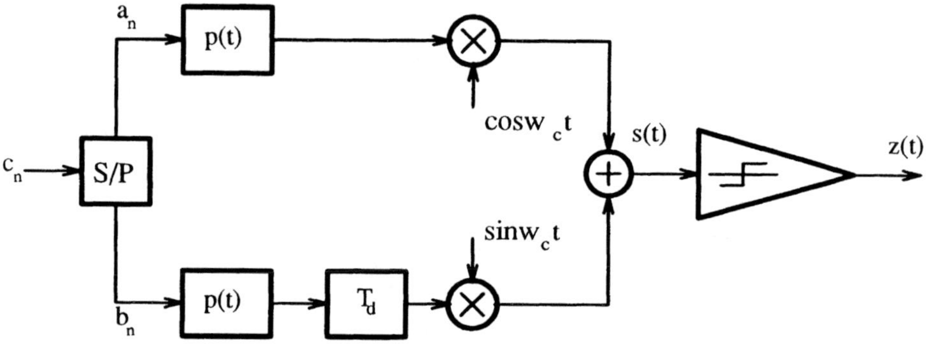


Figure 10.14: Block diagram of a hardlimited QPOM scheme

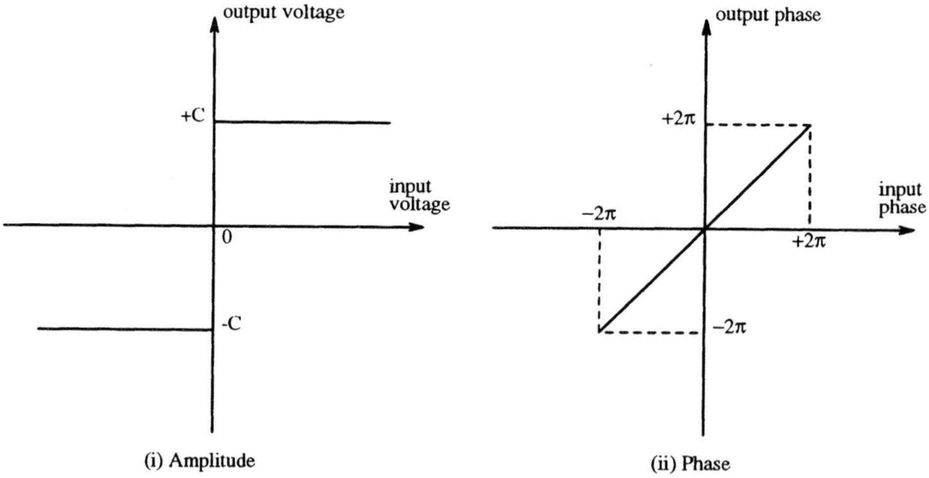


Figure 10.15: Transfer characteristics of an ideal hardlimiter.

and

$$\phi_z(t) = -\tan^{-1}\left(\frac{z_I(t)}{z_Q(t)}\right).$$

It is easy to verify that

$$z_I(t) = \frac{s_I(t)}{\sqrt{s_I^2(t) + s_Q^2(t - T_d)}} \tag{10.25}$$

and

$$z_Q(t) = \frac{s_Q(t)}{\sqrt{s_I^2(t) + s_Q^2(t - T_d)}}, \tag{10.26}$$

where

$$s_I(t) = \sum_{n=-\infty}^{+\infty} A_n p(t - nT)$$

$$s_Q(t) = \sum_{n=-\infty}^{+\infty} B_n p(t - nT).$$

We notice that the two components $z_I(t)$ and $z_Q(t)$ are no longer independent, and any change in $s_I(t)$ (or $s_Q(t)$) will affect both components. In other words, hardlimiting introduces *crosstalk* between the in-phase and quadrature components.

### 10.2.2 Power Spectrum of the Hardlimited QPOM Signals

As its inphase and quadrature components, $z_I(t)$ and $z_Q(t)$, have the same statistical characteristics, the power spectrum of the hardlimited QPOM signal $z(t)$ has the same shape as that of $z_I(t)$ (or $z_Q(t)$).

From (10.25), (10.26), (10.8) and (10.9), it can be seen that in each symbol interval, $nT \leq t \leq (n+1)T$, the component $z_I(t)$ (or $z_Q(t)$) takes one of 16 possible waveforms $v_i(\tau)$,$i = 0, 1, ..., 15$, and $\tau = t - nT$. These waveforms are represented as

$$v_j(\tau) = -v_{15-j}(\tau) = \frac{s_0(\tau)}{\sqrt{s_0^2(\tau) + \acute{s}_j^2(\tau)}} \tag{10.27}$$

and

$$v_{j+4}(\tau) = -v_{11-j}(\tau) = \frac{s_1(\tau)}{\sqrt{s_1^2(\tau) + \acute{s}_j^2(\tau)}}, \tag{10.28}$$

where $j = 0, 1, 2, 3,$
and

$$\acute{s}_0(\tau) = \begin{cases} s_0(\tau + T_d) & \text{for } 0 \leq \tau \leq T_d \\ s_0(\tau - T_d) & \text{for } T_d \leq \tau \leq T \end{cases}$$

$$\acute{s}_1(\tau) = \begin{cases} s_1(\tau + T_d) & \text{for } 0 \leq \tau \leq T_d \\ s_0(\tau - T_d) & \text{for } T_d \leq \tau \leq T \end{cases}$$

$$\acute{s}_2(\tau) = \begin{cases} s_0(\tau + T_d) & \text{for } 0 \leq \tau \leq T_d \\ s_1(\tau - T_d) & \text{for } T_d \leq \tau \leq T \end{cases}$$

$$\acute{s}_3(\tau) = \begin{cases} s_1(\tau + T_d) & \text{for } 0 \leq \tau \leq T_d \\ s_1(\tau - T_d) & \text{for } T_d \leq \tau \leq T \end{cases}$$

The stationary probabilities

$$p_i = Pr\{z_I(t - nT) = v_i(\tau)\} = \frac{1}{16}, \qquad i = 0, 1, ..., 15,$$

and the transition probabilities

$$p_{ik} = Pr\{z_I(t - nT) = v_k(\tau)|z_i(t - (n-1)T) = v_i(\tau)\}, \qquad i, k = 0, 1, 2, \cdots, 15,$$

can be arranged as entries of the $16 \times 16$ transition matrix

$$\mathbf{P} = \{p_{ik}, i, k = 0, 1, 2, \cdots, 15\} = \frac{1}{4} \begin{bmatrix} A & A & 0 & 0 \\ 0 & 0 & B & B \\ A & A & 0 & 0 \\ 0 & 0 & B & B \end{bmatrix},$$

where

$$\mathbf{A} = \begin{bmatrix} 1 & 0 & 1 & 0 \\ 1 & 0 & 1 & 0 \\ 0 & 1 & 0 & 1 \\ 0 & 1 & 0 & 1 \end{bmatrix}, \qquad \mathbf{B} = \begin{bmatrix} 0 & 1 & 0 & 1 \\ 0 & 1 & 0 & 1 \\ 1 & 0 & 1 & 0 \\ 1 & 0 & 1 & 0 \end{bmatrix}$$

and 0 is a $4 \times 4$ null matrix.

Using the results of [142], the power spectral density of $z_I(t)$ is given by

$$\begin{aligned} Z(f) &= \frac{1}{T^2} \sum_{n=-\infty}^{+\infty} \left| \sum_{i=0}^{15} p_i V_i(\frac{n}{T}) \right|^2 \delta(f - \frac{n}{T}) \\ &+ \frac{1}{T} \sum_{i=0}^{15} p_i |V_i(f)|^2 \\ &+ \frac{2}{T} Re \left\{ \sum_{i=0}^{15} \sum_{k=0}^{15} p V_i(f) V_k^*(f) v_{ik} \right\}, \end{aligned} \qquad (10.29)$$

where $V_i(f)$ is the Fourier transform of $v_i(\tau)$, $V_i^*(f)$ is the complex conjugate of $V_i(f)$,

$$v_{ik} = \sum_{n=1}^{+\infty} p_{ik}^{(n)} e^{-j2\pi n f T},$$

$p_{ik}^{(n)}$ is the $ik^{th}$ element of the matrix $\mathbf{P}^n$,

$$\mathbf{P}^n = \underbrace{\mathbf{P}.\mathbf{P}\cdots\mathbf{P}}_{n\ times},$$

and $Re\{.\}$ is the real part of $\{.\}$

Substituting values of $V_i(f)$'s, $p_i$'s, and $z_{ik}$'s into the expression of $Z(f)$, we obtain

$$\begin{aligned} Z(f) &= \frac{1}{4T}\sum_{i=0}^{7}|V_i(f)|^2 \\ &+\frac{1}{8T}Re\{([V_0(f)+V_2(f)+V_4(f)+V_6(f)][V_0(f)+V_1(f)-V_4(f)-V_5(f)]^* \\ &+[V_1(f)+V_3(f)+V_5(f)+V_7(f)][V_2(f)+V_3(f)-V_6(f)-V_7(f)]^*)e^{-j\pi x}\}. \end{aligned}$$

Accordingly, the normalized power spectral density of the hardlimited QPOM signal for a general value of $T_d$ is given as (see Problem 10.3)

$$\begin{aligned} Z(x) &= 2\vartheta\sum_{i=0}^{7}|V_i(x)|^2 \\ &+\vartheta Re\{([V_0(x)+V_2(x)+V_4(x)+V_6(x)][V_0(x)+V_1(x)-V_4(x)-V_5(x)]^* \\ &+[V_1(x)+V_3(x)+V_5(x)+V_7(x)][V_2(x)+V_3(x)-V_6(x)-V_7(x)]^*)e^{-j\pi x}\}\}, \end{aligned}$$

where $\vartheta$ is a constant such that $Z(0) = 1$ and $x = (f - f_c)T$.

**Example 10.7** The spectrum of the hardlimited QPOM signal for a given double-interval pulse $p(t)$ depends on the time offset $T_d$. Figure 10.16 shows the spectrum for various values of $T_d$ with $p(t) = 0.5(1 + \cos\pi t/T)$. It indicates that $T_d = T/2$ yields the best spectrum. As shown in Figure 10.17, the spectral roll-off of the hardlimited QPOM signal with $T_d = T/2$ closely follows that of the original QPOM signal (before hardlimiting).

In the following we will consider two cases, $T_d = 0$ and $T_d = T/2$. For convenience, we use the term hardlimited QPOM for $T_d = 0$ and hardlimited OQPOM for $T_d = T/2$.

Let us first consider the representation of hardlimited QPOM for $T_d = 0$. In this case the number of possible waveforms of $z_I(t)$ (or $z_Q(t)$) in each symbol

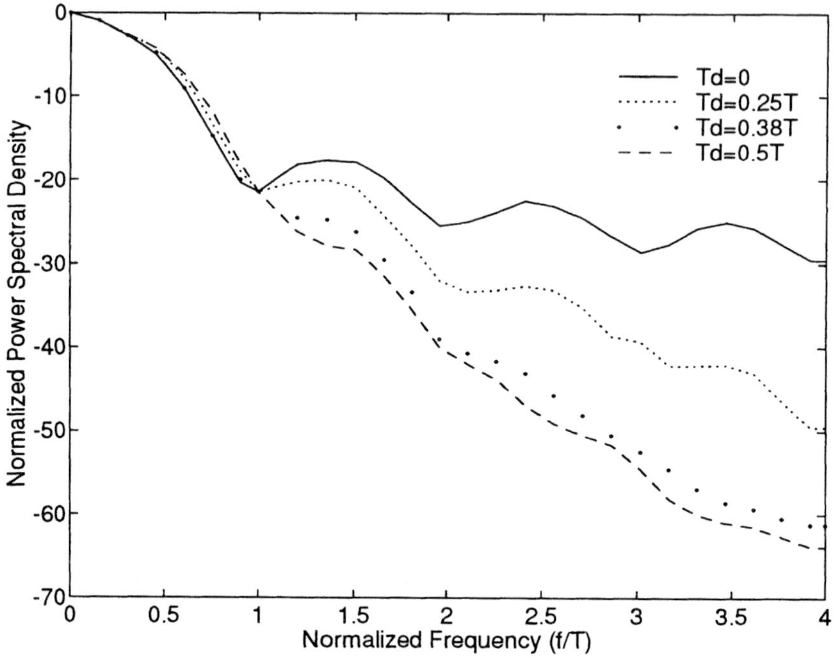


Figure 10.16: Power spectra of hardlimited QPOM signals with different values of $T_d$ ($p(t) = 0.5(1 + \cos(\pi t/T))$).

interval $nT \leq t \leq (n+1)T$ is reduced to 8. These waveforms are derived from (10.25), (10.26) as

$$v_0(\tau) = v_1(\tau) \equiv z_0(\tau) = \frac{1}{\sqrt{2}}, \qquad 0 \leq t \leq T, \tag{10.30}$$

$$v_2(\tau) = v_3(\tau) \equiv z_1(\tau) = \frac{s_0(\tau)}{\sqrt{s_0^2(\tau) + s_1^2(\tau)}}, \qquad 0 \leq t \leq T, \tag{10.31}$$

$$v_4(\tau) = v_5(\tau) \equiv z_2(\tau) = \frac{s_1(\tau)}{\sqrt{s_0^2(\tau) + s_1^2(\tau)}}, \qquad 0 \leq t \leq T, \tag{10.32}$$

$$v_6(\tau) = v_7(\tau) \equiv z_3(\tau) = \begin{cases} \frac{1}{\sqrt{2}}, & 0 \leq \tau \leq T/2 \\ -\frac{1}{\sqrt{2}}, & T/2 \leq \tau \leq T \end{cases} \tag{10.33}$$

The remaining *four* waveforms are antipodal to those given above.

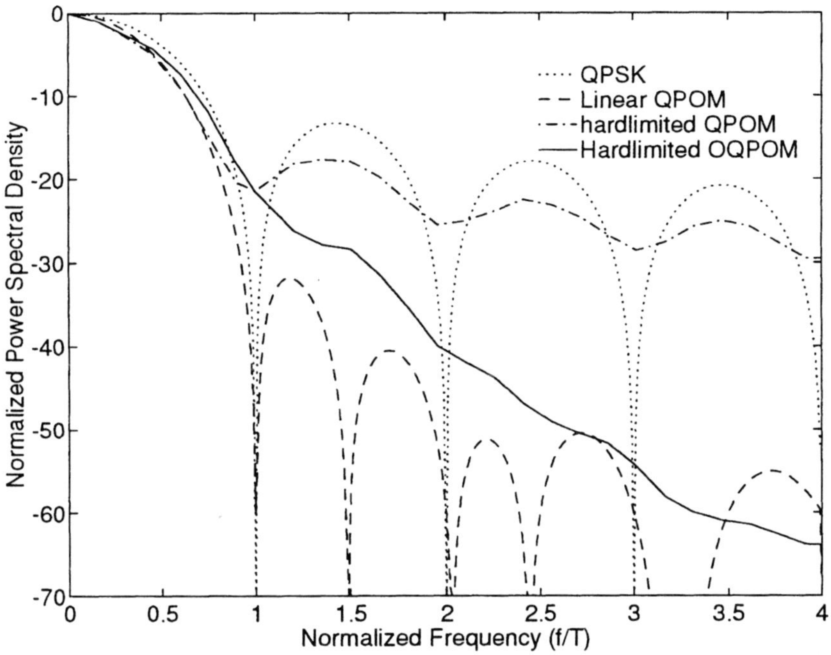


Figure 10.17: Power spectra of the original QPOM signal and the hardlimited QPOM signal with $p(t) = 0.5(1 + \cos(\pi t/T)$.

**Example 10.8** Figure 10.18 shows the plots of the waveforms $z_i(\tau)$'s for the double-interval overlapping raised cosine pulse shape $0.5(1 + \cos \pi t/T)$.

The pair of in-phase and quadrature components $(z_I(t), z_Q(t))$ for $nT \leq t \leq (n+1)T$ becomes a function of $\{a_n, b_n, a_{n-1}, b_{n-1}\}$ as shown in Table 10.1. A hardlimited QPOM scheme with $T_d = 0$ can be modeled as a system with two separate blocks: the hardlimited QPOM encoder followed by a one-to-one signal mapper, as shown in Figure 10.19. The trellis diagram of the QPOM encoder is given in Figure 10.20. This trellis diagram has *four* states and is fully connected. The possible outputs of the encoder are defined by the set $\mathbf{A} = (0, 1, \cdots, 15)$, which is partitioned into four disjoint subsets as shown in Figure 10.20. It can be seen from (10.30), (10.31), (10.32) and (10.33) that the four waveforms are linearly independent. Therefore, an orthonormal basis $\phi_i(\tau), i = 0, 1, 2, 3$, can be found (e.g., using the Gram-Schmidt procedure).

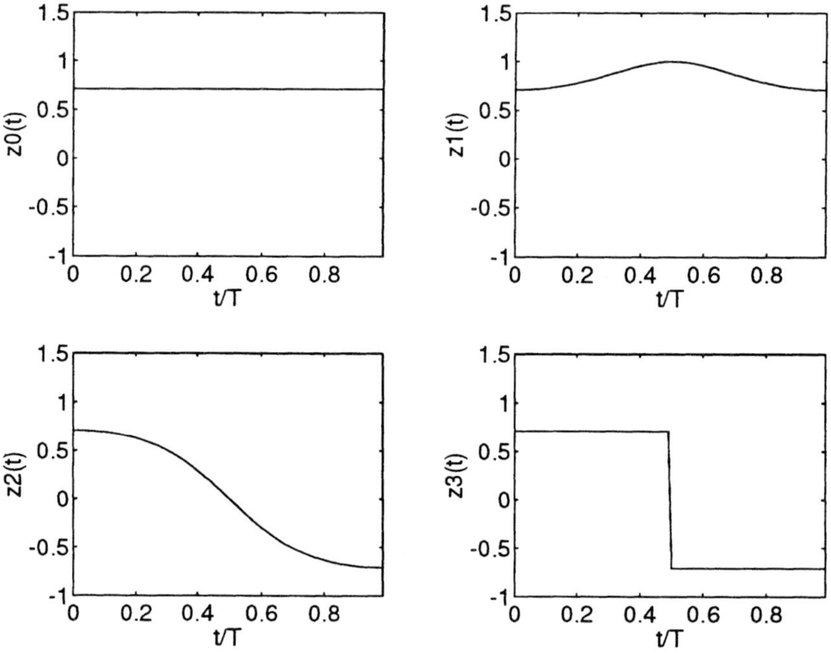


Figure 10.18: Example of transmitted pulse shapes for hardlimited QPOM schemes.

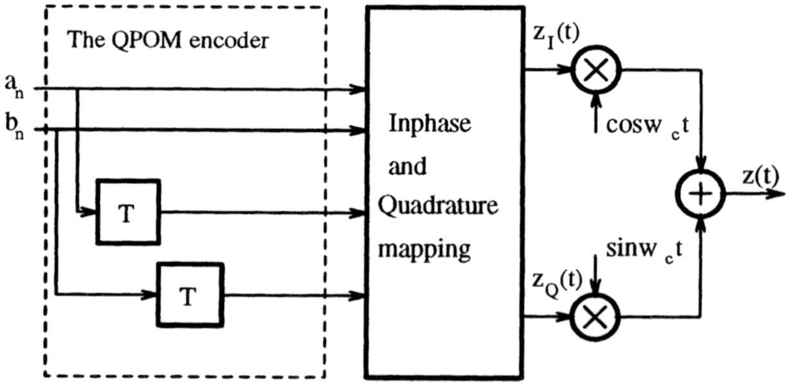


Figure 10.19: An alternative representation of hardlimited QPOM schemes.

| Symbol $a_n b_n a_{n-1} b_{n-1}$ | | Transmitted signals $(s_I(t), s_Q(t))$ |
|---|---|---|
| decimal | binary | $=\mathcal{M}(a_n, b_n, a_{n-1}, b_{n-1})$ |
| 0 | 0 0 0 0 | $(-z_0(t), -z_0(t))$ |
| 1 | 0 0 0 1 | $(-z_1(t), -z_2(t))$ |
| 2 | 0 0 1 0 | $(-z_2(t), -z_1(t))$ |
| 3 | 0 0 1 1 | $(-z_3(t), -z_3(t))$ |
| 4 | 0 1 0 0 | $(-z_1(t), +z_2(t))$ |
| 5 | 0 1 0 1 | $(-z_0(t), +z_0(t))$ |
| 6 | 0 1 1 0 | $(-z_3(t), +z_3(t))$ |
| 7 | 0 1 1 1 | $(-z_2(t), +z_1(t))$ |
| 8 | 1 0 0 0 | $(+z_2(t), -z_1(t))$ |
| 9 | 1 0 0 1 | $(+z_3(t), -z_3(t))$ |
| 10 | 1 0 1 0 | $(+z_0(t), -z_0(t))$ |
| 11 | 1 0 1 1 | $(+z_1(t), -z_2(t))$ |
| 12 | 1 1 0 0 | $(+z_3(t), +z_3(t))$ |
| 13 | 1 1 0 1 | $(+z_2(t), +z_1(t))$ |
| 14 | 1 1 1 0 | $(+z_1(t), +z_2(t))$ |
| 15 | 1 1 1 1 | $(+z_0(t), +z_0(t))$ |

Table 10.1: Mapping rules for the hardlimited QPOM scheme.

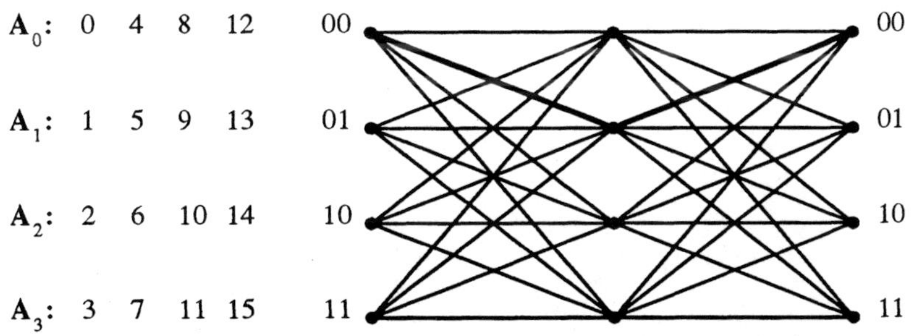


Figure 10.20: Trellis diagram of the hardlimited QPOM encoder.

Accordingly, the waveform $\{z_i(\tau)\}$ can be written as

$$z_i(\tau) = \sum_{j=0}^{3} z_{ij}\phi_j(\tau), \; i = 0, 1, 2, 3.$$

The waveform $z_i(\tau)$ can then be represented by the vector of its coefficients in the signal space $\phi_i(\tau)$, $i = 0, 1, 2, 3$. This vector is a 4-element vector denoted as $\underline{z}_{Ii}$ (if $z_i(\tau)$ represents $z_I(t)$) or $\underline{z}_{Qi}$ (if $z_i(\tau)$ represents $z_Q(t)$), and defined as

$$\underline{z}_{Ii} \text{ or } \underline{z}_{Qi} = (z_{i0}, z_{i1}, z_{i2}, z_{i3}),$$

where

$$z_{il} = \int_0^T z_i(\tau)\phi_l(\tau)d\tau, \qquad l = 0, 1, 2, 3.$$

The optimum MLSE receiver of hardlimited QPOM signals needs to treat both components $z_I(t)$ and $z_Q(t)$ at the same time. The $i$th transmitted pair of waveforms $z_I(t)$ and $z_Q(t)$ can be represented by a unique augmented vector $\underline{z}_i$ defined as

$$\begin{aligned}\underline{z}_i &= (\underline{z}_{Il}, \underline{z}_{Qj}) \\ &= (z_{i0}, z_{i1}, z_{i2}, z_{i3}, z_{i4}, z_{i5}, z_{i6}, z_{i7}),\end{aligned}$$

where

$$z_{ij} = \begin{cases} \int_0^T z_I(t)\phi_j(t)dt & j = 0, 1, 2, 3 \\ \int_0^T z_Q(t)\phi_{j-4}(t)dt & j = 4, 5, 6, 7 \end{cases}$$

### 10.2.3 Performance of Hardlimited QPOM Signals in AWGN Channels

The MLSE receiver for hardlimited QPOM signals is shown in Figure 10.21. The received signal is down-converted to baseband, then each component is passed through a bank of *four* matched filters with impulse response $\{\phi_i(T-t)\}$ and sampled every $T$ seconds giving the elements of the augmented vector $\underline{z}_i$. This vector is then used by the Viterbi processor to estimate the transmitted sequence.

To evaluate the performance of the hardlimited QPOM scheme in an AWGN environment, we follow the similar approach presented in Section 10.1.2. Using the trellis diagram of Figure 10.20 and Table 10.1, the error weight profiles for all possible error vectors are calculated as

$$\mathcal{F}(A_i, \underline{e}_j, D) = \frac{1}{4} \sum_{\underline{a} \in A_i} D^{d_E^2[\mathcal{M}(\underline{a}), \mathcal{M}(\underline{a} \oplus \underline{e}_j)]}.$$

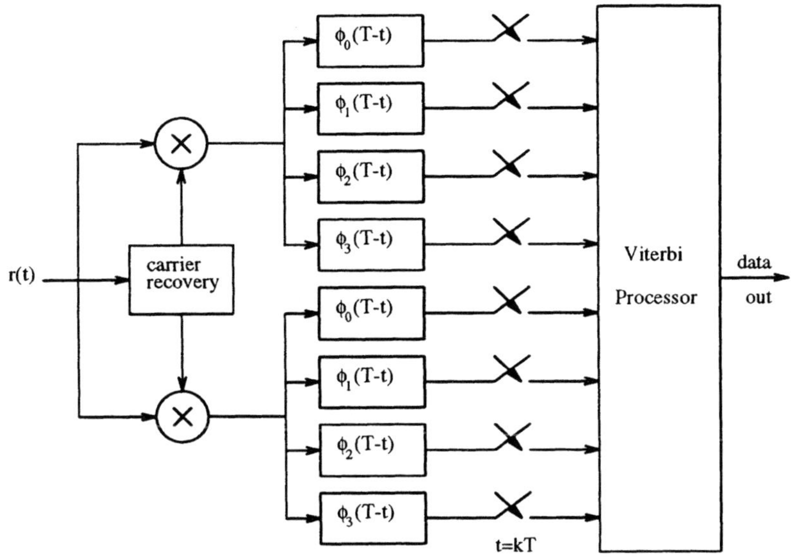


Figure 10.21: MLSE receiver of hardlimited QPOM signals.

It is found that the error weight profiles are identical for all four subsets $A_i$, $i = 1, 2, 3, 4$ (see Problem 10.5),

$$\mathcal{F}(A_i, 0000, D) = 1$$

$$\begin{aligned}\mathcal{F}(A_i, 0001, D) &= \mathcal{F}(A_i, 0010, D) = \mathcal{F}(A_i, 1000, D) = \mathcal{F}(A_i, 0100, D)\\ &= 0.5(D^{\alpha_3} + D^{\alpha_4})\end{aligned}$$

$$\begin{aligned}\mathcal{F}(A_i, 0011, D) &= \mathcal{F}(A_i, 0101, D) = \mathcal{F}(A_i, 0110, D) = \mathcal{F}(A_i, 1001, D)\\ &= \mathcal{F}(A_i, 1010, D) = \mathcal{F}(A_i, 1100, D)\\ &= D^{\alpha_2}\end{aligned}$$

$$\begin{aligned}\mathcal{F}(A_i, 0111, D) &= \mathcal{F}(A_i, 1011, D) = \mathcal{F}(A_i, 1101, D)\\ &= \mathcal{F}(A_i, 1110, D) = 0.5(D^{\alpha_1} + D^{\alpha_5})\end{aligned}$$

$$\mathcal{F}(A_i, 1111, D) = D^{\alpha_0},$$

where $\alpha_i$ represents the squared Euclidean distance between two pairs of waveforms $(z_i(\tau), z_j(\tau))$ and $(z_k(\tau), z_l(\tau))$

$$\alpha_0 = 8E_b, \quad \alpha_1 = 4(1+\beta_0)E_b, \quad \alpha_3 = 4(1-\beta_0)E_b$$

$$\alpha_2 = 4E_b, \quad \alpha_5 = 4(1+\beta_1)E_b, \quad \alpha_4 = 4(1-\beta_1)E_b$$

$E_b$ is the bit energy obtained as

$$E_b = T/2$$

and $\beta_0$ and $\beta_1$ are the coefficients depending on the pulse shape $p(t)$ as

$$\beta_0 = \frac{2}{T}\int_0^{T/2} \frac{p(t)+p(t-T)}{\sqrt{p^2(t)+p^2(t-T)}}\,dt \tag{10.34}$$

$$\beta_1 = \frac{2}{T}\int_0^{T/2} \frac{p(t)-p(t-T)}{\sqrt{p^2(t)+p^2(t-T)}}\,dt. \tag{10.35}$$

The trellis diagram of Figure 10.20 is similar to that of Figure 5.3. It can be shown that its error-state diagram has the same structure as that of Figure 5.13 with the following branch labels (see Problem 10.6):

$$\begin{array}{ll} t_1 = 0.5I(D^{\alpha_3}+D^{\alpha_2}), & t_2 = t_4 = t_5 = t_9 = t_{10} = ID^{\alpha_2} \\ t_3 = I^2 D^{\alpha_2}, & t_6 = I^2 D^{\alpha_0} \\ t_7 = t_{11} = 0.5I^2(D^{\alpha_1}+D^{\alpha_5}), & t_8 = t_{12} = 0.5I(D^{\alpha_1}+D^{\alpha_5}) \\ t_{13} = t_{14} = 0.5(D^{\alpha_3}+D^{\alpha_4}), & t_{15} = D^{\alpha_2}. \end{array}$$

Following the same procedure as Example 5.11, the generating function is derived by solving the nodal equations of the error-state diagram (see Problem 10.7)

$$T(D,I) = \frac{W_0(D)I + W_1(D)I^2}{1 - Q_0(D))I - Q_1(D)I^2}, \tag{10.36}$$

where

$$W_0(D) = 0.5D^{2\alpha_3} + 0.5D^{2\alpha_4} + D^{\alpha_3+\alpha_4} + D^{2\alpha_2}$$

$$\begin{aligned} W_1(D) &= D^{\alpha_1+\alpha_2+\alpha_3} + D^{\alpha_2+\alpha_2+\alpha_5} + D^{\alpha_2+\alpha_4+\alpha_5} \\ &\quad - D^{\alpha_0+\alpha_3+\alpha_4} - 0.5D^{\alpha_0+2\alpha_3} - 0.5D^{\alpha_0+2\alpha_4} - 2D^{2\alpha_2} \end{aligned}$$

$$Q_0(D) = D^{\alpha_0} + 2D^{\alpha_2}$$

$$Q_1(D) = 0.5D^{2\alpha_1} + 0.5D^{2\alpha_5} + D^{\alpha_1+\alpha_5} - 2D^{\alpha_0+\alpha_2}.$$

The minimum squared Euclidean distance of the code is obtained from the transfer function using (10.17)

$$d^2_{free} = 8(1-\beta_m)E_b, \tag{10.37}$$

where $\beta_m$ represents the maximum between $\beta_0$ and $\beta_1$.

Substituting (10.36) and (10.37) into (10.14), the bit error probability of the hardlimited QPOM in an AWGN environment is upper-bounded as

$$P_b \leq \frac{1}{2}\text{erfc}\left(\sqrt{2(1-\beta_m)\frac{E_b}{N_0}}\right)\psi(E_b/N_0, \beta_m), \tag{10.38}$$

where the function $\psi(E_b/N_0, \beta_m)$ is given by

$$\psi(E_b/N_0,\ \beta_m) = D^{d_{free}^2}\frac{W_0(D)(1+Q_1(D)) + W_1(D)(2-Q_0(D))}{[1-Q_0(D)-Q_1(D)]^2},$$

with $D = e^{-1/4N_0}$.

It is noted that at high $E_b/N_0$, the function $\psi(E_b/N_0, \beta_m)$ converges to a constant $C$. In this situation, the upper bound on $P_b$ becomes

$$P_b \leq \frac{C}{2}\text{erfc}\left(\sqrt{2(1-\beta_m)\frac{E_b}{N_0}}\right), \tag{10.39}$$

where

$$C = \begin{cases} C = 1.00, & \beta_m = 0.5 \\ C = 0.25, & \beta_m \neq 0.5 \end{cases}.$$

When $\beta_m = 0.5$, (10.38) becomes

$$P_b \leq \frac{1}{2}\text{erfc}\left(\sqrt{\frac{E_b}{N_0}}\right)\frac{1+0.5e^{-E_b/N_0}}{\left[1-e^{-2E_b/N_0}-2e^{-E_b/N_0}\right]^2},$$

which converges to the performance of ideal coherent QPSK at high signal-to-noise ratio

$$P_b \leq \frac{1}{2}\text{erfc}\left(\sqrt{\frac{E_b}{N_0}}\right), \qquad \frac{E_b}{N_0} \gg 1.$$

It is observed from the above equations that if $p(t)$ is a *single-interval* pulse shape, i.e.,

$$p(t-T) = 0, \qquad t \leq T/2,$$

then the two coefficients $\beta_0$ and $\beta_1$ become equal to 0.5, and the performance of the hardlimited QPOM scheme reduces to that of ideal coherent QPSK at high signal-to-noise ratio.

Consider the case of *double-interval* overlapping pulse shapes satisfying the conditions in (10.2). It is easy to show that if $0 \leq p(t)$ for all $t$, then

$$\sqrt{p^2(t)+p^2(t-T)} \leq p(t)+p(t-T).$$

| Pulse shape $p(t)$ | The coefficient $\beta_m$ |
|---|---|
| $p_1(t) = \cos(\pi t/2T)$ | 0.6366 |
| $p_2(t) = \cos^2(\pi t/2T)$ | 0.59017 |
| $p_3(t) = \cos^4(\pi t/2T)$ | 0.55151 |

Table 10.2: Example of some double-interval overlapping pulse shapes and their coefficients $\beta_m$.

Hence, we conclude that

$$\left.\begin{array}{l}\beta_0 \geq 0.5\\ \beta_1 \leq 0.5\end{array}\right| \Rightarrow \beta_m = \beta_0 \geq 0.5.$$

This indicates that the performance of hardlimited QPOM schemes can reach that of ideal coherent QPSK, without exceeding it.

**Example 10.9** Table 10.2 gives some double-interval overlapping pulse shapes and their corresponding coefficients $\beta_m$. As this table shows, it is possible to select a double-interval pulse shape $p(t)$ which has a coefficient $\beta_m$ as close to 0.5 as possible and which introduces a compact spectrum.

Figure 10.22 gives the upper bound on $P_b$ for hardlimited QPOM signals using the pulse shapes of Table 10.2. Also indicated in this figure is the performance of ideal coherent QPSK. We notice that as $\beta_m$ approaches 0.5, the performance of the QPOM scheme approaches that of ideal coherent QPSK at high signal-to-noise ratios. Therefore, by an appropriate choice of the pulse shape $p(t)$ we can obtain a transmitted QPOM signal with a compact spectrum and a performance similar to that of ideal coherent QPSK.

### 10.2.4 Hardlimited OQPOM in an AWGN Environment

Now let us consider hardlimited QPOM signals with $T_d = T/2$. The performance of these signals can be analyzed by using the same procedure as that used for $T_d = 0$. However, due to the time offset $T_d = T/2$, the two components $z_I(t)$ and $z_Q(t)$ are now functions of the set $\{a_n, b_n, a_{n-1}, b_{n-1}, b_{n-2}\}$ as shown in Table 10.3.

The 16 elements $z_i(t)$, $v_i(t)$, $i = 0, 1, \cdots, 7$, are derived from (10.25), (10.26) as

$$z_0(t) = \begin{cases} \frac{s_0(t)}{\sqrt{s_0^2(t)+s_0^2(t+T/2)}} & 0 \leq t \leq T/2 \\ \frac{s_0(t)}{\sqrt{s_0^2(t)+s_0^2(t-T/2)}} & T/2 \leq t \leq T \end{cases} \tag{10.40}$$

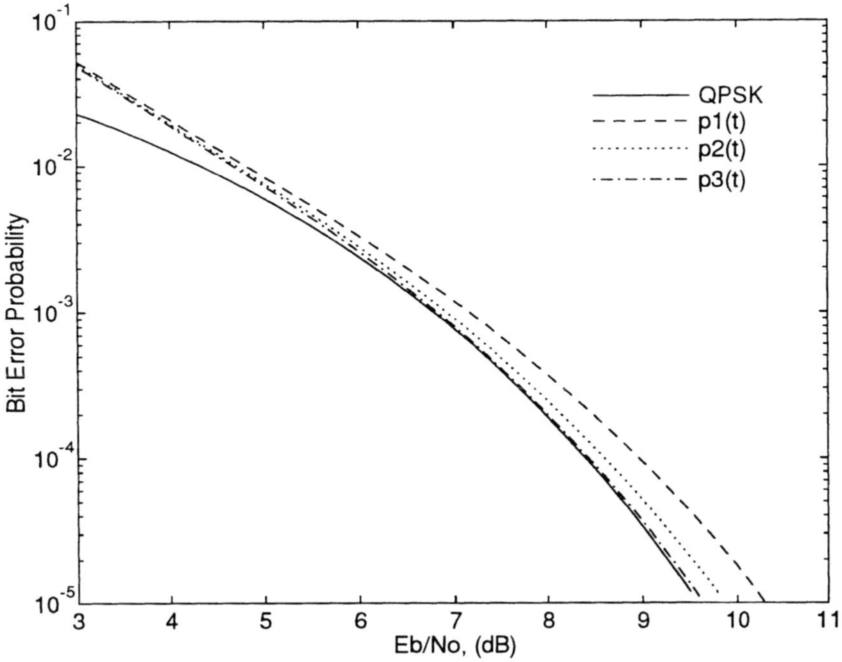


Figure 10.22: Performance of hardlimited QPOM signals in AWGN channels for different pulse shapes (Table 10.2).

$$z_1(t) = \begin{cases} \frac{s_0(t)}{\sqrt{s_0^2(t)+s_1^2(t+T/2)}} & 0 \le t \le T/2 \\ \frac{s_0(t)}{\sqrt{s_0^2(t)+s_0^2(t-T/2)}} & T/2 \le t \le T \end{cases} \tag{10.41}$$

$$z_2(t) = \begin{cases} \frac{s_0(t)}{\sqrt{s_0^2(t)+s_0^2(t+T/2)}} & 0 \le t \le T/2 \\ \frac{s_0(t)}{\sqrt{s_0^2(t)+s_1^2(t-T/2)}} & T/2 \le t \le T \end{cases} \tag{10.42}$$

$$z_3(t) = \begin{cases} \frac{s_0(t)}{\sqrt{s_0^2(t)+s_1^2(t+T/2)}} & 0 \le t \le T/2 \\ \frac{s_0(t)}{\sqrt{s_0^2(t)+s_1^2(t-T/2)}} & T/2 \le t \le T \end{cases} \tag{10.43}$$

$$z_4(t) = \begin{cases} \frac{s_1(t)}{\sqrt{s_1^2(t)+s_0^2(t+T/2)}} & 0 \le t \le T/2 \\ \frac{s_1(t)}{\sqrt{s_1^2(t)+s_0^2(t-T/2)}} & T/2 \le t \le T \end{cases} \tag{10.44}$$

$$z_5(t) = \begin{cases} \frac{s_1(t)}{\sqrt{s_1^2(t)+s_1^2(t+T/2)}} & 0 \le t \le T/2 \\ \frac{s_1(t)}{\sqrt{s_1^2(t)+s_0^2(t-T/2)}} & T/2 \le t \le T \end{cases} \tag{10.45}$$

$$z_6(t) = \begin{cases} \frac{s_1(t)}{\sqrt{s_1^2(t)+s_0^2(t+T/2)}} & 0 \le t \le T/2 \\ \frac{s_1(t)}{\sqrt{s_1^2(t)+s_1^2(t-T/2)}} & T/2 \le t \le T \end{cases} \tag{10.46}$$

$$z_7(t) = \begin{cases} \frac{s_1(t)}{\sqrt{s_1^2(t)+s_1^2(t+T/2)}} & 0 \le t \le T/2 \\ \frac{s_1(t)}{\sqrt{s_1^2(t)+s_1^2(t-T/2)}} & T/2 \le t \le T \end{cases} \tag{10.47}$$

$$v_0(t) = \begin{cases} \frac{s_0(t+T/2)}{\sqrt{s_0^2(t)+s_0^2(t+T/2)}} & 0 \le t \le T/2 \\ \frac{s_0(t+T/2)}{\sqrt{s_0^2(t)+s_0^2(t-T/2)}} & T/2 \le t \le T \end{cases} \tag{10.48}$$

$$v_1(t) = \begin{cases} \frac{-s_1(t+T/2)}{\sqrt{s_0^2(t)+s_1^2(t+T/2)}} & 0 \le t \le T/2 \\ \frac{s_0(t-T/2)}{\sqrt{s_0^2(t)+s_0^2(t-T/2)}} & T/2 \le t \le T \end{cases} \tag{10.49}$$

$$v_2(t) = \begin{cases} \frac{s_0(t+T/2)}{\sqrt{s_0^2(t)+s_0^2(t+T/2)}} & 0 \le t \le T/2 \\ \frac{s_1(t-T/2)}{\sqrt{s_0^2(t)+s_1^2(t-T/2)}} & T/2 \le t \le T \end{cases} \tag{10.50}$$

$$v_3(t) = \begin{cases} \frac{-s_1(t+T/2)}{\sqrt{s_0^2(t)+s_1^2(t+T/2)}} & 0 \le t \le T/2 \\ \frac{s_1(t-T/2)}{\sqrt{s_0^2(t)+s_1^2(t-T/2)}} & T/2 \le t \le T \end{cases} \tag{10.51}$$

$$v_4(t) = \begin{cases} \frac{s_0(t+T/2)}{\sqrt{s_1^2(t)+s_0^2(t+T/2)}} & 0 \le t \le T/2 \\ \frac{s_0(t-T/2)}{\sqrt{s_1^2(t)+s_0^2(t-T/2)}} & T/2 \le t \le T \end{cases} \tag{10.52}$$

$$v_5(t) = \begin{cases} \frac{-s_1(t+T/2)}{\sqrt{s_1^2(t)+s_1^2(t+T/2)}} & 0 \le t \le T/2 \\ \frac{s_0(t-T/2)}{\sqrt{s_1^2(t)+s_0^2(t-T/2)}} & T/2 \le t \le T \end{cases} \tag{10.53}$$

$$v_6(t) = \begin{cases} \frac{s_0(t+T/2)}{\sqrt{s_1^2(t)+s_0^2(t+T/2)}} & 0 \le t \le T/2 \\ \frac{s_1(t-T/2)}{\sqrt{s_1^2(t)+s_1^2(t-T/2)}} & T/2 \le t \le T \end{cases} \tag{10.54}$$

| Symbol $a_n b_n a_{n-1} b_{n-1} b_{n-2}$ decimal | binary | Transmitted signals $(z_I(t), z_Q(t))$ $=\mathcal{M}(a_n b_n a_{n-1} b_{n-1} b_{n-2})$ |
|---|---|---|
| 0 | 00 00 0 | $(-z_0(t), -v_0(t))$ |
| 1 | 00 00 1 | $(-z_1(t), -v_1(t))$ |
| 2 | 00 01 0 | $(-z_2(t), -v_2(t))$ |
| 3 | 00 01 1 | $(-z_3(t), -v_3(t))$ |
| 4 | 00 10 0 | $(-z_4(t), -v_4(t))$ |
| 5 | 00 10 1 | $(-z_5(t), -v_5(t))$ |
| 6 | 00 11 0 | $(-z_6(t), -v_6(t))$ |
| 7 | 00 11 1 | $(-z_7(t), -v_7(t))$ |
| 8 | 01 00 0 | $(-z_3(t), +v_3(t))$ |
| 9 | 01 00 1 | $(-z_2(t), +v_2(t))$ |
| 10 | 01 01 0 | $(-z_1(t), +v_1(t))$ |
| 11 | 01 01 1 | $(-z_0(t), +v_0(t))$ |
| 12 | 01 10 0 | $(-z_7(t), +v_7(t))$ |
| 13 | 01 10 1 | $(-z_6(t), +v_6(t))$ |
| 14 | 01 11 0 | $(-z_5(t), +v_5(t))$ |
| 15 | 01 11 1 | $(-z_4(t), +v_4(t))$ |
| 16 | 10 00 0 | $(+z_4(t), -v_4(t))$ |
| 17 | 10 00 1 | $(+z_5(t), -v_5(t))$ |
| 18 | 10 01 0 | $(+z_6(t), -v_6(t))$ |
| 19 | 10 01 1 | $(+z_7(t), -v_7(t))$ |
| 20 | 10 10 0 | $(+z_0(t), -v_0(t))$ |
| 21 | 10 10 1 | $(+z_1(t), -v_1(t))$ |
| 22 | 10 11 0 | $(+z_2(t), -v_2(t))$ |
| 23 | 10 11 1 | $(+z_3(t), -v_3(t))$ |
| 24 | 11 00 0 | $(+z_7(t), +v_7(t))$ |
| 25 | 11 00 1 | $(+z_6(t), +v_6(t))$ |
| 26 | 11 01 0 | $(+z_5(t), +v_5(t))$ |
| 27 | 11 01 1 | $(+z_4(t), +v_4(t))$ |
| 28 | 11 10 0 | $(+z_3(t), +v_3(t))$ |
| 29 | 11 10 1 | $(+z_2(t), +v_2(t))$ |
| 30 | 11 11 0 | $(+z_1(t), +v_1(t))$ |
| 31 | 11 11 1 | $(+z_0(t), +v_0(t))$ |

Table 10.3: Mapping rules for the hardlimited OQPOM scheme.

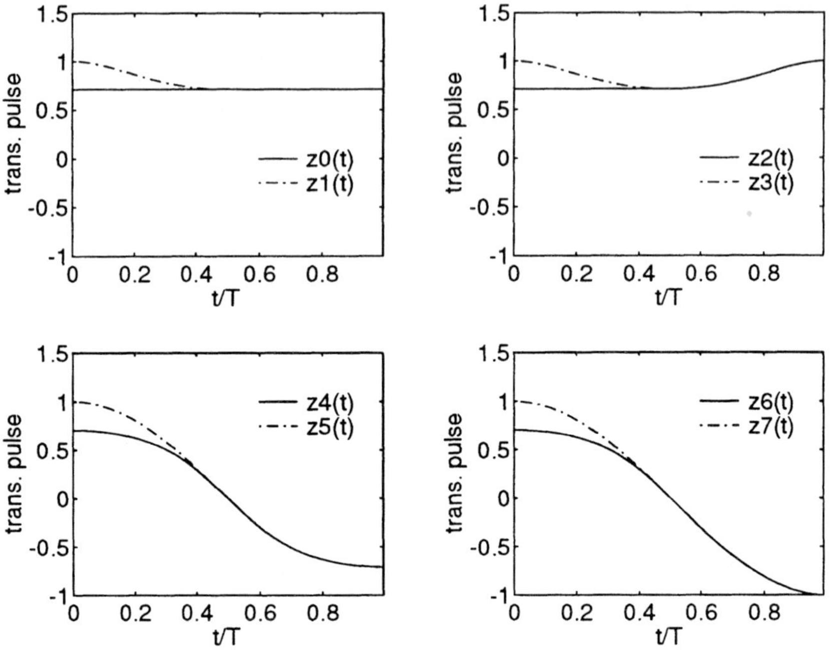


Figure 10.23: Example of transmitted pulses for hardlimited OQPOM signals (inphase component).

$$v_7(t) = \begin{cases} \frac{-s_1(t+T/2)}{\sqrt{s_1^2(t)+s_1^2(t+T/2)}} & 0 \leq t \leq T/2 \\ \frac{s_1(t-T/2)}{\sqrt{s_1^2(t)+s_1^2(t-T/2)}} & T/2 \leq t \leq T \end{cases} . \tag{10.55}$$

**Example 10.10** The waveforms $z_i(t)$, $v_i(t)$, $i = 0, 1, ..., 7$, are shown in Figure 10.23 and Figure 10.24 for the overlapping pulse shape $p(t) = 0.5(1 + \cos \pi t/T)$.

The hardlimited OQPOM scheme is represented by the structure shown in Figure 10.25, and the trellis diagram of its encoder is shown in Figure 10.26. The number of states of this encoder is double that of the QPOM scheme, which effectively increases the complexity of the optimum MLSE receiver. The MLSE receiver is then obtained by first finding two orthonormal bases, the first for the set of waveforms $\{z_i(t)\}$, and the second for the set of waveforms $\{v_i(t)\}$. From (10.40)-(10.55), we notice that

$$z_0(t) + z_3(t) = z_1(t) + z_2(t)$$

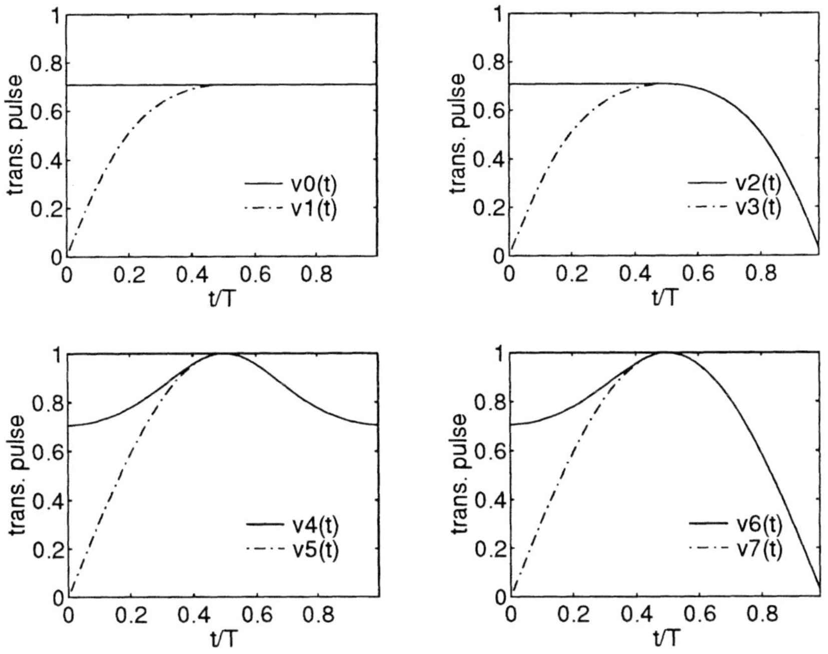


Figure 10.24: Example of transmitted pulses for hardlimited OQPOM signals (quadrature component).

$$\begin{aligned} z_4(t) + z_7(t) &= z_5(t) + z_6(t) \\ v_0(t) + v_3(t) &= v_1(t) + v_2(t) \\ v_4(t) + v_7(t) &= v_5(t) + v_6(t), \end{aligned}$$

indicating that these waveforms are linearly dependent. Only 6 orthonormal waveforms are needed for each basis. The pair of waveforms $(z_I(t), z_Q(t))$ is then represented by an augmented vector $\underline{z}$ of 12-elements. The first 6-elements of this vector are obtained by passing the in-phase component through a bank of 6 matched filters to the first basis, and the last 6-elements are obtained by passing the quadrature component through 6 matched filters to the second basis as shown in Figure 10.27.

To analyze the performance of hardlimited OQPOM signals, we consider the trellis diagram of Figure 10.26. This encoder has 8 subsets $\{\mathbf{A}_i,\ i = 0, ..., 7\}$, with each subset having *four* elements. To check the linearity of this encoder, we have computed the weight profile of these subsets for every error vector $\mathbf{e}$ for the pulse shape $p(t) = 0.5(1+\cos \pi t/T)$. As indicated in Table 10.4, for $\mathbf{e} =$

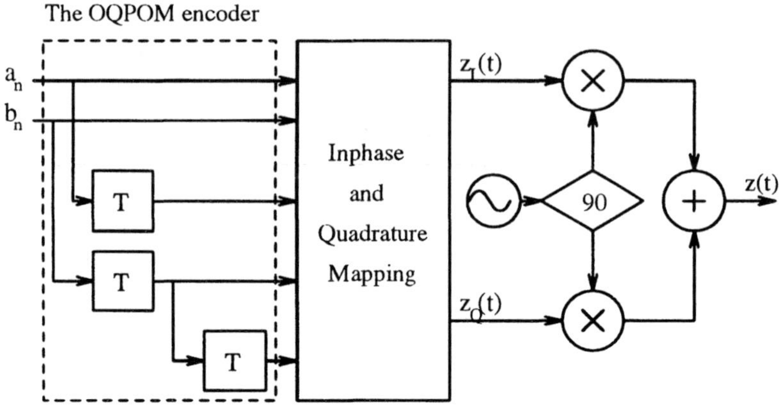


Figure 10.25: An alternative representation of hardlimited OQPOM schemes.

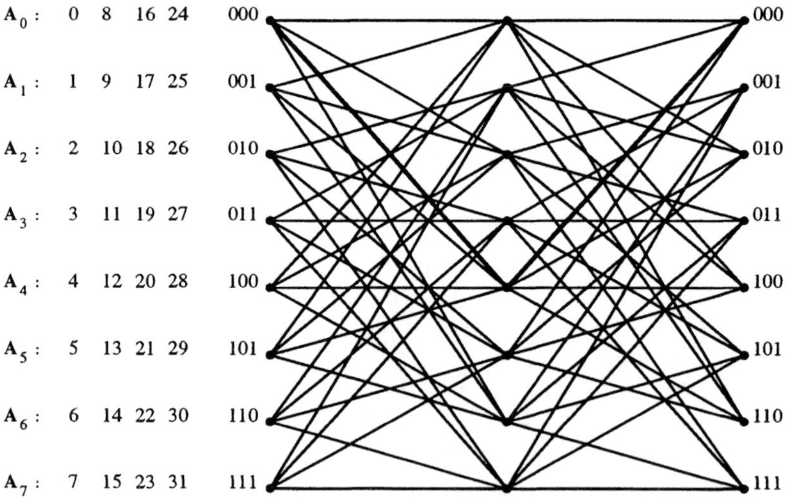


Figure 10.26: Trellis diagram of the hardlimited OQPOM scheme.

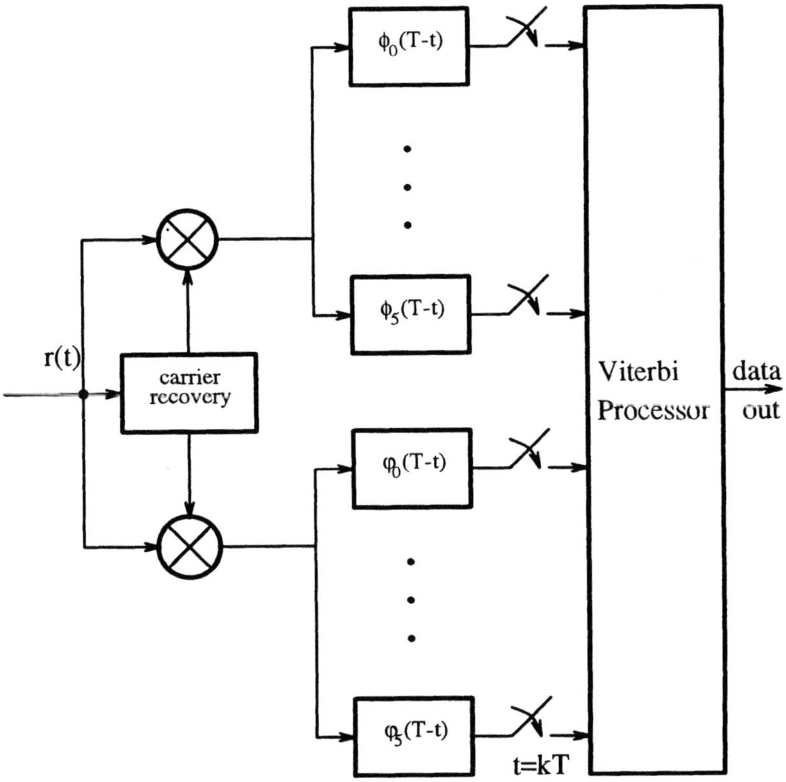


Figure 10.27: MLSE receiver for hardlimited OQPOM signals.

| Subset of Channel Signal | Weight Profile |
|---|---|
| $\mathbf{A}_0$ | $2D^{3.52} + 2D^{5.18}$ |
| $\mathbf{A}_1$ | $2D^{2.69} + 2D^{4.35}$ |
| $\mathbf{A}_2$ | $2D^{2.69} + 2D^{4.35}$ |
| $\mathbf{A}_3$ | $2D^{5.18} + 2D^{3.52}$ |
| $\mathbf{A}_4$ | $2D^{5.18} + 2D^{3.52}$ |
| $\mathbf{A}_5$ | $2D^{4.35} + 2D^{2.69}$ |
| $\mathbf{A}_6$ | $2D^{4.35} + 2D^{2.69}$ |
| $\mathbf{A}_7$ | $2D^{5.18} + 2D^{3.52}$ |

Table 10.4: Hardlimited OQPOM weight profile of subsets $\mathbf{A}_i$'s for $\mathbf{e} = 00011$.

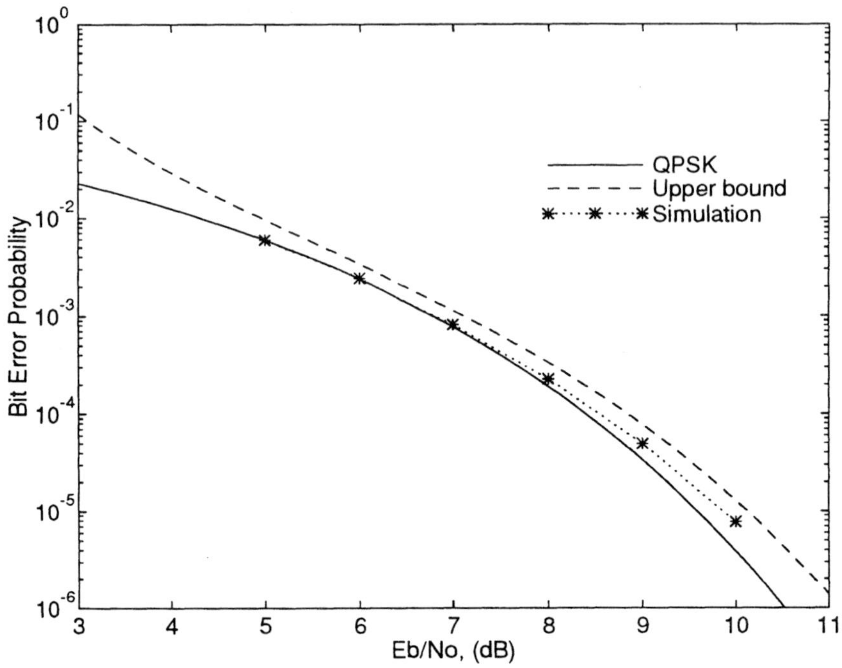


Figure 10.28: Performance of hardlimited OQPOM in AWGN channels.

00011 the subsets $\{\mathbf{A}_i\}$ have different error weight profiles. The hardlimited OQPOM scheme does not have a UEP. In such a situation, the analytical method used for hardlimited QPOM cannot be applied.

One way to derive an upper bound on $P_b$ is to consider the method of pairwise error states which involves all possible correct paths (not only the all-zero path). The modified state diagram then becomes an expanded set of states, namely $2^6$ states. As we can see, in order to derive the transfer function for this system, we need to solve a system of 64 equations. This computation is cumbersome.

To avoid this computational complexity, we consider the weight profile distribution of the code. By using the worst weight profile for each error vector e, it is possible to approximate the transfer function from a state diagram containing only *eight* states (see Problem 10.8). With this assumption, the method of [22] can be applied, and a good approximation for the Chernoff upper bound on $P_b$ can be obtained. Figure 10.28 [135] shows the performance of hardlimited OQPOM signals for the pulse shape $p(t) = 0.5(1 + \cos(\pi t/T))$. As observed from this figure, this scheme performs similarly to the hardlimited

QPOM scheme.

### 10.2.5 Performance of HardLimited QPOM Signals in Fading Channels

Using a similar approach as that of Section 10.1.3, the modified transfer function $T(I)$ of the hardlimited QPOM scheme in a fading channel with Channel State Information (CSI) can be derived (see Problems 10.9 and 10.10)

$$T(I) = \frac{W_0 I + W_1 I^2}{1 - Q_0 I - Q_1 I^2},$$

where

$$\begin{aligned} W_0 &= 0.5D_3^2 + 0.5D_4^2 + D_3 D_4 + D_2^2, \\ W_1 &= D_1 D_2 D_3 + D_2 D_3 D_5 + D_1 D_2 D_4 + D_2 D_4 D_5 \\ &\quad - D_0 D_3 D_4 - 0.5 D_0 D_3^2 - 0.5 D_0 D_4^2 - 2D_2^3, \\ Q_0 &= D_0 + 2D_2, \\ Q_1 &= 0.5D_1^2 + 0.5D_5^2 + D_1 D_5 - 2D_0 D_2, \end{aligned}$$

and for $i = 0, 1, \cdots, 5$,

$$D_i = \begin{cases} \frac{1}{1+\alpha_i/4N_0} & \text{for Rayleigh fading channel} \\ \frac{e^{-C_i/(1+0.2\sigma_\gamma C_i)}}{1+\sigma_a^2 \alpha_i/2N_0}, & \text{for shadowed Rician fading channel} \end{cases}$$

with

$$C_i = \frac{\alpha_i/4N_0}{1 + \sigma_a^2 \alpha_i/2N_0} e^{2m_\gamma}.$$

The Chernoff bound (5.39) on the average bit error probability of hardlimited QPOM signals then becomes

$$P_b \leq \frac{W_0(1 + Q_1) + W_1(2 - Q_0)}{2\left[1 - Q_0 - Q_1\right]^2}. \tag{10.56}$$

The important point to be observed in the above equation is that $P_b$ depends on the pulse shape $p(t)$.

At high signal-to-noise ratios, higher-order terms in the expression of the average bit error probability can be neglected. Hence, for a Rayleigh fading

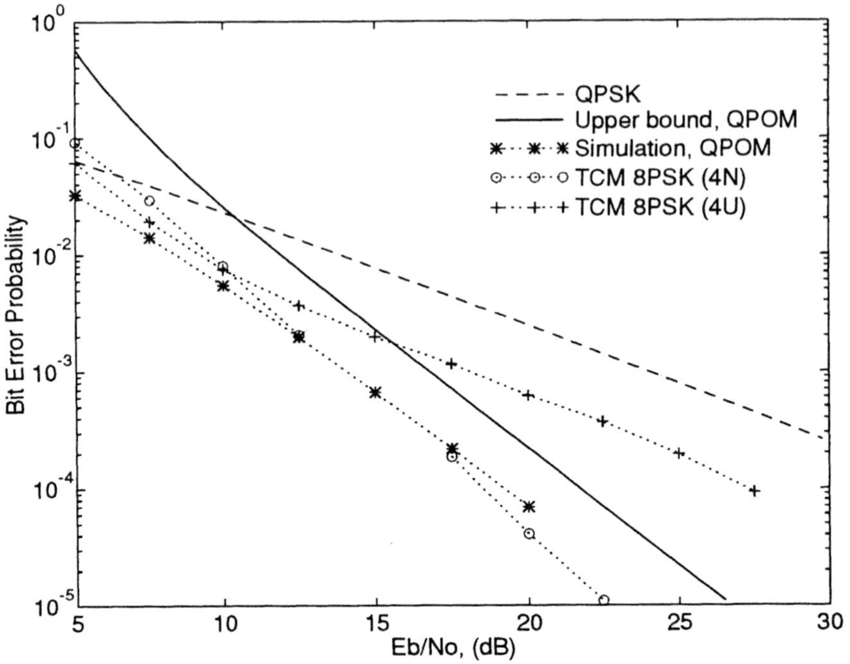


Figure 10.29: Performance comparison of hardlimited QPOM and 4-state 8PSK TCM in Rayleigh fading channels.

channel, $P_b$ can be simplified to

$$\begin{aligned} P_b &\le \frac{(2-\beta_0-\beta_1)^2}{(1-\beta_0)^2(1-\beta_1)^2}\,[E_b/N_0]^{-2} \\ &\le 0.25\frac{(2-\beta_0-\beta_1)^2}{(1-\beta_0-\beta_1)^2}\,[E_b/N_0]^{-2}\,. \end{aligned} \tag{10.57}$$

As we can see the variation of $\beta_0$ and $\beta_1$ does not considerably affect the expression of the upper bound. Therefore, it is possible to select a double-interval overlapping pulse shape which gives a transmitted signal with a compact spectrum and good performance. It is also noted that $P_b$ varies inversely with the square of $E_b/N_0$. Figure 10.29 [136] illustrates the performance of the hardlimited QPOM scheme. Simulation results are obtained for the double-interval overlapping pulse shape $0.5(1+\cos(\pi t/T))$.

The error rate of QPOM decreases inversely with the square of $E_b/N_0$ as shown by (10.57). The results in Figure 10.29 clearly indicate the superior performance of the hardlimited QPOM scheme in Rayleigh fading channels.

Its performance is comparative to that of TCM 8PSK (4N) and better than that of TCM 8PSK (4U).

**Example 10.11 Performance of Hardlimited OQPOM Schemes in a Rayleigh Fading Channel with Perfect CSI**

The trellis diagram of the hardlimited OQPOM encoder is shown in Figure 10.26. We notice that the shortest error event path of this encoder is of length 2. At high $E_b/N_0$, its performance is expected to vary inversely with the square of $E_b/N_0$ (same as hardlimited QPOM schemes).

An upper bound on the performance of this scheme can be obtained by first computing the squared product of Euclidean distances along the shortest path and then using the Chernoff bound. The squared product distance is easily obtained from the trellis diagram of Figure 10.26. The upper bound on $P_b$ can then be approximated by

$$P_b \leq 0.25\left[1+(1-\beta_m)E_b/N_0\right]^{-2},$$

where $\beta_m$ is the maximum of $\beta_0$ and $\beta_1$. We notice that this equation is comparable to (10.57). Both functions vary inversely with the square of $E_b/N_0$. Therefore, the performance of hardlimited OQPOM is similar to that of QPOM, and choosing between the two schemes will depend only on the complexity and the spectral properties of each scheme. Figure 10.30 gives the upper bound on $P_b$ for both schemes. As we can see the performance is comparable. At a high signal-to-noise ratio, hardlimited OQPOM performs better than QPOM due to the structure of its trellis diagram.

**Example 10.12 Tight Upper Bound on Bit Error Probability of Hardlimited QPOM Schemes in a Rayleigh Fading Channel with Perfect CSI**

The effective length of the hardlimited QPOM scheme is $L = 2$. The minimum Euclidean distance between two signal symbols on the shortest error event path is

$$d_{min} = \sqrt{4(1-\beta_m)E_b}.$$

Using a similar approach as that of Example 10.6, we obtain a tighter upper bound on $P_b$ of the hardlimited QPOM scheme in a Rayleigh fading channel with perfect CSI

$$P_b \leq \frac{1}{4}\left[\frac{1}{1+x_{min}} + \frac{1}{(1+x_{min})^2}\right]\frac{W_0(1+Q_1)+W_1(2-Q_0)}{[1-Q_0-Q_1]^2}, \qquad (10.58)$$

where

$$x_{min} = \sqrt{\frac{(1-\beta_m)E_b/N_0}{1+(1-\beta_m)E_b/N_0}}.$$

Figure 10.31 shows the performance of the hardlimited QPOM scheme in a

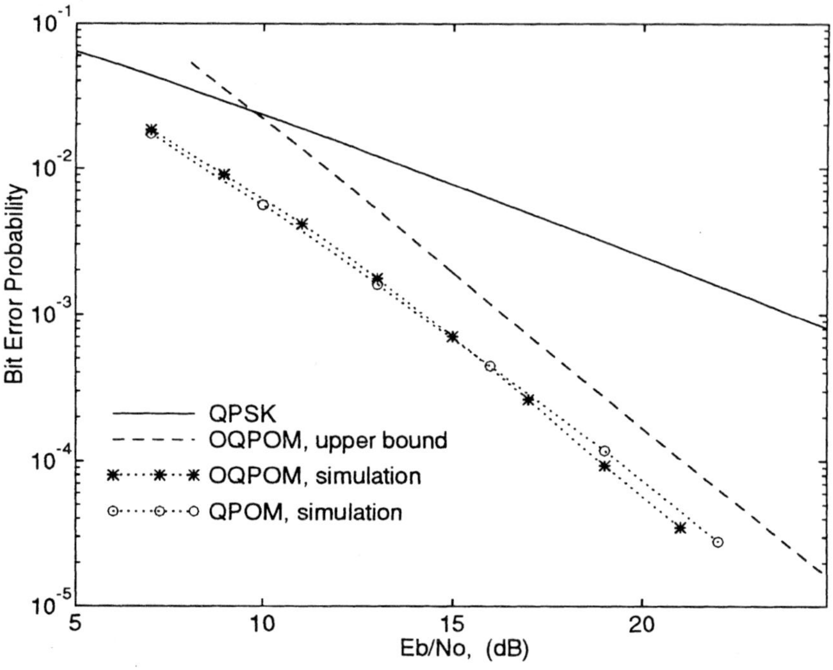


Figure 10.30: Performance comparison of hardlimited QPOM and hardlimited OQPOM signals in Rayleigh fading channels.

Rayleigh fading channel with perfect CSI for the double interval overlapping pulse shape

$$p(t) = \frac{1}{2}(1 + \cos \pi t/T), \qquad |t| \leq T.$$

The new upper bound of (10.58) is much closer to the simulation results than the Chernoff bound of (10.56).

**Example 10.13 Performance of Hardlimited QPOM and OQPOM Schemes in Rician Fading Channels**

Figures 10.32 and 10.33 show simulation results for a lightly shadowed fading channel with $\sigma_a^2 = 0.158$, $\sigma_\gamma = 0.115$ and $m_\gamma = 0.115$ (see Section 1.3). Figure 10.32 shows that hardlimited QPOM and OQPOM have the same performance and that both outperform conventional QPSK. Figure 10.33 shows that the performance of these schemes is comparable to that of 4-state TCM 8PSK schemes (4U is the 4-state Ungerboeck code with parallel paths and 4N is the 4-state code without parallel paths designed for fading channels previously discussed in Chapter 5).

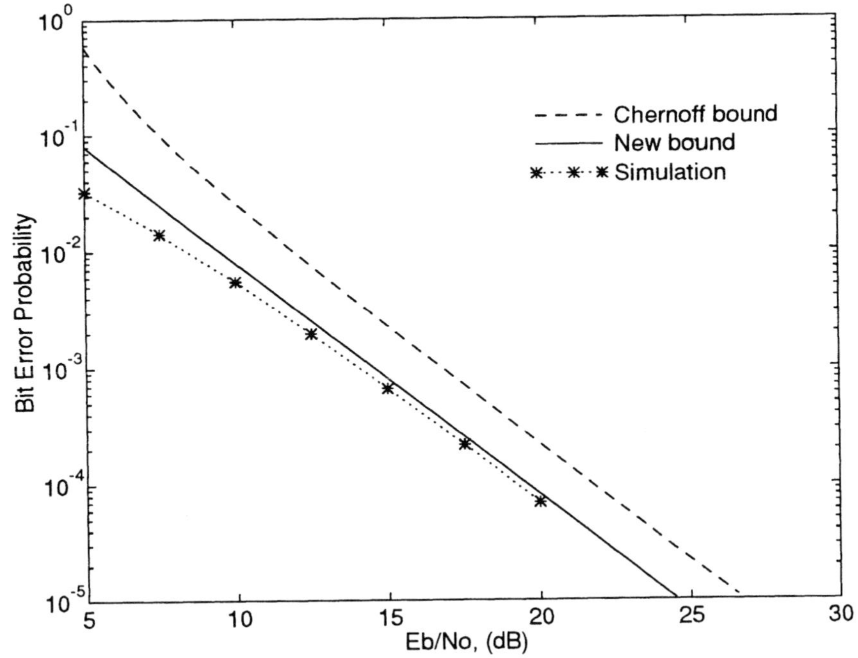


Figure 10.31: Performance of hardlimited QPOM signals in a Rayleigh fading channel with perfect CSI.

**Example 10.14 Performance Comparison and Discussions**

We have considered the average bit error probability of both QPOM and OQPOM signals over fading channels. We have seen that for both schemes, the bit error rate in Rayleigh fading channels decreases inversely with the square of $E_b/N_0$. At high SNR, the performance of hardlimited OQPOM is better than that of hardlimited QPOM due to the structure of its trellis diagram and its signal constellations. Even though the detection of hardlimited OQPOM signals is more complex, this scheme is still attractive due to its spectral characteristics.

Note that in the hardlimited QPOM and OQPOM signals, error correcting coding is not included and the modulation is basically a 4PSK. However the memory caused by the double-interval overlapping baseband pulse shape behaves as an encoder. As previously discussed, the MLSE receiver is constructed by exploiting this inherent *non-redundant* coding structure. Its trellis does not have any parallel transitions. This makes it suitable to combat fading. As shown in Figures 10.29 and 10.33, hardlimited QPOM and OQPOM schemes perform similarly to TCM 8PSK (4N) and both QPOM and OQPOM outperform TCM 8PSK (4U). This is explained by the fact that the hardlimited

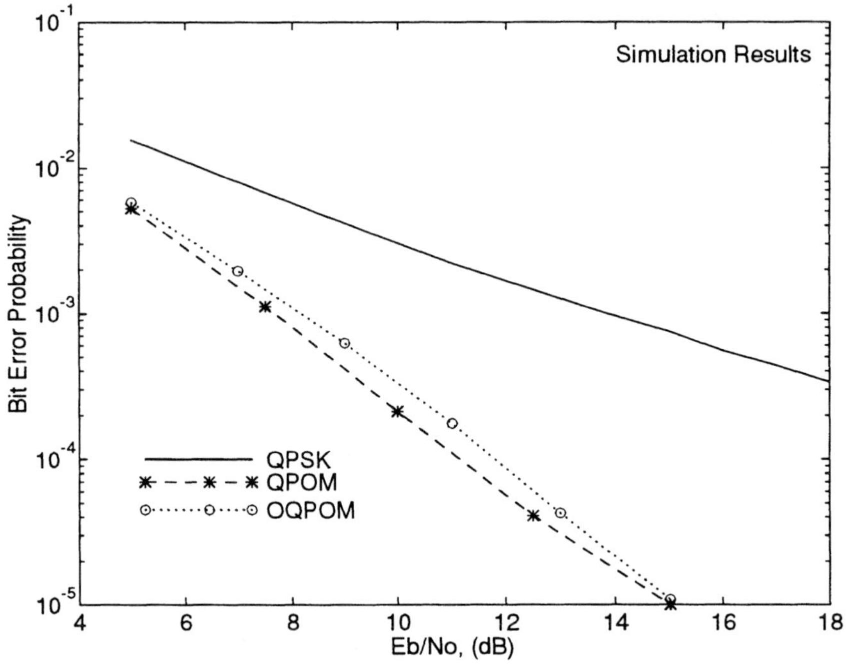


Figure 10.32: Performance comparison of hardlimited QPOM and OQPOM signals in Rician fading channels.

QPOM and TCM 8PSK (4N) exhibit the same diversity $L = 2$ (represented by the shortest error event path), while TCM 8PSK (4U) has a diversity of $L = 1$. As previously stated in Chapter 5, larger diversity yields a better performance in fading channels.

Consider the fact that the hardlimited QPOM or OQPOM is actually a 4PSK scheme. It can be seen that the hardlimited QPOM scheme is less sensitive to phase error than the TCM 8PSK (4N), and both schemes have the same receiver complexity.

In Example 10.11 we assumed an uncorrelated Rayleigh fading channel in which fading effects on adjacent symbols are independent. However, in a practical communication system, this assumption is valid only when long interleaving/de-interleaving is used, as shown in Figure 10.34. For coded modulation schemes with perfect CSI, using a depth of interleaving/de-interleaving larger than one-fifth the duration of a fade cycle, the fading amplitude from one symbol to the next can be considered as independent [146]. As shown in Figure 10.34, since the memory of the hardlimited OQPOM scheme is represented

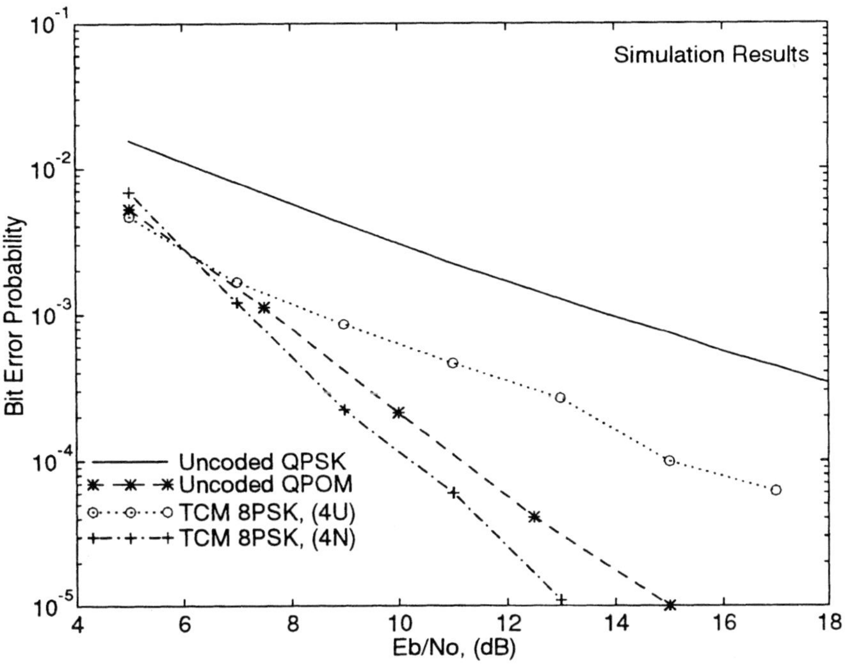


Figure 10.33: Performance comparison of hardlimited QPOM and 4-state 8PSK TCM in Rician fading channels.

by the linear encoder prior to interleaving, it can be successfully reconstructed by the de-interleaver prior to the MLSE.

However, the interleaver/de-interleaver structure should be selected in such a way as to preserve the compact spectrum of the transmitted hardlimited OQPOM signal. To understand this effect, let us examine the power spectral density of hardlimited OQPOM signals. From (10.27) this power spectral density can be derived as (see Problem 10.3)

$$\begin{aligned} Z(x) &= \vartheta \left\{ 2\sum_{i=0}^{7} |Z_i(x)|^2 \right\} \\ &+ \vartheta Re\left([Z_0(x)+Z_3(x)-Z_4(x)-Z_7(x)]^2 e^{-2j\pi x}\right) \\ &+ \vartheta Re\left([Z_1(x)+Z_2(x)-Z_5(x)-Z_6(x)]^2 e^{-j2\pi x}\right), \qquad (10.59) \end{aligned}$$

where $\vartheta$ is a constant such that $Z(0) = 1$ and $x = (f - f_c)T$.

The above expression of $Z(x)$ consists of three terms. The first term represents the sum of the spectra of the individual pulses, and the last two terms

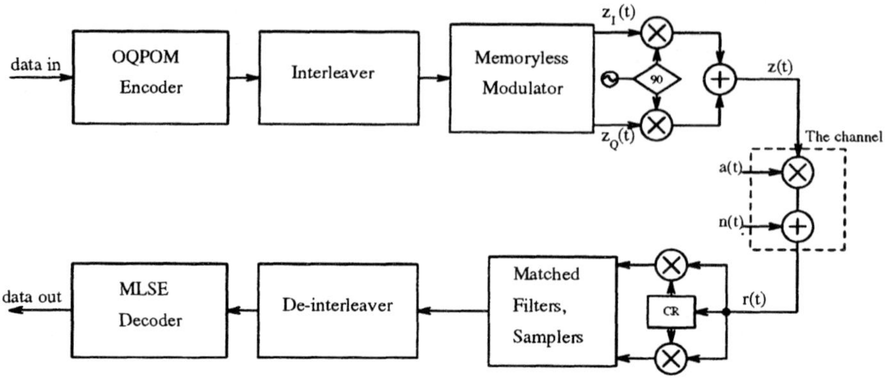


Figure 10.34: Block diagram of an OQPOM scheme in fading channels.

represent the correlation between these pulses. If this correlation is completely removed (i.e., all the pulses are independently generated), it can be shown that the last two terms will be cancelled, and the power spectral density function becomes

$$Z(x) = \vartheta \sum_{i=0}^{7} |Z_i(x)|^2. \tag{10.60}$$

Figure 10.35 shows the plots of (10.59) and (10.60) for the same pulse shape $p(t)0.5(1+\cos(\pi t/T))$. (10.60) represents the *worst-case* power spectrum if the interleaver structure does not at all consider the correlation between pulses. Therefore, the compactness of the transmitted spectrum depends on the structure of the interleaver under consideration. This suggests an interleaver structure design which avoids the occurrence of adjacent pulses with abrupt transitions after interleaving. In this way, the spectra of interleaved OQPOM signals can be made close to that of the original hardlimited OQPOM signal.

Such an interleaver takes a block of $L$ channel symbols, permutes their positions according to a sequence which prevents adjacent pulses with abrupt transitions. With this technique, we can preserve the compact spectrum at the expense of more complexity than a conventional block or convolutional interleaver of the same size.

Let us now consider the case of *no interleaving*. In this case, the fading experienced by different modulation symbols becomes correlated. For a land-mobile radio channel, the correlation coefficient is

$$\xi = J_0(2\pi BT),$$

where $J_0(.)$ is the order zero Bessel function of the first kind and $B$ is the maximum Doppler frequency (see Chapters 1 and 8). Following a similar procedure

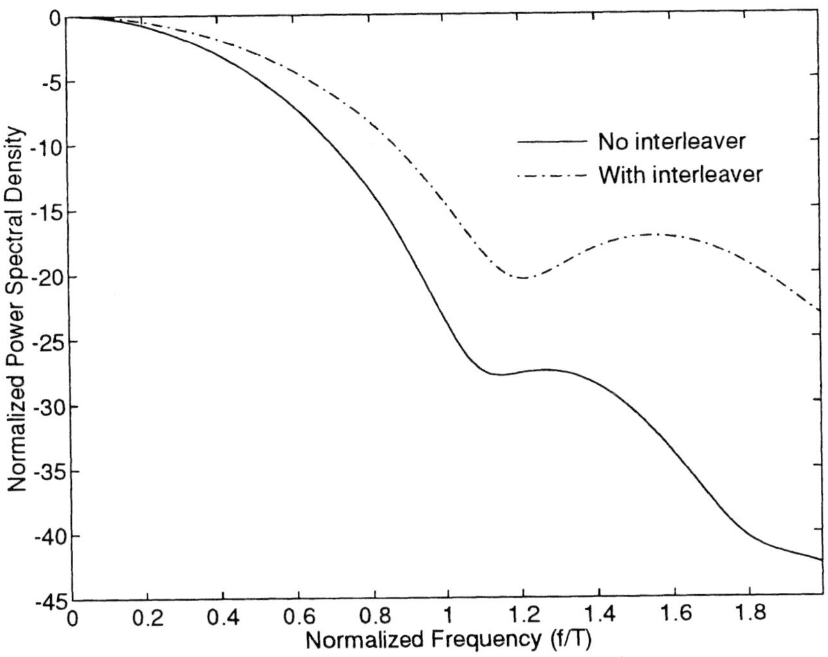


Figure 10.35: Power spectra of hardlimited OQPOM signals with and without interleaving.

as that used in [146] and [147], the performance of hardlimited OQPOM signals over a land-mobile correlated Rayleigh fading channel can be approximated as (see Problem 10.11)

$$P_b \approx \frac{1}{4(\lambda_0 - \lambda_1)} \left[ \lambda_0 \left( 1 - \sqrt{\frac{\lambda_0}{1+\lambda_0}} \right) - \lambda_1 \left( 1 - \sqrt{\frac{\lambda_1}{1+\lambda_1}} \right) \right], \tag{10.61}$$

where

$$\lambda_0 = \frac{E_b}{N_0}(1-\beta_m)[1 - J_0(2\pi BT)] \tag{10.62}$$

$$\lambda_1 = \frac{E_b}{N_0}(1-\beta_m)[1 + J_0(2\pi BT)]. \tag{10.63}$$

The bit error performance of hardlimited OQPOM signals in a correlated Rayleigh fading channel is shown in Figure 10.36 as a function of the product $BT$. Simulation results are also given for the case of an uncorrelated Rayleigh

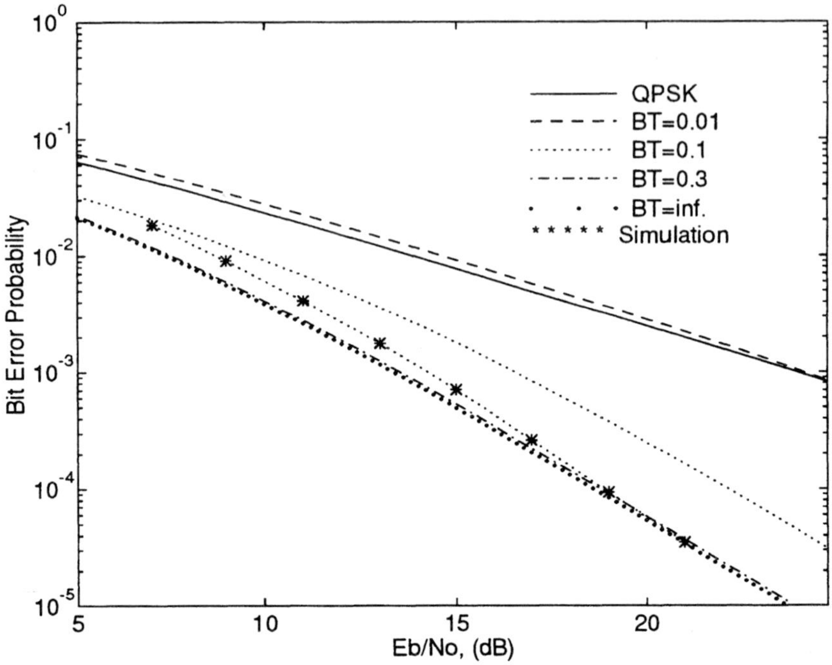


Figure 10.36: Performance of hardlimited OQPOM signals in correlated Rayleigh fading channels with various fading rates.

fading channel. It is observed that for $BT = 0.01$, the performance of hardlimited OQPOM scheme is similar to that of conventional QPSK. The performance of the hardlimited OQPOM gets better as the product $BT$ increases, that is, as the channel becomes less correlated ($\xi \ll 1$). A similar observation for trellis coded modulation schemes was reported in [145]. Performance results for a larger $BT$ are equivalent to those that are obtained when the actual $BT$ is smaller and interleaving is employed. For $BT \geq 0.3$, the obtained performance is similar to that of the uncorrelated fading case. A fading rate of $BT = 0.3$ corresponds to an approximate vehicle speed of 39 Km/h at Ka-band (19.914 GHz) and 4.8 Kb/s channel of the Advanced Communications Technologies Satellite (ACTS) Mobile System [150].

## 10.3 CODED QPOM SCHEMES FOR FADING CHANNELS

In the previous sections, we considered the performance of QPOM signals in fading channels. We observed that the new configuration of the QPOM scheme is suitable to fading channel applications. The use of this new configuration has improved the performance of the QPOM scheme considerably compared to ideal QPSK and this is achieved without any extra redundancy.

The performance of the QPOM scheme can be further improved by combining it with coding schemes. This section presents the structure, performance, and realization of combining hardlimited QPOM schemes and convolutional codes for fading channel applications. As a direct result of the structural analysis, we provide solutions to the problem of how to optimize the combined coding-modulation for hardlimited QPOM schemes. The optimization criterion considered is based on maximizing the length of the shortest error event path and the product of the squared branch distances along that error event path for a given external convolutional code.

### 10.3.1 Background

It is well known that combining coding and modulation improves the performance of the overall system. However, the combination of the two schemes depends on the structure of the modulation scheme. When coding is combined with a modulation scheme with memory, the optimum solution depends on the interaction between the memory of the outer code and that of the modulator. By decomposing the QPOM scheme into an encoder followed by a memoryless modulator, combining this modulation scheme with an additional code is equivalent to a problem of concatenating two codes: the *inner* QPOM code and the *outer* additional code. This concatenation will not affect the operation (mapping rules) of the memoryless modulator. If the outer code is selected as a convolutional code, the resulting concatenated code can be represented as one *single* code with a higher number of states.

An important characteristic in choosing the convolutional code is to reduce the number of states of the overall (concatenated) encoder. The number of states of the concatenated code is minimum when the output symbol of the convolutional code and the input symbol of the QPOM scheme are of equal size. Since the QPOM scheme is a 4-level scheme (2 bits/symbol), the convolutional code of interest is a rate 1/2, convolutional code having the following generators

$$g_i(D) = \sum_{j=0}^{\nu} g_{ij} D^i, \quad i = 1, 2, \tag{10.64}$$

where the coefficients $g_{ij}$'s are either 0 or 1 and $\nu$ represents the total memory of the code.

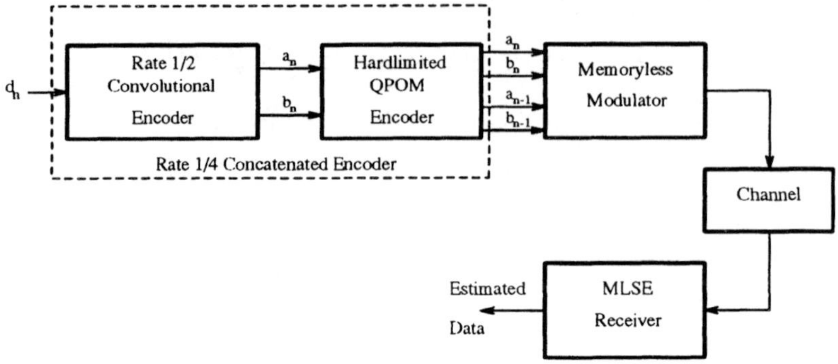


Figure 10.37: Block diagram of coded hardlimited QPOM schemes.

### 10.3.2 Design of Good Coded Hardlimited QPOM Schemes

The system under consideration is shown in Figure 10.37. As we can see the concatenation of the two codes will not affect the operation of the memoryless modulator. The hardlimited QPOM encoder is the rate 2/4 linear encoder shown in Figure 10.20. When this encoder is concatenated with the convolutional code having the generator polynomials of (10.64), the resulting code is a rate 1/4 convolutional code with the following generator polynomials:

$$\begin{aligned} G_1(D) &= g_1(D) \\ G_2(D) &= g_2(D) \\ G_3(D) &= D.g_1(D) \\ G_4(D) &= D.g_2(D). \end{aligned} \tag{10.65}$$

The total memory of the concatenated code is $\nu + 1$.

**Example 10.15 Rate 1/2, 2-State Convolutional Codes**
Consider the general trellis diagram of the rate 1/2 2-state convolutional code given in Figure 10.38. The encoder output is defined by the set of symbols $\{c_0, c_1, c_2, c_3\}$. These symbols are distinct and are chosen from the set of symbols $\{00, 01, 10, 11\}$. Figure 10.39 shows the trellis diagram of the concatenated code. The 4-bit output symbols of the concatenated encoder are derived from (10.65). With an appropriate choice of symbols, $\{c_0, c_1, c_2, c_3\}$, we can control the output of the concatenated encoder in order to optimize the performance of the coded scheme.

The bit error probability of the coded hardlimited QPOM scheme in fading channels can be minimized by maximizing the length of the shortest error event path and the product of the squared branch distances along that path.

From the trellis diagram of Figure 10.39, the length of the shortest error event path is 3. When the all-zero path is taken as a reference, the product of

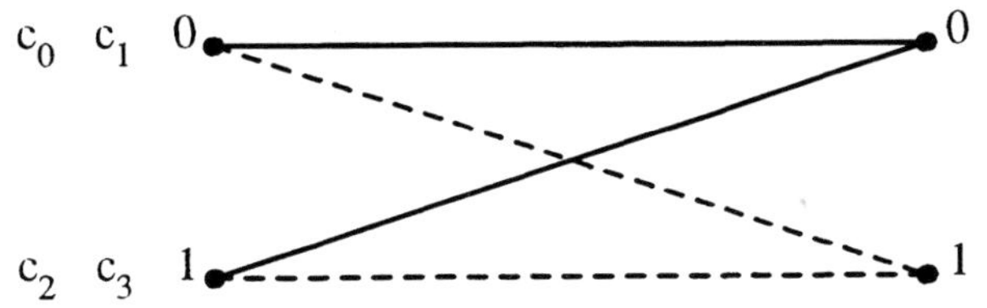


Figure 10.38: Trellis diagram of a rate 1/2, 2-state convolutional code.

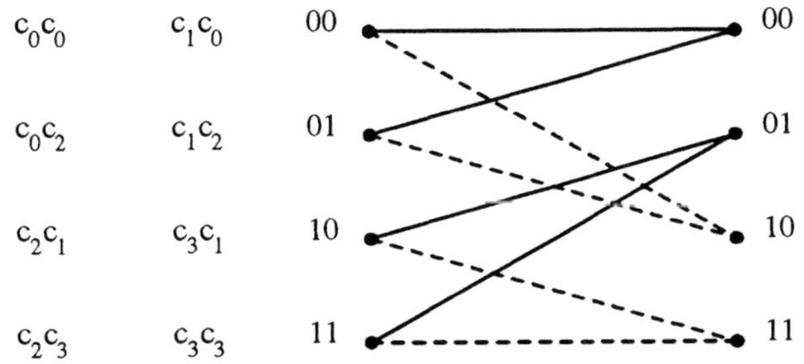


Figure 10.39: Trellis diagram of the code resulting from the concatenation of a 2-state convolutional code and the hardlimited QPOM encoder.

the squared branch distances along the shortest error event path is a function of the following symbols:

$$c_1c_0, \quad c_2c_1, \quad c_0c_2.$$

We start by defining the all-zero path of the code. For linear codes, at the all-zero state an input of 0 produces the all-zero output symbol, i.e.,

$$c_0 = 00.$$

Consider the first subset of the trellis diagram of the hardlimited QPOM encoder (Figure 10.20)

$$\mathbf{A}_0 = \{0, 4, 8, 12\}.$$

The first subset of the concatenated code must be $\{0,4\}$, $\{0,8\}$, or $\{0,12\}$. It can be shown that if this subset is $\{0,12\}$, the obtained code has a UEP and yields the best performance (see Problem 10.12).

It follows that

$$c_1 = 11.$$

The remaining symbols $c_2$ and $c_3$ can be selected as summarized in Table 10.5. It can be shown that the two concatenated codes in Table 10.5 have the same transfer function, and hence the same performance.

| Code | $c_0$ | $c_1$ | $c_2$ | $c_3$ |
|---|---|---|---|---|
| code 1 | 00 | 11 | 01 | 10 |
| code 2 | 00 | 11 | 10 | 01 |

Table 10.5: 2-State Convolutional Codes for Hard-limited QPOM Schemes.

**Example 10.16** Consider the coded hardlimited QPOM scheme using code 1 (Table 10.5) as the outer code. The trellis diagram of the concatenated code is shown in Figure 10.40. The error-state diagram of the concatenated code is similar to that of Figure 5.13 with

$$t_1 = t_3 = t_4 = t_7 = t_5 = t_{14} = t_{12} = t_{15} = 0.$$

Deriving the other branch labels and solving the nodal equations of this error-state diagram yields the following transfer function (see Problem 10.12):

$$T(I) = \frac{0.25ID_2(D_1 + D_5)(D_3 + D_4)}{1 + I[D_2 + 0.25(D_1 + D_5)^2]},$$

where

$$D_i = \begin{cases} e^{-\alpha_i/4N_0} & \text{for AWGN channels} \\ \frac{1}{1+\alpha_i/4N_0} & \text{for Rayleigh fading channels} \\ \frac{e^{-\frac{C_i}{1+0.2\sigma_\gamma C_i}}}{1+\sigma_a^2\alpha_i/2N_0}, & \text{for shadowed Rician fading channels} \end{cases}$$

with

$$C_i = \frac{\alpha_i/4N_0}{1 + \sigma_a^2\alpha_i/2N_0} e^{2m_\gamma},$$

$$\begin{array}{ll} \alpha_0 = 4E_b, & \alpha_2 = 2E_b, \\ \alpha_1 = 2(1+\beta_0)E_b, & \alpha_5 = 2(1+\beta_1)E_b \\ \alpha_3 = 2(1-\beta_0)E_b, & \alpha_4 = 2(1-\beta_1)E_b \end{array} \qquad (10.66)$$

$$\beta_0 = \int_0^{T/2} \frac{p(t) + p(t-T)}{\sqrt{p^2(t) + p^2(t-T)}} dt$$

$$\beta_1 = \int_0^{T/2} \frac{p(t) - p(t-T)}{\sqrt{p^2(t) + p^2(t-T)}} dt.$$

The minimum Euclidean distance of this coded hardlimited QPOM scheme is

$$d_{free} = \sqrt{2(3 + \beta_1 - \beta_0)E_b}.$$

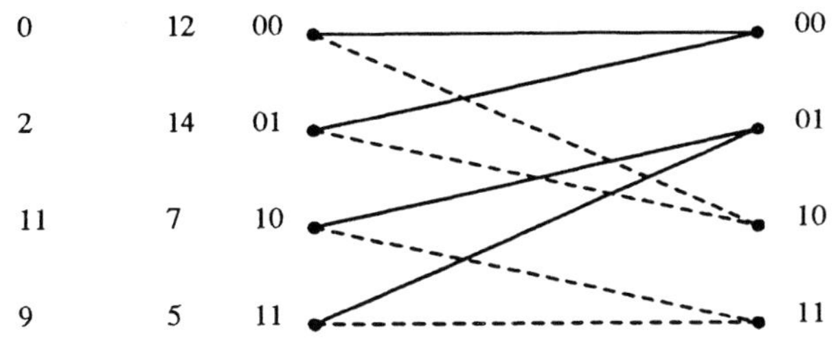


Figure 10.40: Trellis diagram of the code resulting from the concatenation of code 1 and the hardlimited QPOM encoder.

The Chernoff bound on the bit error probability of the coded hardlimited QPOM scheme is

$$P_b \leq \frac{1}{2}\frac{\partial T(I)}{\partial I}\bigg|_{I=1} = \frac{1}{8}\frac{D_2(D_1+D_5)(D_3+D_4)}{[1-D_2-0.25(D_1+D_5)^2]^2}.$$

A tighter upper bound for AWGN and Rayleigh fading channels can be derived by following the procedures given in Examples 10.3 and 10.6 (see Problem 10.14).

We notice that the performance is dependent on the pulse shape $p(t)$. By a proper choice of $p(t)$, both $\beta_0$ and $\beta_1$ become close to 0.5 and a coding gain up 1.76 dB over uncoded BPSK under AWGN conditions can be achieved, which is comparable to 2-state coded QPSK. However, the complexity of 2-state coded QPSK is less than that of coded hardlimited QPOM.

At high SNR, the Chernoff upper bound on $P_b$ in a Rayleigh fading channel can be approximated by

$$P_b \leq 0.125 D_2 D_3 D_5 \approx \frac{1}{(1-\beta_0)(1+\beta_1)}\left[\frac{E_b}{N_0}\right]^{-3}, \quad E_b/N_0 \gg 1,$$

where $\beta_0$ and $\beta_1$ are given by (10.34) and (10.35), respectively. From Figure 10.41, we notice that this coded scheme outperforms the uncoded QPOM by about 3 dB at a bit error rate of $10^{-4}$.

The performance of coded QPOM schemes can further be improved by increasing the number of states of the outer convolutional code. This is illustrated in the next example.

**Example 10.17 Rate 1/2, 4-State Convolutional Codes**

A rate 1/2, 4-state convolutional code can be represented by its trellis diagram as shown in Figure 10.42. In this figure, $c_0$, $c_1$, $c_2$, $c_3$, $c_4$, $c_5$, $c_6$, and $c_7$ represent the output symbols of the code. These symbols are chosen from the set {00, 01, 10, 11}.

When this code is concatenated with the hardlimited QPOM encoder, the overall code has the trellis diagram as shown in Figure 10.43. Its possible outputs

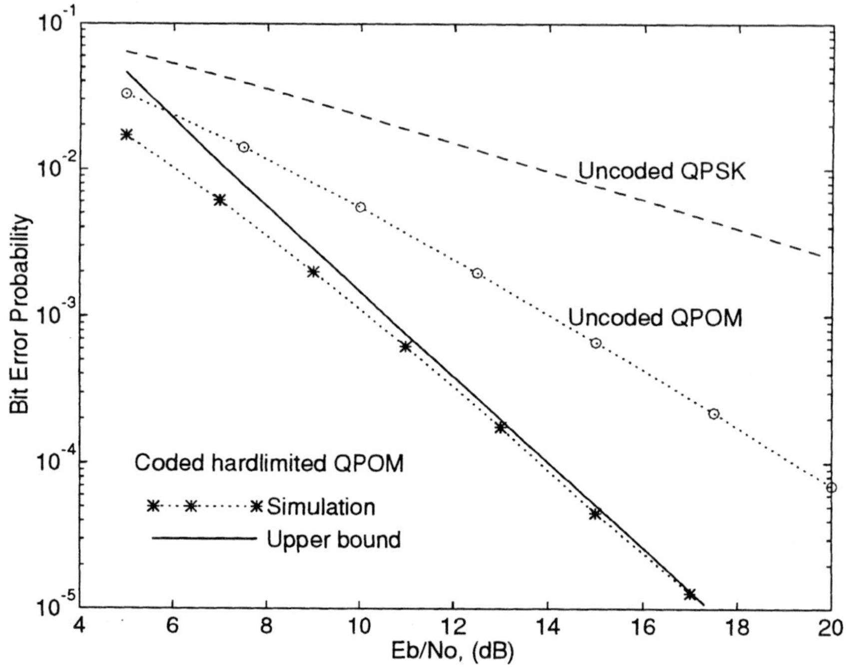


Figure 10.41: Performance of coded hardlimited QPOM schemes over Rayleigh fading channels (using a 2-state convolutional outer code).

are expressed in terms of the possible outputs of the external convolutional code. This code has a total of 8 states and the length of the shortest error event path is 4.

In designing the outer code, our objectives are to maximize *i*) the distance between signals leaving the same state, *ii*) the distance between signals merging into the same state, and *iii*) the product of the squared branch distances along the shortest error event path.

From Figure 10.43, we notice that when the all-zero path is taken as a reference, the product distance along the shortest error event path is a function of the following symbols:

$$c_1 c_0, \quad c_4 c_1, \quad c_2 c_4, \quad c_0 c_2.$$

The all-zero path is defined by the symbol $c_0 c_0$, i.e.,

$$c_0 = 00.$$

Using the same approach as that of Example 10.15 concerning the linearity of

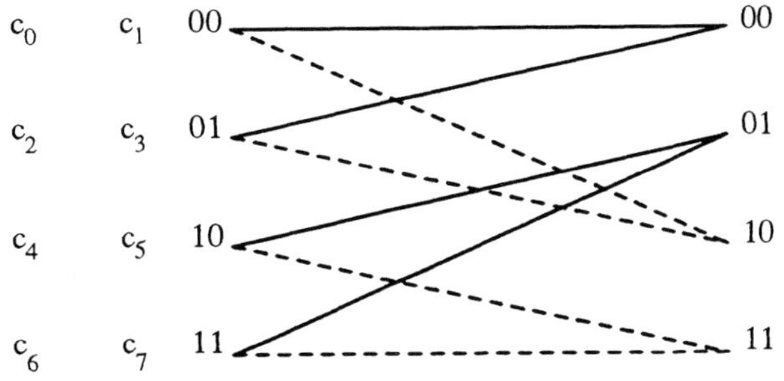


Figure 10.42: Trellis diagram of a rate 1/2, 4-state convolutional code.

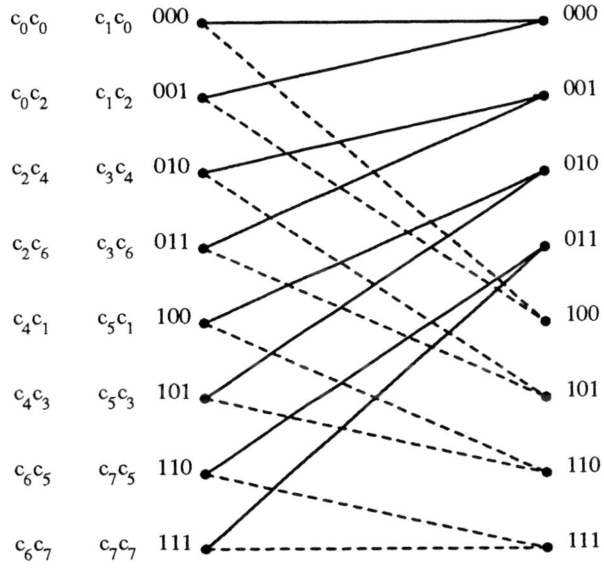


Figure 10.43: Trellis diagram of the code resulting from the concatenation of a 4-state convolutional outer code and the hardlimited QPOM encoder.

| code | $c_0$ | $c_1$ | $c_2$ | $c_3$ | $c_4$ | $c_5$ | $c_6$ | $c_7$ |
|---|---|---|---|---|---|---|---|---|
| code 1 | 00 | 11 | 11 | 00 | 01 | 10 | 10 | 01 |
| code 2 | 00 | 11 | 11 | 00 | 10 | 01 | 01 | 10 |

Table 10.6: 4-state convolutional codes for hardlimited QPOM schemes.

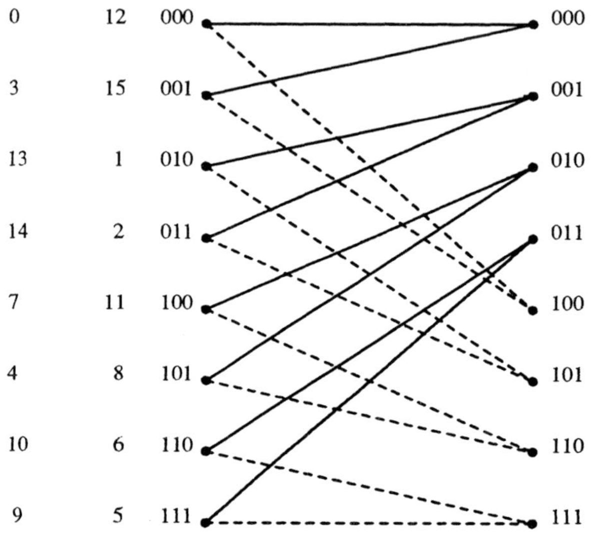


Figure 10.44: Trellis diagram for the concatenated code of Example 10.18.

the concatenated code, the selected values for $c_1$ and $c_2$ are

$$c_1 = 11, \quad c_2 = 11,$$

to avoid the presence of the distance $2(1-\beta_0)$ (always smaller than or equal to one) in the product distance along the shortest error path. The symbol $c_4$ will take one of two values,

$$c_4 = 01 \quad or \quad 10.$$

The selection of the remaining symbols is summarized in Table 10.6.

**Example 10.18** The trellis diagram of the resulting code from the concatenation of code 1 (Table 10.6) and the hardlimited QPOM encoder is given in Figure 10.44. Using the same procedure as that of Example 10.16, its modified transfer function is derived as (see Problem 10.15)

$$T(I) = \frac{0.25 I D_2^2 (D_1 + D_5)^2}{1 - I(D_2 + 0.25[D_3 + D_4]^2 + 0.25[D_1 + D_5]^2)}, \tag{10.67}$$

where $D_i$'s are defined as in Example 10.16.

The minimum normalized squared Euclidean distance of this code is given by

$$d_{free} = \sqrt{4(2+\beta_{min})E_b}, \tag{10.68}$$

with

$$\beta_{min} = min(\beta_0,\ \beta_1).$$

The tight upper bound on $P_b$ of this coded hardlimited QPOM scheme in AWGN, Rayleigh fading and shadowed Rician fading channels can be derived by using the procedures given in Examples 10.3, 10.6 and 10.16 (see Problem 10.17).

In an AWGN channel, the upper bound can be approximated as

$$P_b \le \frac{1}{2}\mathrm{erfc}\left[\sqrt{(2+\beta_{\min})\mathrm{E_b/N_0}}\right].$$

Compared to uncoded QPSK, a coding gain of $10\log(2+\beta_{min})$ dB is achieved. Since $\beta_{min} \le 0.5$, the maximum coding gain obtained with this scheme is 3.98 dB which is comparable to that of 4-state coded QPSK.

In a Rayleigh fading channel, the Chernoff upper bound on the bit error probability is given by

$$P_b \le \frac{0.125D_2^2[D_1+D_5]^2}{(1-D_2-0.25[D_3+D_4]^2-0.25[D_1+D_5]^2)^2},$$

which at high signal-to-noise ratios can be approximated as

$$P_b \le \frac{8(\alpha_1+\alpha_5)^2}{(\alpha_1\alpha_5)^2}\left[\frac{E_b}{N_0}\right]^{-4}, \quad \frac{E_b}{N_0} \gg 1.$$

Figure 10.45 shows the performance of this coded scheme. As indicated, more coding gain is obtained compared to the 2-state code of Example 10.16.

### 10.3.3 Design of Good Coded Hardlimited OQPOM Schemes

The hardlimited OQPOM encoder is a rate 2/5, 8-state encoder (Figure 10.26). When it is concatenated with an outer rate 1/2 convolutional code, the resulting code is a rate 1/5 convolutional code as shown in Figure 10.26. The generator polynomials of this rate 1/5 concatenated code are

$$\begin{aligned}G_1(D) &= g_1(D)\\ G_2(D) &= g_2(D)\\ G_3(D) &= D.g_1(D)\\ G_4(D) &= D.g_2(D)\\ G_5(D) &= D^2.g_2(D),\end{aligned}$$

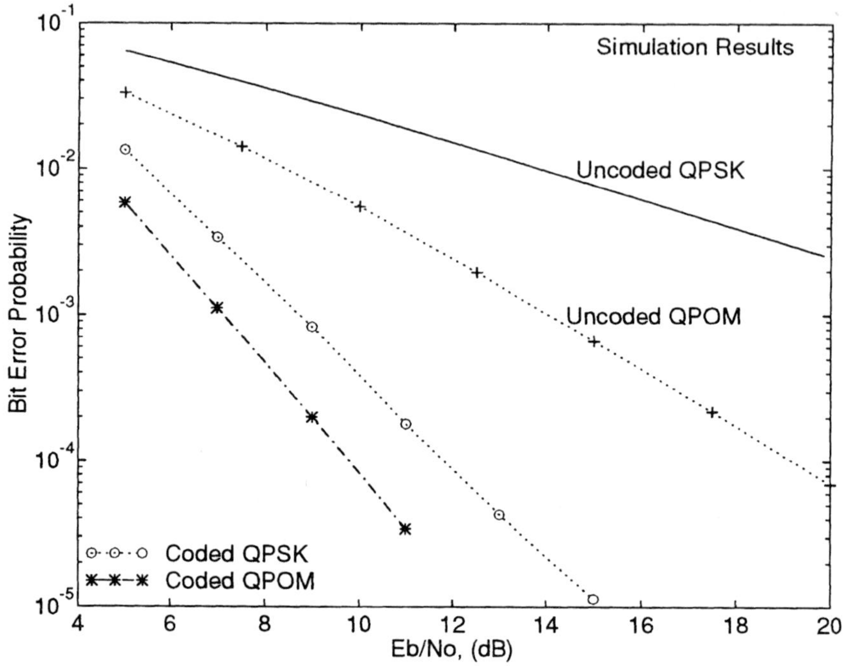


Figure 10.45: Performance of coded QPOM schemes over a Rayleigh fading channel (with a 4-state convolutional outer code).

where $g_1(D)$ and $g_2(D)$ are the generator polynomials of the outer rate 1/2 convolutional code.

Using the trellis diagram of the outer rate 1/2 convolutional code and the above generator polynomials, we can derive the trellis diagram of the rate 1/5 concatenated code. The total memory of this code is $\nu + 2$. Therefore, in general its number of states is *four* times the number of states of the outer convolutional code. Later we will show that by properly choosing the outer code structure, the number of states of the coded hardlimited OQPOM can be kept smaller or equal to that of the uncoded hardlimited OQPOM scheme.

**Example 10.19 Outer rate 1/2 2-State Convolutional Codes**

The general trellis diagram of a rate 1/2 2-state convolutional code is given in Figure 10.38. When this code is concatenated with the hardlimited OQPOM encoder, the resulting code has the trellis diagram of Figure 10.46, where $l_j$ denotes the least significant bit of the 2-bit symbol $c_j$, $j = 0, 1, 2, 3$. Following the same procedure that was used in Example 10.16 we can design convolutional codes suitable to the hardlimited OQPOM scheme [140].

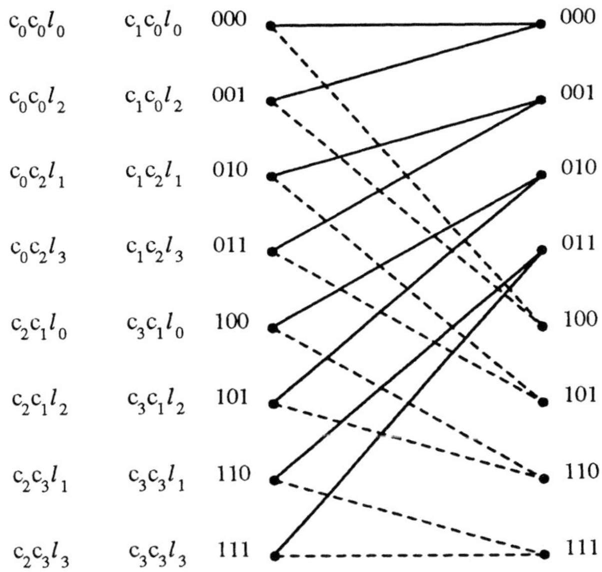


Figure 10.46: Trellis diagram of the code resulting from the concatenation of a 2-state convolutional code and the hardlimited OQPOM encoder.

Let the all-zero path of the code be defined by the symbol $c_0$, i.e., $c_0 = 00$. From Figure 10.46 and (10.64), we obtain the following relations:

$$\begin{aligned} l_1 &= g_{20} \\ l_2 &= g_{21} \\ l_3 &= g_{20} + g_{21}. \end{aligned}$$

If we select $l_2 = 1$ it follows that $l_1$ is the one-complement of $l_3$, i.e. if $l_1 = 1$ then $l_3 = 0$ or vice-versa. The generator polynomials of the outer code (called Code A) are:

$$\begin{aligned} g_1(D) &= 1 \\ g_2(D) &= 1 + D. \end{aligned}$$

If we select $l_2 = 0$ it follows that

$$g_{21} = 0$$

and

$$l_1 = l_3 = g_{20} = 1.$$

The outer code then has the following generator polynomials:

$$\begin{aligned} g_1(D) &= g_{10} + g_{11} D \\ g_2(D) &= 1. \end{aligned}$$

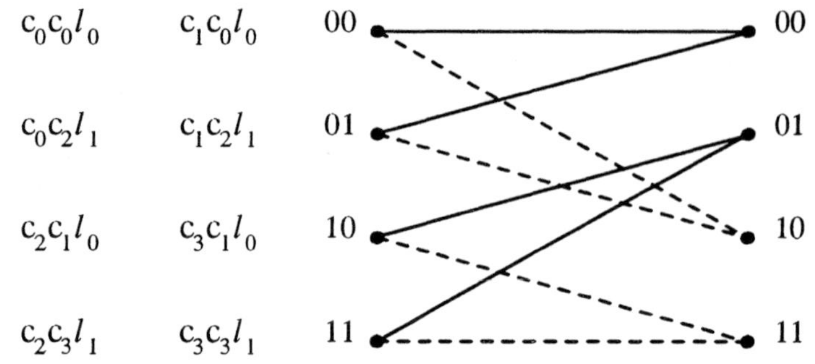


Figure 10.47: Trellis diagram of the reduced state concatenated code resulting from the concatenation of a 2-state convolutional code and the hardlimited OQPOM encoder.

This means that the least significant bit of the output symbols is independent of the memory of the code.

As we can see, the memory of the concatenated code has been reduced from 8 to 4. In this case, the trellis diagram of Figure 10.46 becomes a 4-state diagram as shown in Figure 10.47. The outer code (called Code B) that offers such a structure has the following generator polynomials:

$$g_1(D) = 1 + D$$
$$g_2(D) = 1.$$

By considering the two codes, the obtained results are shown in Figure 10.48. Code A and Code B produce 8-state and 4-state coded OQPOM schemes, respectively. It is observed that both schemes offer comparable performance. Therefore, Code B is a better choice because of its lower complexity (lower than the uncoded scheme).

From Figures 10.41 and 10.48, it is observed that coded hardlimited QPOM and coded hardlimited OQPOM schemes provide similar performance in a Rayleigh fading channel.

Further improvement in the performance of convolutionally coded OQPOM can be achieved by increasing the number of states of the convolutional code. In the following, rate 1/2 4-state convolutional codes are considered. These codes are combined with the hardlimited OQPOM scheme.

**Example 10.20 Rate 1/2 4-State Convolutional Codes**

A 4-state, rate 1/2 convolutional code is represented by its trellis diagram as shown in Figure 10.42. Following the same procedure used in Example 10.19, we can derive the trellis diagram of the rate 1/5 concatenated code as shown in Figure 10.49. We notice that the length of the shortest error event path through the trellis is $L = 5$, which is equal to the length of the shortest error event path of the convolutional outer code plus two.

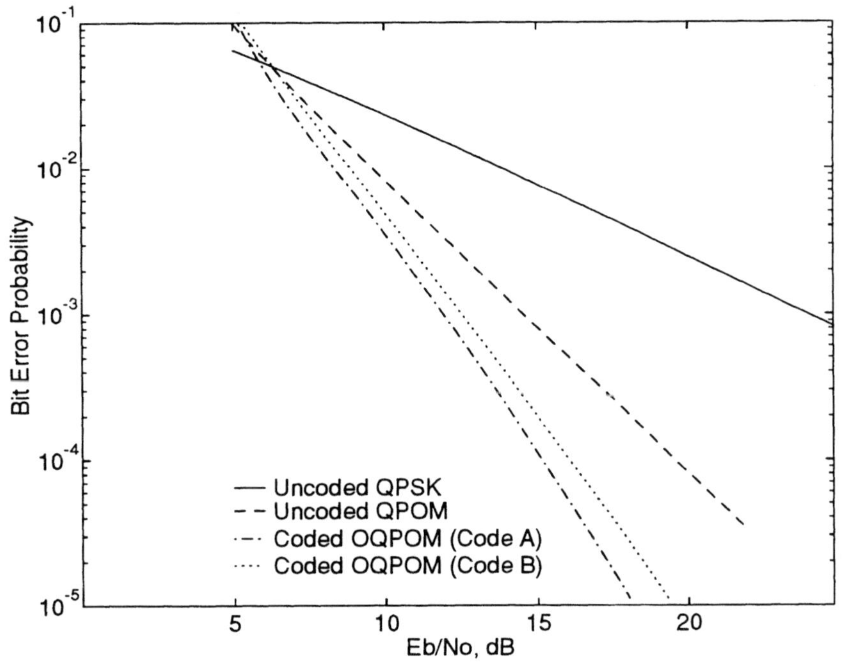


Figure 10.48: Performance of coded hardlimited OQPOM signals over Rayleigh fading channels with a 2-state convolutional outer code.

The design is based on maximizing *i*) the distance between signals leaving the same state, *ii*) the distance between signals merging into the same state, and *iii*) the product of the squared branch distances along the shortest error event path. The distance product along the shortest error event path is a function of the following symbols:

$$c_1c_0l_0, \quad c_0c_0l_2, \quad c_0c_2l_4, \quad c_2c_4l_1, \quad c_4c_1l_0.$$

Let the all-zero path be defined by the symbol $c_0$, i.e., $c_0 = 00$. The least significant bits, $l_i$, of the remaining symbols $c_i$ for $i = 1, 2, \cdots, 7$, are

$$\begin{aligned} & & l_1 &= g_{20} \\ l_2 &= g_{22} & l_3 &= g_{20} + g_{22} \\ l_4 &= g_{21} & l_5 &= g_{20} + g_{21} \\ l_6 &= g_{21} + g_{22} & l_7 &= g_{20} + g_{21} + g_{22}. \end{aligned}$$

If we select $l_2 = 1$ it follows that $g_{22} = 1$ and $l_1$, $l_4$, $l_5$ are the one-complements of $l_3$, $l_6$, $l_7$, respectively. The concatenated code has a total of 16 states (Figure 10.49). Using the optimization criteria the generator polynomials of the 4-state

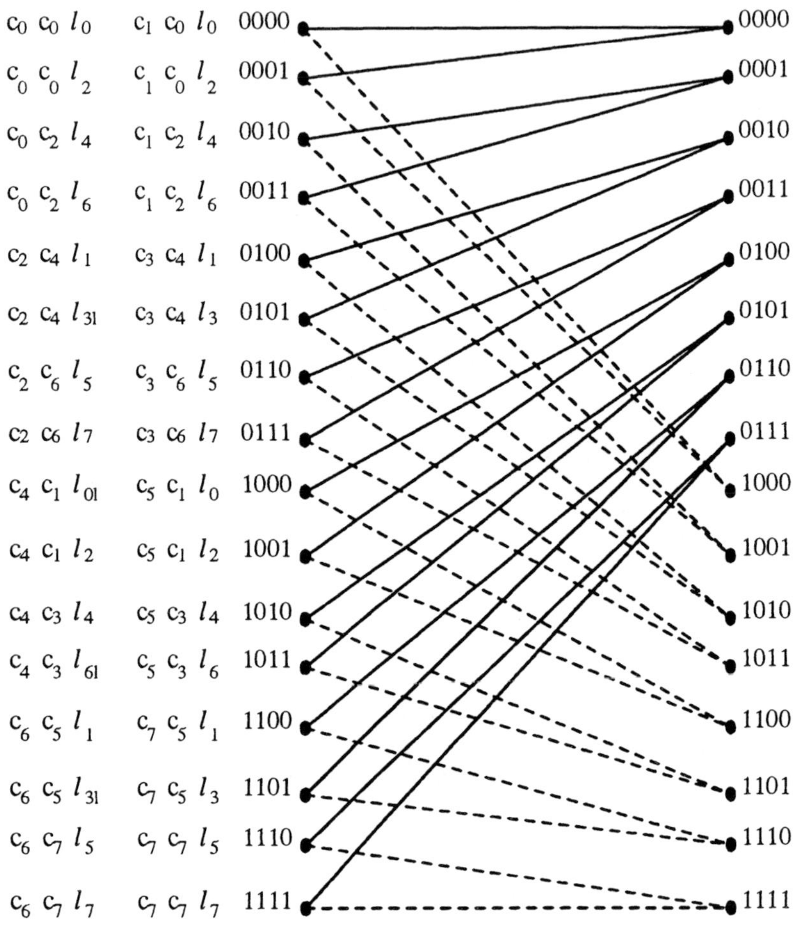


Figure 10.49: Trellis diagram of the concatenated code resulting from the concatenation of a 4-state convolutional code and the hardlimited OQPOM encoder.

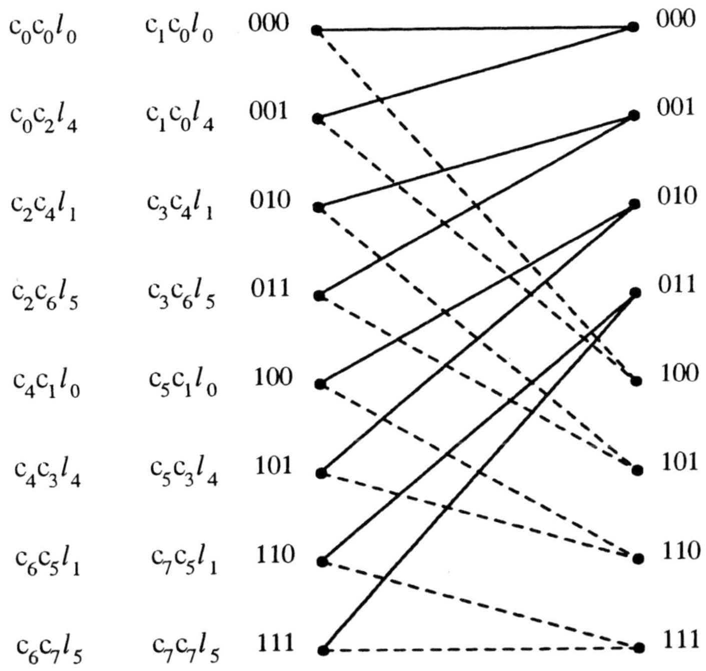

Figure 10.50: Trellis diagram of the reduced concatenated code resulting from the concatenation of a 4-state convolutional code and the hardlimited OQPOM encoder.

outer code are obtained as

$$g_1(D) = 1 + D^2 \quad \text{or} \quad g_1(D) = 1 + D + D^2$$
$$g_2(D) = 1 + D + D^2 \quad\quad\;\; g_2(D) = 1 + D^2.$$

If we select $l_2 = 0$, it follows that $g_{22} = 0$ and $l_1 = l_3$, $l_4 = l_6$, $l_5 = l_7$. The outer code then has the following generator polynomials:

$$g_1(D) = g_{10} + g_{11}D + g_{12}D^2$$
$$g_2(D) = g_{20} + g_{21}D.$$

The concatenated code now has 8 states as shown in Figure 10.50. The generator polynomials of the optimum 4-state outer code are

$$g_1(D) = 1 + D^2$$
$$g_2(D) = 1 + D$$

Figure 10.51 shows the performance comparison of various coded and uncoded hardlimited OQPOM and QPSK schemes. The number of states in parentheses shown in this figure indicates that of the additional (outer) convolutional code

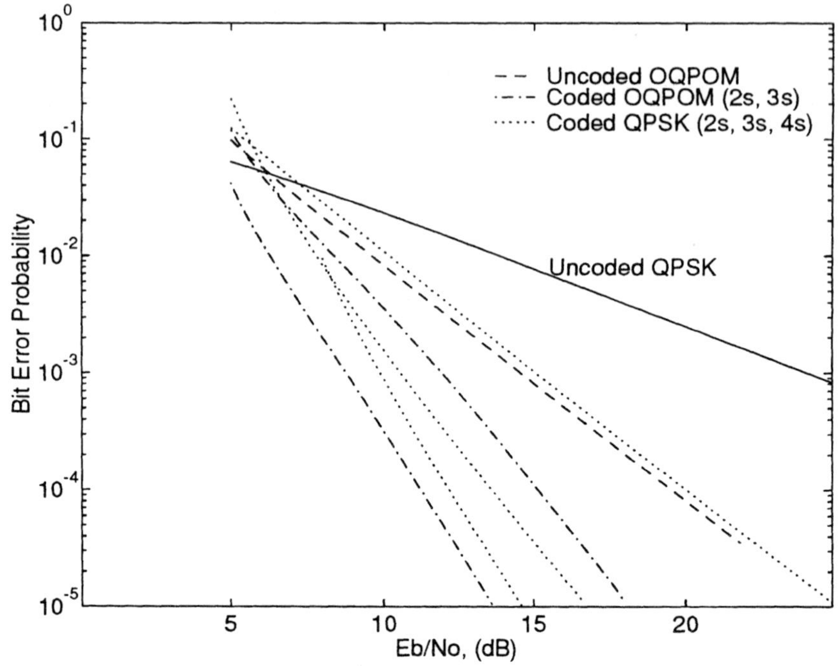


Figure 10.51: Performance of coded hardlimited OQPOM signals over Rayleigh fading channels for different convolutional codes.

in use. Let us first focus on the performance of coded and uncoded hardlimited OQPOM schemes. The coded hardlimited OQPOM schemes provide a significant performance improvement compared to the uncoded hardlimited OQPOM scheme. For instance, at a bit error rate of $10^{-4}$, the coded hardlimited OQPOM schemes using 2-state and 4-state outer codes offer a coding gain of 4.5 dB and 8 dB, respectively, compared to the uncoded scheme. The coding gain is obtained at the expense of bandwidth increase. It is worth noting that the coded hardlimited OQPOM schemes do not increase the number of states and hence the decoding complexity of the uncoded scheme. The *uncoded* hardlimited OQPOM scheme has 8 states (Figure 10.26) while the *coded* hardlimited OQPOM scheme using 2-state and 4-state outer codes has 4 states (Figure 10.47) and 8 states (Figure 10.50), respectively. Therefore, the *coded* hardlimited OQPOM scheme using a 4-state outer code and its *uncoded* counterpart have the same decoding complexity.

Figure 10.51 also indicates that the coded hardlimited OQPOM scheme outperforms the coded QPSK for a same additional convolutional code in use.

# Problems

**10.1** Prove that the bit energy $E_i$ of the waveform $s_i(\tau)$ given by (10.10) is

$$E_0 = (1+\gamma)E_b$$
$$E_1 = (1-\gamma)E_b,$$

where

$$\gamma = \frac{2}{E_b}\int_0^T p(t)p(t-T)dt,$$

and

$$E_b = \int_{-T}^{+T} p^2(t)dt.$$

**10.2** Show that the minimum squared Euclidean distance for a binary overlapping-pulse sequence with a $2K$-interval even pulse shape $p(t)$ is

$$d^2_{free} = 8\int_0^{KT} p^2(t)dt = 4E_b.$$

*Hint:* Following the procedure used in Section 10.1.2, show that the minimum distance is defined by the path $\{10\cdots 0, 010\cdots 0\} = \{n_1, n_2\}$. The squared Euclidean distance is given by

$$d^2 = |\underline{s}_{n_1} - \underline{s}_0|^2 + |\underline{s}_{n_2} - \underline{s}_0|^2,$$

where $\underline{s}_n$ is the vector corresponding to $n$.

**10.3 Power Spectrum of Hardlimited QPOM Signals**

**a)** Prove that the normalized power spectrum of hardlimited QPOM signal for a general value of $T_d$ is

$$\begin{aligned} Z(x) &= 2\vartheta\sum_{i=0}^{7}|V_i(x)|^2 \\ &+\vartheta\,Re\left\{\left([\sum_{i=0}^{3}V_{2i}(x)]\,[V_0(x)+V_1(x)-V_4(x)-V_5(x)]^*\right.\right. \\ &+\left.\left.[\sum_{i=0}^{3}V_{2i+1}(x)]\,[V_2(x)+V_3(x)-V_6(x)-V_7(x)]^*\right)e^{-j2\pi fT}\right\}. \end{aligned}$$

*Hint:* Apply the characteristics of the waveforms $v_i(\tau)$'s and the transition probabilities $p_{ik}$'s.

b) For $T_d = 0$, prove that the normalized power spectrum of a hardlimited QPOM signal is

$$Z(x) = \vartheta \left[ 2 \sum_{i=0}^{3} |V_{2i}(x)|^2 + \mathrm{Re} \left\{ \left( \sum_{i=0}^{3} V_{2i}(x) \right)^2 e^{-j2\pi fT} \right\} \right].$$

*Hint:* By redefining the time origin, $v_0(t)$, $v_2(t)$ become even functions and $v_4(t)$, $v_6(t)$ become odd functions (see Figure 10.18). Also note that $v_{2i+1}(t) = v_{2i}(t), i = 0, 1, 2, 3$.

c) For $T_d = T/2$, prove that the normalized power spectrum of a hardlimited *Offset* QPOM is

$$\begin{aligned} Z(x) &= 2\vartheta \sum_{i=0}^{7} |V_i(x)|^2 + \\ & \vartheta Re \left( e^{-j2\pi fT} \left[ (\sum_{l=0}^{3} V_{2l}(x))^2 + (\sum_{i=0}^{3} V_{2l+1}(x))^2 \right] \right). \end{aligned}$$

*Hint:* By redefining the time origin, $v_0(t)$, $v_3(t)$ become even functions and $v_4(t)$, $v_7(t)$ become odd functions. Also note that $v_1(t) = v_2(-t)$, $v_5(t) = -v_6(-t)$ (see Figure 10.23).

**10.4** Using the results of Problem 10.3, plot the power spectra of the hardlimited QPOM and hardlimited OQPOM signals using a double-interval pulse shape $p(t) = \cos(\pi t/2T), \quad |t| \leq T$.

**10.5** Derive the error weight profiles for all possible error vectors $\mathcal{F}(\mathbf{A}_i, \underline{e}_j, D)$ for the case of a hardlimited QPOM signal and prove that the hardlimited QPOM encoder has a uniform error property (UEP).

**10.6** Following the procedure given in Example 5.11, compute the branch labels of the error-state diagram of the hardlimited QPOM scheme.

**10.7** Derive the generating function of the hardlimited QPOM scheme. (*Hint:* Follow the procedure given in Example 5.11.)

**10.8** Consider the hardlimited OQPOM scheme in an AWGN environment. For each error vector $\mathbf{e}$ compute the worst weight profile. Derive the approximated error-state diagram of the hardlimited OQPOM scheme and its transfer function.

**10.9** Following the procedure given in Examples 5.9-5.11, derive the modified transfer function of the hardlimited QPOM scheme in a Rayleigh fading channel.

**10.10** Repeat Problem 10.9 for a shadowed Rician fading channel.

**10.11** Derive the bit error probability of hardlimited OQPOM schemes over a land-mobile correlated Rayleigh fading channel.

*Hint:* The pairwise error probability can be written as

$$\begin{aligned} P(\mathbf{z} \rightarrow \hat{\mathbf{z}}) &= \text{Prob}(D \leq 0) \\ &= -\sum \text{Residue}[\Phi_D(s)/s]_{\text{RPoles}}, \end{aligned}$$

where $D$ is the decoding metric and $\Phi_D(s)$ is the characteristic function of D given by

$$\Phi_D(s) = \left(\prod_{i=1}^{L} \frac{1}{\lambda_i}\right) \left(\prod_{i=1}^{L} \frac{-1}{(s-p_{2i})(s-p_{1i})}\right).$$

$$p_{ji} = \frac{1}{2} + (-1)^j \sqrt{\frac{1}{4} + \frac{1}{\lambda_i}}, \quad j = 1, 2,$$

where $\lambda_i$ is the eigenvalue of the correlation matrix $\Phi_{\mathbf{z}\hat{\mathbf{z}}}$ of $D$.

$$\Phi_{\mathbf{z}\hat{\mathbf{z}}} = \frac{1}{2} E\{\mathbf{z}\hat{\mathbf{z}}\} = \Delta \Phi \Delta^H$$

$$\Delta = \begin{pmatrix} d_i & & 0 \\ & \ddots & \\ 0 & & d_L \end{pmatrix},$$

where $d_i$ is the Euclidean distance of the $i$th branch. $(.)^H$ represents the Hermitian transpose. $\Phi$ is a square matrix with

$$\phi(i, j) = \rho(i - j) = \frac{E_s}{N_0} R((i - j)T).$$

$R(\tau)$ is the autocorrelation function of the fading process. For a land-mobile fading channel

$$R(\tau) = J_0(2\pi B \tau).$$

**10.12** Compute the error weight profiles for all possible error vectors $\mathcal{F}(\mathbf{A}_i, \underline{e}_i, D)$ for the coded hardlimited QPOM scheme with the trellis diagram shown in Figure 10.40. Verify that this scheme has a uniform error property. Find the transfer function for this scheme.

**10.13** Derive the trellis diagram of the coded hardlimited QPOM scheme using Code 2 of Table 10.5. Find the transfer function for this scheme. Verify that this transfer function is identical to that obtained in Problem 10.12.

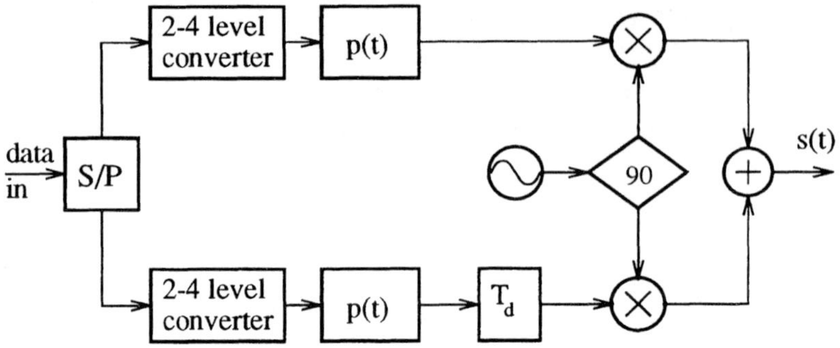

Figure P10.19-1: Block diagram of linear 16-QPOM schemes.

**10.14** Using the results of Examples 10.3, 10.6, and 10.16, derive the tight upper bound on the bit error probability of the coded hardlimited QPOM scheme in AWGN and Rayleigh fading channels.

**10.15** Derive the error-state diagram for the coded hardlimited QPOM scheme with the trellis diagram shown in Figure 10.44. Compute the error weight profiles, the minimum normalized squared Euclidean distance, and the transfer function for this scheme.

**10.16** Repeat Problem 10.15 for the coded hardlimited QPOM scheme using Code 2 of Table 10.6 as a 4-state convolutional outer code.

**10.17** Using the results obtained from Problem 10.15 (or Example 10.18), derive the upper-bound on the bit error probability of the coded hardlimited QPOM scheme using Code 1 of Table 10.6 as a 4-state convolutional outer code over AWGN, Rayleigh fading and shadowed Rician fading channels.

**10.18** Find the upper bound on the bit error performance of the coded hardlimited QPOM scheme using Code 2 of Table 10.6 as a 4-state convolutional outer code over AWGN, Rayleigh fading and shadowed Rician fading channels.

*Hint:* Use the results obtained in Problem 10.16 and follow the procedures given in Examples 10.3, 10.6, and 10.16.

**10.19 16-QPOM Schemes**

The block diagram of a linear 16-QPOM scheme is shown in Figure P10.19-1. The baseband pulse shapes are two identical double-interval pulses satisfying the conditions given by (10.2). The transmitted signal can be written as

$$s(t) = \sum_{n=-\infty}^{+\infty} \left[A_n p(t - nT) \cos 2\pi f_c t + B_n p(t - nT) \sin 2\pi f_c t\right].$$

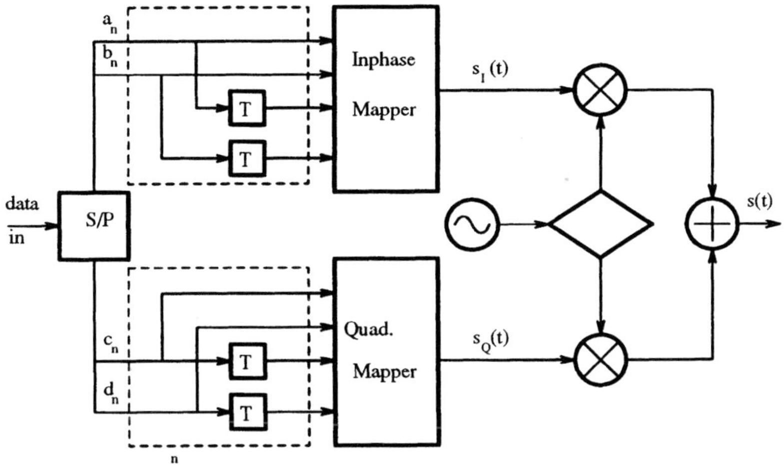


Figure P10.19-2: An alternative representation of 16-QPOM schemes

The 2-to-4 level baseband converter converts its input data stream to 4 level, pulse amplitude modulated (PAM) baseband signals having a symbol duration of

$$T = 4T_b,$$

where $T_b$ represents the bit duration. At time $nT$, the output of each converter can be expressed in terms of its input:

$$A_n = 2(2a_n + b_n) - 3,$$
$$B_n = 2(2c_n + d_n) - 3.$$

**a)** Show that the 16-QPOM scheme can be represented by Figure P10.19-2 with the mapping rules given in Table P10.19-1. Design the MLSE receiver for this 16-QPOM scheme. Derive the trellis diagram of this scheme.

**b)** Consider the transmission of 16-QPOM signals over an AWGN channel. Show that the minimum squared Euclidean distance of this scheme is given by

$$d_{min}^2 = \int_0^T [s_0(t) + s_1(t)]^2 dt = 1.6E_b,$$

where $E_b$ represents the bit energy, $s_0(t)$ and $s_1(t)$ are given in Section 10.1.1. Prove that the bit error probability of 16-QPOM signals over an AWGN channel is tightly upper bounded as

$$P_b \leq \frac{1}{2} erfc\left(\sqrt{0.4E_b/N_0}\right).$$

| Symbol $a_n b_n a_{n-1} b_{n-1}$ | | Transmitted signal $s_I(t)$ or $s_Q(t)$ |
|---|---|---|
| Decimal | Binary | $\mathcal{M}(a_n b_n a_{n-1} b_{n-1})$ |
| 0 | 00 00 | $-3s_0(t)$ |
| 1 | 00 01 | $-2s_0(t) - s_1(t)$ |
| 2 | 00 10 | $-s_0(t) - 2s_1(t)$ |
| 3 | 00 11 | $-3s_1(t)$ |
| 4 | 01 00 | $-2s_0(t) + s_1(t)$ |
| 5 | 01 01 | $-s_0(t)$ |
| 6 | 01 10 | $-s_1(t)$ |
| 7 | 01 11 | $+s_0(t) - 2s_1(t)$ |
| 8 | 10 00 | $-s_0(t) + 2s_1(t)$ |
| 9 | 10 01 | $+s_1(t)$ |
| 10 | 10 10 | $+s_0(t)$ |
| 11 | 10 11 | $+2s_0(t) - s_1(t)$ |
| 12 | 11 00 | $+3s_1(t)$ |
| 13 | 11 01 | $+s_0(t) + 2s_1(t)$ |
| 14 | 11 10 | $+2s_0(t) + s_1(t)$ |
| 15 | 11 11 | $+3s_0(t)$ |

Table P10.19-1: Mapping rules for the inphase (quadrature) component of linear 16-QPOM schemes.

c) Now consider the case of Rayleigh fading channels. From the trellis diagram of the 16-QPOM encoder, show that this scheme offers a time diversity of length 2, and derive the upper bound on $P_b$. Prove that $P_b$ varies inversely with the square of $E_b/N_0$. Verify your results with Figure P10.19-3 and observe that the 16-QPOM outperforms the 16-QAM.

### 10.20 Coded Linear 16-QPOM Schemes

The two components of the linear 16-QPOM scheme are independent (Figure P19.2). A coding scheme can be applied to each one independently. This technique simplifies the design of the code and reduces the complexity of the overall system. The coded scheme is shown in Figure P10.20-1. It uses two identical convolutional codes, one for the in-phase component (I channel), and one for the quadrature component (Q channel). The convolutional code is a rate 1/2 code with a total of $\nu$ memory. As previously discussed in Section 10.3, the rate of this code is chosen in such a way that the number of states of the concatenated code is reduced. After concatenation, each concatenated code is a rate 1/4 code with a total of $\nu + 1$ memory, and the obtained coded scheme has a net

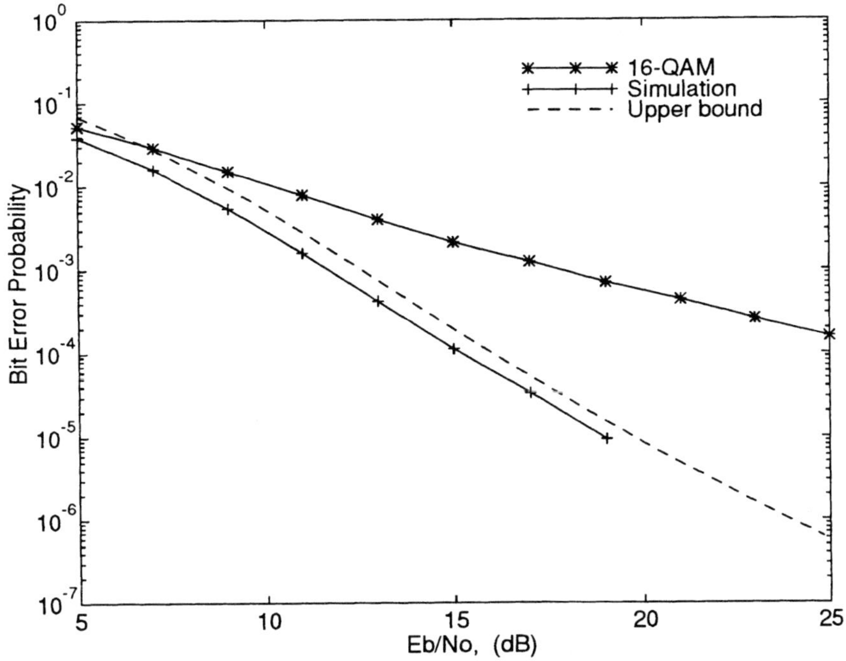


Figure P10.19-3: Performance of 16-QPOM schemes over Rayleigh fading channels.

throughput of 2 bits/symbol.

**a)** We first consider the linear 16-QPOM scheme and 2-state convolutional outer codes. From Section 10.3, each concatenated code will have a total of 4 states. Therefore, the coded scheme will have a total of 16 states. However, the configuration of this coded scheme allows the option of independent decoding for each component. In this case the decoding becomes much simpler, and the total number of states is reduced to 8, 4 states for the in-phase component and 4 states for the quadrature component.

In order to select the optimum convolutional code, we consider the signal constellations of one of the two components of the 16-QPOM scheme shown in Figure P10.20-2. Show that the convolutional code having the following generator polynomials:

$$g_1(D) = 1 + D$$
$$g_2(D) = 1,$$

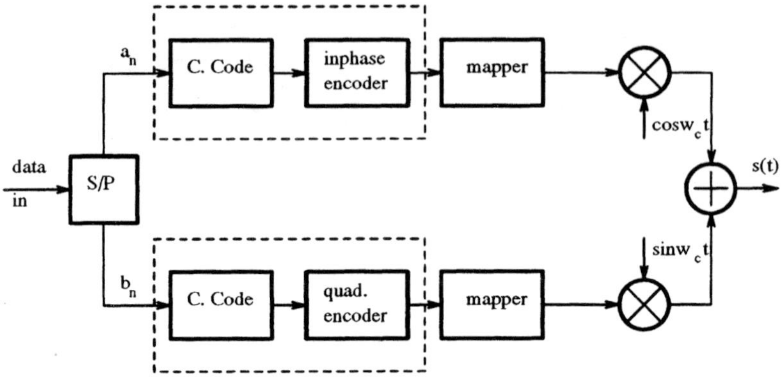


Figure P10.20-1: Convolutionally coded 16-QPOM schemes.

offers the best expected performance.
Find the trellis diagram of the overall coded 16-QPOM scheme.
*Hint:* This can be seen from the obtained signal constellation of the concatenated code (Figure P10.20-2).

**b)** Evaluate the performance of the coded linear 16-QPOM scheme over the Rayleigh fading channel with $p(t) = 0.5[1 + cos(\pi t/2T)]$. Compare your results with those given in Figure P10.20-3. Observe that the performance improvement compared to uncoded QPOM is quite significant (about 4 $dB$ at a bit error rate of $10^{-4}$). When compared to coding schemes of the same complexity, it is observed that for $E_b/N_0 \leq 15\ dB$, coded 16-QPOM performs better than 8-state 8PSK TCM. Even though the two schemes have the same number of states (a total of 8 states), the coded 16-QPOM scheme uses a simpler decoding procedure. This decoding procedure allows the use of two parallel Viterbi processors working on a trellis diagram of 4 states each. This will reduce the time delay introduced by the Viterbi processor and gives the possibility of using higher bit rates. In addition, the 16-QPOM scheme uses pulse shaping in generating the transmitted signal. Therefore, this signal has a more compact spectrum than the uncoded QPSK and coded PSK, making the coded linear 16-QPOM scheme a better choice for band-limited channels.

**c)** With only 2-state convolutional codes, it was possible to generate good coded schemes. By increasing the number of states of the external code, it is possible to design better coding schemes. Next we consider combining the linear 16-QPOM scheme with a 4-state convolutional outer code. Show that the 4-state convolutional code

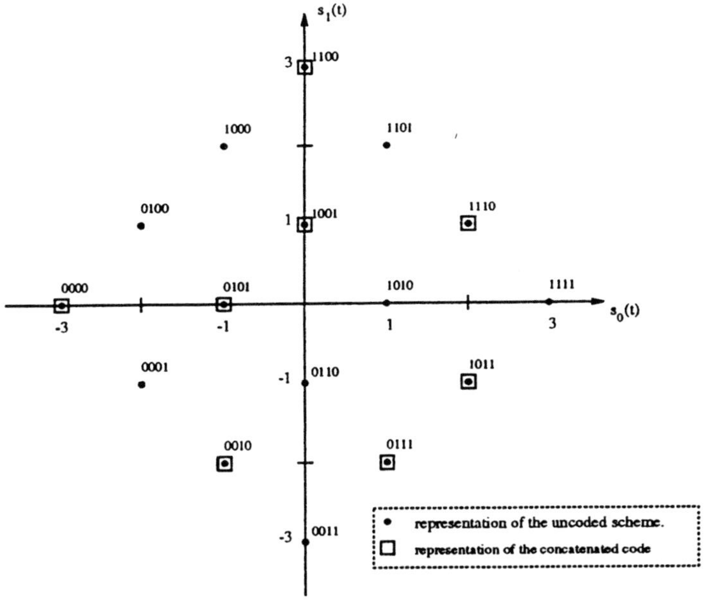


Figure P10.20-2: Signal representation of the inphase (or quadrature) components of the 16-QPOM scheme.

having the following generator polynomials:

$$g_1(D) = 1 + D + D^2$$
$$g_2(D) = 1 + D^2,$$

provides the best performance. Find the trellis diagram of the resulting coded 16-QPOM scheme.

**d)** Derive the upper bound on the bit error probability of the coded scheme over a Rayleigh fading channel. Compare your results with those given in Figure P10.20-4. Observe that increasing the number of states has further improved the performance at the expense of higher decoding complexity.

**10.21** In Figure P10.21-1, the input data stream is split into 4 parallel streams $\{a_n\}$, $\{b_n\}$, $\{c_n\}$, and $\{d_n\}$. The data streams $\{a_n\}$ and $\{b_n\}$ are fed into the inphase and quadrature channels of the first QPOM scheme, respectively, and the data streams $\{c_n\}$ and $\{d_n\}$ are fed into the inphase and quadrature channels of the second QPOM scheme.

The envelope fluctuations of the QPOM signals, $s_1(t)$ and $s_2(t)$, are removed after hardlimiting. Hence, the hardlimited QPOM signals, $z_1(t)$ and $z_2(t)$, suffer no major degradation due to nonlinear amplification.

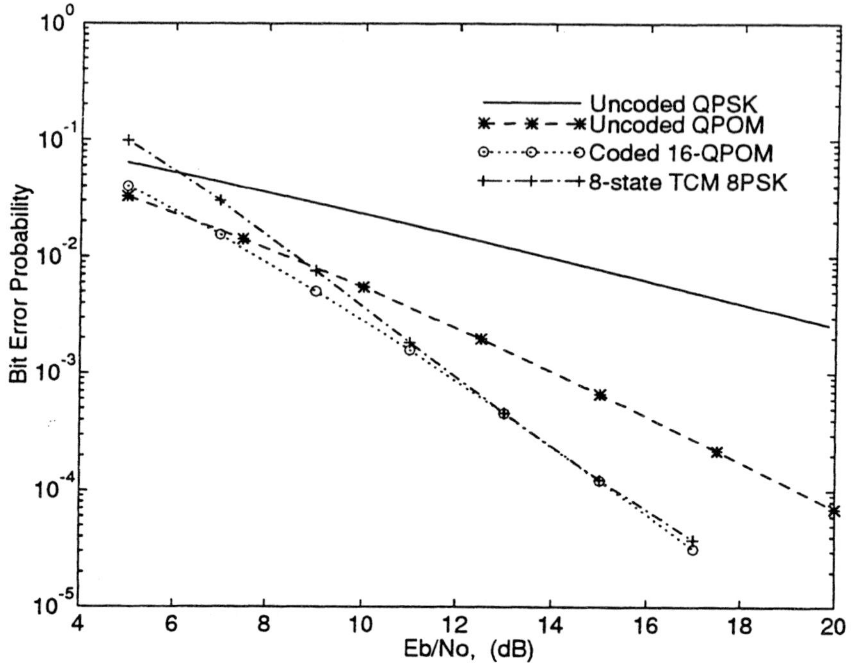


Figure P10.20-3: Performance of coded 16-QPOM signals in a Rayleigh fading channel (with a 2-state convolutional outer code).

The output voltage of the second high power amplifier is arranged to be twice that of the first amplifier. The outputs of the two amplifiers are added to give the 16-QPOM transmitted signal.

Since the shapes of the power spectral density of both $z_1(t)$ and $z_2(t)$ are identical and the 16-QPOM signal $z(t)$ is a linear combination of those two signals, it can be shown that the power spectrum of $z(t)$ has the same shape as that of $z_1(t)$ or $z_2(t)$. In addition, the spectral efficiency of this 16-QPOM scheme is twice that of the QPOM scheme.

a) We have seen in Section 10.2 that at any symbol interval a hardlimited QPOM signal can be represented by a vector $\underline{z}$ in the signal space. This vector is an 8-element vector when hardlimited QPOM signals are considered, or a 12-element vector in the case of OQPOM signals. Show that the hardlimited 16-QPOM signals and hardlimited QPOM signals can be defined over the same signal space, i.e., if at time $nT$, $z_1(t)$ and $z_2(t)$ are represented by the vectors $\underline{z}_{1n}$ and $\underline{z}_{2n}$, respectively, then the 16-QPOM signal, $z(t)$, will be represented

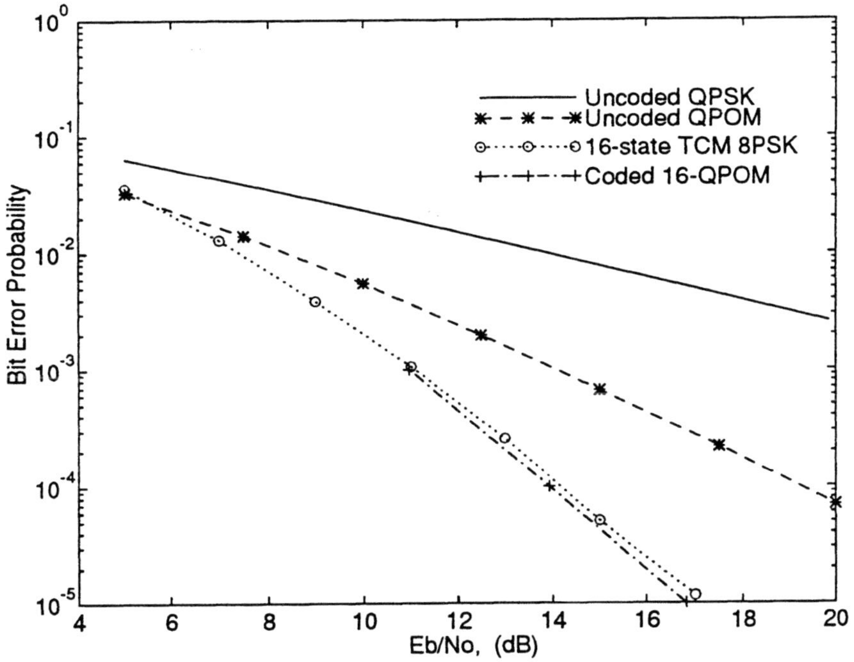


Figure P10.20-4: Performance of coded 16-QPOM signals in a Rayleigh fading channel (with a 4-state convolutional code).

by the vector

$$\underline{z}_n = \underline{z}_{1n} + 2\underline{z}_{2n}.$$

Show that the hardlimited 16-QPOM scheme can be decomposed into an encoder followed by a memoryless modulator. The modulation encoder is a rate 4/8 encoder with an input symbol of length 4 and a total of 16 states. This encoder is constructed using two hardlimited QPOM encoders in parallel. The mapping rules of the memoryless modulator consists of all possible combinations of $z_1(t)$ and $z_2(t)$, which give a total of 256 different pairs of waveforms $(z_I(t),\ z_Q(t))$.

b) Verify that the two components of this scheme cannot be treated independently. (An optimum receiver needs to treat both components simultaneously.)

c) Draw the trellis diagram of this scheme and prove that the minimum

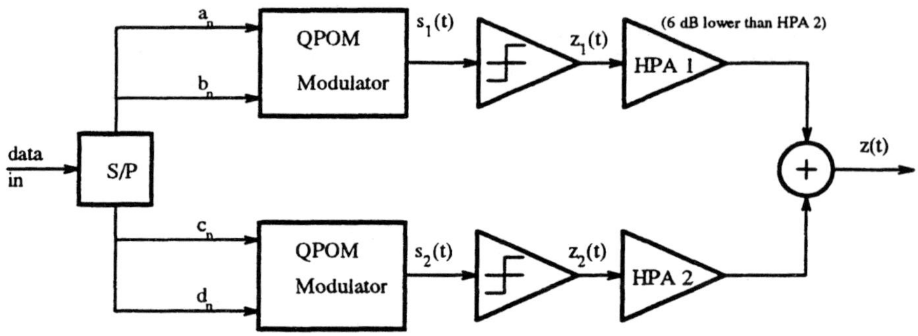


Figure P10.21-1: Block diagram of hardlimited 16-QPOM schemes.

squared Euclidean distance is given by

$$d^2_{min} = 3.2(1 - \beta_0),$$

where $\beta_0$ is a coefficient depending on the pulse shape $p(t)$ and defined by (10.32).

**d)** Derive the upper bound on the bit error probability of this scheme in an AWGN environment and prove that as $\beta_0$ approaches 0.5, the performance of this scheme approaches that of a conventional coherent 16-QAM scheme. Compare your results with the simulation results shown in Figure P10.21-2 for the double-interval pulse $p(t) = 0.5[1 + \cos(\pi t/2T)]$.

**d)** Show that the bit error probability of this scheme over a Rayleigh fading channel with perfect CSI can be upper bounded as

$$P_b \leq \frac{1}{4\left(1 + 0.4(1 - \beta_0)\frac{E_b}{N_0}\right)^2}.$$

Plot this upper bound and compare it with the simulation results shown in Figure P10.21-3.

**10.22** Consider a QPOM scheme using a rectangular double-interval pulse

$$p(t) = \begin{cases} 1 & |t| \leq T \\ 0 & \text{elsewhere} \end{cases}$$

**a)** Show that the signal elements $s_i(\tau)$, $i = 0, 1, 2, 3$, defined in Section 10.1.1, can be represented in a one-dimensional signal space and that the MLSE receiver for this QPOM scheme has a structure as shown in Figure P10.22-1.

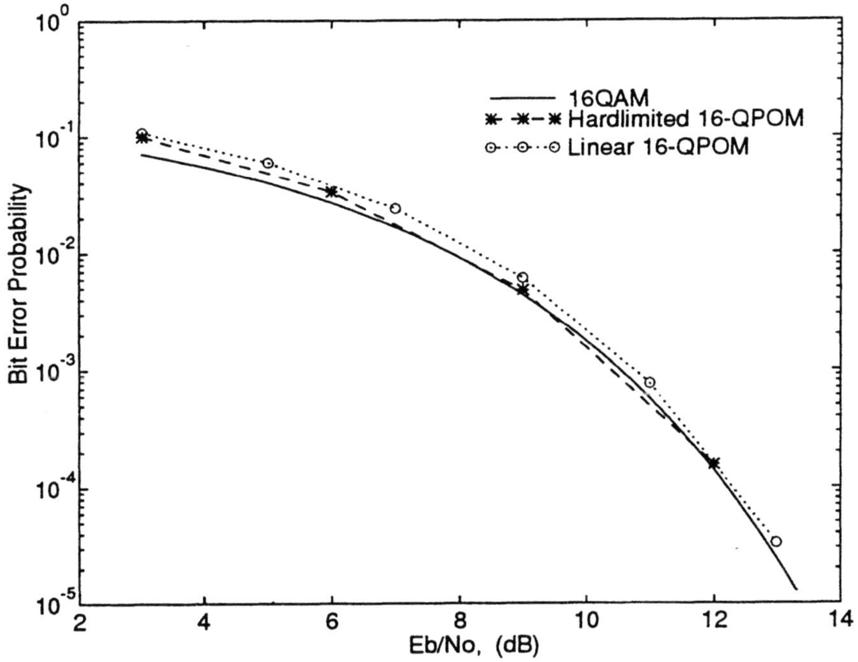


Figure P10.21-2: Performance of hardlimited 16-QPOM signals in AWGN channels.

b) Prove that the bit error probability of this scheme in an AWGN channel is upper-bounded as

$$P_b \leq 2\text{erfc}\left(\sqrt{\frac{E_b}{N_0}}\right)\frac{1}{\left[1 - e^{-2E_b/N_0}\right]^2}.$$

c) Derive the upper-bound on the bit error probability of this sheme in Rayleigh fading and shadowed Rician fading channels.

**10.23 Quadrature Partial Response Systems (QPRS)**

The QPOM scheme described in Problem 10.22 can be extended to the Quadrature Partial Response System (QPRS) shown in Figure P10.23-1. The pulse $p(t)$ in this figure is the inverse Fourier transform of

$$P(f) = \begin{cases} \sqrt{2T}\cos \pi f T e^{-j\pi f T/2}, & |f| \leq 1/2T \\ 0, & |f| \geq 1/2T \end{cases}.$$

The modulator is similar to the QPOM except that the pulse shape $p(t)$ is not time-limited. The transmitted signal $s(t)$ is defined by (10.1).

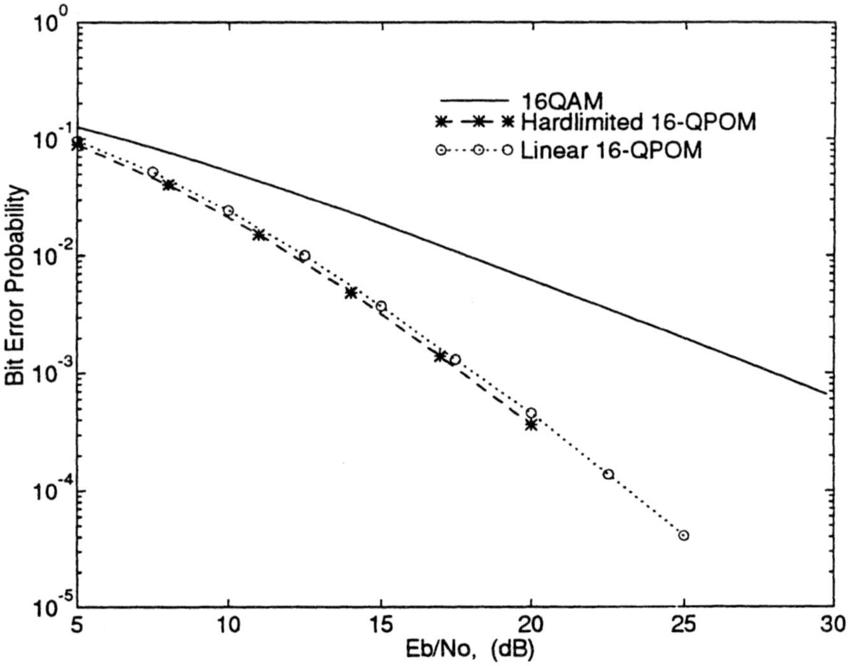


Figure P10.21-3: Performance of hardlimited 16-QPOM signals in Rayleigh fading channels.

a) Show that the average energy per bit is

$$E_b = \frac{4}{\pi}$$

and that the power spectrum of the transmitted signal $s(t)$ has the shape of $|P(f)|^2$. Plot the transmitted power spectrum and observe its compactness.

b) In an AWGN channel with a noise power spectral density of $N_0/2$, the received samples $r_I(kT)$ and $r_Q(kT)$ in Figure P10.23-1 are represented as

$$\begin{aligned} r_I(kT) &= x_I(kT) + n_I(kT) \\ r_Q(kT) &= x_Q(kT) + n_Q(kT). \end{aligned}$$

Prove that $n_I(kT)$ and $n_Q(kT)$ are independent, identically distributed (i.i.d.) Gaussian noise samples with zero mean, variance

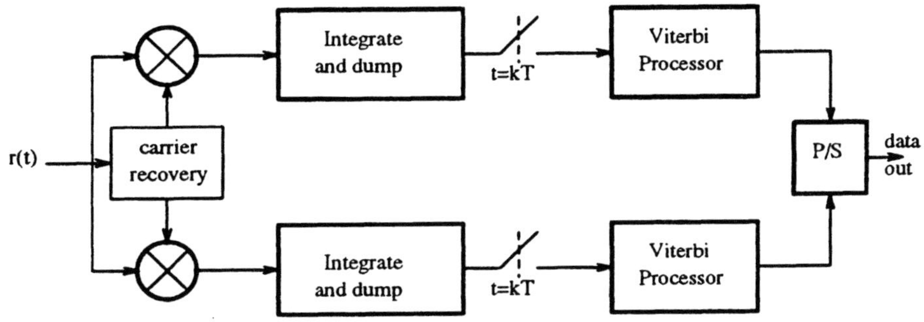


Figure P10.22-1: Block diagram of the MLSE receiver of the QPOM scheme.

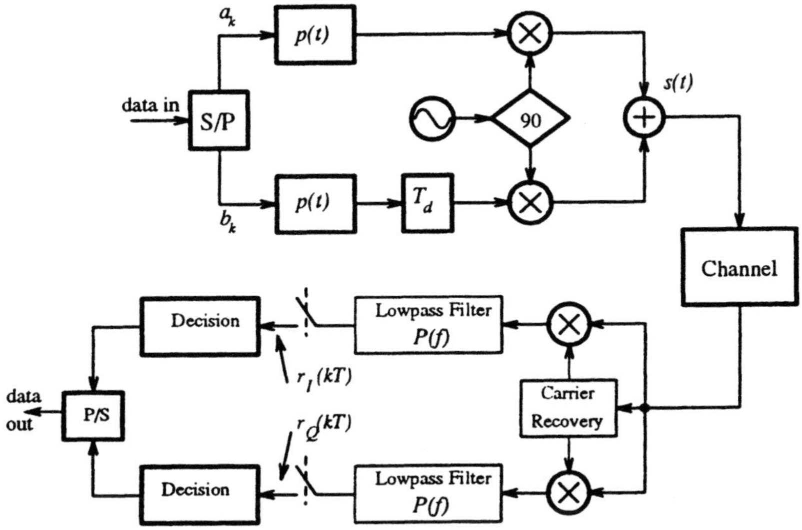


Figure P10.23-1: Block diagram of a QPRS scheme.

of

$$\sigma^2 = N_0 \int_0^{1/2T} |P(f)|^2 df = \frac{2N_0}{\pi},$$

and $x_I(kT)$ (or $x_Q(kT)$) is related to $a_k$ and $a_{k-1}$ (or $b_k$ and $b_{k-1}$) as follows:

$$x_I(kT) = \begin{cases} 2 & \text{if } a_k = a_{k-1} = 1 \\ -2 & \text{if } a_k = a_{k-1} = 0 \\ 0 & \text{if } a_k \neq a_{k-1} \end{cases},$$

where $a_k$ (or $b_k$)$=$ 0 or 1 with a probability of 1/2.

c) Consider the case of symbol-by-symbol decision. In order to avoid error propagation at the receiver, data is precoded before transmission as follows:

$$a_k = a_{k-1} \oplus \acute{a}_k ,$$

where $\acute{a}_k$ is the data symbol ($\acute{a}_k = 0$ or 1 with probability of 1/2) and $\oplus$ is a modulo-2 adder (or an exclusive-or operator). The initial value $a_0$ is 0. Verify that

$$x_I(kT) = \begin{cases} 2 \text{ or } -2 & \text{if } \acute{a}_k = 0 \\ 0 & \text{if } \acute{a}_k = 1 \end{cases} .$$

Show that a symbol-by-symbol decision rule can be

$$\hat{\acute{a}}_k = \begin{cases} 0 & \text{if } |r_I(kT)| > 1 \\ 1 & \text{if } |r_I(kT)| < 1 \end{cases}$$

and that the bit error probability of the QPRS using symbol-by-symbol detection in an AWGN channel is

$$P_b = \frac{1}{2}\text{erfc}\left(\sqrt{\frac{\pi^2}{16}\frac{\text{E}_\text{b}}{\text{N}_0}}\right) .$$

Observe that the QPRS using a symbol-by-symbol detection results in a loss of 2.1 $dB$ in $E_b/N_0$ relative to the QPSK.

**d)** Now consider a decision scheme based on MLSE. Show that the relation between $x_I(kT)$ (or $x_Q(kT)$) and $a_k$ (or $b_k$) can be represented by the trellis diagram in Figure 10.5. Following the procedure given in Section 10.1.2 derive the upper-bound on the bit error probability of the QPRS using an MLSE receiver in an AWGN. Show that the loss of 2.1 $dB$ in $E_b/N_0$ mentioned in **c)** is completely recovered.

**e)** Derive the upper-bound on the bit error probability of the QPRS using an MLSE receiver in Rayleigh fading and shadowed Rician fading channels.

# Appendix A

# CALCULATION OF THE CHERNOFF FACTORS FOR RICIAN AND RAYLEIGH FADING CHANNELS

In this appendix the Chernoff factors are derived for Rician and Rayleigh fading channels with and without the channel state information (CSI).

## A.1 FADING CHANNEL WITH CSI

Assume that the channel state information is available at the receiver as $\mathbf{a} = (a_1, a_2, \cdots, a_N)$ for the transmitted sequence $\mathrm{s} = (s_1, s_2, \cdots, s_N)$ and the received sequence $\mathbf{r} = (r_1, r_2, \cdots, r_N)$. For two-dimensional signals the components of the vectors $\mathbf{a}$, $\mathrm{s}$, and $\mathbf{r}$ are represented as complex numbers. Using (5.2), the ML metric of (5.7) can be written as

$$m(\mathbf{r}, \mathrm{s}; \mathbf{a}) = -|\mathbf{r} - \mathbf{a} \cdot \mathrm{s}|^2,$$

where, for simplicity, the factor $(1/\pi N_0)^{N/2}$ is ignored, since it would be absorbed in the Chernoff bound parameter $\lambda$ anyway.

Using this metric, the Chernoff bound between the codewords s and $\hat{\mathrm{s}}$, defined in (2.31), is written as

$$C(\mathrm{s},\ \hat{\mathrm{s}};\ \lambda) = E_{\mathbf{a}}[E_{\mathbf{r}|\mathrm{s}}[\exp(\lambda\{|\mathbf{r} - \mathbf{a} \cdot \mathrm{s}|^2 - |\mathbf{r} - \mathbf{a} \cdot \hat{\mathrm{s}}|^2\})]],$$

$$= \prod_{i=1}^{N} E_{a_i}[E_{n_i}[\exp(\lambda\{|r_i - a_i s_i|^2 - |r_i - a_i \hat{s}_i|^2\})]], \quad (A.1)$$

where $a_i$ and $n_i$ are the fading amplitude and the additive noise components at time $i$, respectively, and are assumed to be independent of the corresponding components of the other time intervals. Recalling

$$r_i = a_i \cdot s_i + n_i,$$

the statistical expectation with respect to the additive noise component $n_i$ with variance $\frac{N_0}{2}$ in (A.1) can be simplified as

$$E_{n_i}[\exp(\lambda\{|r_i - a_i s_i|^2 - |r_i - a_i \hat{s}_i|^2\})] = \exp(-\lambda a_i^2 |s_i - \hat{s}_i|^2) \cdot E_{n_i}[\exp(-2\lambda a_i \Re\{n_i(s_i - \hat{s}_i)^*\})], \quad (A.2)$$

where $\Re\{\cdot\}$ and $*$ denote the real part and complex conjugate of a complex number, respectively. For a complex Gaussian noise $n_i$ with variance $\sigma_n^2 = \frac{N_0}{2}$, (A.2) can be written as:

$$E_{n_i}[\exp(\lambda\{|r_i - a_i s_i|^2 - |r_i - a_i \hat{s}_i|^2\})] = \exp(-\lambda a_i^2 (1 - N_0\lambda)|s_i - \hat{s}_i|^2). \quad (A.3)$$

Substituting (A.3) into (A.1) yields

$$C(\mathrm{s}, \hat{\mathrm{s}}, \lambda) = \prod_{i=1}^{N} E_{a_i}[\exp(-\lambda a_i^2 (1 - N_0\lambda)|s_i - \hat{s}_i|^2)]. \quad (A.4)$$

The above bound is minimized by setting

$$\lambda = \lambda_{\mathrm{opt}} = \frac{1}{2N_0}, \quad (A.5)$$

which results in

$$\begin{aligned} C(\mathrm{s}, \hat{\mathrm{s}}, \lambda) &= \prod_{i=1}^{N} E_{a_i}[\exp(-\frac{a_i^2}{4N_0}|s_i - \hat{s}_i|^2)], \\ &= \prod_{i=1}^{N} C(s, \hat{s}, \lambda_{\mathrm{opt}}). \end{aligned}$$

Dropping the subscript $i$, for simplicity, the Chernoff factor of signals $s$ and $\hat{s}$ can be written as

$$\begin{aligned} C(s, \hat{s}, \lambda_{\mathrm{opt}}) &= E_a[\exp(-\frac{a^2}{4N_0}|s - \hat{s}|^2)], \\ &= \int_0^{\infty} p_A(a) \cdot \exp(-\frac{a^2}{4N_0}|s - \hat{s}|^2) \cdot da, \quad (A.6) \end{aligned}$$

where $p_A(a)$ is the pdf of the fading amplitude. Now, consider the case of a Rician fading channel. Representing $|s-\hat{s}|^2/4N_0$ by $\delta^2$ in (A.6), we proceed as

$$C(s,\ \hat{s},\ \lambda_{\text{opt}}) = \int_0^\infty 2a(1+K)e^{-[K+a^2(1+K)+a^2\delta^2]}I_0(2a\sqrt{K(1+K)}\ )\cdot da. \tag{A.7}$$

This integral can be evaluated using [92, p. 717]

$$\int_0^\infty te^{-\alpha t^2}\cdot J_0(\beta t)\cdot dt = \frac{1}{2\alpha}e^{\frac{-\beta^2}{4\alpha}}, \tag{A.8}$$

where [92, p. 952]

$$J_0(jt) = I_0(t), \tag{A.9}$$

as

$$C(s,\ \hat{s},\ \lambda_{\text{opt}}) = \frac{1+K}{1+K+\frac{1}{4N_0}|s-\hat{s}|^2}\exp[-\frac{K\frac{1}{4N_0}|s-\hat{s}|^2}{1+K+\frac{1}{4N_0}|s-\hat{s}|^2}]. \tag{A.10}$$

The Chernoff factor for a Rayleigh fading channel can be simplified as

$$C(s,\ \hat{s},\ \lambda_{\text{opt}}) = \frac{1}{1+\frac{1}{4N_0}|s-\hat{s}|^2},$$

by substituting $K=0$ in (A.10).

## A.2 FADING CHANNEL WITHOUT CSI

The ML metric for the case of decoding without the channel state information is given by

$$m(\mathbf{r},\ \mathbf{s}) = -|\mathbf{r}-\mathbf{s}|^2.$$

In such a case, the Chernoff bound between codewords $\mathbf{s}$ and $\hat{\mathbf{s}}$ is expressed as

$$\begin{aligned} C(\mathbf{s},\ \hat{\mathbf{s}};\ \lambda) &= E_{\mathbf{a}}[E_{\mathbf{r}|\mathbf{s}}[\exp(\lambda\{|\mathbf{r}-\mathbf{s}|^2-|\mathbf{r}-\hat{\mathbf{s}}|^2\})]], \\ &= \prod_{i=1}^{N} E_{a_i}[E_{n_i}[\exp(\lambda\{|r_i-s_i|^2-|r_i-\hat{s}_i|^2\})]]. \end{aligned} \tag{A.11}$$

The statistical expectation with respect to $n_i$ can be simplified as

$$\begin{aligned} E_{n_i}[\exp(\lambda\{|r_i-s_i|^2 - |r_i-\hat{s}_i|^2\})] = \\ \exp(\lambda(|s_i|^2-|\hat{s}_i|^2) - 2\lambda a_i(|s_i|^2-\Re\{s_i\hat{s}_i^*\})\ ) \\ \cdot\ E_{n_i}[\exp(-2\lambda\Re\{n_i(s_i-\hat{s}_i)^*\}\ )\ ]. \end{aligned} \tag{A.12}$$

Similar to the case with CSI, the expectation in the right-hand side of (A.12) can be calculated. The result after some simplification is:

$$E_{n_i}[\exp(\lambda\{|r_i - s_i|^2 - |r_i - \hat{s}_i|^2\})] = \exp(\lambda\{|s_i|^2 - |\hat{s}_i|^2 + N_0\lambda|s_i - \hat{s}_i|^2 - a_i\lambda(|s_i|^2 - |\hat{s}_i|^2 + |s_i - \hat{s}_i|^2)\}). \quad (A.13)$$

In deriving the above equation we have used the equality

$$2\Re\{xy^*\} = |x|^2 + |y|^2 - |x - y|^2,$$

where $x$ and $y$ are two complex numbers. Substituting (A.13) into (A.11) yields

$$C(\mathrm{s}, \hat{\mathrm{s}}; \lambda) = \prod_{i=1}^{N} \exp(\lambda\{|s_i|^2 - |\hat{s}_i|^2 + N_0\lambda|s_i - \hat{s}_i|^2\}) \cdot E_{a_i}[\exp(-a_i\lambda\{|s_i|^2 - |\hat{s}_i|^2 + |s_i - \hat{s}_i|^2\})]. \quad (A.14)$$

Unlike the case with CSI, (A.14) cannot be optimized over $\lambda$ to yield a constant value for this parameter. The optimization procedure should be done after averaging over the fading amplitude distribution. Therefore, replacing the Chernoff parameter $\lambda$ by $\frac{\lambda}{N_0}$ (denormalization), the Chernoff factor of signals $s$ and $\hat{s}$ is obtained as

$$C(s, \hat{s}; \lambda) = \exp(\frac{\lambda}{N_0}\{|s|^2 - |\hat{s}|^2 + \lambda|s - \hat{s}|^2\}) \cdot E_a[\exp(-a\frac{\lambda}{N_0}\{|s|^2 - |\hat{s}|^2 + |s - \hat{s}|^2\})].$$

The above expression cannot be evaluated in a closed form for a Rician fading channel. In such a case, similar to [43], this is expressed as

$$C(s, \hat{s}; \lambda) = \exp(\frac{\lambda}{N_0}(|s|^2 - |\hat{s}|^2 + \lambda|s - \hat{s}|^2))\frac{e^{-K}}{\pi}\int_0^{\pi}(1 - \sqrt{\pi}\mathrm{erfc}(\nu)e^{\nu^2}\nu)d\theta,$$

where

$$\nu = \frac{\frac{\lambda}{2N_0}(|s|^2 - |\hat{s}|^2 + |s - \hat{s}|^2)}{\sqrt{1+K}} - \sqrt{K}\cos(\theta). \quad (A.15)$$

For a Rayleigh fading channel $C(s, \hat{s}, \lambda)$ is expressed in a closed form. For $K = 0$ the integral in (A.15) is evaluated as

$$\int_0^{\pi}(1 - \sqrt{\pi}\mathrm{erfc}(\nu)e^{\nu^2}\nu)d\theta = 1 - \sqrt{\pi}\mathrm{erfc}(\beta)e^{\beta^2}\beta,$$

where

$$\beta = \frac{\lambda}{2N_0}(|s|^2 - |\hat{s}|^2 + |s - \hat{s}|^2),$$

and

$$C(s, \hat{s}; \lambda) = \exp(\frac{\lambda}{N_0}(|s|^2 - |\hat{s}|^2 + \lambda|s - \hat{s}|^2))(1 - \sqrt{\pi}\mathrm{erfc}(\beta)e^{\beta^2}\beta).$$

# Appendix B

# CALCULATION OF THE TRANSITION PROBABILITIES FOR THE MPSK SIGNAL SET OVER A RAYLEIGH FADING CHANNEL

In this appendix the transition probabilities for an MPSK signal set over a Rayleigh fading channel are derived.

In general, the decision regions for an MPSK signal set include a null zone for the erasure decoding as shown in Figure 2.15. Due to the symmetry of the MPSK signal set the transition probabilities $\{q_{li}\}$ are independent of the $i$th signal. Thus, dropping the subscript $i$, these probabilities are evaluated for signal $s_0$. Then $q_l$ is defined as

$$q_l = E_a[\int_{\Lambda_l} p(r|s_0, a)dr], \qquad l = 0, 1, \cdots, q-1, \tag{B.1}$$

where $\Lambda_l$ shows the $l$th decision region. Recalling that $r, n$, and $s_0$ are complex numbers and using (2.7), (B.1) can be written as

$$\begin{aligned} q_l &= E_a\left[\int\int_{\Lambda_l} \frac{1}{\pi N_0} e^{-\frac{1}{N_0}[(r_1-as_1)^2+(r_2-as_2)^2]} dr_1 dr_2\right] \\ &= E_a\left[\int\int_{\Lambda_l} \frac{E_s}{\pi N_0} e^{-\frac{E_s}{N_0}\,[R_1^2+R_2^2-2aR_1+a^2]} dR_1\, dR_2\right], \end{aligned} \tag{B.2}$$

where we use $r = r_1 + jr_2 = \sqrt{E_s}(R_1 + jR_2)$, $s_0 = s_1 + js_2$ and $|s_0| = \sqrt{E_s}$. Introducing the polar coordinates as

$$\begin{aligned} R_1 &= \rho\cos\theta \\ R_1 &= \rho\sin\theta \\ dR_1 dR_2 &= \rho d\rho d\theta \end{aligned}$$

(B.2) can be simplified as

$$\begin{aligned} q_l &= E_a\left[\int_{a_T}^{\infty}\int_{\alpha_1}^{\alpha_2}\frac{E_s}{\pi N_0}\, e^{-\frac{E_s}{N_0}[\,\rho^2-2a\rho\cos\theta+a^2\,]}\rho\, d\theta\, d\rho\right], \\ &= \int_0^{\infty}(2ae^{-a^2}\int_{a_T}^{\infty}\int_{\alpha_1}^{\alpha_2}\frac{E_s}{\pi N_0}\, e^{-\frac{E_s}{N_0}[\,\rho^2-2a\rho\cos\theta+a^2\,]}\rho\, d\theta\, d\rho\,)\, da, \end{aligned} \tag{B.3}$$

where $\alpha_1 = (2l-1)\pi/q$ and $\alpha_2 = (2l+1)\pi/q$. By interchanging the order of integrations and after some simplification, (B.3) can be written as

$$q_l = \frac{2E_s}{\pi N_0}\int_{\alpha_1}^{\alpha_2}(\int_{a_T}^{\infty}\rho\, e^{-\frac{E_s}{N_0}\rho^2}(\int_0^{\infty} a\, e^{-(1+E_s/N_0)\,[\,a^2-\frac{2a\rho E_s/N_0}{1+E_s/N_0}\cos\theta\,]}da\,)\, d\rho\,)\, d\theta. \tag{B.4}$$

The integration with respect to $a$ can be expressed as

$$\begin{aligned} \int_0^{\infty} a\, e^{-(1+E_s/N_0)\,[\,a^2-\frac{2a\rho E_s/N_0}{1+E_s/N_0}\cos\theta\,]}da &= \frac{1}{2(1+E_s/N_0)} \\ \cdot\left[1+\frac{\rho E_s/N_0\sqrt{\pi}\cos\theta}{\sqrt{1+E_s/N_0}}\, e^{\frac{\rho^2(E_s/N_0)^2\cos^2\theta}{1+E_s/N_0}}\operatorname{erfc}(-\frac{\rho \mathrm{E_s/N_0}\cos\theta}{\sqrt{1+\mathrm{E_s/N_0}}})\right]. \end{aligned} \tag{B.5}$$

Substituting (B.5) into (B.4) results in

$$\begin{aligned} q_l = &\frac{E_s/N_0}{\pi(1+E_s/N_0)}\left[\frac{\pi}{qE_s/N_0}\, e^{-a_T^2\frac{E_s}{N_0}} + \int_{(2l-1)\pi/q}^{(2l+1)\pi/q} d\theta\,\left(\int_{a_T}^{\infty}\rho^2\,\frac{E_s}{N_0}\sqrt{\pi}\cos\theta\right.\right. \\ &\left.\left.\cdot e^{-\rho^2E_s/N_0+\frac{\rho^2(E_s/N_0)^2\cos^2\theta}{1+E_s/N_0}}\operatorname{erfc}(\frac{-\rho\mathrm{E_s/N_0}\cos\theta}{\sqrt{1+\mathrm{E_s/N_0}}})\, \mathrm{d}\rho\right)\right], \end{aligned} \tag{B.6}$$

The probability of the received signal being in the null zone can be obtained from $\{q_l\}$ as

$$q_e = 1 - \sum_{l=0}^{q-1} q_l.$$

Note that the transition probability for hard-decision decoding can simply be obtained from (B.6) by setting $a_T = 0$. It is obvious that in this case $q_e = 0$.

# Appendix C

# ENCODING AND DECODING TECHNIQUES FOR RS CODES USING FOURIER TRANSFORM

This appendix is devoted to the encoding and decoding techniques of RS codes using Fourier transform. We begin with a definition of the finite-field Fourier transform. Then we discuss the transform encoding and decoding algorithms.

## C.1 FINITE-FIELD FOURIER TRANSFORM

Let $\mathbf{a} = (a_{n-1}, \cdots, a_1, a_0)$ be an $n$-element vector with elements chosen from GF($2^m$). The (finite-field) *Fourier transform* of $\mathbf{a}$ is another $n$-element vector $\mathbf{A}$ in GF($2^m$) defined as

$$\mathbf{A} = \mathcal{F}\{\mathbf{a}\} = (A_{n-1}, A_{n-2}, \cdots, A_1, A_0),$$

where

$$A_j = a(\alpha_j) = \sum_{i=0}^{n-1} a_i \alpha^{ij}, \qquad j = 0, 1, \cdots, (n-1).$$

$A_j$ is called the $j$th *spectral component* of $\mathbf{a}$. $A(x)$ is known as the *spectrum polynomial* or *associated polynomial* of $a(x)$.

Equivalently, **a** is the *inverse* (finite field) *Fourier transform* of **A** which is represented as

$$\mathbf{a} = \mathcal{F}^{-1}\{\mathbf{A}\}.$$

In this case the elements of **a** are related to the elements of **A** as

$$a_i = \frac{1}{n \bmod 2} A(\alpha^{-i}) = A(\alpha^{-i}) = \sum_{j=0}^{n-1} A_j \alpha^{-ij}, \qquad i = 0,\ 1,\ \cdots,\ (n-1).$$

It can be seen that the Fourier transform is a linear operation. **A** and **a** can be expressed as $\mathbf{A} = \mathbf{Fa}$ and $\mathbf{a} = \mathbf{F'A}$, where the *forward transform matrix* $\mathbf{F} = [f_{ij} = \alpha^{ij}]$ and the *inverse transform matrix* $\mathbf{F}' = [f'_{ij} = \alpha^{-ij}]$ are $n \times n$ symmetric matrices.

Let **a**, **b**, and **c** be three time-domain $n$-element vectors in GF($2^m$) and **A**, **B**, and **C** be their corresponding Fourier transforms. The following properties can be shown.

| **Time-Domain** | **Frequency-Domain** |
|---|---|
| 1) *Linearity* | |
| $c(x) = \beta_1 a(x) + \beta_2 b(x)$, | $C(x) = \beta_1 A(x) + \beta_2 B(x)$ |
| $\beta_1, \beta_2 \in \mathrm{GF}(2^m)$ | $\beta_1, \beta_2 \in \mathrm{GF}(2^m)$ |
| 2) *Convolution* | |
| $c(x) = \sum_{i=0}^{n-1} c_i x^i$ | $C(x) = \sum_{j=0}^{n-1} C_j x^j$ |
| $c_i = a_i b_i$ | $C_j = \sum_{l=0}^{n-1} A_l B_{j-l}$ $= \sum_{l=0}^{n-1} A_{j-l} B_l$ |
| $c_i = \sum_{l=0}^{n-1} a_l b_{i-l}$ $= \sum_{l=0}^{n-1} a_{i-l} b_l$ | $C_j = A_j B_j$ |
| 3) $l$th *Cyclic Shift* | |
| $a'' = x^l a(x) \bmod (x^n + 1)$ | $A''(x) = \sum_{j=0}^{n-1} (\alpha^{lj} A_j) x^j$ |
| $b(x) = \sum_{i=0}^{n-1} (\alpha^{-li} a_i) x^i$ | $B(x) = x^l A(x) \bmod (x^n + 1)$ |
| 4) *Delta Function* | |
| $a(x) = \beta x^i$ | $A(x) = \beta \sum_{j=0}^{n-1} \alpha^{ij} x^j$ |
| $b(x) = \beta \sum_{i=0}^{n-1} \alpha^{-ij} x^i$ | $B(x) = \beta x^j$ |

## C.2 TRANSFORM ENCODING

Let a valid code polynomial $c(x)$ be represented as an $n$-element vector $\mathbf{c}$

$$\mathbf{c} = (c_{n-1},\ c_{n-2},\ \cdots,\ c_1, c_0).$$

The $j$th spectral element of $\mathbf{c}$ is

$$C_j = c(\alpha^j), \qquad j = 0,\ 1,\ \cdots,\ (n-1).$$

As $c(x)$ is a valid code polynomial

$$c(\alpha^{h+i}) = 0, \qquad i = 0, 1, \cdots, (n-k-1),$$

it follows that

$$C_{(h+i) \bmod n} = 0, \qquad i = 0, 1, \cdots, (n-k-1).$$

In other words, the Fourier transform $\mathbf{C}$ of a valid code vector $\mathbf{c}$ contains $(n-k)$ consecutive zero components.

Let us make the remaining $k$ components of $\mathbf{C}$ equal to the $k$ coefficients of the information polynomial $A(x)$, i.e.,

$$C_{(h+i) \bmod n} = A_{[i-(n-k)] \bmod n}, \qquad i = (n-k), (n-k+1), \cdots, (n-1). \quad \text{(C.1)}$$

The components of the valid code vector $\mathbf{c}$ are generated by taking the inverse Fourier transform of $\mathbf{C}$, defined in (C.1), i.e.,

$$c_i = \sum_{j=0}^{n-1} C_j \alpha^{-ij} = \sum_{l=0}^{k-1} A_l (\alpha^{-i})^{(h+l-k) \bmod n}.$$

It is obvious that transform encoding produces non-systematic code vectors. Transform encoding procedure provides an encoder structure suitable for a universal $(n,\ k)$ code.

For convenience, for transform encoding, $h$ is usually chosen to be $k$ so that

$$C_j = \begin{cases} A_j, & \text{if } j = 0, 1, \cdots, (k-1), \\ 0, & \text{if } j = k,\ (k+1),\ \cdots,\ (n-1). \end{cases}$$

## C.3 TRANSFORM DECODING TECHNIQUES

An alternative decoding procedure for RS codes that can be computationally more effective than the syndrome-based procedure for large values of $(n-k)$, presented in Chapter 6, is based on the relationship between the received vector and its Fourier transform.

Consider first the errors-and-erasures transform decoding procedure for RS $(n,k)$ codes. The received polynomial $r(x)$ is represented as an $n$-component vector $\mathbf{r}$ in the time-domain as

$$\mathbf{r} = (r_{n-1},\ r_{n-2},\ \cdots,\ r_1,\ r_0) = \mathbf{c} + \mathbf{e},$$

where $\mathbf{c} = (c_{n-1},\ \cdots,\ c_1,\ c_0)$ and $\mathbf{e} = (e_{n-1},\ \cdots,\ e_1,\ e_0)$ are the transmitted codeword and the error vector, respectively. The $n$ components of these vectors are the $n$ coefficients of their corresponding polynomial, that is, the codeword and error polynomials, introduced in Chapter 6.

The components of the errata vector are zero at locations containing no error or erasure, i.e., $e_q = 0$ for $q \neq i_l, l = 1, 2, \cdots, \epsilon$ and $q \neq j_p,\ p = 1, 2, \cdots, s$ where $\epsilon$ and $s$ denote the number of errors and the number of erasures, respectively. Similarly, the errata-locator polynomial $\Lambda(x)$ in (6.23) can also be represented as $n$-component *errata-locator vector* in the *frequency domain* as

$$\mathbf{\Lambda} = (\Lambda_{n-1},\ \Lambda_{n-2},\ \cdots,\ \Lambda_0),$$

where $\Lambda_i$'s are the coefficients of $\Lambda(x)$. As the degree of $\Lambda(x)$ is $(s+\epsilon)$, it is obvious that $\Lambda_j = 0$ for $j = (s+\epsilon+1), (s+\epsilon+2), \cdots, (n-1)$. Also note that $\Lambda_0 = 1$.

The inverse Fourier transform of the errata-locator vector is

$$\lambda = (\lambda_{n-1}, \lambda_{n-2}, \cdots, \lambda_0),$$

where $\lambda_i = \Lambda(\alpha^{-i})$. As $\alpha^{-j_p}$'s, $p = 1, 2, \cdots, s$ and $\alpha^{-i_l}$'s, $l = 1, 2, \cdots, \epsilon$ are roots of $\Lambda(x)$, it is obvious that $\lambda_q = 0$ for $q = i_l, l = 1, 2, \cdots, s$ and $q = j_p,\ p = 1, 2, \cdots, \epsilon$. Using the properties of $\lambda_i$'s and $e_i$'s it can be shown that

$$\lambda_i e_i = 0, \qquad i = 0,\ 1,\ \cdots,\ (n-1).$$

This leads to the following relation in the frequency domain

$$\sum_{l=0}^{n-1} \Lambda_l E_{j-l} = \sum_{l=0}^{n-1} \Lambda_{j-l} E_l = 0, \qquad j = 0,\ 1,\ \cdots,\ (n-1), \tag{C.2}$$

where $E_l$'s are the spectral components of $\mathbf{e}$, i.e., the Fourier transform of $\mathbf{e}$ is

$$\mathbf{E} = (E_{n-1},\ E_{n-2},\ \cdots,\ E_0)$$

and

$$E_l = e(\alpha^l), \qquad l = 0, 1, 2, \cdots, (n-1).$$

As $\alpha^{h+i}$'s are roots of $c(x)$ we obtain

$$\mathbf{E}_{h+i} = R_{h+i} = r(\alpha^{h+i}) = S_{i+1}, \qquad i = 0, 1, \cdots, (n-k-1).$$

Note that(6.24) has the same form as (C.2) except that the index $j$ in (C.2) varies from 0 to $(n-1)$ rather than 1 to $(n-k)$ in (6.24). Using (6.10), (6.24), and (C.2) the spectral components of $\mathbf{e}$ can be found as follows.

**(i)** Compute $(n-k)$ spectral components $E_{h+i}$'s from (6.10) for $i = 0, 1, \cdots, (n-k-1)$.

**(ii)** Using $(n-k)$ spectral components $E_{h+i}$'s obtained in (i), and (6.24) solve for $(s+\epsilon)$ spectral components of the error-locator vector $\Lambda_l, l = 1, 2, \cdots, (s+\epsilon)$. The Berlekamp-Massey algorithm shown in Figure 6.3, can be used for this purpose.

**(iii)** Using $(n-k)$ spectral components $E_{h+i}$'s, obtained in *(i)*, and $(s+\epsilon)$ components $\Lambda_l$'s, obtained in *(ii)*, and (C.2), compute the $k$ remaining spectral components of the error vector **e**.

After the evaluation of **E** the remaining steps for the transform decoding procedure are as follows:

**For systematic codes**

**(iv)** Compute the errata vector **e** from **E** by using the inverse Fourier transform, i.e.,

$$\mathbf{e} = \mathcal{F}^{-1}\{\mathbf{E}\}.$$

**(v)** Correct the received vector as

$$\mathbf{c} = \mathbf{r} + \mathbf{e}.$$

**For Transform Codes**

**(iv)** Compute the remaining spectral components of **r** by using Fourier transform

$$\mathbf{R} = \mathcal{F}\{\mathbf{r}\}.$$

**(v)** Produce the $k$ *information* symbols as

$$A_{i-(n-k)} = C_{h+i} = R_{h+i} + E_{h+i}, \quad i = (n-k),\ (n-k+1),\ \cdots,\ (n-1),$$

by using the linearity of the Fourier transform.

The transform decoding flowchart is shown in Figure C.1.

To compare the syndrome-based and transfer decoding algorithms, we can make the following comments with reference to Figure 6.2 and Figure C.1.

**(a)** Block 1 of Figure 6.2 and Figure C.1 uses the same computation. However, the transfer decoding procedure requires $k$ additional components.

**(b)** Block 2 of Figure 6.2 and Figure C.1 is identical.

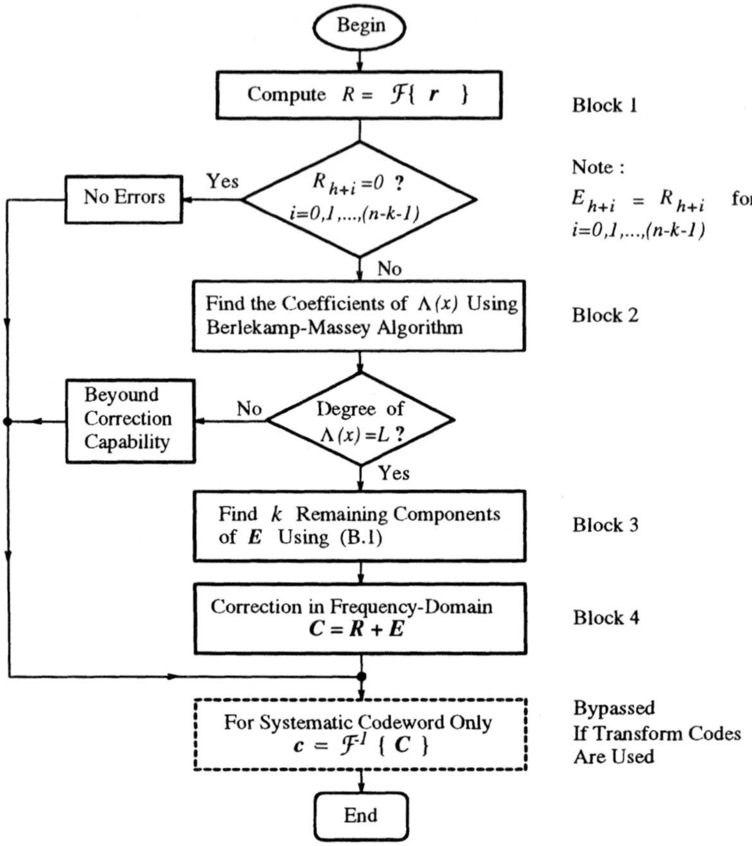


Figure C.1: Errors-and-erasures transform decoding algorithm.

**(c)** Block $3A$ and $3B$ of Figure 6.2 are replaced by block 3 of Figure C.1. In the transform decoding procedure, block 3 of Figure C.1 needs $k$ iteratives. In the syndrome-based decoding procedure, blocks $3A$ and $3B$ of Figure 6.2 need a number of iterations proportional to $(n-k)$. For a given $\mathrm{GF}(2^m)$ we can find a threshold value for $(n-k)$ at which the amount of computations for block 3 of Figure C.1 is similar to that of blocks $3A$ and $3B$ of Figure 6.2. In general, as $(n-k)$ increases, block 3 of Figure C.1 becomes more and more computationally efficient compared to blocks $3A$ and $3B$ of Figure 6.2.

**(d)** Block 3 of Figure C.1 shares the basic structure of (C.2) with block 2 of Figure 6.2 , while blocks 2 and $3A$ of Figure 6.2 have different structures, i.e., the transform decoding procedure has a more regular structure than the syndrome-based decoding procedure. Note that regularity is important for the VLSI implementation of RS decoders.

For singly-extended RS $(n+1,\ k)$ codes, the approach to compute the components of $\mathbf{\Lambda}$, presented in Chapter 6, is completely applicable to transform decoding. In other words, the transform decoding procedure shown in Figure C.1 can be used to correct both errors and erasures for a singly-extended RS $(n+1,\ k)$ code by using the modified BMA of Figure 6.4 to replace that of Figure 6.3.

# C.4 TIME-DOMAIN DECODING TECHNIQUES

The transform decoding procedure, shown in Figure C.1, provides a more regular structure compared to the syndrome-based decoding procedure. However, it still needs a Fourier transform block. The *time-domain decoding* procedures, introduced by Blahut [74], [93] provide a more regular structure which is important for VLSI implementation of ***universal*** RS decoders. By universal, we mean that the decoder parameters such as $n$ and $k$ can be programmed for RS codes defined in GF($2^m$).

With reference to the transform decoding procedure, shown in Figure C.1, the time-domain procedures aim to avoid the Fourier transform block by developing an equivalent BMA in time domain.

The time-domain Berlekamp-Massey algorithm produces components of time-domain vectors $\lambda$ and $\mathbf{e}$ rather than their frequency-domain counterparts $\mathbf{\Lambda}$ and $\mathbf{E}$. The time- domain BMA for errors-and-erasures decoding of RS $(n,\ k)$ codes is developed as follows.

Referring to Figure 6.3, the errata polynomial $\Lambda(x)$ is initialized with $s$ erasure-locators in the following manner.

$$\begin{aligned} \Lambda^{(0)}(x) &= 1, & i&=0, \\ \Lambda^{(i)}(x) &= \Lambda^{(i-1)}(x)\,(1+X_i^* x), & i&=1,\ 2,\ \cdots,\ s, \end{aligned}$$

where $X_i^*$'s are erasure-locators.

At each step $i$, $\Lambda^{(i)}(x)$ can be represented as an $n$-component vector in frequency domain as

$$\mathbf{\Lambda}^{(i)} = (\Lambda_{n-1}^{(i)},\ \Lambda_{n-2}^{(i)},\ \cdots,\ \Lambda_1^{(i)},\ \Lambda_0^{(i)}),$$

where

$$\begin{aligned} \Lambda_0^{(i)} &= 1, \\ \Lambda_j^{(i)} &= \Lambda_j^{(i-1)} + X_i^* \Lambda_{j-1}^{(i-1)}, \qquad j=1,2,\cdots,i, \end{aligned}$$

and $\Lambda_j^{(i)} = 0$ for $j=(i+1),(i+2),\cdots,(n-1)$.

The inverse Fourier transform of $\mathbf{\Lambda}^{(i)}$ is $\lambda^{(i)}$, where

$$\lambda^{(i)} = (\lambda_{n-1}^{(i)}, \lambda_{n-2}^{(i)}, \cdots, \lambda_1^{(i)}, \lambda_0^{(i)}),$$

and

$$\lambda_j^{(i)} = \Lambda^{(i)}(\alpha^{-j}).$$

The time-domain vector $\lambda^{(i)}$ can be represented by the time-domain polynomial $\lambda^{(i)}(x)$.

Using the Fourier transform properties discused in Section C.1, we obtain (see Problem 6.30)

| **Frequency-Domain** | **Time-Domain** |
|---|---|
| $\Lambda^{(0)}(x) = 1$ | $\lambda^{(0)}(x) = \sum_{j=0}^{n-1} x^j$ |
| $\Lambda^{(i)}(x) = \Lambda^{(i-1)}(x) + X_i^* x \Lambda^{(i-1)}(x)$ | $\lambda^{(i)}(x) = \sum_{j=0}^{n-1}(\lambda_j^{(i-1)} + X_i^* \alpha^{-j} \lambda_j^{(i-1)}) x^j,$ |

implying that $\lambda_j^{(0)}$'s=1 and $\lambda_j^{(i)} = (1 + X_i^* \alpha^{-j})\lambda_j^{(i-1)}$.

For the $(n-k-s)$ following steps, the errata polynomial $\Lambda(x)$ is updated by the BMA. The structure of the flowchart in Figure 6.3 is unchanged for the time-domain BMA but the computations involve the time-domain vectors $\mathbf{a}$, $\lambda$, $\mathbf{r}$, and $\mathbf{b}$ rather than their counterparts $\mathbf{A}$, $\mathbf{\Lambda}$, $\mathbf{R}$, and $\mathbf{B}$.

Let $\mathbf{A}^{(i)}$, $\mathbf{B}^{(i)}$, and $\mathbf{R}^{(i)}$ be the $n$-component vectors in frequency domain at step $i$ for $i = (s+1), (s+2), \cdots, (n-k)$, i.e.,

$$\begin{aligned} \mathbf{A}^{(i)} &= (A_{n-1}^{(i)}, A_{n-2}^{(i)}, \cdots, A_1^{(i)}, A_0^{(i)}), \\ \mathbf{B}^{(i)} &= (B_{n-1}^{(i)}, B_{n-2}^{(i)}, \cdots, B_1^{(i)}, B_0^{(i)}), \end{aligned}$$

where $A_j^{(i)}$'s and $B_j^{(i)}$'s are the coefficients of polynomials $A(x)$ and $B(x)$, respectively, at step $i$ and

$$\mathbf{R}^{(i)} = (R_{n-1}^{(i)}, R_{n-2}^{(i)}, \cdots, R_1^{(i)}, R_0^{(i)}),$$

where $R_{(h+l) \bmod n}^{(i)} = S_{l+1} = r(\alpha^{h+l})$ for $l = 0, 1, \cdots, (i-1)$ and the remaining components of $\mathbf{R}^{(i)}$ are zero.

In Figure 6.3 we update

$$A^{(i)}(x) = x A^{(i-1)}(x)$$

or equivalently

$$\begin{aligned} A_0^{(i)} &= 0, \\ A_j^{(i)} &= A_{j-1}^{(i-1)}. \end{aligned}$$

The inverse Fourier transform of $\mathbf{A}^{(i)}$ is an $n$-component vector in time domain $\mathbf{a}^{(i)} = (a_{n-1}^{(i)}, \cdots, a_1^{(i)}, a_0^{(i)})$ where the components of $\mathbf{a}^{(i)}$ are updated as

$$a_j^{(i)} = \alpha^{-j} a_j^{(i-1)}.$$

In Figure 6.3 the discrepancy $\delta$ at step $i$ is computed as

$$\begin{aligned}\delta_i &= \sum_{j=0}^{L} \Lambda_j^{(i-1)} S_{i-j} \\ &= \sum_{j=0}^{n-1} \Lambda_j^{(i-1)} R_{(h+i-j-1) \bmod n}^{(i)},\end{aligned}$$

where

$$\Lambda_j^{(i-1)} = \sum_{l=0}^{n-1} \lambda_l^{(i-1)} \alpha^{lj},$$

and

$$R_{(h+i-j-1) \bmod n}^{(i)} = \sum_{p=0}^{n-1} r_p \alpha^{p(h+i-j-1)}$$

and $r_p$'s are the components of the received vector $\mathbf{r}$. Therefore, $\delta_i$ at step $i$ can be computed in terms of $\lambda_l$'s and $r_p$'s as

$$\begin{aligned}\delta_i &= \sum_{j=0}^{n-1}\sum_{l=0}^{n-1}\sum_{p=0}^{n-1} \lambda_l^{(i-1)} r_p \alpha^{lj+ph+pi-pj-p} \\ &= \sum_{l=0}^{n-1}\sum_{p=0}^{n-1} \lambda_l^{(i-1)} r_p \alpha^{p(i+h-1)} \sum_{j=0}^{n-1} \alpha^{(l-p)j}.\end{aligned}$$

Since

$$\sum_{j=0}^{n-1} \alpha^{j(l-p)} = \begin{cases} 1, & \text{if } l = p \\ 0, & \text{if } l \neq p \end{cases}$$

we obtain

$$\delta_i = \sum_{l=0}^{n-1} \alpha^{l(i+h-1)} \lambda_l^{(i-1)} r_l. \tag{C.3}$$

After the evaluation of $\lambda$ by using the time-domain BMA, the errata vector $\mathbf{e}$ should be found. Therefore, our next step is to develop the procedure to generate $\mathbf{e}$ from $\lambda$ in the time domain.

For $s$ erasures and $\epsilon$ errors with $2\epsilon+s \leq (n-k)$, at the end of step $(n-k)$ the frequency-domain errata-locator is $\Lambda = (\Lambda_{n-1}, \cdots, \Lambda_1, \Lambda_0)$ where $\Lambda_0 = 1$ and $\Lambda_j = 0$ for $j = (s+\epsilon+1), (s+\epsilon+2), \cdots, (n-1)$. The frequency-domain errata-locator $\Lambda$ remains unchanged for the following steps $(n-k+1), (n-k+2), \cdots, n$.

The Fourier transform of the errata vector at the end of step $(n-k)$ is

$$\mathbf{E}^{(n-k)} = (E_{n-1}^{(n-k)},\ E_{n-2}^{(n-k)},\ \cdots,\ E_1^{(n-k)},\ E_0^{(n-k)}),$$

where

$$E_{(h+j) \bmod n}^{(n-k)} = R_{(h+j) \bmod n} = S_{j+1}, \quad j = 0,\ 1,\ \cdots,\ (n-k-1),$$

and the $k$ remaining components of $\mathbf{E}^{(n-k)}$ are zero.

At each following step $i$, $i = (n-k+1), (n-k+2), \cdots, n$ one new component $E_{h+i-1}^{(i)}$ is produced by using (6.16) as

$$E_{(h+i-1)}^{(i)} = \sum_{l=1}^{s+\epsilon} \Lambda_l E_{(h+i-1-l) \bmod n}^{(i-1)}.$$

As $E_{(h+i-1) \bmod n}^{(i-1)} = 0$ and $\Lambda_l = 0$ for $l = (s+\epsilon+1), \cdots, (n-1)$, we can write

$$E_{h+i-1}^{(i)} = \sum_{l=0}^{n-1} \Lambda_l E_{(h+i-1-l) \bmod n}^{(i-1)}.$$

In other words, the frequency-domain errata-locator at step $i$, for $i = (n-k+1),\ (n-k+2),\ \cdots,\ n$, is updated as

$$\mathbf{E}^{(i)} = \mathbf{E}^{(i-1)} + \mathbf{F},$$

where

$$\mathbf{F} = (F_{n-1},\ F_{n-2},\ \cdots,\ F_1,\ F_0),$$

and

$$F_j = \begin{cases} 0, & j \neq (h+i-1) \bmod n, \\ \delta_i = \sum_{l=0}^{n-1} \Lambda_l E_{(h+i-l) \bmod n}^{(i-1)}, & j = (h+i-1) \bmod n. \end{cases}$$

Equivalently, the time-domain errata vector $\mathbf{e}^{(i)}$ at step $i$, for $i = (n-k+1), (n-k+2), \cdots, n$ is updated as

$$\mathbf{e}^{(i)} = \mathbf{e}^{(i-1)} + \mathbf{f},$$

where

$$\begin{aligned} \mathbf{f} &= \mathcal{F}^{-1}\{\mathbf{F}\} \\ &= (f_{n-1},\ f_{n-2},\ \cdots,\ f_1,\ f_0) \end{aligned}$$

with

$$f_l = \alpha^{-l(h+i-1)} \cdot \delta_i$$

and

$$\delta_i = \sum_{j=0}^{n-1} \alpha^{j(h+i-1)} \lambda_j e_j^{(i-1)}.$$

In the foregoing equation $e_j^{(i-1)}$'s are components of $\mathbf{e}^{(i-1)}$ and $\mathbf{e}^{(n-k)} = \mathbf{r}$, and $\lambda_j$'s are components of $\lambda$, obtained at the end of step $(n-k)$. The vector $\mathbf{e}^{(n)}$, obtained at step $n$ , is the derived errata vector $\mathbf{e}$ which is used to correct $\mathbf{r}$ as $\mathbf{c} = \mathbf{r} + \mathbf{e}$.

The complete errors-and-erasures time-domain decoding algorithm for RS $(n,\ k)$ codes is shown in Figure C.2.

# C.5 TIME-DOMAIN DECODING TECHNIQUES FOR SINGLY-EXTENDED RS $(n+1, k)$ CODES

For singly-extended RS $(n+1,\ k)$ codes the additional parity-check symbol $c_+$ (or $c_-$) is actually the $(h+n-k)$th (or $(h-1)$th) *spectral component* of the time domain interior $\mathbf{c}$. Following the discussion in Chapter 6, if the discrepancy at step $(n-k)$ is not zero, then the received symbol $r_+$ is assumed to be error-free (i.e., $r_+ = c_+ = C_{h+n-k}$) and is used in step $(n-k+1)$ to find the coefficients of the errata-locator polynomial $\Lambda(x)$. We will study the equivalent time-domain expression when $\delta_{n-k} \neq 0$ in the following paragraphs.

From Section 6.8, the discrepancy at step $(n+1-k)$ is computed as

$$\begin{aligned}
\delta_{n+1-k} &= (S_{n+1-k} + r_+) + \sum_{l=1}^{d_{n-k}} \Lambda_l^{(n-k)} S_{n+1-k-l} \\
&= r_+ + \sum_{l=0}^{d_{n-k}} \Lambda_l^{(n-k)} S_{n+1-k-l}.
\end{aligned}$$

As $\Lambda_l^{(n-k)} = 0$ for $l = (d_{n-k}+1), (d_{n-k}+2), \cdots, (n-1)$ and

$$S_{n+1-k-l} = R_{h+n-k-l} = R_{h+n-k-l}^{(n-k-1)},$$

we can rewrite $\delta_{n+1-k}$ as

$$\delta_{n+1-k} = \sum_{l=0}^{n-1} \Lambda_l^{(n-k)} R_{h+n-k-l}^{(n+1-k)} + r_+ \tag{C.4}$$

Using (C.3) and (C.4), we obtain

$$\delta_{n+1-k} = \sum_{l=0}^{n-1} \alpha^{l(h-1+i)} \lambda_l^{(i-1)} r_l + r_+,$$

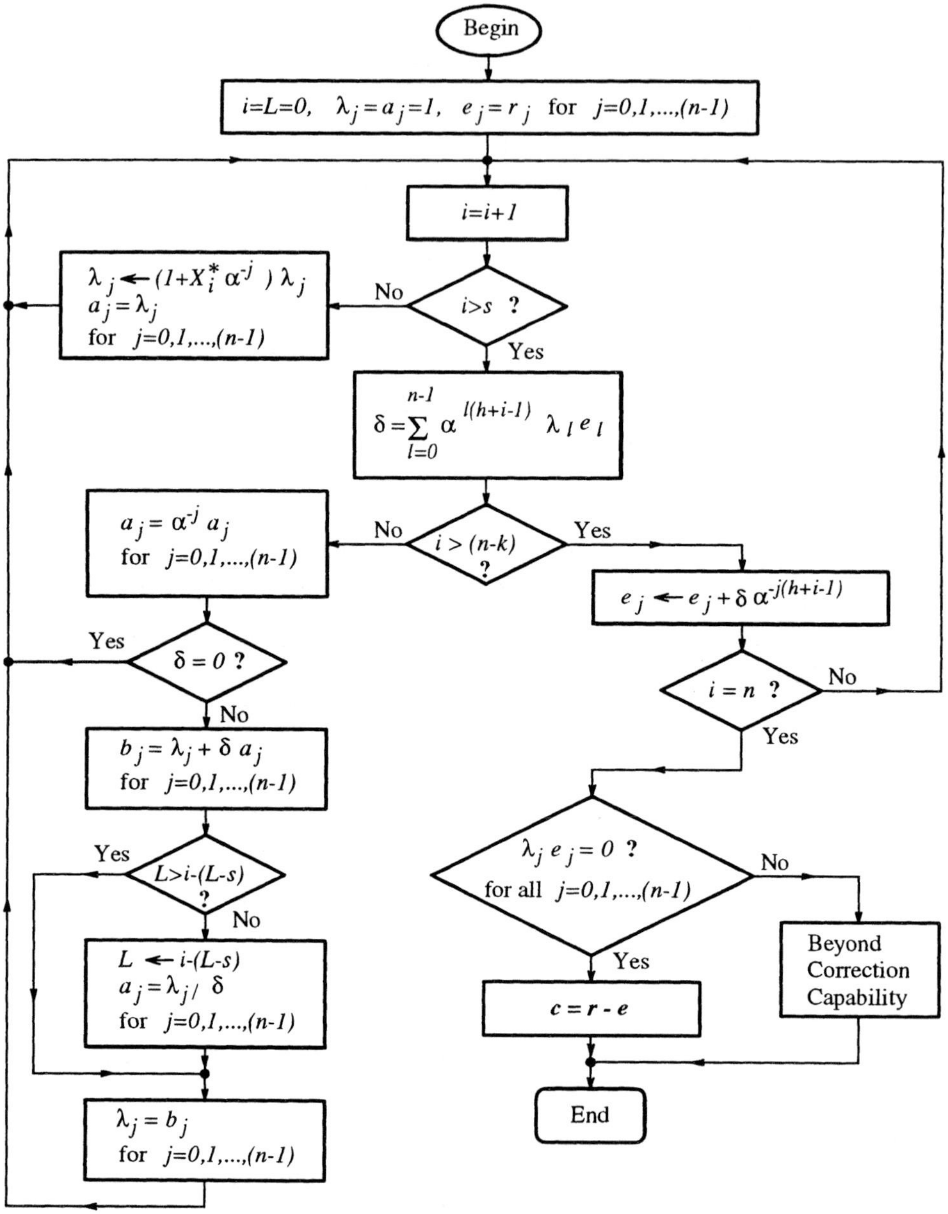


Figure C.2: An errors-and-erasures time-domain decoding algorithm for RS $(n, k)$ codes with $s < (n - k)$.

where $i = n + 1 - k$.

In this case ($\delta_{n-k} \neq 0$), the frequency-domain errata vector at step $(n+1-k)$ is modified accordingly as

$$\mathbf{E}^{(n+1-k)} = (E_{n-1}^{(n+1-k)}, E_{n-2}^{(n+1-k)}, \cdots, E_1^{(n+1-k)}, E_0^{(n+1-k)})$$

where $E_{(h+j) \bmod n}^{(n+1-k)} = R_{(h+j) \bmod n}$ for $j = 0, 1, \cdots, (n-k-1)$,

$$E_{(h-k) \bmod n}^{(n+1-k)} = R_{(h-k) \bmod n} + r_+,$$

and the $(k-1)$ remaining components of $\mathbf{E}^{(n+1-k)}$ are zero. In other words, the frequency-domain errata vector $\mathbf{E}^{(n+1-k)}$ is modified as

$$\mathbf{E}^{(n+1-k)} = \mathbf{E}^{(n-k)} + \mathbf{R}^+,$$

where

$$\begin{aligned} \mathbf{R}^+ &= (R_{n-1}^+, R_{n-2}^+, \cdots, R_1^+, R_0^+), \\ \mathbf{R}_{(h-k) \bmod n}^+ &= r_+, \end{aligned}$$

and the other components of $\mathbf{R}^+$ are zero.

Equivalently, the time-domain errata vector at step $(n+1-k)$, $\mathbf{e}^{(n+1-k)}$, is modified as

$$\mathbf{e}^{(n+1-k)} = (e_{n-1}^{(n+1-k)}, e_{n-2}^{(n+1-k)}, \cdots, e_1^{(n+1-k)}, e_0^{(n+1-k)})$$

where

$$e_j^{(n+1-k)} = e_j^{(n-k)} + r_+\alpha^{-j(h-k)}.$$

The time-domain decoding algorithm for singly-extended RS $(n+1, k)$ codes is shown in Figure C.3. The procedure in Figure C.3 also includes the verification for the case $s = n - k + 1$.

The decoding time of a time-domain decoding procedure is independent of $k$, and hence, independent of the number of correctable errors and erasures. As shown in Figure C.3 many operations require repetitive computations for all elements of involved vectors.

The time-domain decoding algorithm is therefore less computationally efficient than the syndrome-based and transform decoding algorithms. However, it provides a more regular structure suitable to VLSI implementation of universal decoders. Other time-domain decoding algorithms can be found in [93].

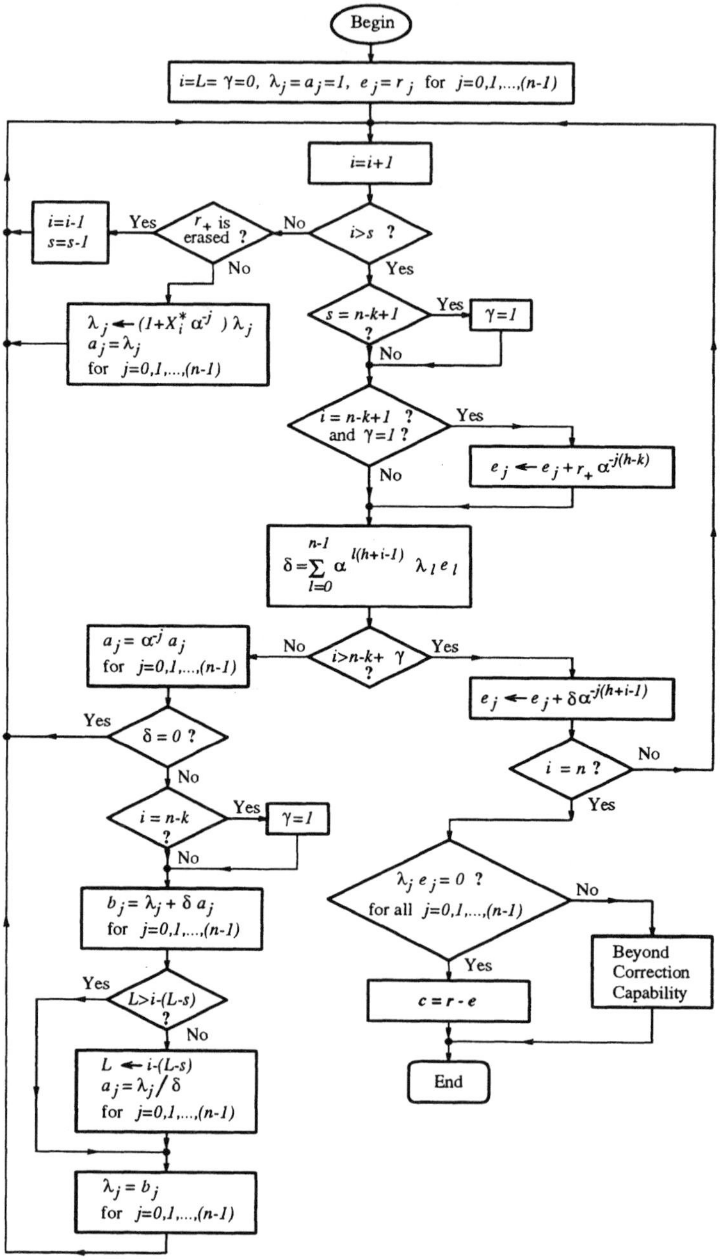


Figure C.3: An errors-and-erasures time-domain decoding algorithm for singly-extended RS $(n+1, k)$ codes $(s < n-k)$.

# Appendix D

# DERIVATION OF THE PROBABILITY OF CORRECT DECISION FOR RELIABLE AND UNRELIABLE MPSK SYMBOLS

In this appendix we evaluate the probability of correct decision for a reliable and an unreliable MPSK symbol where a circular region with radius $\rho_T$ in the MPSK signal space is considered for erasures (erasure generation without CSI). Also, the probability of a reliable MPSK symbol is calculated.

## D.1 CALCULATION OF THE PROBABILITY OF THE CORRECT DECISION OF A RELIABLE MPSK SYMBOL

The probability of the correct decision of a reliable MPSK symbol, $Q_i^{\overline{E}}$, is defined as (7.8). For the equiprobable transmitted MPSK symbols this probability is independent of the $i$th signal. Hence, dropping the subscript $i$, the probability of correct decision of a reliable MPSK signal is evaluated for the decision region $\Lambda_0$. This evaluation can be simplified by interchanging the integration and averaging operations. Further simplification can be obtained by introducing the polar coordination $\rho$ and $\theta$ related to $r_1$ and $r_2$ as

$$\begin{aligned} r_1 &= \rho \cos\theta \\ r_2 &= \rho \sin\theta \\ dr_1\, dr_2 &= \rho\, d\rho\, d\theta. \end{aligned} \tag{D.1}$$

Representing the radius of the erasure region by $\rho_T$ (see Figure 2.15), after some mathematical manipulation (7.8) can be written as

$$\begin{aligned} Q^{\overline{E}} &= \frac{e^{-\rho_T^2/2\sigma^2}}{M(1+\gamma_s)} + \frac{\sqrt{2\gamma_s}}{2\pi\sigma^3(1+\gamma_s)^{3/2}} \int_{\rho_T}^{\infty} \rho^2\, e^{-\frac{\rho^2}{2\sigma^2(1+\gamma_s)}} \\ &\quad \cdot \left( \int_{-\pi/M}^{\pi/M} \cos\theta\, e^{-\frac{\gamma_s \rho^2 \sin^2\theta}{2\sigma^2(1+\gamma_s)}} \int_{-\frac{\sqrt{\gamma_s}\,\rho\cos\theta}{\sqrt{2\sigma^2(1+\gamma_s)}}}^{\infty} e^{-x^2}\, dx\, d\theta \right) d\rho, \end{aligned} \tag{D.2}$$

where $\gamma_s \equiv \frac{E_s}{N_0}$ is the symbol energy-to-noise ratio and $\sigma^2 = N_0/2$. In general, this expression should be calculated numerically. However, applying the approximation

$$\mathrm{erfc(x)} \equiv \frac{2}{\pi} \int_{\mathrm{x}}^{\infty} \mathrm{e}^{-\mathrm{t}^2} \mathrm{dt} \approx \frac{2}{\sqrt{\pi}\mathrm{x}}\, \mathrm{e}^{-\mathrm{x}^2}, \tag{D.3}$$

the expresion inside the parentheses in (D.2) can be simplified as

$$\begin{aligned} &\int_{-\pi/M}^{\pi/M} \cos\theta\, e^{-\frac{\gamma_s \rho^2 \sin^2\theta}{2\sigma^2(1+\gamma_s)}} \int_{-\frac{\sqrt{\gamma_s}\rho\cos\theta}{\sqrt{2\sigma^2(1+\gamma_s)}}}^{\infty} e^{-x^2} \cdot dx\, d\theta \\ &\qquad = \frac{\sigma\pi\sqrt{2(1+\gamma_s)}}{\rho\sqrt{\gamma_s}} \left( 1 - \mathrm{erfc}\Big(\frac{\sqrt{\gamma_s}\rho\ \sin(\pi/\mathrm{M})}{\sqrt{2\sigma^2(1+\gamma_s)}}\Big) \right) \\ &\qquad\qquad - \frac{2\pi\sigma\sqrt{1+\gamma_s}}{\rho M\sqrt{2\gamma_s}}\, e^{-\frac{\gamma_s\rho^2}{2\sigma^2(1+\gamma_s)}}. \end{aligned} \tag{D.4}$$

Substituting (D.4) into (D.2) and performing some calculations yield

$$Q^{\overline{E}} = e^{-\frac{\rho_T^2}{2\sigma^2(1+\gamma_s)}} - \frac{1}{\sigma^2(1+\gamma_s)} \int_{\rho_T}^{\infty} \rho\, e^{-\frac{\rho^2}{2\sigma^2(1+\gamma_s)}}\, \mathrm{erfc}\left( \frac{\sqrt{\gamma_s}\ \rho\sin(\frac{\pi}{\mathrm{M}})}{\sqrt{2\sigma^2(1+\gamma_s)}} \right) \mathrm{d}\rho. \tag{D.5}$$

Using integral by parts the integral part of (D.4) can be calculated as

$$\int_{\rho_T}^{\infty} \rho\, e^{-\frac{\rho^2}{2\sigma^2(1+\gamma_s)}}\, \mathrm{erfc}\left( \frac{\sqrt{\gamma_s}\rho\sin(\pi/\mathrm{M})}{\sqrt{2\sigma^2(1+\gamma_s)}} \right) \mathrm{d}\rho$$

$$= \frac{1}{2}\sigma^2(1+\gamma_s)\, e^{-\frac{\rho_T^2}{2\sigma^2(1+\gamma_s)}} \operatorname{erfc}\left(\frac{\sqrt{\gamma_s}\,\rho_T \sin(\pi/M)}{\sqrt{2\sigma^2(1+\gamma_s)}}\right)$$

$$-\frac{\sqrt{\gamma_s}\sigma^2(1+\gamma_s)\,\sin(\frac{\pi}{M})}{2\sqrt{1+\gamma_s \sin^2(\pi/M)}} \operatorname{erfc}\left(\frac{\rho_T\sqrt{1+\gamma_s \sin^2(\pi/M)}}{\sqrt{2\sigma^2(1+\gamma_s)}}\right). \tag{D.6}$$

The final result can be obtained by substituting (D.6) into (D.5) as

$$Q^{\overline{E}} = e^{-\frac{\rho_T^2}{2\sigma^2(1+\gamma_s)}} + \frac{\sqrt{\gamma_s}\,\sin(\pi/M)}{\sqrt{1+\gamma_s \sin^2(\pi/M)}} \operatorname{erfc}\left(\frac{\rho_T\sqrt{1+\gamma_s \sin^2(\pi/M)}}{\sqrt{2\sigma^2(1+\gamma_s)}}\right)$$
$$-e^{-\frac{\rho_T^2}{2\sigma^2(1+\gamma_s)}} \operatorname{erfc}\left(\frac{\rho_T\sqrt{\gamma_s}\sin(\pi/M)}{\sqrt{2\sigma^2(1+\gamma_s)}}\right). \tag{D.7}$$

Defining the normalized erasure threshold region radius $\rho_{T_n}$ as $\rho_{T_n} \equiv \rho_T/\sqrt{E_s}$ and using the approximation $\gamma_s \approx \gamma_s + 1$ at high signal-to-noise ratios, (D.7) can be written as

$$Q^{\overline{E}} = e^{\rho_{T_n}^2} + \frac{\sqrt{\gamma_s}\,\sin(\pi/M)}{\sqrt{1+\gamma_s \sin^2(\pi/M)}} \operatorname{erfc}\left(\sqrt{1+\gamma_s \sin^2(\frac{\pi}{M})}\,\rho_{T_n}\right)$$
$$-e^{\rho_{T_n}^2} \operatorname{erfc}\left(\sqrt{\gamma_s}\rho_{T_n}\,\sin(\frac{\pi}{M})\right).$$

# D.2 CALCULATION OF THE PROBABILITY OF CORRECT DECISION OF AN UNRELIABLE MPSK SYMBOL

To find the probability of correct decision of MPSK symbol when the received signal falls into the erasure region we may follow the same steps as in the previous part. In this case the integration with respect to $r$ in (D.2) should be carried out from 0 to $\rho_T$. Following the approach of Section D.1 the probability of correct decision of an unreliable MPSK symbol, $Q^E$, at high signal-to-noise ratios, is found to be

$$Q^E = e^{-\rho_{T_n}} \operatorname{erfc}\left(\sqrt{\gamma_s}\,\rho_{T_n}\,\sin(\frac{\pi}{M})\right) + \frac{\sqrt{\gamma_s}\,\sin(\frac{\pi}{M})}{\sqrt{1+\gamma_s\,\sin^2(\frac{\pi}{M})}} - e^{-\rho_{T_n}^2}$$
$$-\frac{\sqrt{\gamma_s}\,\sin(\frac{\pi}{M})}{\sqrt{1+\gamma_s\,\sin^2(\frac{\pi}{M})}} \operatorname{erfc}\left(\sqrt{1+\gamma_s\,\sin^2(\frac{\pi}{M})}\,\rho_{T_n}\right).$$

## D.3 CALCULATION OF THE PROBABILITY OF A RELIABLE MPSK SYMBOL

To find the conditional probability $P_{sM|\overline{E}}$ and $P_{sM|E}$ the integration in (7.9) has to be evaluated. Interchanging the integration and averaging operations in (7.9) and introducing the polar coordinates as (D.1) this equation can be written as

$$\begin{aligned} P_{\overline{E}M} &= \int_{\rho_T}^{\infty}\int_{-\pi/2}^{3\pi/2}\Big[\frac{1}{2\pi\sigma^2(1+\gamma_s)}\,e^{-\frac{\rho^2}{2\sigma^2}} + \frac{\sqrt{2\gamma_s}\,\rho^2\cos\theta}{2\pi\sigma^3(1+\gamma_s)^{3/2}}\,e^{-\frac{\rho^2}{2\sigma^2}+\frac{\gamma_s\rho^2\cos^2\theta}{2\sigma^2(1+\gamma_s)}} \\ &\quad\cdot\int_{-\frac{\sqrt{\gamma_s}\rho\cos\theta}{\sqrt{2\sigma^2(1+\gamma_s)}}}^{+\infty} e^{-x^2}\cdot dx\,\Big]\,\rho\,d\rho\,d\theta. \end{aligned}$$

The interval of the integration with respect to phase, i.e., $(-\frac{\pi}{2},\ \frac{3\pi}{2})$ can be divided into two intervals $(-\frac{\pi}{2},\ \frac{\pi}{2})$ and $(\frac{\pi}{2},\ \frac{3\pi}{2})$. After some mathematical manipulation the following expression results

$$\begin{aligned} P_{\overline{E}M} &= \frac{1}{1+\gamma_s}\,e^{-\frac{\rho_T^2}{2\sigma^2}} + \sqrt{\pi}\int_{\rho_T}^{\infty}\frac{\rho^2\sqrt{2\gamma_s}}{2\pi\sigma^3(1+\gamma_s)^{3/2}}\,e^{-\frac{\rho^2}{2\sigma^2}+\frac{\gamma_s\rho^2}{2\sigma^2(1+\gamma_s)}} \\ &\quad\cdot\int_{-\pi/2}^{\pi/2}\cos\theta\,e^{-\frac{\gamma_s\rho^2\sin^2\theta}{2\sigma^2(1+\gamma_s)}}\left(1-\mathrm{erfc}\Big(\frac{\cos\theta}{\sigma\sqrt{1+\gamma_s}}\Big)\right)\,d\theta\,d\rho. \end{aligned} \tag{D.8}$$

Using the approximation of (D.3) and ignoring 1 with respect to $\gamma_s$ at high signal-to-noise ratios, (D.8) can be simplified as

$$P_{\overline{E}M} = e^{-\rho_{T_n}^2}\left(1-\mathrm{erfc}(\sqrt{\mathrm{mR_c}\gamma_b}\,\rho_{T_n})\right) + \mathrm{erfc}(\sqrt{\mathrm{mR_c}\gamma_b}\,\rho_{T_n}).$$

The probability of being in unreliable (erasure) region, $P_{EM}$ is obtained as

$$P_{EM} = 1 - P_{\overline{E}M}.$$

# REFERENCES


[1] S. Lin and D. J. Costello, Jr., *Error Control Coding.* Englewood Cliffs, NJ: Prentice-Hall, 1983.

[2] D. G. Forney, Jr., "Convolutional codes I: Algebraic structure," *IEEE Trans. Inform. Theory*, vol. IT-16, pp. 720–738, Nov. 1970.

[3] A. J. Viterbi and J. K. Omura, *Principles of Digital Communication and Coding.* New York: McGraw-Hill, 1979.

[4] R. G. Gallager, *Information Theory and Reliable Communication.* New York: John Wiley & Sons, 1968.

[5] E. A. Bucher and J. A. Heller, "Error probability bounds for systematic convolutional codes," *IEEE Trans. Inform. Theory*, vol. IT-16, pp. 219–224, Mar. 1970.

[6] J. L. Massey and M. K. Sain, "Codes, automata, and continuous systems: Explicit interconnections," *IEEE Trans. Automatic Control*, vol. AC-12, pp. 644–650, Dec. 1967.

[7] J. M. Wozencraft and B. Reiffen, *Sequential Decoding.* Cambridge, Mass.: MIT Press, 1961.

[8] R. M. Fano, "A heuristic discussion of probabilistic decoding," *IEEE Trans. Inform. Theory*, vol. IT-9, pp. 64–74, Apr. 1963.

[9] K. Zigangirov, "Some sequential decoding procedures," *Problemy Peredachi Informatsii*, vol. 2, pp. 13–25, 1966.

[10] F. Jelenik, "A fast sequential decoding algorithm using a stack," *IBM J. Res. Develop.*, vol. 13, pp. 675–685, Nov. 1969.

[11] J. L. Massey, *Threshold Decoding.* Cambridge, Mass.: MIT Press, 1963.

[12] A. J. Viterbi, "Error bounds for convolutional codes and an asymptotically optimum decoding algorithm," *IEEE Trans. Inform. Theory*, vol. IT-13, pp. 260–269, Apr. 1967.

[13] D. G. Forney, Jr., "The Viterbi algorithm," *Proc. IEEE*, vol. 61, pp. 268–278, Mar. 1973.

[14] A. J. Viterbi, "Convolutional codes and their performance in communication systems," *IEEE Trans. Commun.*, vol. COM-19, pp. 751–772, Oct. 1971.

[15] J. A. Heller and I. M. Jacobes, "Viterbi decoding for satellite and space communications," *IEEE Trans. Commun.*, vol. COM-19, pp. 835–848, Oct. 1971.

[16] L. V. D. Meeberg, "A tighter upper bound on the error probability of binary convolutional codes with Viterbi decoding," *IEEE Trans. Inform. Theory*, vol. IT-20, pp. 389–391, May 1974.

[17] J. M. Wozencraft and I. M. Jacobs, *Principles of Communication Engineering*. New York: John Wiley & Sons, 1965.

[18] G. Ungerboeck, "Channel coding with multilevel/phase signals," *IEEE Trans. Inform. Theory*, vol. IT-28, pp. 55–67, Jan. 1982.

[19] A. R. Calderbank and J. E. Mazo, "A new description of trellis codes," *IEEE Trans. Inform. Theory*, vol. IT-30, pp. 784–791, Nov. 1984.

[20] G. Ungerboeck, "Trellis-coded modulation with redundant signal sets-Part II: State of the Art," *IEEE Communications Mag.*, vol. 25, pp. 12–21, Feb. 1987.

[21] J. E. Porath, "Algorithms for converting convolutional codes from feedback to feed-forward form and vice versa," *Electron. Letters*, vol. 25, pp. 1008–1009, July 1989.

[22] E. Biglieri and P. J. McLane, "Uniform distance and error probability properties of TCM schemes," *IEEE Trans. Commun.*, vol. COM-39, pp. 41–53, Jan. 1991.

[23] E. Zehavi and J. K. Wolf, "On the performance evaluation of trellis codes," *IEEE Trans. Inform. Theory*, vol. IT-33, pp. 196–202, Mar. 1987.

[24] J. K. Omura, "Generalized transfer function bounds," *Proc. IEEE Int'l Conf. Commun.*, June 1981.

[25] D. Divsalar, *Performance of Mismatched Receivers on Bandlimited Channels*. PhD thesis, Univ. of California, Los Angeles, 1978.

[26] Y. -J. Liu, I. Oka, and E. Biglieri, "Error probability for digital transmission over nonlinear channels with application to TCM," *IEEE Trans. Inform. Theory*, vol. IT-36, pp. 1101–1110, Sept. 1990.

[27] E. Biglieri, D. Divsalar, P. J. McLane, and M. K. Simon, *Introduction to Trellis-Coded Modulation with Applications.* New York: Macmillan, 1991.

[28] H. Imai and S. Hirakawa, "A new multilevel coding method using error correcting codes," *IEEE Trans. Inform. Theory*, vol. IT-23, pp. 371–376, May 1977.

[29] E. L. Cusack, "Error control codes for QAM signaling," *Electron. Letters*, vol. 20, pp. 62–63, Jan. 1984.

[30] S. I. Sayegh, "A class of optimum block codes in signal space," *IEEE Trans. Commun.*, vol. COM-34, pp. 1043–1045, Oct. 1986.

[31] D. G. Forney, Jr., R. G. Gallager, G. R. Lang, F. M. Longstaff, and S. U. Qureshi, "Efficient modulation for band-limited channels," *IEEE J. Select. Areas Commun.*, vol. SAC-2, pp. 632–647, Sept. 1984.

[32] D. G. Forney, Jr., "Coset codes - Part I: Introduction and geometrical classification," *IEEE Trans. Inform. Theory*, vol. IT-34, pp. 1123–1151, Sept. 1988.

[33] D. G. Forney, Jr., "Coset codes - Prat II: Binary lattices and related codes," *IEEE Trans. Inform. Theory*, vol. IT-34, pp. 1152–1187, Sept. 1988.

[34] F. R. Kschischang, P. G. du Buda, and S. Pasupathy, "Block coset codes for M-ary phase shift keying," *IEEE J. Select. Areas Commun.*, vol. SAC-7, pp. 900–913, Aug. 1989.

[35] T. Kasami, T. Takata, T. Fujiwara, and S. Lin, "A concatenated coded modulation scheme for error control," *IEEE Trans. Commun.*, vol. COM-38, pp. 752–763, June 1990.

[36] J. L. Massey, "Coding and modulation in digital communications," in *Proc. 1974 Int. Zurich Seminar on Digital Commun.*, pp. E2(1)–(4), March 1974.

[37] A. R. Calderbank and N. J. A. Sloane, "Four-dimensional modulation with an eight-state trellis code," *AT&T Tech. J.*, vol. 64, pp. 1005–1018, 1985.

[38] A. R. Calderbank and N. J. A. Sloane, "An eight-dimensional trellis code," *Proc. IEEE*, vol. 74, pp. 757–759, 1987.

[39] A. R. Calderbank and N. J. A. Sloane, "New trellis codes based on lattices and cosets," *IEEE Trans. Inform. Theory*, vol. IT-33, pp. 177–195, Mar. 1987.

[40] L.-F. Wei, "Trellis-coded modulation with multidimensional constellations," *IEEE Trans. Inform. Theory*, vol. IT-33, pp. 483–501, July 1987.

[41] D. Divsalar and J. H. Yuen, "Asymmetric MPSK for trellis codes," *Proc. IEEE Globecom.*, pp. 20.6.1–20.6.8, Nov. 1984.

[42] D. Divsalar, M. K. Simon, and J. H. Yuen, "Trellis coding with asymmetric modulation," *IEEE Trans. Commun.*, vol. COM-35, pp. 130–141, Feb. 1987.

[43] D. Divsalar and M. K. Simon, "Multiple trellis coded modulation (MTCM)," *IEEE Trans. Commun.*, vol. COM-36, pp. 410–419, Apr. 1988.

[44] L.-F. Wei, "Rotationally invariant convolutional channel coding with expanded signal space–Part I:180°," *IEEE J. Select. Areas Commun.*, vol. SAC-2, pp. 659–671, Sept. 1984.

[45] L.-F. Wei, "Rotationally invariant convolutional channel coding with expanded signal space–Part II: Nonlinear codes," *IEEE J. Select. Areas Commun.*, vol. SAC-2, pp. 672–686, Sept. 1984.

[46] CCITT Study Group XVII, "Recommendation V.32 for a family of 2-wire, duplex modems operating at data signaling rates of up to 9600 bits/s for use on the general switched telephone network and on leased telephone-type circuits." Document AP VIII-43-E, May 1984.

[47] D. Divsalar and M. K. Simon, "The design of trellis coded MPSK for fading channels: Performance criteria," *IEEE Trans. Commun.*, vol. COM-36, pp. 1004–1012, Sept. 1988.

[48] S. G. Wilson and Y. S. Leung, "Trellis-coded phase modulation on Rayleigh fading channels," *Proc. IEEE Int'l Conf. Commun.*, pp. 21.3.1–21.3.5, June 1987.

[49] J. K. Cavers and P. Ho, "Analysis of the error performance of trellis-coded modulations in Rayleigh-fading channels," *IEEE Trans. Commun.*, vol. COM-40, pp. 74–83, Jan. 1992.

[50] J. Hagenauer, "Viterbi decoding of convolutional codes for fading and burst-channels," in *Proc. 1980 Int. Zurich Seminar on Digital Commun.*, pp. G2(1)–(7), March 1980.

[51] D. Divsalar and M. K. Simon, "Trellis coded modulation for 4800-9600 bits/s transmission over a fading mobile satellite channel," *IEEE J. Select. Areas Commun.*, vol. SAC-5, pp. 162–175, Feb. 1987.

[52] C. Schlegel and D. J. Costello, Jr., "Bandwidth efficient coding for fading channels: Code construction and performance analysis," *IEEE J. Select. Areas Commun.*, vol. SAC-7, pp. 1356–1368, Dec. 1989.

[53] S. H. Jamali and T. Le-Ngoc, "A new 4-state 8PSK TCM scheme for fast fading, shadowed mobile radio channels," *IEEE Trans. Vehic. Technol.*, vol. VTC-40, pp. 216–222, Feb. 1991.

[54] J. M. Turgeon and P. J. McLane, "Minimal transmitter complexity design of analytically described trellis codes," *IEEE Trans. Commun.*, vol. COM-38, pp. 1352–1358, Sept. 1990.

[55] C. Schlegel, *Bandwidth Effective Coding for Non-Gaussian Channels.* PhD thesis, Univ. of Notre Dame, Notre Dame, Indiana, 1988.

[56] J. Du and B. Vucetic, "New M-PSK trellis codes for fading channels," *Electron. Letters*, vol. 26, pp. 1267–1269, Aug. 1990.

[57] D. Divsalar and M. K. Simon, "The design of trellis coded MPSK for fading channels: Set partitioning for optimum code design," *IEEE Trans. Commun.*, vol. COM-36, pp. 1013–1021, Sept. 1988.

[58] M. K. Simon, J. K. Omura, R. A. Scholtz, and B. K. Levitt, *Spread Spectrum Communications, Vol. 1.* Rockville, Maryland: Computer Science Press, 1985.

[59] M. Schwartz, W. Bennett, and S. Stein, *Communication Systems and Techniques.* New York: McGraw-Hill, 1966.

[60] J. G. Proakis, *Digital Communications.* New York: McGraw-Hill, 1983.

[61] J. H. Lodge and M. L. Moher, "Time diversity for mobile satellite channels using trellis coded modulations," *Proc. IEEE Globecom.*, pp. 8.7.1–8.7.5, Nov. 1987.

[62] M. L. Moher and J. H. Lodge, "TCMP- A modulation and coding strategy for Rician fading channels," *IEEE J. Select. Areas Commun.*, vol. SAC-7, pp. 1347–1355, Dec. 1989.

[63] E. Zehavi, "8-PSK trellis codes for a Rayleigh channel," *IEEE Trans. Commun.*, vol. COM-40, pp. 873–884, May 1992.

[64] R. G. McKay, "Error bounds for trellis coded modulation on fading channels," Master's thesis, Queen's University, Kingston, Canada, 1989.

[65] R. G. McKay, P. J. McLane, and E. Biglieri, "Error bounds for trellis-coded MPSK on a fading mobile satellite channel," *IEEE Trans. Commun.*, vol. COM-39, pp. 1750–1761, Dec. 1991.

[66] K.-Y. Chan and A. Bateman, "The performance of reference-based M-ary PSK with trellis coded modulation in Rayleigh fading," *IEEE Trans. Vehic. Technol.*, vol. VTC-41, pp. 190–198, Apr. 1992.

[67] I. S. Reed and G. Solomon, "Polynomial codes over certain finite fields," *J. Soc. Ind. Appl. Math.*, vol. 8, pp. 300–304, June 1960.

[68] W. W. Peterson and E. J. Weldon, Jr., *Error-Correcting Codes.* Cambridge, Mass.: MIT Press, 1972.

[69] A. Hocquenghem, "Codes correcteurs d'erreurs," *Chiffres*, vol. 2, pp. 147–156, 1959.

[70] R. C. Bose and D. K. Ray-Chaudhuri, "On a class of error-correcting binary group codes," *Inform. Contr.*, vol. 3, pp. 68–79, Mar. 1960.

[71] W. W. Peterson, *Error-Correcting Codes.* Cambridge, Mass.: MIT Press, 1961.

[72] A. R. Michelson and A. H. Levesque, *Error-Control Techniques for Digital Communication.* New York: John Wiley & Sons, 1985.

[73] D. C. Gorenstein and N. Zierler, "A class of cyclic linear error-correcting codes in $p^m$ symbols," *J. Soc. Ind. Appl. Math.*, vol. 9, pp. 207–214, June 1961.

[74] R. E. Blahut, *Theory and Practice of Error Control Codes.* Reading, Mass.: Addison-Wesley, 1983.

[75] J. K. Wolf, "Adding two information symbols to certain nonbinary BCH codes," *Bell Syst. Tech. J.*, vol. 48, pp. 2405–2404, 1969.

[76] R. E. Blahut, "On extended BCH codes," in *Proc. 18th Allerton Conf., Commun., Contr., Comput.*, (Univ. of Illinois, Monticello, I11.), pp. 50–59, 1980.

[77] W. W. Peterson, "Encoding and error-correction procedures for the Bose-Chaudhuri codes," *IEEE Trans. Inform. Theory*, vol. IT-6, pp. 459–470, 1960.

[78] E. R. Berlekamp, *Algebraic Coding Theory.* New York: McGraw-Hill, 1968.

[79] J. L. Massey, "Shift-register synthesis and BCH decoding," *IEEE Trans. Inform. Theory*, vol. IT-15, pp. 122–127, Jan. 1969.

[80] Y. Sugiyama, M. Kasahara, S. Hirasawa, and T. Namekawa, "A method for solving key equations for decoding Goppa codes," *Inform. Contr.*, vol. 27, pp. 87–99, Jan. 1975.

[81] D. M. Mandelbaum, "Decoding beyond the designed distance for certain algebraic codes," *Inform. Contr.*, vol. 29, pp. 207–228, 1977.

[82] L. R. Welch and R. A. Scholtz, "Continued fractions and Berlekamp's algorithms," *IEEE Trans. Inform. Theory*, vol. IT-25, pp. 19–27, Jan. 1979.

[83] R. T. Chien, "Cyclic decoding procedures for Bose-Chaudhuri-Hocquenghem codes," *IEEE Trans. Inform. Theory*, vol. IT-10, pp. 357–363, Oct. 1964.

[84] D. G. Forney, Jr., "On decoding BCH codes," *IEEE Trans. Inform. Theory*, vol. IT-11, pp. 549–557, Oct. 1965.

[85] D. G. Forney, Jr., "Generalized minimum distance decoding," *IEEE Trans. Inform. Theory*, vol. IT-12, pp. 125–131, Apr. 1966.

[86] E. J. Weldon, Jr., "Decoding binary block codes on Q-ary output channels," *IEEE Trans. Inform. Theory*, vol. IT-17, pp. 713–718, Nov. 1971.

[87] D. Chase, "A class of algorithms for decoding block codes with channel measurement information," *IEEE Trans. Inform. Theory*, vol. IT-18, pp. 170–182, Jan. 1972.

[88] S. Wainberg and J. K. Wolf, "Algebraic decoding of block codes over a q-ary input, on Q-ary output channel, Q>q," *Inform. Contr.*, vol. 22, pp. 232–247, Apr. 1973.

[89] J. K. Wolf, "Efficient maximum likelihood decoding of linear block codes using a trellis," *IEEE Trans. Inform. Theory*, vol. IT-24, pp. 76–81, 1978.

[90] D. G. Forney, Jr., *Concatenated Codes.* Cambridge, Mass.: MIT Press, 1966.

[91] G. Einersson and C. E. Sundberg, "A note on soft decision decoding with successive erasures," *IEEE Trans. Inform. Theory*, vol. IT-22, pp. 88–96, Jan. 1976.

[92] I. S. Gradshteyn and I. M. Ryzhik, *Tables of Integrals, Series, and Products.* New York: Academic Press, 1980.

[93] R. E. Blahut, "A universal Reed-Solomon decoder," *IBM J. Res. Develop.*, vol. 28, pp. 150–158, 1984.

[94] D. G. Forney, Jr., "Generalized minimum distance decoding," *IEEE Trans. Inform. Theory,* vol. IT-12, pp. 125–131, Apr. 1966.

[95] J. Hagenauer and E. Lutz, " Forward error correction coding for fading compensation in mobile satellite channels," *IEEE J. Select Areas Commun.,* vol. SAC-5, no. 2, pp. 215–225, Feb. 1987.

[96] M. K. Simon and D. Divsalar, "The performance of trellis coded multilevel DPSK on a fading mobile satellite channel," *IEEE Trans. Vehic. Technol.*, vol. VTC-37, no. 2, pp. 78–91, May 1988.

[97] M. Schwartz, W. R. Bennet, and S. Stein,*Communication Systems and Techniques.* New York : McGraw-Hill, 1966.

[98] D. L. Johnston and S. K. Jones, "Spectrally efficient communication via fading channels using coded multilevel DPSK," *IEEE Trans. Commun.*, vol. COM-29, no. 3, pp. 276–284, March 1981.

[99] J. G. Proakis, "Probabilities of error for adaptive reception of M-phase signals," *IEEE Trans. Commun. Technol.*, vol. COM-16, pp. 71–81, Feb. 1968.

[100] L. J. Mason, " Error probability evaluation for systems employing differential detection in a Rician fast fading environment and Gaussian noise", *IEEE Trans. Commun.*, vol. COM-35, no. 1, Jan. 1987.

[101] D. Rainish and J. M. Perl, "Generalized cutoff rate of time- and frequency-selective fading channels," *IEEE Trans. Commun.*, vol. COM-37, no. 5, pp. 449–467, May 1989.

[102] D. Divsalar and M. K. Simon, "Performance of trellis-coded MDPSK on fast fading channels, " *Proc. IEEE Int'l Conf. Commun.*, pp. 9.1.1–9.1.7, June 1989.

[103] J. Huang and L. L. Campbell, "Trellis coded MDPSK in correlated and shadowed Rician fading channels," *IEEE Trans. Vehic. Technol.*, vol. VTC-40, no. 4, pp. 786–797, Nov. 1991.

[104] E. Edbauer, "Performance of interleaved trellis-coded differential 8-PSK modulation over fading channels." *IEEE J. Selec. Areas Commun.*, vol. SAC-7, no. 9, pp. 1340–1346, Dec. 1989.

[105] K. Brayer, Editor, *Data Communications via Fading Channels.* New York: IEEE Press, 1975.

[106] S. Stein, "Fading Channel Issues in System Engineering," *IEEE J. Select. Areas Commun.*, vol. SAC-5, no. 2, pp. 68–89, Feb. 1987.

[107] A. Papoulis, *Probability, Random Variables, and Stochastic Processes.* New York: McGraw-Hill 1965.

[108] F. Davarian, "Channel simulation to facilitate mobile-satellite communication research," *IEEE Trans. Commun.*, vol. COM-35, no. 1, pp. 47–56, Jan. 1987.

[109] C. Loo, "A statistical model for a land mobile satellite link," *IEEE Trans. Vehic. Technol.,* vol. VT-34, no. 3, pp. 122–127, Aug. 1985.

[110] C. Loo, "Measurements and models for a land mobile satellite channel and their applications to MSK signals," *IEEE Trans. Vehic. Technol.,* vol. VT-35, no. 3, pp. 114–121, Aug. 1987.

[111] C. Loo, and N. Secord, "Computer models for fading channels with application to digital transmission," *IEEE Trans. Vehic. Technol.,* vol. VT-40, no. 4, pp. 700–707, Nov. 1991.

[112] E. Lutz, W. Papke, and E. Plochinger, "Land mobile satellite communications channel model, modulation and error control," *Seventh International Conference on Digital Satellite Communications,* pp. 537–543, May 1986.

[113] E. Lutz, D. Cygan, M. Dippold, F. Dolainsky, and W. Papke, "The land mobile satellite communication channel-recording, statistics, and channel model," *IEEE Trans. Vehic. Technol.,* vol. VT-40, no. 2, pp. 375–386, May 1991.

[114] R. M. Barts, W. L. Stutzman, W. T. Smith, R. C. Schmier, and C. W. Bostian, "Land mobile satellite propagation modeling," *Proc. of the 1987 IEEE A&P Society International Symposium,* vol. 1, pp. 20–23, 1987.

[115] J. Goldhrish, and W. J. Vogel, *Propagation Effects for Land Mobile Satellite Systems: Overview of Experimental and Modeling Results.* SATCOM Report No. 203 (NASA Refrence Publication 1274), Feb. 1992.

[116] C. Loo, E. E. Matt, J. S. Butterworth, and M. Duffer, "Measurements and modeling of land-mobile satellite signal statistics," *presented at 1986 Vehic. Technol. Conf.*, Dallas, TX, May 20–22, 1986.

[117] G. A. Arredondo, W. H. Chriss, and E. H. Walker, "A multipath fading simulator for mobile radio," *IEEE Trans. Vehic. Technol.*, vol. VT-22, no. 4, pp. 241–244, Nov. 1973.

[118] J. R. Ball, "A real-time fading simulator for mobile radio," *The Radio and Electronic Engineer,* vol. 52, no. 10, pp. 475–478, Oct. 1982.

[119] L. Ehrman, L. B. Bates, J. F. Eschle, and J. M. Kates, "Real-time software simulation of the HF radio channel," *IEEE Trans. Commun.* vol. COM-30, no. 8, pp. 1809–1817, Aug. 1982.

[120] E. Arthurs and H. Dym, "On the optimum detection of digital signals in the presence of white Gaussian noise - A geometric interpretation and a study of three basic data transmission systems," *IRE Trans. Commun. Systems*, vol. CS-10, pp. 336-372, Dec. 1962.

[121] C. E. Shannon, "A mathematical theory of communication," *Bell System Tech. J.*, vol. 27, (pt. I) pp. 379-423, (pt. II) pp. 623-656, 1948.

[122] S. B. Slimane and T. Le-Ngoc, "Performance of 16-QPOM signaling on a fading mobile satellite channel", *GLOBECOM'93*, Houston, Texas, Nov. 29-Dec. 2, 1993.

[123] S. B. Slimane, T. Le-Ngoc, "A tight upper bound on the error probability of coded modulation schemes in Rayleigh fading channels", *PIMRC'93*, Yokohama, Japan, Sept. 8-11, 1993.

[124] P. J. McLane, P. H. Wittke, P. K.-M. Ho and C. Loo, "PSK and DPSK trellis codes for fast fading, shadowed mobile satellite communication channels," *IEEE Trans. Commun.*, vol. COM-36, pp. 1242–1246, Nov. 1988.

[125] A. C. Lee and P. J. McLane, "Convolutional interleaved PSK and DPSK trellis codes for shadowed, fast fading mobile satellite communication channels", *IEEE Trans. Vehic. Technol.*, vol. VTC-39, PP. 37–47, Feb. 1990.

[126] S. H. Jamali, and T. Le-Ngoc, "Performance comparison of different decoding strategies for a bandwidth-efficient block-coded scheme on mobile channels," *IEEE Trans. Vehic. Technol.*, vol. VTC-41, no. 4, pp. 505–515, Nov. 1992.

[127] S. H. Jamali, and T. Le-Ngoc, " Performance of bandwidth-efficient RS-coded multilevel DPSK schemes on a mobile fading channel," *Proc. of the 1992 Vehic. Technol. Conf.*, Denver, Colorado, pp. 451–454, May 1992.

[128] M. C. Austin and M. U. Chang, "Quadrature overlapped raised-cosine modulation," *IEEE Trans. Commun.*, vol. COM-27, pp. 1616-1624, Oct. 1979.

[129] P. A. Bello, "Aeronautical channel characterization", *IEEE Trans. Commun.*, vol. COM-21, pp. 548-563, May 1973.

[130] F. Amoroso, "The use of quasi-bandlimited pulses in MSK transmission," *IEEE Trans. Commun.*, vol COM-27, pp. 1616-1624, Oct., 1979.

[131] M. Rabzel and S. Pasupathy, "Spectral shaping in minimum shift keying (MSK)-type signals," *IEEE Trans. Commun.*, vol. COM-26, pp. 189-195, January 1978.

[132] T. Le-Ngoc, "Power and bandwidth efficient ISI and jitter free (IJF) signals," *Journal, IERE,* vol. 55, NO. 7, pp. 268-274, July 1985.

[133] Y. Akaiwa and Y. Nagata, "Highly efficient digital mobile communications with a linear modulation method", *IEEE J. Select. Areas Commun.*, vol. SAC-5, pp. 890-895, June 1987.

[134] S. B. Slimane and T. Le-Ngoc, "Performance of quadrature pulse-overlapping modulated signals over fading channels," *Globecom'91*, Phoenix, Arizona, Dec. 2-5, 1991.

[135] T. Le-Ngoc and S. B. Slimane, "IJF-OQPSK modulation schemes with MLSE receivers for portable/mobile communications," *VTC'92*, Denver, Colorado, May 11-13 1992.

[136] S. B. Slimane and T. Le-Ngoc, "Maximum likelihood sequence estimation of quadrature pulse-overlapping modulated signals for portable/mobile satellite communications," *IEEE J. Select. Areas Commun.*, vol. SAC-10, pp. 1278-1288, October 1992.

[137] D. Chase, "A combined coding and modulation approach for communication over dispersive channels," *IEEE Trans. Commun.*, vol. COM-21, pp. 455-470, March 1973.

[138] C. Loo, "A statistical model for a land mobile satellite link," *IEEE Trans. Vehic. Techn.*, vol. 34, pp. 122-127, August 1985.

[139] C. Loo, E. E. Matt, J. S. Butterworth, and M. Dufour, "Measurements and modeling of land-mobile satellite signal statistics," *1986 Vehic. Techn. Conf.*, Dallas, TX, May 20-22, 1986.

[140] S. B. Slimane and T. Le-Ngoc, "Coded offset-quadrature pulse-overlapping modulation techniques," *Globecom'92*, Orlando, Florida, December 6-9 1992.

[141] T. Le-Ngoc, "A class of power and bandwidth efficient transmission techniques for digital radio communications systems," *Ph.D. dissertation*, U. of Ottawa, Ottawa, Canada, February 1983.

[142] R. C. Titsworth and L. R. Welch, "Power spectra of signals modulated by random and pseudo-random sequences," *Technical Report No. 32-440*, Jet Propulsion Laboratory, October 1961.

[143] M. Nakagami, "The m-distribution-ageneral formula of intensity distribution of rapid fading," in *Statistical Methods of Radio Wave Propagation*, W. C. Hoffman, Ed. Elmsford, NY: Pergamon, pp. 3-36, 1960.

[144] R. Bracewell, *The Fourier Transform and its Applications,* McGraw Hill, NewYork, 1965.

[145] A. C. M. Lee and P. J. McLane, "Convolutionally interleaved PSK and DPSK trellis codes for shadowed, fast fading mobile satellite communication channels," *IEEE Trans. Veh. Techn.*, vol. 39, pp. 37-47, February 1990.

[146] P. Ho and D. K. P. Fung, "Error performance of interleaved trellis-coded PSK modulations in correlated Rayleigh fading channels," *IEEE Trans. Commun.*, vol. COM-40, NO. 12, pp. 1800-1809, December 1992.

[147] F. Gagnon and D. Hacoun, "Bounds on the performance of coding for nonindependent Rician fading channels," *IEEE Trans. Commun.*, vol. COM-40, pp. 351-360, February 1992.

[148] W. C. Y. Lee, *Mobile Communications Engineering,* New York, McGraw Hill, 1982.

[149] J. K. Cavers and P. Ho, "Analysis of the error performance of trellis coded modulations in Rayleigh fading channels," *IEEE Trans. Commun.*, vol. COM-40, pp. 74-80, Jan. 1992.

[150] K. Dessouky and T. Jedrey, "The ACTS mobile terminals", *SATCOM Quarterly,* Jet Propulsion Laboratory, No. 8, pp. 16-23, January 1993.

[151] S. B. Slimane, "Maximum likelihood sequence estimation of quadrature pulse-overlapping modulated signals", *Ph.D. dissertation*, Concordia University, Montréal, Canada, October 1992.

# Index